THERMAL STRESSES IN SEVERE ENVIRONMENTS

m̃ £76.07

This book is to be returned on or before
below.

THERMAL STRESSES IN SEVERE ENVIRONMENTS

Edited by

D. P. H. Hasselman

and

R. A. Heller

Virginia Polytechnic Institute and State University
Blacksburg, Virginia

PLENUM PRESS · NEW YORK AND LONDON

Library of Congress Cataloging in Publication Data

International Conference on Thermal Stresses in Materials and Structures in Severe Thermal
Environments, Virginia Polytechnic Institute and State University, 1980.
Thermal stresses in severe environments.

Includes index.
1. Thermal stresses—Congresses. I. Hasselman, D. P. H., 1931-   II. Heller, Robert A.
III. Title.
TA418.58.I55 1980   620.1'121   80-17767
ISBN 0-306-40544-X

Proceedings of the International Conference on Thermal Stresses in Materials and Structures
in Severe Thermal Environments, held at Virginia Polytechnic Institute and State University,
Blacksburg, Virginia, March 19—21, 1980.

PREFACE

This volume of Thermal Stresses in Materials and Structures in Severe Thermal Environments constitutes the proceedings of an international conference held at Virginia Polytechnic Institute and State University in Blacksburg, Virginia, USA, on March 19, 20 and 21, 1980.

The purpose of the conference was to bring together experts in the areas of heat transfer, theoretical and applied mechanics amd materials science and engineering, with a·common interest in the highly interdisciplinary nature of the thermal stress problem. It is the hope of the program chairmen that the resulting interaction has led to a greater understanding of the underlying principles of the thermal stress problem and to an improved design and selection of materials for structures subjected to high thermal stresses.

The program chairmen gratefully acknowledge the financial assistance for the conference provided by the Department of Energy, the National Science Foundation, the Army Research Office and the Office of Naval Research as well as the Departments of Engineering Science and Mechanics and Materials Engineering at Virginia Polytechnic Institute and State University. A number of professional societies also provided mailing lists for the program at no nominal cost

The Associate Director, Mr. R. J. Harshberger and his staff at the Conference Center for Continuing Education at VPI and SU should be recognized especially for their coordination of the conference activities, lunches and banquet. Provost John D. Wilson gave a most enlightening and provocative after-dinner speech. Finally, we also wish to thank our secretaries: Mrs. B. Johnson, Mrs. P. Epperly and Mrs. M. Taylor for their patience in helping us with typing of many of the manuscripts and the almost endless stream of correspondence.

Blacksburg, Virginia   D. P. H. Hasselman
March 21, 1980   R. A. Heller

CONTENTS

CONTENTS

THERMAL STRESSES: A SURVEY

Bruno A. Boley

The Technological Institute
Northwestern University
Evanston, Illinois 60201

I want to thank the organizers of this Congress for asking me
to present the opening lecture. I thought it would be appropriate
if I were to start with a rapid review of the field of thermal
stresses and give some thoughts regarding an assessment (although
necessarily a very personal one) of the present state of the field,
and finally include guesses as to what the future might bring.

The history of the development of the field of thermal stresses
forms an interesting study of a growth of a scientific discipline.
It may thus be useful not only in judging what has happened and is
likely to happen in the field itself, but may provide some broad
observations in a general way upon progress in a scientific field.

A glance at the number of publications in this field (Table1,
[1]) shows that a remarkable growth has taken place in the field of
thermal stresses, but that it has taken place relatively recently.
For example, one may note that only 17 papers were published in the
first 65 years of the subject (i.e., from 1835-1900), and an equal
number in the next 20 years. But an admittedly incomplete listing
for the two years 1972 and 1973 alone [2] contains roughly ten times
that number of papers, and there is no doubt that the rate of publications has continued to increase ever since.

Rather than to dwell on the number of papers published, however,
it is probably more instructive to consider a few of the landmarks in
the long history of the subject, and thus to see whether the
scientific advance followed a similarly increasing rate of growth.

The preparation of this paper was supported by the Office of Naval
Research.

TABLE 1
Thermal Stress Publications [1]

| | Number of Publications | |
	During the five-year period	Cumulative
before 1836	4	4
1836-1840	2	6
1841-1845	1	7
1846-1850	0	7
1851-1855	0	7
1856-1860	0	7
1861-1865	0	7
1866-1870	1	8
1871-1875	1	9
1876-1880	1	10
1881-1885	2	12
1886-1890	0	12
1891-1895	2	14
1896-1900	3	17
1901-1905	5	22
1906-1910	5	27
1911-1915	2	29
1916-1920	5	34
1921-1925	15	49
1926-1930	15	64
1931-1935	21	85
1936-1940	30	115
1941-1945	30	145
1946-1950	83	228
1951-1955	204	432

Aside from some preliminary experimental investigations on the effect of temperature and deformations of solids, the first paper on thermoelasticity is that of Duhamel [3]. With this paper, published only fifteen years after the appearance of Fourier's treatise on heat conduction [4] and thirteen years after the publication of Navier's classical paper on the foundations of the theory of elasticity [5], the subject seems to have emerged full grown. In this paper the various boundary-value problems in linear thermoelasticity are formulated (including the effect of thermomechanical coupling), they are simplified and solved for a number of special cases. Little advance occurred for almost 60 years or so, that is until the work of the Italian elasticians (notably Almansi [6] and Tedone [7]), in the decade from about 1897 to 1907. Their researches paved the way for Muskhelishvili's later studies of two dimensional problems, including the introduction of the dislocation analogy [8] and the use of complex variables [9]; foreshadowed by this early work was also Goodier's method, published in 1937 [10].

In addition to the above rather basic advances, other papers were published in the first century of thermoelasticity, dealing with a variety of solutions to specific problems. Among these might be mentioned those relating to stresses, in spheres and cylinders (the first ones being respectively those of Hopkinson in 1879 [11] and of Leon in 1904 and 1905 [12 and 13]), Timoshenko's analysis of bimetallic strips [14], and some work on thermoelastic plates.

The foundations of thermoelasticity were thus well laid, and led to the solution in the 30's and 40's of a large number of problems of practical interest. Particular emphasis was given to the study of problems with common practical geometries, such as cylinders spheres and plates, so that the practical solution of problems related to many important engineering applications was then well understood and within the reach of the engineering community. We might perhaps mention two papers of a somewhat different cast, published during this period, namely Signorini's [15] on finite thermoelastic deformations, and Biot's [16] study of the stresses arising under steady-state temperature distributions.

This activity can be said to culminate in the appearance of the first text on the subject, namely that of Melan and Parkus [17] in 1953; although restricted to steady-state temperature distributions, the appearance of this book may be said to mark the first recognition that thermoelasticity was indeed a field in its own right, distinct from elasticity (or inelasticity) on the one hand, and heat conduction on the other.

At the same time, substantial work on the inelastic behavior of solids under thermal effects was begun, with consequent appearance of important generalizations of the theories of viscoelasticity (e.g. [18], where the viscoelastic-elastic analogy was first given

for incompressible materials, later generalized [19,20],and plasticit
(e.g.[21,22]).Simultaneously, advances were being made in the
extension of the domain of thermoelasticity: important topics of
interest were the establishment of thermoelastic energy theorems
[23,24] and the analysis of thermally-induced waves in solid bodies
[25] and of thermally-induced vibration of beams [26]. Coupled to
these developments was the establishment of theorems on overall
thermoelastic deformations [27] and the introduction into thermoelasticity the well known structural use of the principle of virtual
work [28]. These were accompanied by considerable effort at noting
both the similarities and the differences between elastic and thermoelastic behaviors and theories. Perhaps the feeling at the time
could best be summarized by noting that the similarities were useful
in providing a direct extension of known procedures into the new
field, while the differences had to be kept in mind so as not to be
led astray. Examples of this sort of work are given by [28,29,30].

It is well to note that the great interest in thermal effects
in structures, illustrated by the works which have been mentioned,
did not spring alone from a spontaneous burst of intellectual
curiosity, but was greatly spurred by problems arising in engineering
practice, primarily in the then rapidly developing fields of rocket-
powered high-speed flight and by the nascent exploitation of nuclear
sources of energy. It seemed an appropriate time to collect the
diverse strands of the subject in a single book, which would hopefull
make available to researchers, practicing engineers, and students
the physical and mathematical foundations of the field, as well as
provide a compendium of previous work, and thus a sound basis for
future advances, all in a single source and from a unified point of
view. Indeed, this is what J.H. Weiner and I attempted to do at
just about that time [31], and I must say that it has always been a
source of comfort to us that the appearance of the book did not
immediately result in an immediate regression of all work in the
field. Indeed, the study of thermal effects in solids and structures
continued to grow unabated in a number of directions. Among these
we might mention investigations of many specific and practical
problems on the basis of the theory that was by then well understood,
but which nevertheless presented considerable difficulty when actual
solutions were sought, because of such factors as complicated geometries, complicated heating or cooling conditions, and complicated
elastic material behavior in extreme temperature regimes. It goes
without saying that developments in these directions were accompanied
by extensive application and adaptation of numerical methods.

At the same time, other problems were arising which lay at the bounda
of validity of what may be called standard thermoelastic theory.
These therefore required more careful assessment of the basis of that
theory, and particularly the validity of some of the usual simplifyir
assumptions common in thermal stress analysis. Thus research
shedding new insight into the effect of thermoelastic coupling

was undertaken, including such fundamental work as a proof of uniqueness [32], analysis of thermoelastic waves [25,33,34,35], the solution of boundary value problems in coupled thermoelasticity [36], and considerations of electromagnetic coupling [37]. A summary of the results of these researches in the general field of dynamic thermoelasticity appeared in Nowacki's book published in 1966 [38]. Examination of the phenomenon of the so called second speed of sound, necessitating the inclusion of hyperbolic terms in the heat conduction equation, emerged shortly thereafter [39].

Another subject which began to attain prominence in the early sixties dealt with the study of mechanical behavior accompanying changes of phase [40], including the determination of transient and residual stresses and deformations. Application of these were important in traditional areas, such as the solidification of castings, but it was not long before the possibility of accurate thermo-mechanical analyses of ablating bodies [41] and of nuclear reactor meltdown accidents [42] began to be seriously examined. Allied to these works is the analysis of thermoelastic waves during change of phase contained in [43].

Another emerging technology of the sixties was concerned with composite materials; here too the behavior under non-isothermal conditions needed to be examined [44]. Also of growing importance was the theory of anisotropic thermoelasticity [45]. In parallel with the work which has been mentioned, more fundamental efforts at simplifying the practical solution of thermoelastic problems were continuing. Some of these were directed as the development of approximate analytical procedures (and, of course, at establishing the underlying theory; e.g., [45]),at constructing general bounds on the thermal stresses due to arbitrary temperature distributions [46], and at estimating errors in approximate calculations [47]. Not to be overlooked is the considerable body of analytical results pertaining to particular problems, especially in Japan, for example [48]).

The preceding overview up to about a decade or so ago has attempted to give a rapid chronological summary of the subject of thermal stresses. The references listed are those which appear to be the earliest in the particular topic mentioned, with no effort being made at providing a comprehensive bibliography in any one topic. The latter would be a tremendous task, particularly because of the great volume of publications which, as has been mentioned, continues to appear in print. It is nevertheless hoped that the above discussion has helped in putting in some historical perspective the origins of much of today's research. One word of demurral must however be added, because of the restriction here solely to advances of an analytical or theoretical nature. It must therefore be remembered that in many cases considerable experimental research had to be carried out in order to identify either the physical phenomena (for example, the second speed of sound) or the material properties in the specific thermal environment, and of course to verify theoretical results. No discussion of these aspects is presented here, but their essential role must not go unnoticed. Th reader will also note that coverage of numerical methods has likewise been omitted here.

I think it is now at any rate clear that, by any reasonable definition, the field of thermal stresses can be considered to have reached full maturity. It might be of interest, in fact, to note what characteristics may be thought of as typical of a mature field and to see whether a framework appropriate for examining present research and possible future directions might not thus emerge. Perhaps the following four closely related features may be identified as pertaining to a field that has reached maturity:

1. A tendency to take for granted some of the fundamental assumptions underlying the commonly used theory, and not to examine carefully the conditions in which particular equations might be valid. Another way of stating this might be to say that the field is ready for "handbook engineering" exploitation. Many instances o this can be found: in addition to examples previously mentioned, it appears that some of the early concerns with the validity of thermo elastic beam theory [24] and thermoelastic theory [48,49] are now virtually unnoticed, as are the caveats which have been issued concerning the use of conventional thermoelastic theory and formula in the analysis of beams of arbitrary cross-section [50] and rings [51]. Similarly, the careful assessment of criteria for uncoupling [48] is no longer a cause of great concern. The reasons for this are evident: experience has taught that reliance on the accepted formulas is generally justified, and it would be both unhealthy and inefficient to indulge in protracted worry. Nevertheless, the earl: studies point to special, but by no means pathological, cases, in which analyses more accurate than the conventional ones are needed, and it would be well if the former found their place in the courses on thermal stresses and thus at least in the subconscious of the practicing engineers.

2. An increased separation between applied and theoretical work. There is of course considerable interaction between practicir engineers and researchers, but, perhaps as a partial corollary of it 1 above, fundamental work is often regarded as irrelevant to practic solutions, while the need for the latter is often of little interest to the more mathematically inclined researcher. There is, in other words, a tendency to dismiss mathematical theorems by practicing engineers as mere curiosities, while simultaneously they may be quit correctly regarded by the mathematicisns as providing deeper insight theoretically. Attempts to bring the exponents of the two extremes together have been made, and continue to be made, such as, for examp two recent symposia [52,53] in the field of problems involving movin boundaries or change of phase (depending on whether one is mathemati cally or physically inclined), but much more must be done before a

real dialogue can be said to be taking place. The present Conference appears to be in fact an excellent move in this direction. One reason for the separation is easy to understand: researchers are increasingly drawn to more and more to unconventional problems, which though of course still physically meaningful and important in certain advanced applications, do not fall within the usual experience of most practitioners. A plea can be entered here for more understanding of the problems experiences by both camps.

3. Research on more esoteric problems. This has already been referred to as a partial cause of the increased separation of the practical from the theoretical, but it must be understood, in a mature but not stagnant field, as a case of continuous interchange between the two: as more extreme conditions become analytically tractable, applications for them are found. New applications in turn give rise to new needs, and then further, and more advanced or complex work must take place. This is an unending interaction, and there are numerous indications that in the field of thermal stresses they are indeed occurring in such diverse areas as the effect of extremely rapid intense heating or in change-of-phase problems, including the melting of reactor elements.

4. The appearance of a journal devoted to the field. A need for this is perceived, not necessarily to accommodate the very advanced work, but to provide an outlet for the increased number of workers concerned with a variety of problems of increasing complexity.

I conclude from an examination of the above characteristics that the field of thermal stresses is indeed mature and lively, and certainly the present symposium gives evidence of both. One more question might nevertheless be asked, namely what direction in the field would I deem most interesting and fruitful for future research. The answer to that question must of necessity be a subjective one, since the outlook of any one person is bound to be affected by his own previous work and interests. It might nevertheless be of some value to attempt to give my own views, with the idea that they might provide a useful starting point for discussion and further research.

The topics covered by the present Conference can certainly be taken to form a reasonable overview of current interests, and in fact each paper presented could be taken as representative of a problem that will receive additional future attention. In fact, almost everyone of the broad areas identified below as future research topics finds an illustrative counterpart on the Conference's program.

The following topics are, in my opinion, worthy of the reader's attention:
1. The propagation of thermoelastic waves. Studies should include all coupling effects, consideration of finite propagation speeds, anisotropic and inelastic effects, and identification of the various discontinuities and wave-front characteristics (e.g. [54].
2. Further development of general theorems in thermoelasticity.

This area is not only an intellectually satisfying one, but also
still can disclose useful and important basic information. As
examples, one might note Dundurs' work [55] on the a priori prediction of thermoelastic distorsions in steady-state problems, and some
useful extensions and applications of thermoelastic energy and virt
work theorems in [56]. Some general theorems in thermopiezoelectri
city should also be noted [57].

    3. Thermomechanical behavior accompanying changes of phase,
and generally in moving-boundary problems. In part no doubt because
of the complexity of problems of this type, and of the necessity to
incorporate consideration of inelastic behavior of materials in a
regime where little accurate information is available, comparatively
little has as yet been done in this field.

    4. Much more work will no doubt be needed in all aspects of
viscoelasticity, plasticity, and combined inelastic effects, ranging
from the establishment of accurate constitutive equations, to the
solution of practical problems. Similarly, the analysis of anisotropic media and composite materials will need further attention.

    5. Thermal effects in fracture are obviously part of a still
developing subject.

    6. To all of the above must of course be added the development
of numerical and experimental methods, essential for the solution of
actual complicated problems. I would add to these, as a useful
comparison the establishment of approximate rapid analytical methods
of calculation of stresses and deflection, which are extremely
useful in the verification of numerical results and in preliminary
design.

    I believe the above categories contain most of the important
topics which require consideration, and are likely to continue to
do so for some time to come. What is clear is that there is a
great deal of work to be done, and I can probably be most useful at
this point by cutting my presentation short and, with renewed thanks
to the organizers of this Conference, get out of the way so that wor
can proceed.

REFERENCES

1. B.A. Boley, J.H. Weiner, and I.S. Tolins, "Thermal Stress Analys
    for Aircraft Structures-Part II: Bibliography", WACD Tech. Rep.
    pp. 56-102, Part II, November 1955.
2. B.A. Boley, "Thermal Stresses Today", Proc. 7th U.S. Nat. Congres
    of Applied Mechanics, Boulder, Colorado, June 3-7, 1974; publ.
    by ASME, New York (1974) pp. 99-107.
3. Duhamel, J.M.C., Second mémoire sur les phénomènes thermo-
    mécaniques, Journal de l'École Polytechnique, tome 15, cahier 2!
    1837, pp. 1-57. This paper was read before the Academy of
    Sciences on February 23, 1835.
4. Fourier, J. Théorie analytique de la chaleur, Paris, 1822.
5. Navier, C.L.M.H., Mémoire sur les lois de l'équilibre et du

mouvement des corps solides élastiques", Mém. Inst. Nat. (1824). This paper was presented before the Academy of Sciences on May 14, 1921.

6. Almansi, E., "Use of the stress function in thermo-elasticity", Mem. Reale Accad. Sci. Torino, Series 2, vol. 47, 1897.

7. Tedone, O. "The General Theroems of the Mathematical Theory of Elasticity",Encyklopadie d. math. Wiss., Bd. IV, 4 Teilband, 1907, p. 68.

8. Muskhelishvili, N., "Relation between Thermal Stress in Steady Heat Flow and Dislocational Stress", Bull. Elec. Tech. Inst., St. Petersburg, Vol. 13, 1916, p. 23.

9. Muskhelishvili, N., "Sur l'integr ation de l'equation biharmonique", Bulletin de l'Academie des Sciences de Russie, VI Serie, Tome 13, 1919, pp. 663-686.

10. Goodier, J.N., Integration of thermoelastic equations, Phil. Mag. Vol. 23, 1937, p. 1017.

11. Hopkinson, J., Thermal Stresses in a Sphere, whose temperature is a function of r only, Messenger of Math., Vol. 8, 1879, p. 168.

12. Leon, A., On thermal stresses, Der Bautechniker, Bd. 26, 1904, p. 968.

13. _____ , Stresses and deformations of a hollow cylinder under the assumption of a linear temperature distribution law, Zeit. f. Math u. Physik, Bd. 52, 1905, pp. 174-190.

14. Timoshenko, S., Bending and buckling of bi-metallic strips, J. Optical Society of America, vol. 11, 1925, p. 233.

15. Signorini, A., Finite Thermo-elastic deformation, Proc. 3rd Int. Cong. Applied Mechanics, Stockholm, Vol. 2, 1930, pp. 80-89.

16. Biot, M.A., Propriété générale tensions thermiques en régime stationare dans corps cylindrique, Soc. Scientifiques de Bruxelles, serie B, Vol. 54, 1934, pp. 14-18.

17. Melan, E. and Parkus, H., Wärmespannungen infolge stationärer Temperaturfelder, Springer, Vienne, 1953.

18. Alfrey, T., Mechanical Behavior of High Polymers, Interscience Publ., New York, 1948.

19. Tsien, H.S., "A Generalization of Alfrey's Theorem for Viscoelastic Media", Quart. Appl. Math., Vol. 8, 1950, pp. 104-106.

20. Read, W.T., "Stress Analysis for Compressible Viscoelastic Materials", J. Appl. Physics, Vol. 21, 1950, pp. 671-674.

21. Freudenthal, A.M., On inelastic thermal stresses in flight structures, J. of the Aero. Sci., Vol. 21, 1954, pp. 772-778.

22. Prager, W., "Non-Isothermal Plastic Deformations", Proc. Kon. Nederl, Adad van Wetenschappen, Series B, Vol. 61, no. 3, 1958, pp. 176-182.

23. Argyris, J.H., Thermal stress analysis and energy theorems, Great Britain Air Ministry, Aeronautical Research Council, Vol. 16, Dec. 1953, p. 489.

24. Hemp, W.S., Fundamental principles and methods of thermoelasticity Aircraft Engrng., Vol. 26, no. 302, Apr. 1954, pp. 126-127.

25. Danilovskaya, V.I., On a dynamical problem of thermoelasticity

Prikl. Mat. Mekh., Vol. 16, no. 3, May/June 1952, pp. 341-344.

26. Boley, B., Thermally induced vibrations of beams, Journ. of the Aero. Sci., Vol. 23, No. 2, Feb. 1956, pp. 179-181.

27. Goodier, J.N., "Thermal Stresses and Deformations", J. Appl. Me Vol. 24, No. 3, 1957, pp. 467-474.

28. Boley, B.A., "The Calculation of Thermoelastic Beam Deflections by the Principle of Virtual Work", J. Aero. Sci., Vol. 24, no. 2, 1957, pp. 139-141.

29. _____, The determination of temperature, stresses and deflections in two-dimensional thermoelastic problems, Journ of the Aero. Sci., Vol. 23, No. 1, January 1956, pp. 67-75. (Originally published as WADC Tech. Rept. 54-424, March 1954

30. Boley, B.A. and Tolins, I.S., Thermoelastic stresses and deflections in thin-walled beams, WADC TR 54-426, December 1954.

31. Boley, B.A. and Weiner, J.H., Theory of Thermal Stresses, Wiley New York, 1960.

32. Weiner, J.H., " A Uniqueness Theorem for Coupled Thermoelastic Problems", Quart. Appl. Math., Vol. 15, no. 1, 1957, pp. 102-105.

33. Mura, T., "Thermal Strains and Stresses in Transient State", Prof. Second Jap. Congress Appl. Mech., 1952.

34. Ignaczak, J., "Thermal Displacements in an Elastic Semispace Due to Sudden Heating of the Boundary Plane", Nedb. Arch. Mech. Stos., Vol. 9, 1957, p. 395.

35. Deresiewicz, H., "Plane Waves in a Thermoelastic Solid", J. Acoust. Soc. Am., Vol. 29, no. 2, 1957, pp. 204-209.

36. Boley, B.A. and Tolins, I.S., "Transient Coupled Thermoelastic Boundary-Value Problems in the Half-Space", J. Appl. Mech., Vol. 29, no. 4, 1962, pp. 637-646.

37. Mindlin, R.D., "Investigations in the Mathematical Theory of Vibrations of Anisotropic Bodies", U.S. Army Signal Corps Eng. Labs., Fort Monmouth, N.J., 1956, pp. 84ff.

38. Nowacki, W., Dynamiczne zagadniena termosprezystości, Państ. Wydawn. Nauk., Warsaw, 1966. (English translation as Dynamic Problems of Thermoelasticity, Nordhoff, 1975.

39. Lord, H.W. and Shulman, U., "A generalized Dynamical Theory of Thermoelasticity", J. Mech. Phys. Solids, Vol. 15, no. 5, 1967, pp. 299-309.

40. Weiner, J.H. and Boley, B.A., "Elasto-Plastic Thermal Stresses i a Solidifying Slab", J. Mech. Phys. Solids, Vol. 11, 1963, pp. 145-154.

41. Friedman, E. and Boley, B.A., "Stresses and Deformations in Melting Plates", J. Spacecraft and Rockets, Vol. 7, no. 3, 1970, pp. 324-333.

42. Boley, B.A. "Temperature and Deformations in Rods and Plates Melting under Internal Heat Generation", Prof. First Int. Cong. Struct. Mech. in Reactor Tech. (SMiRT-1), Berlin, 1971, vol. 5, Part 1, Paper L2/3.

43. Tadjbakhsh, I., "Thermal Stresses in an Elastic Half-Space with

a Moving Boundary", AIAA Journ., Vol. 1, no. 1, 1963, pp. 214-215.

44. Testa, R.B. and Boley, B.A., "Basic Thermoelastic Problems in Fiber-Reinforced Materials", paper in Mechanics of Composite Materials, Pergamon Press, 1970. (Prof. of Fifth Symp. of Naval Struct. Mech., Philadelphia, Pennsylvania, 1967).

45. Biot, M.A., "Thermoelasticity and Irreve·sible Thermodynamics", J. Appl. Phys., Vol. 27, no. 3, 1956, pp. 240-253.

46. Boley, B.A., "Bounds on the Maximum Thermoelastic Stress and Deflection in a Beam or Plate", J. Appl. Mech., Vol. 23, no. 4, 1966, pp. 881-887.

47. _____, "Estimate of Errors in Approximate Temperature and Thermal Stress Calculations". Proc. Eleventh Int. Congress Appl. Mech., Springer, 1964.

48. Takeuti, Y., "Approximate Formulae for Temperature Distribution in a Heat-Generating Polygonal Cylinder with Central Circular Hole", Nucl. Eng. Des., Vol. 8, 1968, pp. 241-246.

49. Boley, B.A., "Thermal Stresses" in Structural Mechanics, Edited by J.N. Goodier and J.J. Hoff, Pergamon Press, 1960, pp. 378-406 (Proc. fo First Symp. on Naval Struct. Mech., Stanford Univ., August 11-14, 1958).

50. _____, "On Thermal Stresses in Beams", Int. J. Solids and Struct., Vol. 8, no. 4, 1972, pp. 571-579.

51. _____, "On Thermal Stresses and Deflections in Thin Rings", Int. J. Mech. Sci., Vol. 11, 1969, pp. 781-789.

52. Ockendon, J.R. and Hodgkins, W.R., Moving Boundary Problems in Heat Flow and Diffusion, Clarendon Press, Oxford, 1975.

53. Wilson, D.G., Solomon, A.D. and Boggs, P.T., Moving Boundary Problems, Acad. Press, 1978.

54. Boley, B.A. and Hetnarski, R.B., "Propagation of Discontinuities in Coupled Thermoelastic Problems", J. Appl. Mech., Vol. 35, no. 3, 1968, pp. 489-494.

55. Dundurs, J., "Distorsion of a Body Caused by Free Thermal Expansion", Mech. Res. Communications, Vol. 1, no. 3, 1974, pp. 121-124.

56. Tabakman, H.D. and Lin, Y.J., "Stresses and Displacements in Simply Supported Curved Plates Subjected to an Arbitrary Temperature Distribution", J. of Thermal Str., Vol. 1, no. 2, 1978, pp. 183-191.

57. Nowacki, W., "Some General Theorems of Thermopiezoelectricity", J. Th. Str., Vol. 1, no. 2, 1978, pp. 171-182.

THERMOELASTICITY IN POLYMERIC AND CRYSTALLINE SOLIDS FROM THE

ATOMISTIC VIEWPOINT*

J. H. Weiner

Division of Engineering
Brown University
Providence, RI   02912

ABSTRACT

On the macroscopic level, the difference in the thermoelastic behavior between crystalline solids and amorphous polymeric solids such as rubber is striking. While crystalline solids expand with increase in temperature, stretched rubber contracts upon heating. On the atomic level this difference may be traced to the greater degree of disorder in amorphous polymeric solids and consequently to the central role played by entropy in their mechanical behavior.

In this paper we present two simple linear chain atomistic models with prescribed nearest-neighbor interactions, one which represents, in highly idealized form, a crystalline solid and the second a long-chain molecule under tension. For these models, it is possible to calculate the thermoelastic relation by the application of the basic principles of classical equilibrium statistical mechanics. The model[1] for the long-chain molecule is found to exhibit a transition in behavior, over a narrow range of temperature, from that expected for a harmonic crystal at low temperature levels, to that characteristic of a polymer at higher temperature levels.

REFERENCES

1. J. H. Weiner and M. R. Pear, "Computer Simulation of Conformational Transitions in an Idealized Polymer Model," Macromolecules Vol. 10 (1977) page 317.

*Research supported by the Gas Research Institute (Grant No. 5009-362-0015).

2. L. R. G. Treloar, The Physics of Rubber Elasticity, Third Edition, Clarendon Press, Oxford, 1975.

3. P. J. Flory, Principles of Polymer Chemistry, Cornell University Press, Ithaca, NY, 1953.

4. J. H. Weiner and M. R. Pear, <u>Macromolecules</u>, Vol. 10, 317-325, 1977.

THERMOELASTICITY WITH FINITE WAVE SPEEDS - A SURVEY

Józef Ignaczak

Polish Academy of Sciences
Swietokrzyska 21, 00-049
Warsaw, Poland

ABSTRACT

At present there are at least two different generalizations of the classical linear thermoelasticity. The first one proposed by Green and Lindsay (1972) (G-L Theory) involves two relaxation times of a thermoelastic process, while the second theory due to Lord and Shulman (1967) (L-S Theory) admits only one relaxation time. Both theories have been developed in an attempt to eliminate the paradox of an infinite velocity of thermoelastic propagation inherent in the classical case. In this paper a number of general results concerning these theories for a homogeneous isotropic solid are presented. They include (G-L Theory): 1. Domain of Influence Theorem, 2. Decomposition Theorem, 3. Uniqueness Theorem, and 4. Variational Principle. Theorem 1 asserts that in the new theory thermoelastic disturbances produced by the data of bounded support propagate with a finite velocity only. Theorem 2 which is quite similar to a Boggio result in linear elastodynamics shows that a thermoelastic disturbance can be split into two fields each of which propagates with a different finite velocity. Theorem 3 covers uniqueness for a stress-temperature boundary value problem with arbitrary initial tensorial data. Finally, the variational principle gives an alternative description of the theory in terms of a thermoelastic process (S,q), where S and q represent the stress tensor and the heat flux vector, respectively. In the L-S theory uniqueness of a stress-flux initial-boundary value problem is discussed. A number of suggestions concerning those areas of the theory that are critically in need of further investigation are also given.

1. INTRODUCTION

Classical linear thermoelasticity is a well developed theory of solid mechanics today. Recently published monographs in this field[1] offer the derivation of the basic equations, the fundamental theorems, and the general solutions of the governing equations. The theory is described by hyperbolic-parabolic equations that imply an infinite speed of thermoelastic disturbances in a solid. Since such implication is not acceptable from a physical point of view, some new thermoelastic theories aiming at elimination of the infinite speed paradox have been put forward in the technical literature. In the present paper two such new theories are discussed. The first one proposed by Green and Lindsay in 1972 (G-L Theory) is a generalization of classical thermoelasticity in the sense that it involves two relaxation times in a thermoelastic process: a relaxation time τ_0 modifies the classical energy equation and another relaxation time τ_1 changes the Duhamel-Neumann relation. The second theory due to Lord and Shulman in 1967 (L-S Theory), being also a generalization of the classical theory, admits only one relaxation time: a time τ_0^* that modifies the Fourier law of heat conduction. For both these theories a number of general results analogous to the basic theorems of classical thermoelasticity are presented.

The results are restricted to a homogeneous and isotropic solid, when the material functions are independent of temperature. One-dimensional problems or vibrations harmonic in time, which have been published in a scattered form in the technical literature, are not discussed.[2] The paper is confined to smooth thermoelastic processes, i.e., thermoelastic singular surfaces are not discussed.[3] Most of the article is devoted to the results obtained recently by the author for the G-L Theory. They include:
1. Domain of Influence Theorem, 2. Decomposition Theorem,
3. Uniqueness Theorem, and 4. Variational Principle. In the L-S Theory uniqueness for a natural-stress-flux problem is presented. The survey ends up with a number of suggestions concerning those areas of thermoelasticity with finite wave speeds, that are in critical need of further investigation.

2. BASIC EQUATIONS IN G-L THEORY

The fundamental system of field equations for a homogeneous isotropic linear thermoelastic solid in the G-L theory can be

---

[1]See, e.g., Carlson [1], Nowacki [2], and Kupradze et al. [3].
[2]See, e.g., a survey article by Francis [6], and the papers [7]-[9].
[3]For a discontinuous thermoelastic process in the L-S Theory see, e.g., [10].

deduced from Eqs. 4.1 - 4.16 of Ref. 5, and it consists of the strain-displacement relation

$$E = \frac{1}{2} (\nabla u + {}^{\cdot}\nabla u^T) \equiv \hat{\nabla} u \tag{1}$$

and the equation of motion

$$\text{div } S + b = \rho \ddot{u} \tag{2}$$

the energy equation

$$- \text{div } q + r = c_E(\dot{\delta} + \tau_o \ddot{\delta}) + (3\lambda + 2\mu)\alpha\theta_o \text{ tr}\dot{E} \tag{3}$$

the stress-strain-temperature relation

$$S = 2\mu E + \lambda(\text{tr}E)1 - (3\lambda + 2\mu)\alpha(\delta + \tau_1\dot{\delta})1 \tag{4}$$

and the heat conduction equation

$$q = - k\nabla\delta. \tag{5}$$

Here u, E, S, δ, q, b and r are the displacement, strain, stress, temperature difference, heat flux, body force, and heat supply fields, respectively; while ρ, λ and μ, α, k and c_E are the density, the Lame moduli, the coefficient of thermal expansion, the conductivity, and the specific heat for zero deformation, respectively; finally τ_o and τ_1 are the relaxation times,[4] and θ_o is the fixed uniform reference temperature ($\theta_o > 0$).[5]

The material constants and the relaxation times obey the inequalities

$$\rho > 0 \quad , \quad k > 0 \quad , \quad c_E > 0 \quad , \quad \alpha > 0 \tag{6}$$

$$\mu > 0 \quad , \quad 3\lambda + 2\mu > 0 \quad , \tag{7}$$

$$\tau_1 \geq \tau_o \geq 0 . \tag{8}$$

If $\tau_o = \tau_1 = 0$, Eqs. 1-5 reduce to the basic field equations of classical thermoelasticity [see, e.g., [1] p. 327, and [2] p. 13].

---

[4]The relaxation times τ_o and τ_1 are not introduced explicitly by Green and Lindsay in [5]. However, they can be easily defined in terms of the constitutive constants of Green and Lindsay by comparing their linear energy equation and the constitutive relation with Eqs. 3 and 4, respectively.

[5]Eq. 1-5 are written in Carlson's notation [1]. However, there is one to one correspondence between the G-L notation and that of Carlson. A superposed dot denotes $\partial/\partial t$, where $\hat{t}$ is the time.

Let B denote an open regular domain of euclidean three space and $(0,\infty)$ be the time interval. A thermoelastic solution in the G-L Theory is an ordered array $[u,E,S,\delta,q]$ that satisfies Eqs. 1-5 on $B \times (0,\infty)$ and complies with suitable initial and boundary conditions. Of particular importance is a thermoelastic solution in which

$$u = \nabla\hat{\phi}$$ (9)

and $\hat{\phi} = \hat{\phi}(\hat{x},\hat{t})$ is a scalar field.

For convenience, it is assumed that $b = 0$ in Eq. 2, and $r = 0$ in Eq. 3. Eqs. 1-5 subject to Eq. 9 are then satisfied provided $\hat{\phi}$ and δ meet the equations

$$\nabla^2\hat{\phi} - \frac{\rho}{\lambda + 2\mu} \ddot{\hat{\phi}} = \frac{3\lambda + 2\mu}{\lambda + 2\mu} \alpha(\delta + \tau_1\dot{\delta})$$ (10)

$$\nabla^2\delta - \frac{c_E}{k} (\dot{\delta} + \tau_0\ddot{\delta}) - \frac{1}{k} (3\lambda + 2\mu)\alpha\theta_0\nabla^2\dot{\hat{\phi}} = 0$$ (11)

These equations can be put in a more convenient form by using non-dimensional variables

$$x = \frac{\hat{x}}{\hat{x}_0} \quad , \quad t = \frac{\hat{t}}{\hat{t}_0} \quad ,$$ (12)

$$\theta = \frac{\delta}{\theta_0} \quad , \quad \phi = \frac{(\lambda + 2\mu)\hat{\phi}}{(3\lambda + 2\mu)\alpha\theta_0\hat{x}_0^2} \quad ,$$ (13)

where

$$\hat{x}_0 = \frac{k}{c_E c_1} \quad , \quad \hat{t}_0 = \frac{k}{c_E c_1^2} \quad ,$$ (14)

and

$$c_1^2 = \frac{\lambda + 2\mu}{\rho} .$$ (15)

These variables, when substituted into Eqs. 10 and 11, yield the following non-dimensional relations

$$\cdot \ \nabla^2\phi - \ddot{\phi} = \theta + t_1\dot{\theta} \ ,$$ (16)

$$\nabla^2\theta - (\dot{\theta} + t_0 \ddot{\theta}) = e\nabla^2\dot{\phi} \ ,$$ (17)

where ∇^2 is the Laplacian with respect to x and the dots denote differentiation with respect to t. Moreover, e is the thermoelastic constant

$$e = \frac{(3\lambda + 2\mu)^2\alpha^2\theta_0}{(\lambda + 2\mu)c_E} \quad , \quad (e > 0) \quad ,$$ (18)

and the dimensionless relaxation times t_0 and t_1 are given by

$$t_0 = \tau_0/\hat{t}_0 \quad , \quad t_1 = \tau_1/\hat{t}_0 . \tag{19}$$

Elimination of θ from Eqs. 16 and 17 leads to

$$\left[\left(\nabla^2 - \frac{\partial^2}{\partial t^2}\right)\left(\nabla^2 - t_0 \frac{\partial^2}{\partial t^2} - \frac{\partial}{\partial t}\right)\right.$$

$$\left. - \epsilon\nabla^2 \frac{\partial}{\partial t}\left(1 + t_1 \frac{\partial}{\partial t}\right)\right]\phi = 0 . \tag{20}$$

This is a central equation of the G-L Theory.[6] If ϕ is a solution
of Eq. 20, then θ is easily obtained from Eq. 16. Now suppose
that $\phi = \phi(x,t)$ and $\theta = \theta(x,t)$ satisfy Eqs. 16 and 17 on
$B \times [0,\infty)$ subject to the conditions

$$\phi(x,0) = \phi_0(x) \quad , \quad \dot{\phi}(x,0) = \phi_1(x)$$

$$\theta(x,0) = \theta_0(x) \quad , \quad \dot{\theta}(x,0) = \theta_1(x) \quad \text{on } \overline{B} \tag{21}$$

and

$$n \cdot \nabla\phi = f \quad , \quad \theta = g \quad \text{on } \partial B \times [0,\infty) , \tag{22}$$

where $n(x)$ is the unit outward normal to ∂B at x and ϕ_0, ϕ_1, θ_0,
θ_1, f and g are given functions.

The conditions, Eq. 21, define the initial state of the solid
$\overline{B}$, while f and g are the boundary displacement and temperature,
respectively for $t \geq 0$.

Also, suppose that B is a bounded or unbounded regular
region, and introduce the positive constant v as

$$v \geq \max\left[(1 + 2\epsilon)\frac{t_1}{t_0} , \frac{1}{t_1} , 2\right] . \tag{23}$$

Given a pair (ϕ,θ) satisfying Eqs. 16 and 17 and conditions, Eqs.
21 and 22, and a given time $t > 0$, let D_t^0 denote the set

$$D_t^0 = \{x\epsilon\overline{B} : \text{if } x\epsilon B \text{ then } \phi_0(x) \neq 0 \quad \text{or} \quad \phi_1(x) \neq 0$$

$$\text{or} \quad \theta_0(x) \neq 0 \quad \text{or} \quad \theta_1(x) \neq 0 ;$$

---

[6]Eq. 20 is important not only for discussion of the particular
solution generated by (ϕ,θ), but also for analysis of the general
solution in which $u = \nabla\phi + \text{curl } \psi$, where ψ is a vector field.

if $(x,\tau)\varepsilon\ \partial B\times[0,t]$ then

$$f(x,\tau)\neq 0\quad\text{or}\quad g(x,\tau)\neq 0\}\ . \tag{24}$$

A domain of influence of the data at time t for the initial-boundary value problem, Eqs. 16, 17, 21 and 22, is the set

$$D_t = \{x\varepsilon\bar{B}\ :\ D_t^o\ \cap\ \overline{\Sigma_{vt}(x)}\neq 0\}\ , \tag{25}$$

where $\Sigma_{vt}(x)$ denotes the open ball of radius vt and center at x. The set D_t^o is the support of the initial and boundary data, while D_t covers a domain of thermoelastic disturbances produced by the data at time t. This interpretation is motivated by the following:

Theorem (Domain of Influence Theorem in the G-L Theory). If (ϕ,θ) is a pair of functions satisfying Eqs. 16, 17, 21 and 22, and if D_t is the domain of influence of its data at time t, then

$$\phi = \theta = 0\quad\text{on}\ (\bar{B} - D_t)\times[0,t]\ . \tag{26}$$

The proof of this theorem, based on a lemma analogous to Zaremba's result (1915), is given by Ignaczak in [11].

The theorem implies that the thermoelastic disturbances generated by the pair (ϕ,θ) propagate with a finite speed only, and for this reason the G-L theory can be called thermoelasticity with finite wave speeds. The speed is bounded from above by the parameter v. If $t_o \rightarrow 0$ and t_1 is finite then, by the definition of v, $v \rightarrow \infty$. Also, if $t_1 \rightarrow 0$ then, by Eq. 8, $t_o \rightarrow 0$. Thus, if $t_1 \rightarrow 0$ and the ratio t_1/t_o is finite, one obtains $v \rightarrow \infty$. Therefore, for the vanishing relaxation times of the G-L theory the thermoelastic disturbances described by (ϕ,θ) propagate with an infinite speed, as expected, since in this case the G-L theory reduces to classical thermoelasticity.

3.   DECOMPOSITION THEOREM IN G-L THEORY

In this section an important property of the thermoelastic process, described by the pair (ϕ,θ) [see Section 2] is discussed. As a result the process is decomposed into two simpler processes. The analysis will be restricted to the homogeneous initial conditions only.

Elimination of the temperature initial data, using Eqs. 16, 17 and 21 with $\phi_o = \phi_1 = \theta_o = \theta_1 = 0$, leads to the four homogeneous initial conditions for ϕ:

$$\frac{\partial^k}{\partial t^k}\ \phi(x,0) = 0\quad\text{for}\ k = 0,1,2,3\ . \tag{27}$$

The following theorem holds true:

Theorem.   Let $\phi = \phi(x,t)$ be a scalar field satisfying Eq. 20, i.e.,

$$\left[\left(\nabla^2 - \frac{\partial^2}{\partial t^2}\right)\left(\nabla^2 - t_0 \frac{\partial^2}{\partial t^2} - \frac{\partial}{\partial t}\right)\right.$$

$$\left. - \epsilon\nabla^2 \frac{\partial}{\partial t}\left(1 + t_1 \frac{\partial}{\partial t}\right)\right]\phi = 0 \tag{28}$$

subject to the homogeneous initial conditions, Eq. 27, and define the function $K = K(t)$ on $[0,\infty)$ through the relation

$$K(t) = 2 \frac{d}{dt}\left[e^{\alpha t} \frac{J_1(\beta t)}{\beta t}\right] , \tag{29}$$

where $J_1 = J_1(x)$ is the Bessel function of order 1 of the first kind, and

$$\alpha = [1 - \epsilon - (1 + \epsilon)(t_0 + \epsilon t_1)]\Delta^{-1} ,$$

$$\beta = 2\sqrt{\epsilon} [1 + (1 + \epsilon)(t_1 - t_0)]^{1/2} \Delta^{-1} , \tag{30}$$

with

$$\Delta = (1 - t_0 + \epsilon t_1)^2 + 4\epsilon t_0 t_1 > 0 . \tag{31}$$

Then

$$\phi(x,t) = \phi_1(x,t) + \phi_2(x,t) , \tag{32}$$

where ϕ_1 and ϕ_2 satisfy the equations

$$\left(\nabla^2 - \frac{1}{v_1^2} \frac{\partial^2}{\partial t^2} - k_1 \frac{\partial}{\partial t} - \lambda - \lambda K*\right)\phi_1 = 0 , \tag{33}$$

$$\left(\nabla^2 - \frac{1}{v_2^2} \frac{\partial^2}{\partial t^2} - k_2 \frac{\partial}{\partial t} + \lambda + \lambda K*\right)\phi_2 = 0 . \tag{34}$$

Here

$$v_{1,2}^{-2} = \frac{1}{2} (1 + t_0 + \epsilon t_1 \pm \Delta^{1/2}) , \tag{35}$$

$$k_{1,2} = \frac{1}{2} (1 + \epsilon \mp \alpha \Delta^{1/2}) , \tag{36}$$

$$\lambda = \frac{1}{4} \beta^2 \Delta^{1/2} , \tag{37}$$

and * denotes the operation of convolution with respect to time t, that is

$$K*\phi_k = \int_o^t K(t - \tau)\phi_k(x,\tau)d\tau , \quad k = 1,2 .$$ (38)

Clearly, since $t_1 > t_o > 0$ and $\epsilon > 0$,

$$v_2 > v_1 > 0 \quad \text{and} \quad \lambda > 0 .$$ (39)

Moreover, if $(t_o,t_1) \to (0,0)$ for $t_1 \geq t_o \geq 0$, then

$$(v_1,v_2) \to (1,\infty) \quad , \quad (k_1,k_2) \to (\epsilon,1) \quad ,$$

$$\lambda \to \epsilon \quad , \quad \alpha \to 1 - \epsilon \quad , \quad \beta \to 2\sqrt{\epsilon} ,$$ (40)

and the theorem reduces to that of Ref. 12. The proof of this theorem is given by Ignaczak [13].

The decomposition theorem implies that a thermoelastic disturbance corresponding to the homogeneous initial data and non-homogeneous boundary conditions is a sum of two disturbances one of which propagates with the velocity v_1 and another with the velocity v_2. Since $v_2 > v_1$, a domain of influence for the thermoelastic body with a quiescent past can be constructed in terms of v_2 only.

4.   UNIQUENESS FOR STRESS-TEMPERATURE EQS. IN G-L THEORY

In this section we prove a uniqueness theorem for a natural stress-temperature formulation of the G-L theory.[7]

By eliminating u, E and q from the system of Eqs. 1-5 the equations involving S and δ are obtained as follows

$$\rho^{-1}\hat{\nabla}(\text{div } S) - \frac{1}{2\mu}\left[\ddot{S} - \frac{\lambda}{3\lambda + 2\mu}(\text{tr } \ddot{S})1\right]$$

$$- \alpha(\ddot{\delta} + \tau_1\ddot{\delta})1 + \rho^{-1}\hat{\nabla}b = 0 \quad ,$$ (41)

$$c_S^{-1}(k\nabla^2\delta - \alpha\theta_o \text{ tr } \dot{S} + r) - \tau\ddot{\delta} - \dot{\delta} = 0 \quad ,$$ (42)

where

---

[7]A conventional uniqueness theorem for the G-L theory generalizing Weiner's result [14] was established in [5].

$$c_S = c_E + 3(3\lambda + 2\mu)\theta_0\alpha^2 \quad , \tag{43}$$

$$\tau = \left(1 - \frac{c_E}{c_S}\right)\tau_1 + \frac{c_E}{c_S}\tau_0 \quad . \tag{44}$$

learly, $c_S > 0$ represents the specific heat for zero stress. It lso follows from the restrictions, Eqs. 6-8, and from the defini- ions, Eqs. 43 and 44, that

$$0 < 1 - \frac{c_E}{c_S} \leq \frac{\tau}{\tau_1} \leq 1 \quad . \tag{45}$$

o the system of field Eqs. 41 and 42 to be satisfied in the region $\mathcal{B} \times [0,\infty)$, the initial conditions are adjoined:

$$S(x,0) = S^0(x) \quad , \quad \dot{S}(x,0) = S^1(x) \quad ,$$

$$\delta(x,0) = \delta^0(x) \quad , \quad \dot{\delta}(x,0) = \delta^1(x) \quad \text{on B} \; , \tag{46}$$

the stress boundary condition

$$Sn = p \quad \text{on } \partial B \times [0,\infty), \tag{47}$$

and the temperature boundary condition

$$\delta = g \quad \text{on } \partial B \times [0,\infty) \; . \tag{48}$$

Here S^0, S^1, δ^0 and δ^1 are the prescribed initial stresses, initial stress velocities, initial temperature, and initial temperature velocity; p and g are the given surface traction and temperature, respectively.[8]

The problem of finding a pair of functions (S,δ) on $\overline{B} \times [0,\infty)$ that satisfies the field Eqs. 41 and 42, the initial conditions, Eq. 46, and the boundary conditions, Eqs. 47 and 48, will be called a natural stress-temperature problem of the G-L theory. A solution of this problem is an ordered array (S,δ) that satisfies Eqs. 41-48.

Uniqueness Theorem. If a solution (S,δ) of the natural stress- temperature problem, Eqs. 41-48, exists, then this solution is unique. Proof of this theorem follows from the integral relation corresponding to the homogeneous data [see [15]]:

---

[8]Eqs. 41-48 are written in the dimensional form of Sec. 2. For convenience we omit superposed "hut" from x and t.

$$\frac{1}{2} \int_B dB \int_0^t dv \left\{ \frac{1}{\rho} (\text{div } S)^2 + \frac{1}{2\mu} \left(\dot{S} - \frac{1}{3} (\text{tr } \dot{S}) 1 \right)^2 \right.$$

$$+ \frac{1}{3(3\lambda + 2\mu)} \left[1 - \left(1 - \frac{c_E}{c_S} \right) \frac{\tau_1}{\tau} \right] (\text{tr } \dot{S})^2 \right\}$$

$$+ \frac{1}{2\theta_o} \int_B dB \int_0^t dv \left\{ \frac{k^2}{c_S} \frac{\tau_1}{\tau} (\nabla^2 \delta)^2 + k\tau_1 (\nabla \dot{\delta})^2 + \frac{\tau_1 - \tau}{\tau} c_S \, \dot{\delta}^2 \right\}$$

$$+ \frac{1}{\theta_o} \int_B dB \int_0^t dv(t - v) [\tau_1 - \tau) c_S \, \ddot{\delta}^2 + k(\nabla \dot{\delta})^2]$$

$$+ \frac{k}{2\theta_o} \frac{\tau_1 - \tau}{\tau} \int_B dB(\nabla \delta)^2 = 0 \tag{49}$$

in which the integrand is a sum of non-negative terms. This relation implies the uniqueness in the following cases:
1. $\tau_1 > \tau_o > 0$, 2. $\tau_1 > 0$, $\tau_o = 0$, 3. $\tau_1 = \tau_o = 0$. In the last case Eq. 49 reduces to

$$\frac{1}{2} \int_B dB \left\{ \frac{1}{\rho} (\text{div } S)^2 + \frac{1}{2\mu} \left(\dot{S} - \frac{1}{3} (\text{tr } \dot{S}) 1 \right)^2 \right.$$

$$+ \frac{1}{3(3\lambda + 2\mu)} \frac{c_E}{c_S} (\text{tr } \dot{S})^2 + \frac{k^2}{\theta_o c_S} (\nabla^2 \delta)^2 \right\}$$

$$+ \frac{k}{\theta_o} \int_B dB \int_0^t dv(\nabla \dot{\delta})^2 = 0 . \tag{50}$$

In the general case the LHS of Eq. 49 can also be used in constructing a linear space of pairs (S, δ) in which there exists a generalized solution to the problem defined by Eqs. 41-48.

5. VARIATIONAL CHARACTERIZATION OF STRESS AND HEAT FLUX IN G-L THEORY

In this section a variational principle corresponding to a natural stress-flux problem of the G-L theory is examined. Discussion is confined to the case of homogeneous initial data, zero body forces and zero heat supply.

By eliminating u, E and δ from Eqs. 1-5 with b = 0 and r = 0, and by incorporating the homogeneous initial data into the field equations one arrives at equations involving S and q only:

$$\frac{1}{\tau_1} e^{-\frac{t}{\tau_1}} * \left[\frac{t}{\rho} * \hat{\nabla}(\text{div } S) - \frac{1}{2\mu}\left(S - \frac{\lambda}{3\lambda + 2\mu}(\text{tr } S)1 \right) \right]$$

$$+ \frac{\alpha}{c_S}\left(1 - e^{-\frac{t}{\tau}} \right) * (\text{div } q + \alpha\theta_o \text{ tr } \dot{S})1 = 0 \quad ,$$

$$\frac{1}{c_S}\left(1 - e^{-\frac{t}{\tau}} \right) * \nabla(\text{div } q + \alpha\theta_o \text{ tr } \dot{S}) - \frac{1}{k}q = 0 \quad . \tag{51}$$

To these field equations which are to be satisfied on $B \times [0,\infty)$, we adjoin the stress boundary conditions

$$Sn = p \quad \text{on } \partial B \times [0,\infty) \quad , \tag{52}$$

and the heat-flux boundary conditions

$$q \cdot n = Q \quad \text{on } \partial B \times [0,\infty) \quad , \tag{53}$$

where p has the same meaning as in Section 4 and Q is a prescribed scalar field.

A solution (S,q) satisfying Eqs. 51-53 can be obtained by using the following.

Variational Principle. Let K be the set of all pairs (S,q) that satisfy the boundary conditions, Eqs. 52 and 53. Let $R = (S,q) \epsilon K$, and for each $t \epsilon [0,\infty)$ define the functional $\Omega_t\{\cdot\}$ on K through the relation

$$\Omega_t\{R\} = \frac{1}{2}\frac{1}{\tau}e^{-\frac{t}{\tau}} * \int_B \left\{ \frac{e^{-\frac{t}{\tau_1}}}{\tau_1} * \frac{t}{\rho} * (\text{div } S) * (\text{div } S) \right.$$

$$+ \frac{e^{-\frac{t}{\tau_1}}}{\tau_1} * \frac{1}{2\mu}\left[S * S - \frac{\lambda}{3\lambda + 2\mu}(\text{tr } S) * (\text{tr } S) \right]$$

$$- \frac{e^{-\frac{t}{\tau}}}{\tau} * \frac{\alpha^2\theta_o}{c_S}(\text{tr } S) * (\text{tr } S)$$

$$\left. - 2\frac{\alpha}{c_S}\left(1 - e^{-\frac{t}{\tau}} \right) * (\text{div } q) * (\text{tr } S) \right\}dB$$

$$- \frac{1}{2\theta_o} \left(1 - e^{-\frac{t}{\tau}}\right) * \int_B \left\{ \frac{1}{k} q * q \right.$$

$$+ \frac{1}{c_S} \left(1 - e^{-\frac{t}{\tau}}\right) * (\text{div } q) * (\text{div } q) \left. \right\} dB \ . \tag{54}$$

Then $\delta\Omega_t\{R\} = 0$ over K $(t \geq 0)$ if and only if (S,q) is a solution of the problem, Eqs. 51–53. Proof of this principle, which is quite analogous to a Nickell-Sackman result [16], is given by Ignaczak in [17].

The variational principle can be used to find (S,q) satisfying Eqs. 51–53 numerically.

6. BASIC EQUATIONS IN L-S THEORY

The L-S theory proposed in 1967 has been discussed by various authors more extensively than the G-L theory, and for this reason only two topics are discussed here: one concerning a particular solution (ϕ,θ) similar to that of Section 2, and another dealing with uniqueness for a stress-flux problem.

The governing equations of the L-S theory for a homogeneous isotropic thermoelastic body consist of the strain-displacement relation

$$E = \frac{1}{2} (\nabla u + \nabla u^T) \equiv \hat{\nabla} u \tag{55}$$

the equation of motion

$$\text{div } S + b = \rho\ddot{u} \tag{56}$$

the energy equation

$$- \text{div } q + r = c_E \dot{\delta} + (3\lambda + 2\mu)\alpha\theta_o \text{ tr } \dot{E} \tag{57}$$

the stress-strain-temperature relation

$$S = 2\mu E + \lambda(\text{tr } E)1 - (3\lambda + 2\mu)\alpha\delta1 \tag{58}$$

and the heat conduction equation

$$q + \tau_o^* \dot{q} = - k\nabla\delta \ , \qquad (\tau_o^* > 0) \ . \tag{59}$$

All symbols in Eqs. 55–59 have the same meaning as in Section 2. Note that Eqs. 55–58 are identical to corresponding equations of classical thermoelasticity. Equation 59 is often called the Maxwell-Cattaneo equation.[9]

If the dimensionless variables of Section 2 [cf. Eqs. 12 and 14] are introduced, a particular solution of Eqs. 55-59 is described by a dimensionless pair (ϕ, θ) satisfying the equations

$$\nabla^2 \phi - \ddot{\phi} = \theta \quad ,$$

$$\nabla^2 \theta - t_o^* \ddot{\theta} - \dot{\theta} = \epsilon \nabla^2 (\dot{\phi} + t_o^* \ddot{\phi}) \quad , \tag{60}$$

where

$$t_o^* = \tau_o^* / \hat{t}_o \quad . \tag{61}$$

By elimination of θ from Eq. 6 the central equation of the L-S theory is obtained:

$$\left[\left(\nabla^2 - \frac{\partial^2}{\partial t^2} \right) \left(\nabla^2 - \frac{t_o^*}{1} \frac{\partial^2}{\partial t^2} - \frac{\partial}{\partial t} \right) \right.$$

$$\left. - \epsilon \nabla^2 \frac{\partial}{\partial t} \left(1 + t_o^* \frac{\partial}{\partial t} \right) \right] \phi = 0 \quad . \tag{62}$$

Comparison of Eqs. 62 and 20 shows that Eq. 62 can be obtained from Eq. 20 by substitution of $t_1 = t_o = t_o^*$.[10] Therefore, for $t_1 = t_o = t_o^*$ the central equation of both theories coincide. However, the temperatures computed from ϕ in both cases are different. This proves that even in this particular case thermoelastic disturbances described by both theories take different forms.

For a pair (ϕ, θ) in the L-S theory one can establish a domain of influence theorem and a decomposition theorem similar to the results of Sections 2 and 3, respectively.

7.   UNIQUENESS FOR THE STRESS-FLUX PROBLEM IN L-S THEORY

By eliminating u, E and δ from Eqs. 55-59 one arrives at the system involving S and q only:

---

[9]A one-dimensional form of the governing Eqs. 55-59 is given in [18]. The energy equation in terms of δ and tr E, derived by Lord and Shulman in [4], is obtained by applying the operator $(1 + \tau_o^* \partial/\partial t)$ to Eq. 57 and taking Eq. 59 into account [see Eq. 23 of [4]]. A version of the L-S theory valid at law reference temperatures is given in [20].

[10]This fact was proved by Agarwal in [9].

$$\rho^{-1} \nabla (\text{div } S) - \frac{1}{2\mu} \left[\ddot{S} - \frac{\lambda}{3\lambda + 2\mu} (\text{tr } \ddot{S}) 1 \right]$$

$$+ \frac{\alpha^2 \theta_o}{c_S} (\text{tr } \ddot{S}) 1 + \frac{\alpha}{c_S} (\text{div } \dot{q}) 1 + \rho^{-1} \hat{\nabla} b = 0 \quad ,$$

$$\frac{1}{c_S} \nabla [\text{div } q + \alpha \theta_o (\text{tr } \dot{S})] - \frac{1}{k} (\dot{q} + \tau_o^* \ddot{q}) - \frac{1}{c_S} \nabla r = 0 \quad , \quad (63)$$

where c_S is given by Eq. 43.

(S,q) is a solution of the stress-flux-initial-boundary value problem in the L-S theory if (S,q) satisfies Eq. 63 on B × $[0,\infty)$, and if

$$S(x,0) = S_o(x) \quad , \quad \dot{S}(x,0) = S_1(x) \quad ,$$

$$q(x,0) = q_o(x) \quad , \quad \dot{q}(x,0) = q_1(x) \quad \text{on } \overline{B} \quad\quad\quad (64)$$

and

$$Sn = p \quad , \quad q \cdot n = Q \quad \text{on } \partial B \times [0,\infty) \quad , \quad\quad\quad (65)$$

where S_o, S_1, q_o and q_1 are prescribed initial stresses, initial stress velocities, initial heat-flux, and initial heat-flux velocity.

Uniqueness Theorem. If a classical solution (S,q) of the stress-flux-initial-boundary-value problem, Eqs. 63-65, exists, then this solution is unique. Proof of this theorem follows from the integral relation corresponding to the homogeneous data [see [19]]:

$$\frac{1}{2} \frac{1}{\rho} \int_B (\text{div } S) \cdot (\text{div } S) dB$$

$$+ \frac{1}{2} \int_B \left\{ \frac{1}{2\mu} \left[\dot{S} - \frac{1}{3} (\text{tr } \dot{S}) 1 \right] \cdot \left[\dot{S} - \frac{1}{3} (\text{tr } \dot{S}) 1 \right] \right.$$

$$+ \frac{1}{3(3\lambda + 2\mu)} \frac{c_E}{c_S} (\text{tr } \dot{S})^2 \Big\} dB + \frac{1}{\theta_o} \left[\frac{1}{2c_S} \int_B (\text{div } q)^2 dB \right.$$

$$+ \frac{\tau_o^*}{2k} \int_B \dot{q} \cdot \dot{q} \, dB + \frac{1}{k} \int_0^t d\tau \int_B \dot{q} \cdot \dot{q} \, dB \Big] = 0 \quad . \quad\quad\quad (66)$$

This relation can also be used to construct a linear space of pairs (S,q) in which a generalized solution to the problem, Eqs. 63-65, exists.

CLOSURE

   This survey presents a number of general theorems of the linear thermoelasticity with finite wave speeds which have their counterparts in the classical three-dimensional thermoelasticity. Emphasis was placed on the domain of influence and on uniqueness theorems. The results were obtained for smooth thermoelastic processes, and their extension to singular thermoelastic surfaces requires further research into this field. The following problems are in need of further investigation in the G-L and L-S theories:
1. Decomposition Theorems with Nonhomogeneous Initial Data,
2. General Domain of Influence Theorems, 3. Existence Theorems,
4. Green Functions, 5. Analysis of the Wave Operators with Convolution [see Eqs. 33 and 34].

   Moreover, both theories are critically in need of experimental verification. Francis [6] gives the following values of τ_0^* [in sec]:

   a. $\tau_0^* = 6.6 \times 10^{-14}$ for uranium dioxide at 25° C

   b. $\tau_0^* = 1.5 \times 10^{-12}$ for uranium silicate at 25° C

   c. $\tau_0^* = 2.0 \times 10^{-9}$ for liquid He II at 0.25 K

   d. $\tau_0^* = 8.0 \times 10^{-12}$ for aluminum alloys at 25° C

   e. $\tau_0^* = 1.6 \times 10^{-12}$ for carbon alloys at 25° C

These relaxation times can be used when a solution of the L-S theory is compared to corresponding experimental results. For the G-L theory existence of the relaxation times τ_0 and τ_1 should be proved in an experiment at least for some materials.

REFERENCES

[1] D. E. Carlson, Linear thermoelasticity, Encyclopedia of Physics, Mechanics of Solids II, vol. 6a/2, Springer, Berlin, 1972.
[2] W. Nowacki, "Dynamic Problems of Thermoelasticity," PWN, Warsaw, and Noordhoff, Leyden, 1975.
[3] V. D. Kupradze, T. G. Gegelia, M. O. Basheleishvili, and T. V. Burchuladze, "Three-dimensional Problems of the Mathematical Theory of Elasticity and Thermoelasticity," North-Holl. Publ. Co., Amsterdam, New York, Oxford, 1979.
[4] H. W. Lord and Y. Shulman, A generalized dynamical theory of thermoelasticity, J. Mech. Phys. Solids, vol. 15, pp. 299-309, 1967.

[5]   A. E. Green and K. A. Lindsay, Thermoelasticity, J.
       Elasticity, vol. 2, pp. 1-7, 1972.

[6]   P. H. Francis, Thermo-mechanical effects in elastic wave
       propagation - a survey, J. Sound and Vibration, vol. 21,
       no. 2, pp. 181-192, 1972.

[7]   A. Nayfeh and S. Nemat-Nasser, Thermoelastic waves in solids
       with thermal relaxation, Acta Mechanica, vol. 12, pp. 35-
       69, 1971.

[8]   P. Puri, Plane waves in generalized thermoelasticity, Int. J.
       Engng. Sci., vol. 11, pp. 735-744, 1973. Errata: Int. J.
       Engng. Sci., vol. 13, pp. 339-340, 1975.

[9]   V. K. Agarwal, On plane waves in generalized thermo-
       elasticity, Acta Mechanica, vol. 31, pp. 185-198, 1979.

[10] J. D. Achenbach, The influence of heat conduction on propa-
       gating stress jumps, J. Mech. and Physics of Solids,
       vol. 16, pp. 273-282, 1968.

[11] J. Ignaczak, Domain of influence theorem in linear thermo-
       elasticity, Int. J. Engng. Sci., vol. 16, pp. 139-145,
       1978.

[12] J. Ignaczak, Thermoelastic Counterpart to Boggio's Theorem
       of Linear Elastodynamics, Bull. Acad. Polon. Sci., Ser.
       Tech., vol. 24, no. 3, pp. 129-137, 1976.

[13] J. Ignaczak, Decomposition theorem for thermoelasticity with
       finite wave speeds, J. Thermal Stresses, vol. 1, no. 1,
       pp. 41-52, 1978.

[14] J. H. Weiner, A uniqueness theorem for the coupled thermo-
       elastic problem, Q. Appl. Math., vol. 15, pp. 102-105,
       1957.

[15] J. Ignaczak, A uniqueness theorem for stress-temperature
       eqs. of dynamic thermoelasticity, J. Thermal Stresses,
       vol. 1, no. 2, pp. 163-170, 1978.

[16] R. E. Nickell and J. L. Sackman, Variational principles for
       linear coupled thermoelasticity, Q. Appl. Math., vol. 26,
       pp. 11-26, 1968.

[17] J. Ignaczak, Variational characterization of stress and heat
       flux in dynamic thermoelasticity, in Polish, Matematyka,
       WSP, Kielce 1979, in print.

[18] E. B. Popov, Dynamic coupled problem of thermoelasticity
       for a semi-space taking into account finite speed of
       heat propagation, in Russian, Prikl, Mat. Mekh., vol. 31,
       pp. 328-334, 1967.

[19] J. Ignaczak, Uniqueness in generalized thermoelasticity,
       J. Thermal Stresses, vol. 2, no. 2, pp. 171-175, 1979.

[20] Y. H. Pao and D. K. Banerjee, A theory of anisotropic
       thermoelasticity at law reference temperature, J. Thermal
       Stresses, vol. 1, no. 1, pp. 99-112, 1978.

THERMAL STRESSES: TRANSIENT AND PERMANENT

Robert Gardon

Research Staff
Ford Motor Company
Dearborn, Michigan

ABSTRACT

Under certain circumstances and in certain materials, (transient) thermal stresses can give rise to permanent stresses. The mechanisms involved in the genesis of such "frozen-in" thermal stresses are briefly reviewed in the context of the tempering of glass. The main purpose of the paper is to trace the development of theoretical treatments of this process. Some well developed computational schemes are now available, which may well find wider application.

INTRODUCTION

Taking the name of this conference to reflect a concern for the continued performance or even survival of objects that, in use, are subjected to severe thermal environments, I could not see how my interests would fit in. Then I realized that one man's severe environment may be another's deliberate choice for some industrial process. Thus, in the manufacture of tempered safety glass, the glass is first heated almost to the point of losing its shape and then severely quenched to "freeze in" the stresses that give it the desired strength and fracture characteristics. I will make our progress in understanding this process a vehicle for discussing thermal stresses in a non-elastic material and methods of their mathematical analysis.

First, let us see how, in tempering glass, transient thermal stresses are converted into permanent stresses. In order to concentrate on the physics involved, we will look at a very simple geometry: namely, the central region of a large plate that is cooled from both surfaces, so that the temperature distribution of interest is one-dimensional and symmetric about the midplane. The resulting stresses will be uniform plane stresses, parallel to

31

the plate, and two-dimensionally isotropic, varying only in the x-direction, normal to the plate.

Initially at a uniform temperature T_o, the plate is quenched in air at room temperature and with a constant, high heat transfer coefficient. The temperature-time histories of its surface and midplane are shown schematically in Fig. 1a. Figure 1b shows the variation of the temperature difference ΔT between the mid-plane and surface. Note that this temperature difference at first increases rapidly, reaches a maximum and then decays to zero as the glass returns to isothermal conditions at room temperature.

If we were dealing with a purely elastic material, thermal stresses in the plate would parallel the temperature differences causing them: Colder regions would tend to contract, but be partially restrained by warmer regions; so that regions below the average temperature of the plate would be in tension, those above in compression. Except for a brief transient immediately following the onset of quenching, temperature distributions in the plate are parabolic. It follows that stresses in the plate also have a parabolic distribution, with tensile stresses in the colder surface layers being balanced by compression in the interior. Stresses will increase while ΔT increases, and decrease as ΔT decreases. They will vanish when, at ambient conditions, ΔT shall have vanished. Indeed, the plot of ΔT vs. time - to an appropriate scale - may also be regarded as the stress-time history of the mid-plane of an elastic plate, as indicated by curve 1 of Fig. 1c.

Such simple thermal stresses are probably familiar to most. For this particular geometry they are given by

$$\sigma(x,\ t) = \frac{E\beta}{1-\nu}\ \left(\overline{T(x,t)} - T(x,t)\right) \tag{1}$$

where $\overline{T(x,t)}$ = average temperature (averaged over x), corresponding to the instantaneous temperature distribution $T(x,t)$

             E = Young's modulus

             β = coefficient of linear expansion

             ν = Poissons's ratio.

Let us turn next to a glass plate quenched from an initial temperature high enough for stress relaxation by viscous flow to come into play. As quenching begins, the surfaces tend - as in the elastic case - to go into tension, the mid-plane into compression. Now, however, this is partly offset by simultaneous stress relaxation, so that stresses rise less rapidly than in the elastic material. The maximum transient compression in the mid-plane is, correspondingly, also lower, as shown by curve 2 in Fig. 1c. Stresses in this initially heat-softened material are thus seen to be lower than they would be in an otherwise identical, but always elastic material having the same temperature distribution. At lower temperatures the glass becomes elastic again. From here on, further changes in stresses are determined solely by changes in the temperature distribution, so that curve 2 for the glass plate runs parallel to curve 1 for the elastic plate. In a material that had been elastic throughout this process, the stresses produced by the decay of temperature gradients would be equal and opposite to those produced by their growth. In case 2, these two are not equal, since some of the stresses generated during the early phase of cooling were relaxed by viscous flow. As a result, the glass ends up with some stresses in it even after its temperature has become uniform. These are the desired temper stresses: desired because they compressively prestress the surfaces of the glass, thus raising its effective strength in a manner analogous to that of prestressed concrete. Note also that the frozen-in, permanent stresses are opposite in sign to the transient stresses in an elastic body subjected to the same heat treatment.

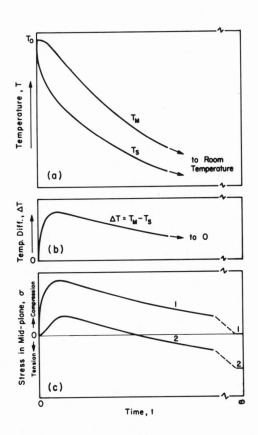

Fig. 1   Schematic temperature- and stress-histories during the quenching of a (glass) plate

So much for a very simple picture of the basic idea of tempering. In fact, stress distributions in glass during tempering are not always parabolic, so that the mid-plane stress cannot always be taken as a measure of stresses elsewhere. The great variety of transient stress distributions that can arise is illustrated by Fig. 2, drawn from selected frames of movies of photoelastic fringe patterns observed during tempering. [1]

The problem I propose to address is how such transient stresses and the resulting "frozen-in" thermal stresses may be calculated, given the process parameters and relevant physical properties of our material. Clearly, the material must be "heat softenable", and we will consider different levels of sophistication in describing - and treating - this particular property.

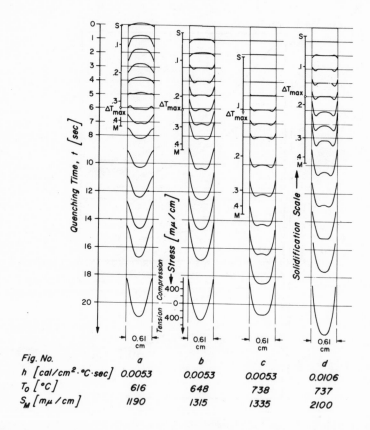

Fig. No.	a	b	c	d
h [cal/cm$^2 \cdot °C \cdot sec$]	0.0053	0.0053	0.0053	0.0106
T_0 [°C]	616	648	738	737
S_M [mμ/cm]	1190	1315	1335	2100

Fig. 2 Stress distributions through the thickness of glass plates at various times during quenching

GENERAL CONSIDERATIONS

Any theoretical treatment of transient thermal stresses must start with a knowledge of the temperature-position-time history of the object in question. The thermal history of our glass plate during quenching is determined by the initial temperature T_0 of the plate, the temperature of the quenching medium and the heat transfer coefficient governing the rate of heat exchange between the two. With these known, the calculation of glass temperatures is essentially an exercise in transient thermal conduction, thermal effects of viscous flow in the glass being wholly negligible and radiative effects only secondary.[1]

Once the temperature-position-time history of our plate is known, its mechanical response to heat treatment is governed by three perfectly general stress-analytical considerations:

(i)   The condition of (geometrical) <u>compatibility</u> of strains: Since the plate must expand or contract as a whole, this requires that strains be independent of x. They can therefore vary only with time. Thus,

$$\epsilon_y = \epsilon_z = \epsilon = \epsilon(t) \tag{2}$$

(ii)   The <u>equilibrium</u> of the plate requires that stresses be balanced, i.e., that the average stress through the thickness be zero. Thus,

$$\overline{\sigma(x,t)} = \frac{1}{\ell} \int_0^\ell \sigma(x,t)\, dx = 0 \tag{3}$$

(iii) The law governing stress relaxation, which is also the <u>stress-strain relation</u> for the glass.

Most commonly, we deal with elastic materials, for which the stress-strain relation is simply Hooke's law; stress relaxation is not involved and time, as such, is not a significant variable. I want to review some approaches to incorporating more complex and time-dependent stress-strain relations into the handling of a geometrically relatively simple problem in stress analysis. To do this requires either a model of stress relaxation simple enough to be used in conjunction with a more-or-less conventional analysis of stresses in a body subjected to an arbitrary temperature-time history; or, alternatively, computational methods in stress analysis that are capable of also handling more realistic and more complicated models of stress relaxation. I will give an example of each of these two approaches.

"INSTANT FREEZING" THEORY

The first attempt at a quantitative treatment of stress relaxation in glass during tempering was that of Bartenev[2], who developed the first of what I will call "instant freezing" theories. He postulated that - in view of the rapid variation of the viscosity of glass with temperature and of temperature with time during tempering - glass be treated as "solidifying" at a single "solidification temperature" T_g. In this simple model, glass above T_g is treated as a fluid incapable of supporting any stress, while glass below T_g is regarded as an elastic solid in which no flow can occur and no stress relaxation. Accordingly, the solidification of a glass plate progresses simply as the T_g isotherm travels from the surface to the mid-plane.

Bartenev's formulation of the problem was ingenious, but his subsequent treatment had flaws that make it of no interest to this audience. However, a stress-analytically rigorous solution - still based on Bartenev's formulation - was obtained a few years later by Indenbom [3]. His point of departure was the recognition that proper application of the compatibility condition to a plate solidifying in accordance with the "instant freezing" model requires separate consideration of the elastic, viscous and thermal components of strain. Each of these is a function of x as well as t, even while the total strain (as per Eq. 2) is a function of t only. For details of this treatment, interested readers are referred to references 3 or 4. The point to be made here is that, using little more than the tools of conventional elastic stress analysis, Bartenev and Indenbom were quite successful in describing the end results of a process that also involves stress relaxation. Admittedly, their treatment applies only to a somewhat special and highly idealized case. I also advisedly stressed "end results" for - while their calculations yield remarkably good predictions of the temper stresses ultimately produced - they say nothing at all about the transients involved. The use of this approach to calculating transient stresses in the glass during tempering yields meaningless results. The reason for this is that the instant freezing hypothesis is too crude a model of stress relaxation in glass to account for details of events in the critical temperature range. Its principal virtue lay in its simplicity that allowed it to be coupled to the other considerations involved in the analysis of thermal stresses - even without a high-speed computer.

In fact - and unlike crystalline materials - a glass changes from a liquid to an elastic solid not at a single temperature but over an extended temperature interval, in which it exhibits both viscous and elastic characteristics.

"VISCOELASTIC" THEORY

The next important advance in tempering theory flowed from the development by Morland and Lee [5] and Lee and Rogers [6] of computational methods of stress analysis in viscoelastic materials having temperature-dependent stress relaxation functions. In particular, they treated thermo-rheologically simple, linear viscoelastic materials. These developments were applied to the tempering of glass by Lee, Rogers and Woo. [7]

A viscoelastic material is said to be thermorheologically simple if the change with temperature of its relaxation modulus can be allowed for by a change of the time-scale only. [8] This characteristic is illustrated by Figs. 3

and 4.[9]   The first of these shows the decay with time of shear stresses evoked in glass - first brought to equilibrium at various temperatures - by a stepwise application of a shear strain at time t=0.   The normalized stress shown is the actual stress at time t, divided by the instantaneous (elastic) stress produced at t=0 upon straining the glass. These curves are drawn on the basis of log(time), and it may be noted that they all have the same shape.   This circumstance allows them to be brought into coincidence by shifting them along the log(time)-axis. The result is a "master" relaxation curve, such as that shown in Fig. 4, drawn for an arbitrarily selected base temperature of $T_B=473°C$.

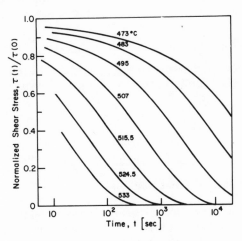

Fig. 3 Stress relaxation in glass as a function of stabilization temperature
        and time

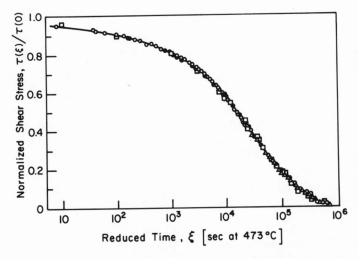

Fig. 4 Stress relaxation in stabilized glass as a function of reduced time

That the relaxation curves can be shifted in this manner shows that a stabilized glass, i.e. one brought to equilibrium at a given temperature, is indeed a thermorheologically simple viscoelastic material. That tempered glass, which is rapidly chilled, is not stabilized need not concern us for the moment.

The magnitude of the shift required to bring the curves of Fig. 3 into coincidence is called the shift function, $\phi(T)$, which depends on the ratio of the viscosities at the temperatures T and T_B, respectively. Making the shift is equivalent to counting time faster at temperatures higher than T_B, and more slowly at temperatures lower than T_B. Time measured at the base temperature is called "reduced time", ξ, and it is defined by

$$d\xi = \phi(T) \cdot dt \tag{4}$$

The (time-dependent) shear stress $\tau(t)$ produced in a linear viscoelastic material by a step-wise applied shear strain of magnitude γ is - analogously to the elastic case -

$$\tau(t) = G(t) \cdot \gamma \tag{5}$$

only that now G(t) is a time dependent shear modulus, or "relaxation modulus", obtainable from Fig. 4 by multiplying its ordinates by G(0), the instantaneous - or elastic - shear modulus.

Since Eq. (5) is linear in γ, the response of glass to a succession of strain increments may be obtained by superposition. This yields a more general stress-strain relation in integral form

$$\tau(t) = \int_0^t G(t-t') \frac{d\gamma(t')}{dt'} dt' \tag{6a}$$

where   t   is the time of interest;
          t' is a running time variable, from 0 to t, during which
              strain increments are accumulated at a rate $d\gamma(t')/dt'$;
and   G(t-t'), the "memory kernel," represents a measure of the
              remembered part of the corresponding stress increments.

If, in addition to varying strains, temperatures also vary, the reduced time must be substituted for the real time t. This leads to the relation

$$\tau(\xi) = \int_0^\xi G(\xi-\xi') \frac{d\gamma(\xi')}{d\xi'} d\xi' \tag{6b}$$

A similar relation may be written to relate tensile stresses $[\sigma(\xi)]$ to the tensile strain-rate $[d\epsilon(\xi')/d\xi']$ and a time-dependent Young's modulus $[E(\xi)]$. In fact, the stresses produced in a plate by tempering are neither pure shear nor simple tension, but two-dimensionally isotropic tension. Muki and Sternberg [10] derived an auxiliary relaxation modulus $R(\xi)$, which is for such a stress system what $G(\xi)$ and $E(\xi)$ are for pure shear and tension, respectively.

Using this relaxation modulus $R(\xi)$, Lee et al.[7] derived the following expression for transient stresses in tempering

$$\sigma(x,t) = 3\int_0^t R\Big(\xi(x,t) - (\xi(x,t'))\Big) \frac{\partial}{\partial t'}\Big[\varepsilon(t') - \beta T(x,t')\Big]\partial t' \qquad (7)$$

Note that in this equation the reduced time ξ is a function of position x as well as of time t, reflecting the x-dependence of temperatures and therefore also of the shift function. Eq. 7 incorporates both the stress-strain relation and compatibility condition (Eq. 2). From Eq. (7) and the equilibrium condition (Eq. 3) the two unknowns, $\sigma(x,t)$ and $\varepsilon(t)$, can be obtained. For details of the derivation and numerical solution the reader is referred to Ref. 7. All that should be noted here is that Eq. 7 can be solved only numerically and step-by-step in time. The need to proceed step-by-step, rather than directly to any desired point in time, arises because the solution for any given layer within the plate cannot be carried out independently of that for all other layers. Rather, the total strain $\varepsilon(t)$ must be determined at each intermediate time-step, and stresses in all layers redistributed to satisfy the equilibrium condition at that instant. The individual stress increments thus calculated therefore represent stresses generated by two simultaneous mechanisms:

(i)   The direct generation of stress increments in response is small, step-wise changes in thermal strain, produced by the changing temperature distribution across the thickness of the glass; and

(ii)   The redistribution of the sum-total of "remembered" stresses in order that equilibrium be maintained across the thickness of the plate, even though stresses in different layers are relaxing at different rates.

These calculations necessarily require a high-speed computer. Even so, the step-by-step procedure outlined above is followed only as long as any part of the glass is still hot enough to relax. As, with decreasing temperatures, the effects of stress relaxation diminish to where they may be regarded as negligible, these "viscoelastic" calculations can advantageously be stopped. The further increment of stresses, from here to isothermal conditions at room temperature, are "temperature equalization stresses" in a wholly elastic plate, that can be calculated very simply by Eq. 1.

With a minor modification of Lee et al.'s treatment, bearing only on computational technique to improve the precision of integrating Eq. 7, this theory was found to be in good agreement with experimental data - but still only over a limited range of conditions.[11] Thus, the "viscoelastic" theory appeared to be superior to the "instant freezing" theory only in that it could predict transient stresses as well as final temper stresses. In quenching from low initial temperatures neither theory fitted the experimental facts. It was believed that the reason for this was that, in quenching from low temperatures, some aspects of the glass transition that are not comprehended by the viscoelastic theory also become significant. Indeed, it is known that stress relaxation in glass occurs in a temperature interval in which the structure of the glass is also changing, and with that its physical properties. In other words, tempered glass, unlike stabilized glass, is not a thermorheologically simple viscoelastic material.

"STRUCTURAL" THEORY

The next, and for the present, last step in refining the theory of tempering was to take account of structural changes in the glass as it is being cooled through the transformation range. Such changes were thought likely to affect tempering in two ways: Structure-related changes in viscosity would modulate the rates of stress relaxation already comprehended in the preceding "viscoelastic" theory, and structure-related changes in density would have to be considered along with thermal strains as a potential source of frozen-in stresses.

First, a word about such structure-dependent property changes: When glass-forming liquids are cooled, their rapidly increasing viscosity inhibits their crystallization and causes them to retain the structure of a liquid, even while they become progressively more rigid. At low temperatures, i.e., in the "glassy" state, the structural arrangement of atoms is frozen, and thermal expansion or contraction reflect only changes in interatomic distances. These changes are instantaneous, reversible and governed by the coefficient of expansion β_g for the glass. At much higher temperatures, the liquid is mobile enough for its volume changes also to be instantaneous and reversible. The corresponding "liquid" coefficient of expansion, β_ℓ, is larger than β_g for the reason that, in the liquid, temperature changes affect the arrangement of atoms as well as distances between them. At intermediate temperatures, in what is called the transformation range of glass, viscosities are so high that structural rearrangements cannot keep pace with temperature changes, but lag behind. The structure is then dependent on time (or cooling rate) as well as temperature; and the same goes for structure-related properties, such as density and viscosity.

Structural relaxation - i.e., the change with time of structure and structure-dependent properties - is inherently more complicated than stress relaxation. While the classic experiment on stress relaxation - namely observation of the decay of stresses following a step change in strain - can be done isothermally (cf. Fig. 3), the equivalent approach to studying structural relaxation necessarily entails a step function in temperature. Thus, even if one wishes to observe the relaxation only of, say, density, one cannot avoid simultaneous changes in other properties, including viscosity, which itself determines the rate of all relaxation processes, including that of density as well as stress. As a result, structural relaxation is inherently non-linear, an effect first accounted for in this manner by Tool[12].

Following Tool, it is helpful to characterize the structural state of the glass by its fictive temperature T_f. This may be thought of as that temperature from which glass at equilibrium must be rapidly quenched in order to produce its particular structural state. The physical properties of a glass are thus functions not only of the actual temperature T, but also of the fictive temperature T_f, which, in the transformation range, also depends on time (cf. Fig. 18 of Ref. 4). Thus, the "glassy" component of volume changes can reasonably be expected still to follow temperature changes instantly, while the structural rearrangement is delayed, tending - more or less exponentially - to an equilibrium volume governed by the "structural" expansivity $\beta_s = \beta_\ell - \beta_g$.

To allow for these effects in tempering requires a model of the relaxation - i.e., change with time - of the glass structure in a form that permits prediction of the evolution of structure-dependent properties for arbitrary temperature-time histories. Such a model was developed by Narayanaswamy[13], using the mathematical formalism that Lee and Rogers had developed for handling stress relaxation in viscoelastic materials with temperature-dependent properties.

Accordingly, the scheme for computing stresses by the "structural" model of tempering is basically the same as that for the viscoelastic model, only that now the evolution of structure-dependent properties must be followed as well as that of stresses.[14] Structural changes in density are taken into account along with ordinary thermal expansion by re-writing Eq. 7 of the viscoelastic theory with an additional term:

$$\sigma(x,t) = 3\int_0^t R\big(\xi(x,t) - \xi(x,t')\big)\frac{\partial}{\partial t'}\Big[\epsilon(t') - \beta_g T(x,t') - \beta_s T_f(x,t')\Big]dt' \quad (8)$$

This is the last term in the square bracket and represents the local strain due to structural volume changes in the glass. The effect of structural changes in viscosity on the stress-time history is allowed for by expanding or contracting the time scale of the modulus R not only in response to temperature changes but also in response to changes in the fictive temperature T_f.

Given the temperature history T(x,t) of the glass, the required history of fictive temperatures $T_f(x,t)$ must therefore be calculated first of all. This requires an iterative solution, by successive approximations, of a set of coupled equations, relating $T_f(x,t)$ and $\xi(x,t)$, both of which are initially unknown. Once both T(x,t) and $T_f(x,t)$ are known for all times, Eq. 8 can be solved for $\sigma(x,t)$, following the step-by-step procedure touched upon before in connection with the viscoelastic case.

That this approach works rather well is shown in Ref. 14, from which Fig. 5 is reproduced. This figure shows measured and calculated transient stress distributions at various times during tempering. Note that the structural model matches experimental observations quite well, while - for the rather low initial temperatures of this particular experiment (cf. Fig. 1) - the simpler viscoelastic model yields an altogether false picture of transient stress distributions. That, nevertheless, its prediction of the residual stress (σ_{ct}) matches the experimental observation better than that of the structural model will be touched upon below.

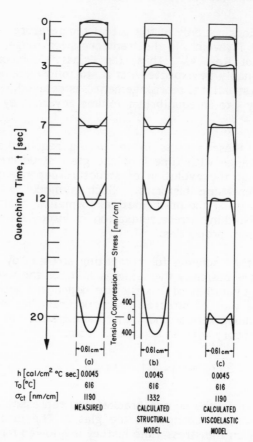

Fig. 5 Comparison of measured transient stress distributions with those predicted by the structural and viscoelastic models of tempering

DISCUSSION

Perhaps the first point to be made in comparing the three theories of tempering reviewed above is that they are closely related. Thus the "instant freezing" hypothesis can be recognized as a special case of the "viscoelastic" theory; which, in turn, is a special case of the "structural" theory. Starting with the last of these, the viscoelastic theory may be derived by suppressing all structure-related changes in physical properties, i.e. by setting $T_f \equiv T$ or $\beta_s \equiv 0$. Indenbom's theory can be derived from that of Lee et al. by setting $R(\xi(x,t) - \xi(x,t'))$ in Eq. 7 identically equal to zero for $t' < t_g$, and equal to $R(0) = E(0)/3(1-\nu)$ for $t' > t_g$, where t_g is the time at which $T(x,t_g) = T_g$.

Discrepancies between experimental observations and the predictions of earlier theories of tempering could be ascribed to incompleteness of the theories.   Thus it was expected that shortcomings of the instant freezing theory would be corrected by the viscoelastic theory; and its shortcomings, in turn, by allowance for structural effects.   To some extent these expectations have been fulfilled, notably as regards capability to calculate transient stresses in the temperature domain where properties are rapidly changing (cf. Fig. 5). However, in some other respects predictions of the later theories are no better than those of the first. One reason for this is that tests of these theories also constitute tests of the available data on relevant physical properties of glass.   Shortcomings of these data can be more clearly recognized now, with the availability of a comprehensive theory and attempts to interpret the results of tempering experiments, using data on physical properties obtained from other, independent experiments.

Even if all relevant physical properties were known, they could, of course, be used only with the most comprehensive theory, since this is the only one that refers to a "real" glass. The other theories necessarily require some "effective" physical properties. In the light of the structural theory one can recognize, for example, that the coefficient of expansion β to be used with the viscoelastic theory (Eq. 7) should lie between β_g and β_ℓ, the latter being typically three times as large as the former. Predictions of the viscoelastic theory can evidently be no better than one's estimate of the applicable value of β.   On the other hand, predictions can always be improved by a judicious choice of "physical properties", notably by deriving these from the same, process-related experiments. Thus, if one treats β as an adjustable parameter, the viscoelastic theory can be made to fit experimental data on temper stresses perfectly.   The effective values of β range from 1.5 β_g for relatively slow cooling to only 1.2 β_g for severe quenching.   These figures reflect structural contributions to permanent temper stresses of from 34 to 17%, respectively.

CONCLUSION

The aim of this paper has been to illustrate some approaches to calculating thermal stresses in non-elastic materials. For this we reviewed three progressively more sophisticated models of stress relaxation in glass and their quantitative development to allow prediction of the thermal stresses that can be "frozen into" glass.

The first model was ingenuously simple in that it allowed the problem to be treated as one in elastic stress analysis.

The second example introduced a method of stress analysis in linear viscoelastic materials, with temperature-dependent properties, that are subjected to arbitrary temperature-time histories.   This method was developed for thermorheologically simple materials, for which an equivalence of temperature and time exists, such that a change in the former can be allowed for by a change in the scale of the latter. This is probably the most general topic I have dealt with here; and the most concrete idea I hope to leave with those of you not already familiar with it is some acquaintance with this very powerful development by Professor Lee and his coworkers.

The third example dealt with a "real" glass - a viscoelastic material that is still linear as far as its stress-strain relation is concerned, but not thermorheologically simple. Its physical properties vary - in an inherently non-linear manner - not only with temperature but also with its changing (i.e. time-dependent) structure. It is noteworthy that the method employed to follow the structural relaxation of glass was built on the mathematical formalism of Lee's treatment of stress relaxation in viscoelastic materials.

Finally, of course, glass is not everybody's cup of tea and viscoelasticity not the only possible departure from elasticity. Even so, I hope that - within an interdisciplinary conference such as this - a review of these developments may stimulate an exploration of their applicability to other fields.

REFERENCES

1.   R. Gardon, "Tempering Flat Glass by Forced Convection," Paper No. 79 in Proc. 7th Internat. Congress on Glass, Bruxelles, 1965; Inst. Nat. du Verre, Charleroi, Belgium.

2.   G. M. Bartenev, "A Study of the Tempering of Glass," Zh. Tekh. Fiz. 19, 1423-33 (1949).

3.   V. L. Indenbom, "Theory of Tempering Glass," Zh. Tekh. Fiz. 24, 925-8 (1954).

4.   R. Gardon, "Thermal Tempering of Glass," in Vol. 5 of "Glass Science & Technology," D. R. Uhlman and N. J. Kreidl, editors, Academic Press, New York (1980)

5.   L. W. Morland and E. H. Lee, "Stress Analysis for Linear Viscoelastic Materials with Temperature Variation," Trans. Soc. Rheology 4, 233-63, (1960).

6.   E. H. Lee and T. G. Rogers, "Solution of Viscoelastic Stress Analysis Problems Using Measured Creep or Relaxation Functions," J. Appl. Mech. 30, 127-33 (1963).

7.   E. H. Lee, T. G. Rogers and T. C. Woo, "Residual Stresses in a Glass Plate Cooled Symmetrically from both Surfaces," J. Am. Ceram. Soc. 48, 480-87 (1965).

8.   F. Schwarzl and A. J. Staverman, "Time-Temperature Dependence of Linear Viscoelastic Behavior," J. Appl. Phys. 23, 838-43 (1952).

9.   C. R. Kurkjian, "Relaxation of Torsional Stress in the Transformation Range of a Soda-Lime-Silica Glass," Phys. Chem. Glasses 4, 128-36 (1963).

10. R. Muki and E. Sternberg, "Transient Thermal Stresses in Viscoelastic Materials with Temperature-Dependent Properties, J. Appl. Mech. 28, 193-207 (1961).

11.   O. S. Narayanaswamy and R. Gardon, "Calculation of Residual Stresses
       in Glass," J. Am. Ceram. Soc. 52, 554-58 (1969).
12.   A. Q. Tool, "Relation Between Inelastic Deformation and Thermal
       Expansion of Glass in Its Annealing Range," J. Am. Ceram.
       Soc. 29, 240-53 (1946).
13.   O. S. Narayanaswamy, "A Model of Structural Relaxation in Glass,"
       J. Am. Ceram. Soc. 54, 491-98 (1971).
14.   O. S. Narayanaswamy, "Stress and Structural Relaxation in Tempering
       Glass," J. Am. Ceram. Soc. 61, 146-52 (1978).

A COUPLED, ISOTROPIC THEORY OF THERMOVISCOPLASTICITY AND ITS PREDICTION FOR STRESS AND STRAIN CONTROLLED LOADING IN TORSION

E.P. Cernocky

E. Krempl

Dept. of Mechanical Eng.
University of Colorado
Boulder, CO 80301

Dept. of Mechanical Eng.
Aeronautical Eng. & Mechanics
Rensselaer Polytechnic Institute
Troy, NY 12181

ABSTRACT

A coupled, isotropic, infinitesimal theory of thermoviscoplasticity was developed in [32,33] and applied to a variety of loadings including thermal monotonic and cyclic straining [32]. In this paper we rederive the coupled equations using the first law of thermodynamics. The predictions of the theory in torsion are examined qualitatively and by numerical experiments. They simulate monotonic loading at loading rates differing by four orders of magnitude. Jumps in loading rate are also included. The theory exhibits initial linear elastic response followed by nonlinear, rate-dependent plastic behavior. The adiabatic temperature changes are initially isothermal followed by heating. The theory exhibits rate-dependence, a difference in strain and stress controlled loading and deformation induced temperature changes which are qualitatively in agreement with recent experiments.

INTRODUCTION

Mechanical deformation of materials is accompanied by deformation induced temperature changes. At low loading rates these temperature changes go unnoticed due to the equilibrating effect of heat conduction. However in rapid monotonic loading or in rapid large-amplitude cycling, inelastic deformation and self-heating of materials can become important; this may even result in melting of the material [1]. In particular situations it may be necessary to accurately calculate the temperature changes generated during deformations; examples are given in [2].

Experiments [3-21] have indicated significant strain (stress)-rate dependence in the mechanical behavior of metals both at static (10^{-8} to 10^{-2} sec^{-1}) and dynamic (10^2 sec^{-1} or more) strain rates; rate-dependence is significant at low, room, and elevated temperatures. The theory proposed here is intended to represent both this rate-dependent mechanical behavior and upon augmentation [22,32,33] the hysteretic behavior observed in metals, including cyclic hardening, cyclic softening, and the Bauschinger effect. The constitutive equation of heat conduction is proposed in order to qualitatively represent deformation-induced temperature changes measured in experiments [23-31].

The constitutive equations of this theory are proposed first. Then the predictions of the proposed equations are studied qualitatively under adiabatic and spatially-homogeneous conditions. Torsional constant strain (stress) -rate loading and loading with sudden jumps in strain (stress) -rate are examined. It is shown that large instantaneous increases in the strain (stress) -rate immediately result in thermoelastic material response. The theory is shown to predict a difference between material responses to displacement-controlled loading and to load-controlled loading.

CONSTITUTIVE EQUATIONS

The Mechanical Constitutive Equation

We propose this theory within the context of infinitesimal deformations, and we designate ε as the infinitesimal strain tensor and σ as the stress tensor. The absolute temperature is θ while the reference temperature is θ_0. A superimposed dot denotes differentiation with respect to time, and the usual comma-notation is employed to represent differentiation with respect to position coordinates. Following [32] we propose the isotropic mechanical constitutive equation

$$E\dot{\psi}_{ij} - \dot{\sigma}_{ij} = \frac{1}{k[\ ]} (\sigma_{ij} - G_{ij}[\underset{\sim}{\varepsilon},\theta]) \qquad (1)$$

with

$$\psi_{ij} = \frac{\nu\varepsilon_{nn}}{(1+\nu)(1-2\nu)} \delta_{ij} + \frac{\varepsilon_{ij}}{(1+\nu)} - \frac{\alpha}{(1-2\nu)} (\theta-\theta_0)\delta_{ij} \qquad (2)$$

and

$$\Gamma = \{(\sigma_{ij} - G_{ij})(\sigma_{ij} - G_{ij})\}^{1/2} \qquad (3)$$

In the above, square brackets denote a function of the indicated arguments, α is the coefficient of thermal expansion, ν is Poisson's ratio, and E is the elastic modulus. Here for brevity E, α, and ν will be approximated as constant; in reality these material parameters depend slowly upon temperature and ν also varies with deformation; such cases are considered in [32,33].

In this paper $k[\ ]$ is a positive and bounded function of Γ only. We identify $\underset{\sim}{G}[\ ]$ as the equilibrium stress-strain response of the material. This corresponds to the limit of the stress-strain response in constant strain (stress) -rate loading as the strain (stress) -rate approaches zero [32]. The equilibrium response represents a lower bound upon the mechanical response which the theory predicts.

The difference between flow stress and the equilibrium stress is $(\underset{\sim}{\sigma}-\underset{\sim}{G})$, and this is referred to as the overstress. The invariant gamma represents the magnitude of the overstress-tensor, and the nonlinear dependence of $k[\ ]$ upon overstress permits (1) to represent the nonlinear rate-sensitivity of real metal behavior [22].

We represent $\underset{\sim}{G}[\ ]$ by writing

$$G_{ij}[\underset{\sim}{\varepsilon}, \theta] = \psi_{ij} \frac{g[\phi]}{\phi} \quad * \tag{4}$$

The symbol ϕ denotes the "effective strain"

$$\phi = \left(\frac{1.5\ e_{ij}e_{ij}}{(1+\nu)^2}\right)^{1/2} \tag{5}$$

where $\underset{\sim}{e}$ is the deviatoric strain tensor. The function $g[\ ]$ is required to have the appearance of a uniaxial stress-strain curve, and

$$\frac{g[\phi]}{\phi} \approx E \quad \text{when } \phi \approx 0 \tag{6}$$

Thermomechanical Coupling

Thermoelasticity. In the context of infinitesimal deformations the first law of thermodynamics may be written as

$$\rho\dot{e} = \sigma_{ij}\dot{\varepsilon}_{ij} - q_{i,i} + \rho R \tag{7}$$

where ρ is the constant mass density, e is the internal energy per unit mass, $q_{i,i}$ is the divergence of the heat flux vector, and R is an internal heat supply per unit mass. In thermoelastic deformations the equation of heat conduction is [35]

$$\rho C\dot{\theta} = -q_{i,i} + \rho R - \frac{E\alpha\theta}{(1-2\nu)}\dot{\varepsilon}_{kk} \tag{8}$$

Here C is the constant specific heat of the material. The thermoelastic heat equation (8) corresponds to a particular choice for the thermoelastic internal energy rate

---

*An explicit dependence of g on θ is considered in [32 and 33].

$$\rho\dot{e} = \rho C\dot{\theta} + \sigma_{ij}\dot{\varepsilon}_{ij} + \frac{E\alpha\theta}{(1-2\nu)}\dot{\varepsilon}_{kk} \tag{9}$$

The combination of (7) and (9) furnishes (8). Equation (7) represents a law of nature but the constitutive assumption (9) is necessary to obtain the heat equation (8).

The thermoelastic internal energy rate (9) may be rewritten in terms of stress by using

$$\sigma_{ij} = E\psi_{ij} \tag{10}$$

and from (9) using (10) we obtain

$$\rho\dot{e} = \rho\hat{C}\dot{\theta} + \sigma_{ij}(\frac{1+\nu}{E}\dot{\sigma}_{ij} - \frac{\nu}{E}\dot{\sigma}_{kk}\delta_{ij} + \alpha\dot{\theta}\delta_{ij}) + \alpha\theta\dot{\sigma}_{kk} \tag{11}$$

where

$$\hat{C} = C + \frac{3E\alpha^2\theta}{\rho(1-2\nu)}$$

Here the thermoelastic internal energy rate was postulated as a constitutive assumption. This combined with the first law of thermodynamics results in the generally accepted thermoelastic heat conduction equation. Instead of proposing $\dot{e}$ a constitutive assumption may be made for the form of the entropy function. In thermoelasticity the entropy depends on the present strain and temperature, and this assumption combined with the second law of thermodynamics and the thermoelastic mechanical equation (10) results in the heat conduction equation (9); [35].

Thermoviscoplasticity. For inelastic material behavior, the entropy may be postulated to depend on a number of variables. It is not definitely known whether the entropy in a viscoplastic material should depend upon the temperature, strain, and strain rate, or the strain and stress; upon internal (hidden) variables, or in some manner upon the past history of deformation. Any choice for the form of the entropy function produces, through the usual derivation, restrictions upon the constitutive equations of the theory; such restrictions may result in theoretical predictions which are not observed experimentally.

Here we pursue an alternative approach by directly postulating a choice for the inelastic internal energy rate $\dot{e}$. We make this assumption about $\dot{e}$ rather than making an assumption about the entropy. The proposed $\dot{e}$ has the following properties:

i) The heat conduction equation should predict initial thermoelastic response.
ii) It should predict inelastic self-heating when mechanical behavior is inelastic.
iii) In a cycle of loading starting and ending at zero stress, all mechanical work should be converted into temperature change.

(Experiments [36-37] indicate that more than 90% of the mechanical work in such a cycle goes to temperature change, while a remaining small portion goes to microstructure change. Here we neglect the diversion of mechanical work to microstructure change.)

The internal energy rate

$$\rho\dot{e} = \rho\hat{C}\dot{\theta} + \frac{d}{dt}\left(\frac{1+\nu}{2E}\sigma_{ij}\sigma_{ij} - \frac{\nu}{2E}\sigma_{kk}^2 + \alpha\theta\sigma_{kk}\right) \tag{12}$$

fulfills these conditions.

This equation is the <u>elastic internal energy rate</u> (11) in terms of stress, but is <u>now</u> assumed to be valid for inelastic <u>deformation</u> <u>as well</u>. When (12) is combined with (1) we have our coupled equations on thermoviscoplasticity which reduce to thermoelasticity in a neighborhood of the stress origin by virtue of the properties of (1), see [32].

This property can be easily demonstrated by defining a thermo-viscoplastic strain

$$\dot{\varepsilon}_{ij}^{pl} \equiv \dot{\varepsilon}_{ij} - \dot{\psi}_{ij}^{-1}/E \tag{13}$$

Using (7) and (13) the heat conduction equation is obtained as

$$\rho\hat{C}\dot{\theta} = \sigma_{ij}\dot{\varepsilon}_{ij}^{pl} - \alpha\theta\dot{\sigma}_{kk} - q_{i,i} + \rho R \tag{14}$$

In the neighborhood of the origin (1) predicts $\dot{\varepsilon}_{ij}^{pl} \approx 0$ and with $\dot{\sigma}_{kk}$ obtained from (10) and (13),(14) reduces to (8).

When mechanical behavior is inelastic (1) and (14) predict inelastic heating. A more general discussion of the theory, including examination of the predictions in cycling and in fast (slow) loading is presented in [32,22].

PURE TORSIONAL DEFORMATIONS

Basic Equations

We assume adiabatic conditions and the absence of an internal heat source, and we consider homogeneous torsional deformation of a thin walled tube. The tube has traction-free lateral surfaces and is not axially constrained. The nontrivial torsional stresses are $\sigma_{12} = \sigma_{21} = \tau$. The corresponding nontrivial components of $\underset{\sim}{\psi}$ are $\psi_{12} = \psi_{21} = \frac{\gamma}{1+\nu}$ where γ is the tensorial shear strain. The mechanical constitutive equation (1) reduces to

$$\frac{E}{1+\nu} \dot{\gamma} - \dot{\tau} = \frac{\tau - G_{12}[\gamma]}{k[\tau - G_{12}[\gamma]]} \tag{15}$$

where

$$G_{12}[\gamma] = \frac{1}{\sqrt{3}} g[\sqrt{3} \frac{\gamma}{1+\nu}] . \tag{16}$$

From (14) we obtain with $q_{i,i} = R = 0$

$$\rho \hat{C} \dot{\theta} = \frac{2(1+\nu)}{E \; k[\;\;]} \tau \; (\tau - G_{12}) \tag{17}$$

or equivalently

$$\rho \hat{C} \frac{d\theta[\gamma]}{d\gamma} = 2\tau(1 - \frac{1+\nu}{E} \frac{d\tau}{d\gamma}) \tag{18}$$

or

$$\rho \hat{C} \frac{d\theta[\tau]}{d\tau} = 2\tau(\frac{d\gamma}{d\tau} - \frac{1+\nu}{E}) \tag{19}$$

Equations (15),(16) and (17) are the coupled nonlinear differential equations of a thin-walled tube free to move axially and subjected either to a strain controlled (γ is prescribed) or to a stress controlled loading (τ is prescribed). Once the loading is specified as a function of time (15),(16) and (17) are nonlinear, nonautonomous differential equations.

The general qualitative properties of these equations are discussed in [22,32,33]. Our purpose here is to demonstrate specifically the influence of loading rate, the difference in the behavior predicted by our theory in strain and stress controlled loading and the responses upon a jump change in loading rate. In each of these instances the corresponding adiabatic temperature changes are also obtained.

Strain Rate Change Behavior

Consider a jump in strain rate with $\dot{\gamma}^{+}$ and $\dot{\gamma}^{-}$ designating the strain rate before and after the jump, respectively, and let $\dot{\gamma}^{+} = \delta \dot{\gamma}^{-}$ where δ is some number. If $\delta < 0$ a reversal in strain rate occurs.

Following [22,32,33] we obtain from (15) and (18)

$$\frac{d\tau}{d\gamma}^{+} = \frac{E}{1+\nu}(1 - \frac{1}{\delta}) + \frac{1}{\delta} \frac{d\tau}{d\gamma}^{-} \tag{20}$$

and

$$\frac{d\theta^+}{d\gamma} = \frac{1}{\delta}\frac{d\theta^-}{d\gamma}$$ (21)

where the superscripts + and - denote the value of the super-scripted quantity before and after the jump, respectively.

For a large strain rate increase $\delta \gg 1$ (typically $\delta > 10^2$) $\frac{d\tau^+}{d\gamma} = \frac{E}{1+\nu}$ and $\frac{d\theta^+}{d\gamma} = 0$, the material responds instantaneously in a thermoelastic fashion.

For a large strain rate decrease on the other hand with $0 < \delta \ll 1$ (typically $\delta < 10^{-2}$) $\frac{d\tau^+}{d\gamma}$ may become negative with concurrent heating since $\frac{d\theta^+}{d\gamma} > \frac{d\theta^-}{d\gamma}$.

For strain rate reversal ($\delta = -1$) $\frac{d\tau^+}{d\gamma} < \frac{E}{1+\nu}$ and $\frac{d\theta^+}{d\gamma} = -\frac{d\theta^-}{d\gamma}$. We note that $\dot\theta$ remains unchanged since $\dot\gamma^+ = -\dot\gamma^-$.

Stress Rate Change Behavior

A jump in stress rate is considered next. We let $\dot\tau^+ = \delta\dot\tau^-$. Then we obtain from (15) and (19)

$$\frac{d\gamma^+}{d\tau} = \frac{1+\nu}{E}(1-\frac{1}{\delta}) + \frac{1}{\delta}\frac{d\gamma^-}{d\tau}$$ (22)

and

$$\frac{d\theta^+}{d\tau} = \frac{1}{\delta}\frac{d\theta^-}{d\tau} .$$ (23)

A large stress rate increase ($\delta \gg 10^2$) causes again elastic response. We note, however, that the last term in (22) may not be negligible since $\frac{d\gamma^-}{d\tau}$ can be large.

For $0 < \delta \ll 1$, a stress rate decrease occurs but no unloading results. In this case $\frac{d\gamma^+}{d\tau}$ becomes very large and the stress-strain curve departs the point of stress rate change with a very small positive slope. At the same time heating sets in.

For a reversal ($\delta = -1$) $\frac{d\tau}{d\gamma}$ is negative if the reversal occurs in the plastic region where usually $\frac{d\gamma}{d\tau} > \frac{(1+\nu)}{E}$. The temperature rate $\dot\theta$ remains unchanged and heating continues.

Numerical Experiments

To illustrate the behavior of our model the coupled equations (15),(16) and (17) are solved using a nonlinear differential equation integration routine [38] together with some hypothetical but realistic material properties given in Table 2 of [32].

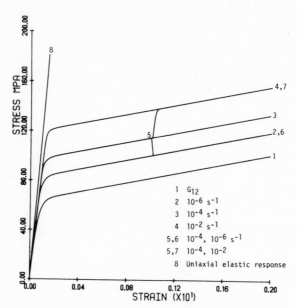

Fig. 1.   Shear stress versus tensorial shear strain for strain loading at various strain rates.

Figures 1 and 2 show the prediction of the theory in a shear strain controlled loading at various strain rates including a jump increase and decrease in strain rates.

Figures 3 and 4 are for stress controlled loading.

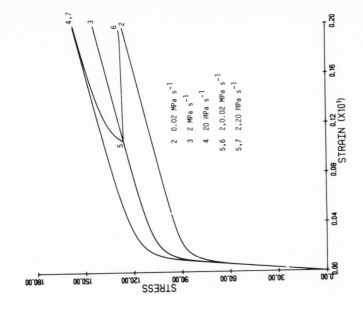

Fig. 3. Shear stress (MPa) vs. tensorial shear strain for stress controlled loading at various stress rates. (G_{12} is not graphed).

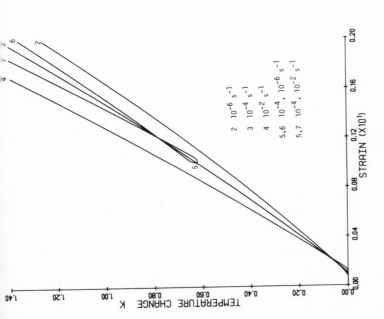

Fig. 2. Adiabatic temperature change vs. tensorial shear strain resulting from the mechanical loading depicted in Fig. 1.

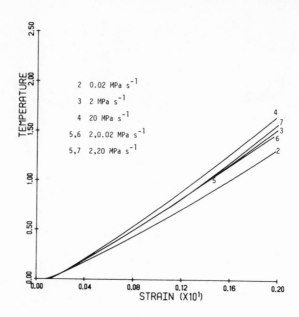

Fig. 4. Adiabatic temperature change (°K) vs. tensorial shear
         strain resulting from the mechanical loading shown in
         Fig. 3.

DISCUSSION

In the initial linear region of Figs. 1 and 3 no rate effects
are noticed until a departure from linearity occurs. Then a nonlinear dependence of the overstress on loading rate is observed.
A change of two orders of magnitude in loading rate causes a small
change in the overstress.

Figures 2 and 4 show initial isothermal behavior followed by
heating which increases with increasing loading rate. Overall
little temperature change occurs.

A comparison of Figs. 1 and 3 demonstrates that the transition
from the initial linear behavior to "plastic" deformation is sharp
for the strain controlled case but gradual for stress control.

A similar difference can be observed for the responses after
a jump in loading rates. The transition is rapid in Fig. 1 but very
gradual in Fig. 3. Ultimately all curves with the same loading rate
merge. The theory exhibits fading memory for the mechanical variables. The behavior immediately after the jump corresponds to the
predictions in (20) and (22).

The temperature vs. strain graphs do not exhibit a fading memory. Curves with the same loading rate do not appear to merge. The behavior immediately after the strain rate jump shown in Fig. 2 is isothermal (rate increase, curve 5,7) and heating (rate decrease, curve 5,6) as predicted by (21). However, the behavior at the point of a jump change in stress rate shown at point 5 in Fig. 4 does not appear to conform to the predictions of (23). However, the slope in Fig. 4 is $\frac{d\theta}{d\gamma}$ and not $\frac{d\theta}{d\tau}$ as required by (23). Using the chain rule (23) can be rewritten as

$$\frac{d\theta}{d\gamma}^{+} = \frac{1}{\delta} \frac{d\theta}{d\gamma}^{-} \frac{\frac{d\tau}{d\gamma}^{+}}{\frac{d\tau}{d\gamma}^{-}} \tag{24}$$

The predictions of (24) are in agreement with the results of Fig. 4.

The qualitative behaviors predicted by the theory correspond to observed mechanical behavior of metals [3-5]. The temperature change, however, was not measured in these experiments. Our theory predicts a very small temperature effect under these comparatively slow loadings. This prediction is in agreement with other experiments [2] and general expectations.

In this paper we have concentrated on thermomechanical coupling and resulting deformation induced temperature changes. The same theory with temperature dependent material properties was used to predict the behavior under thermal heating and thermal cycling [32,33].

ACKNOWLEDGEMENT

The support of the National Science Foundation and of the Office of Naval Research is gratefully acknowledged.

REFERENCES

1. J. F. Tormey and S. C. Britton, Effect of Cyclic Loading on Solid Propellant Grain Structures, AIAA J., 1, 1763-1770 (1963).
2. O. W. Dillon, Jr., Some Experiments in Thermoviscoplasticity, in "Constitutive Equations in Viscoplasticity, Phenomenological and Physical Aspects," K. C. Valanis, editor, AMD V.21, ASME, New York, NY (1976).
3. E. Krempl, An Experimental Study of Room-temperature Rate-sensitivity, Creep and Relaxation of Aisi Type 304 Stainless Steel, J. Mech. Phys. Solids, 27, (Oct. 1979).
4. D. Kujawski, V. Kallianpur and E. Krempl, Uniaxial Creep, Cyclic Creep and Relaxation of Aisi Type 304 Stainless Steel at Room Temperature. An experimental study. Rensselaer Polytechnic Institute Report RPI CS 79-4 (August 1979).

5.   D. Kujawski and E. Krempl, The Rate(Time)-Dependent Behavior of
     the Ti-7Al-2Cb-1Ta Titanium Alloy at Room Temperature under
     Monotonic and Cyclic Loading, Rensselaer Polytechnic Insti-
     tute Report RPI CS 79-5 (December 1979).

6.   J. A. Bailey and A.R.E. Singer, Effect of Strain Rate and Tem-
     perature on the Resistance of Deformation of Aluminum, Two
     Aluminum Alloys, and Lead, J. Inst. Metals, 92, 404-408
     (1964).

7.   E. J. Sternglass and D. A. Stuart, An Experimental Study of the
     Propagation of Transient Longitudinal Deformations in Elast
     plastic Media, Trans. ASME, J. Appl. Mech., 20, 427-434
     (1953).

8.   B.E.K. Alter and C. W. Curtis, Effect of Strain Rate on the
     Propagation of Plastic Strain Pulse Along a Lead Bar,
     J. Appl. Phys., 27, 1079-1085 (1956).

9.   E. Convery and H.Ll.D. Pugh, Velocity of Torsional Incremental
     Waves in Metals Stressed Statically into the Plastic Range,
     J. Mech. Eng. Sci., 10, 153-164 (1968)

10. Z. S. Basinski and J. W. Christian, The Influence of Tempera-
     ture and Strain Rate on the Flow Stress of Annealed and
     Decarburized Iron at Subatomic Temperatures, Australian J.
     Phys., 13, 243-377 (1960).

11. C. H. Karnes and E. A. Ripperger, Strain Rate Effects in Cold
     Worked High-Purity Aluminum, J. Mech. Phys. Solids, 14,
     75-88 (1966).

12. D. L. Holt, S. G. Babcock, S. J. Green and C. J. Maiden, The
     Strain-Rate Dependence of the Flow Stress in Some Aluminum
     Alloys, Trans. ASM, 60, 152-159 (1967).

13. U. S. Lindholm, Some Experiments with the Split Hopkinson
     Pressure Bar, J. Mech. Phys. Solids, 12, 317-335 (1964).

14. J. Klepaczko, Effects of Strain-Rate History on the Strain
     Hardening Curve of Aluminum, Arch. Mech. Sto., 19,
     211-228 (1967).

15. K. Hoge, Influence of Strain Rate on Mechanical Properties of
     6061-T6 Aluminum under Uniaxial and Biaxial States of Stres
     Experimental Mechanics, 6, 204-211 (1966).

16. F. E. Hauser, J. A. Simmons and J. E. Dorn, Strain Rate Effect
     in Plastic Wave Propagation in "Response of Metals to High
     Velocity Deformation," P. G. Shewmon and V. F. Zackay, ed.,
     Interscience, 93-114 (1961).

17. T. Nicholas, Strain-rate and Strain-rate-history Effects in
     Several Metals in Torsion, Experimental Mechanics, 11,
     370-374 (1971).

18. J. D. Campbell and A. R. Dowling, The Behaviour of Materials
     Subjected to Dynamic Incremental Shear Loading, J. Mech.
     Phys. Solids, 18, 43-63 (1970).

19. R. A. Frantz, Jr. and J. Duffy, The Dynamic Stress-strain
     Behavior in Torsion of 1100-0 Aluminum Subjected to a Sharp
     Increase in Strain Rate, Trans. ASME, J. Appl. Mech., 39,
     939-945 (1972.

0. R. N. Orava, G. Stone and H. Conrad, The Effects of Temperature
    and Strain Rate on the Yield and Flow Stresses of Alpha-
    Titanium, Trans. ASM, 59, 171-184 (1966).

1. P. E. Senseny, J. Duffy and R. H. Hawley, Experiments on Strain
    Rate History and Temperature Effects During the Plastic De-
    formation of Close-packed Metals, Trans. ASME, J. Appl. Mech.
    45, 60-66 (1978)

2. E. P. Cernocky and E. Krempl, A Theory of Viscoplasticity Based
    on Infinitesimal Total Strain, to appear in Acta Mechanica.

3. O. W. Dillon, Jr., The Heat Generated During the Torsional
    Oscillations of Copper Tubes, Int. J. of Solids and Struc-
    tures, 2, 181-204 (1966).

4. O. W. Dillon, Jr. Coupled Thermoplasticity, J. Mech. Phys.
    Solids, 11, 21-33 (1963).

5. S. L. Adams, Ph.D. Thesis, Rensselaer Polytechnic Institute,
    forthcoming.

6. O. W. Dillon, Jr., An Experimental Study of the Heat Generated
    During Torsional Oscillations, J. Mech. Phys. Solids, 10,
    235-244 (1962).

7. O. W. Dillon, Jr., Temperature Generated in Aluminum Rods
    Undergoing Torsional Oscillations, J. Appl. Mech., 10,
    3100-3105 (1962).

8. T. R. Tauchert, The Temperature Generated During Torsional
    Oscillation of Polyethylene Rods, Int. J. Eng. Sci., 5,
    353-365 (1967).

9. T. R. Tauchert and S. M. Afzal, Heat Generated During Tor-
    sional Oscillations of Polymethylmethacrylate Tubes, L.
    Appl. Phys., 38, 4568-4572 (1967).

30. G. R. Halford, Stored Energy of Cold Work Changes Induced by
    Cyclic Deformation, Ph.D. Thesis, University of Illinois,
    Urbana, Illinois (1966).

31. J.D. Campbell, Material Behavior Governing Plastic Wave Prop-
    agation in Metals, Proceedings of the Workshop on Nonlinear
    Waves in Solids, ed. T.C.T. Ting, R. J. Clifton and T. Bel-
    ytschko, Chicago, 547-568 (1977).

32. E. P. Cernocky and E. Krempl, A Theory of Thermoviscoplasti-
    city Based on Infinitesimal Total Strain, to appear in
    Int. J. of Solids and Structures.

33. E. P. Cernocky and E. Krempl, A Theory of Thermoviscoplasti-
    city for Uniaxial Mechanical and Thermal Loading, Rensselaer
    Polytechnic Institute Report No. RPI CS 79-3 (July 1979).

34. E. P. Cernocky and E. Krempl, Construction of Nonlinear Mono-
    tonic Functions with Selectable Intervals of Almost Constant
    or Linear Behavior, Trans. ASME, J. Appl. Mech., 45, 780-784
    (1978).

35. B. A. Boley and J. H. Weiner, "Theory of Thermal Stresses,"
    Wiley, New York, NY (1960).

36. F. H. Müller,"Thermodynamics of Deformation: Calorimetric In-
    vestigations of Deformation Processes,"Rheology, 5, edited
    by F. R. Eirich, Academic Press, New York, NY (1969).

37. G. I. Taylor and H. Quinney, The Latent Energy Remaining in a
    Metal after Cold Working, Proc. Roy. Soc., A, 143, 307-326
    (1934).
38. R. Bulrisch and J. Stoer, Numerical Treatment of Ordinary Dif-
    ferential Equations by Extrapolation Methods, Num. Math.,
    8, 1-13 (1966).

THE EFFECTS OF THE TEMPERATURE-DEPENDENCE OF PROPERTIES ON THE THERMAL STRESSES IN CYLINDERS

P. Stanley

F. S. Chau

Simon Engineering
Laboratories
University of
Manchester
Oxford Road
Manchester M13 9PL

Unilever Research
Colworth House
Sharnbrook
Bedford MK44 1LQ

INTRODUCTION

It is usual to assume in thermal stress calculations that material properties are independent of temperature. Significant variations do however occur over the working temperature range of the "engineering ceramics", particularly in the coefficient of thermal conductivity, k. (Godfrey[1] has reported decreases of up to 45 per cent in the thermal conductivity of various samples of silicon nitride between 0 and 400 ^{O}C.) The question arises: what are the effects of these variations on the stress distributions in ceramic components?

Ganguly et al.,[2] have looked at the particular case of an unrestrained flat plate with a temperature-dependent conductivity of the form $k(T) = k_o + k_1/T$ and constant Young's modulus, E, and expansion coefficient α, subjected to a steady-state heat flux. Since the thermal stresses in such a plate are zero for the case of a constant conductivity, there is some difficulty in interpreting, in general terms, the finite stresses obtained. A fuller investigation has been carried out using the infinitely long, unrestrained, hollow cylinder, with a steady radial heat flow. The relevant temperature range is 0-1000 ^{O}C. In contrast with the flat plate [2], this "component" experiences non-zero thermal stress for the case of constant k, and the effects of temperature-dependence can be assessed therefore in general terms.

COVERAGE

The hollow cylinder considered has been assumed homogeneous, isotropic, elastic and unrestrained in both radial and axial directions. In the initial work with the BCRA polynomial and the related linear variations (see below) a thickness/diameter ratio (t/D) of 0.25 was used; with the generalised linear form of conductivity-dependence a t/D ratio range of 0.05-0.25 was covered. (The absolute dimensions of the cylinder are immaterial.)

Following the constant k case, four forms of conductivity dependence have been studied:-

i)   the polynomial form[1] proposed for RBSN by the British Ceramic Research Association (BCRA), viz.

$$k = k_o(1-0.48 \times 10^{-3}T + 0.32 \times 10^{-6}T^2 - 0.14 \times 10^{-9}T^3 + 0.24 \times 10^{-13}T^4)$$

ii) a form varying linearly between the $T = 0$ ^{o}C ($k_o = 14.6$ Wm^{-1} $^{o}C^{-1}$) and $T = 1000^{o}C$ ($k_{1000} = 10.4$ Wm^{-1} $^{o}C^{-1}$) values from the BCRA polynomial (referred to subsequently as "linear 1"),

iii) a form varying linearly over the BCRA temperature range with both the value and the rate of change given by the polynomial at 0 ^{o}C ($k_o = 14.6$ Wm^{-1} $^{o}C^{-1}$, $k_{1000} = 8.4$ Wm^{-1} $^{o}C^{-1}$) (referred to subsequently as "linear 2"),

iv) a generalised linear form given by

$$k = k_o(1 + \beta T) \tag{1}$$

where k_o is the thermal conductivity at 0 ^{o}C and β, the coefficient of temperature variation, is the rate of change of k with temperature T. (The range $-10^{-3} < \beta \leqslant 10^{-2}$ has been studied.)

The first three conductivity variations are depicted in Fig. 1.

The Young's moduli, E, of the common engineering ceramics appear not to vary markedly with temperature up to 1000 ^{o}C [3,4] (e.g. RBSN shows a reduction of 5 per cent). The coefficient of thermal expansion, α, however, increases considerably with temperature[1]. The theory (see later) allows for the independent treatment of temperature-dependence in E and α.

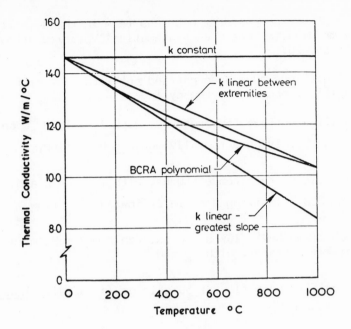

Fig. 1.   The BCRA polynomial k-T variation and the related linear
variations.

In the present work calculations were first completed for the first
three k variations (i-iii on page 2) with both E and α constant;
they were then repeated with E asssumed constant and α assumed to
vary linearly between $2.50 \times 10^{-6} \, ^{o}C^{-1}$ at $0 \, ^{o}C$ and $3.25 \times 10^{-6} \, ^{o}C^{-1}$
at $1000 \, ^{o}C$. (These figures were based on the "Lucas RBSN data"
given in reference[1]). Calculations for the fourth k variation were
with E and α assumed constant only.

The strength variability of brittle materials has been successfully modelled using the Weibull probability distribution, and a
design procedure based on this distribution has been developed[5,6,7],
in which the failure probability of a stressed component is derived
as a function of four non-dimensional terms involving the component
volume and load, and the material properties. A crucial quantity
in the calculation is a stress integral taken over the entire volume
of the component and referred to as the "stress-volume integral",
$\Sigma(V)$. (It is noteworthy that $\Sigma(V)$ is not a function of the maximum
stress alone. The relevance of $\Sigma(V)$ is that, other things being
equal, it is directly proportional to the logarithm of the reciprocal of the survival probability of the cylinder. It is also
inversely proportional to the logarithm of the mean failure load of
nominally identical specimens from a large batch[6].) $\Sigma(V)$ has been
calculated in the present work and since it depends upon the Weibull

modulus (i.e. the exponent of the stress term in the Weibull expression), this modulus, m, has been systematically varied with values
of 5, 10, 15 and 20.

The calculations have been carried out with four different
thermal conditions assumed in the cylinder:-

  i)   specified surface temperatures, with T_i = 1000 ^{o}C and
       T_o = 0 ^{o}C, giving a radially <u>outward</u> heat flow,

  ii) specified surface temperatures, with T_i = 0 ^{o}C and
       T_o = 1000 ^{o}C, giving a radially <u>inward</u> heat flow,

  iii) a specified heat flux of 50 cals/cm^2/s (2.1 MJ/m^2/s)
       radially <u>outwards</u>, with T_o = 0 ^{o}C,

  iv) a specified heat flux of 50 cals/cm^2/s radially <u>inwards</u>,
       with T_i = 0 ^{o}C.

The necessity to consider both inward and outward heat flow for both
conditions arises because with temperature-dependent properties the
effect of reversing the heat flow is not simply to change the signs
of the stresses; moreover, as a result of the greater compressive
strength of ceramic materials, the signs of the stresses are
particularly important in evaluating the stress-volume integral
$\Sigma(V)$ (see later).

The Weibull modulus, m, and Poisson's ratio, ν, of the material
have been assumed to be independent of temperature throughout.
Further work dealing with i) temperature-dependent strength, ii) more
complex temperature-dependence of thermal conductivity and
iii) transient thermal conditions will be described elsewhere.

THEORY AND METHODS OF SOLUTION

For the cases of specified surface temperatures, the following
equation is derived from the Fourier heat conduction equation:

$$\int_{T_i}^{T} k\, dT = \frac{\log (r/a)}{\log (b/a)} \int_{T_i}^{T_o} k\, dT \qquad (2)$$

where T is the temperature at a radius r, T_i and T_o are inner and
outer surface temperatures respectively, and a and b are the
inner and outer radii respectively of the cylinder.

If k is expressed in the polynomial form

$$k = k_o(1 + \beta_1 T + \beta_2 T^2 + \beta_3 T^3 + \ldots \beta_n T^n) \tag{3}$$

where k_o is the zero temperature value and β_1, β_2 .... β_n are temperature coefficients, then the following general polynomial expression can be derived from equation (2):-

$$T + \beta_1 \frac{T^2}{2} + \beta_2 \frac{T^3}{3} + \cdots \beta_n \frac{T^{(n+1)}}{n+1}$$

$$= \left(T_i + \beta_1 \frac{T_i^2}{2} + \beta_2 \frac{T_i^3}{3} + \cdots \beta_n \frac{T_i^{(n+1)}}{n+1} \right)$$

$$- \frac{\log(r/a)}{\log(b/a)}\left[(T_i - T_o) + \frac{\beta_1}{2}\left(T_i^2 - T_o^2 \right) + \cdots \right.$$

$$\left. \cdots \frac{\beta_n}{n+1}\left(T_i^{(n+1)} - T_o^{(n+1)} \right)\right] \tag{4}$$

When the coefficients β_1 ..... β_n are known, the temperature T at any radius r can be obtained from equation (4) using analytical or numerical techniques. (Note. In this case the temperature distribution is independent of k_o).

Using the same polynomial for k (equation (3)), corresponding general expressions are readily obtained for the cases of a specified heat flux, F, in the form:-

$$T + \beta_1\frac{T^2}{2} + \beta_2\frac{T^3}{3} + \cdots \beta_n \frac{T^{(n+1)}}{n+1}$$

$$= \frac{F}{2\pi k_o}\log(b/r) + T_o + \beta_1\frac{T_o^2}{2} + \cdots \beta_n \frac{T_o^{(n+1)}}{n+1} \tag{5}$$

Equation (5) is for the case of <u>outward</u> heat flow with T_o fixed; for <u>inward</u> heat flow with T_i fixed, T_i replaces T_o on the right-hand side of the equation and the log term becomes $\log(r/a)$.

Having obtained the temperature distribution from the appropriate equation, the stresses for cases of constant $E\alpha$ follow readily from the standard thermal stress equations[8]:-

$$\sigma_r = \frac{E\alpha}{1-\nu} \frac{1}{r^2} \left[\frac{r^2-a^2}{b^2-a^2} \int_a^b T r \, dr - \int_a^r T r \, dr \right]$$

$$\sigma_\theta = \frac{E\alpha}{1-\nu} \frac{1}{r^2} \left[\frac{r^2+a^2}{b^2-a^2} \int_a^b T r \, dr + \int_a^r T r \, dr - T r^2 \right] \qquad (6)$$

$$\sigma_z = \frac{E\alpha}{1-\nu} \left[\frac{2}{b^2-a^2} \int_a^b T r \, dr - T \right]$$

where ν is the Poisson's ratio of the material and the subscripts r, θ and z denote the radial, circumferential and axial coordinate directions respectively.

If E and α are temperature-dependent and therefore functions of the radius r, the stresses can be obtained numerically using for example, as in the present work, finite element techniques[9].

The stress-volume integral[5,6,7] is obtained from the appropriate stress solution in the form:-

$$\Sigma(v) = \oint_{VOL} \left[\left(\frac{\sigma_r}{\sigma_{nom} H(\sigma_r)} \right)^m + \left(\frac{\sigma_\theta}{\sigma_{nom} H(\sigma_\theta)} \right)^m + \left(\frac{\sigma_z}{\sigma_{nom} H(\sigma_z)} \right)^m \right] \frac{dV}{V} \qquad (7)$$

where σ_{nom} is a nominal stress given by

$$\sigma_{nom} = E\alpha (T_i - T_o)/(1-\nu)$$

for specified boundary temperature cases and by

$$\sigma_{nom} = \left[F \log (b/a) \right]/2\pi k_o$$

for specified heat flux cases, V is the volume (taken arbitrarily as unity) and H is a step function with a value of 1 when the stress is tensile and 8 when the stress is compressive. (H is defined in this way so that the increased compressive strength of typical ceramic materials, relative to their tensile strength, can be allowed for.)

Computer programs were written or modified for the solutions and evaluations required in the foregoing equations. In the computa-tions the cylinder wall was divided into 200 equal-radial-width annular elements and temperatures and stresses were determined at the mid-point of each element. For convenience, the stresses were obtained as multiples of the relevant nominal stress (i.e. as stress indices, I.)

The constant E and α values were $200 \times 10^9 \text{Nm}^{-2}$ and $3.0 \times 10^{-6} \; {}^\circ\text{C}^{-1}$ respectively throughout (and in the first σ_{nom} expression); ν was taken as 0.3.

RESULTS AND DISCUSSION

Specified Surface Temperatures ($T_i > T_o$)

Values of the maximum tensile stress index and the stress-volume integrals for m values of 5, 10, 15 and 20 in a cylinder of t/D = 0.25 are given in Table 1 overleaf for constant k, the BCRA polynomial variation and the associated linear variations (Fig. 1), with E and α constant throughout. (Bracketed figures are percentage differences relative to the corresponding value for constant k.)

These k variations give rise to small reductions in the maximum tensile stress index, which occurs at the outer surface; the maximum reduction of 11 per cent occurs for the linear variation with the greater gradient. Considerably greater reductions are obtained in the stress-volume integrals, and these reductions increase with increasing m. They range from 23-37 per cent for m = 5 to 73-88 per cent for m = 20, the maximum of 88 per cent occurring again for the second linear variation (Fig. 1). These figures gave an early indication of the importance of allowing for k variations where accurate failure probability assessments are required; it was clear also that errors in the assessment may occur unless an accurate version of the k - T variation is used.

The exercise was repeated for the linear α - T variation detailed above, with E constant. The results are given in Table 2, in which bracketted figures are percentage differences relative to the corresponding value in Table 1. Further reductions of about 2.5 per cent occur in the maximum tensile stress index in all cases. The differences in $\Sigma(V)$ are small for m = 5 but increase with increasing m to further reductions of 23-31 per cent for m = 20. The work suggests that the effects of typical $E\alpha$ - T variations should be taken into account when accurate assessments are required.

The effects of the generalised linear k - T variations represented in equation (1) were assessed for a range of t/D values (0.02 to 0.25) and m values of 5, 10, 15 and 20, with E and α assumed constant. Positive and negative values of β, the coefficient of temperature variation, within the approximate range -10^{-3} to 10^{-2} were included. For negative values of β the radial temperature gradient at the outer surface of the cylinder (where the tensile stresses occur) is less than for the non-varying k case, and vice versa.

Table 1. Maximum tensile stress index and stress-volume integrals
for four k - T variations.
Specified surface temperatures $(T_i > T_o)$.
E, α constant. t/D = 0.25.

k - T VARIATIONS	MAXIMUM TENSILE STRESS INDEX	STRESS-VOLUME INTEGRAL, $\Sigma(V)$			
		m = 5 $\times 10^{-2}$	m = 10 $\times 10^{-4}$	m = 15 $\times 10^{-7}$	m = 20 $\times 10^{-9}$
CONSTANT	0.387	0.224	0.112	0.691	0.467
BCRA POLYNOMIAL	0.361 (−6.7%)	0.172 (−23.2%)	0.061 (−45.5%)	0.263 (−61.9%)	0.124 (−73.5%)
LINEAR 1	0.360 (−7.0%)	0.169 (−24.6%)	0.060 (−46.4%)	0.261 (−62.2%)	0.124 (−73.5%)
LINEAR 2	0.344 (−11.1%)	0.141 (−37.1%)	0.040 (−64.3%)	0.139 (−79.9%)	0.053 (−88.7%)

Table 2. Maximum tensile stress index and stress-volume integrals
for four k - T variations.
Specified surface temperatures $(T_i > T_o)$.
E, α varying. t/D = 0.25.

k - T VARIATIONS	MAXIMUM TENSILE STRESS INDEX	STRESS-VOLUME INTEGRAL, $\Sigma(V)$			
		m = 5 $\times 10^{-2}$	m = 10 $\times 10^{-4}$	m = 15 $\times 10^{-7}$	m = 20 $\times 10^{-9}$
CONSTANT	0.378 (−2.3%)	0.230 (+2.7%)	0.106 (−5.4%)	0.590 (−14.6%)	0.359 (−23.1%)
BCRA POLYNOMIAL	0.352 (−2.5%)	0.173 (+0.6%)	0.055 (−9.8%)	0.214 (−18.6%)	0.090 (−27.4%)
LINEAR 1	0.352 (−2.2%)	0.170 (+0.6%)	0.055 (−8.3%)	0.211 (−19.2%)	0.089 (−28.2%)
LINEAR 2	0.336 (−2.3%)	0.139 (−1.4%)	0.036 (−10.0%)	0.109 (−21.6%)	0.037 (−30.2%)

Longitudinal stress indices for a series of β values and a particular
t/D (0.25) are shown in Fig. 2. (Hoop stress indices varied in a
generally similar manner but the maximum tensile stress values at the
outer surface were somewhat smaller than those in Fig. 2. Radial
stress indices were much smaller in magnitude but were included in
the $\Sigma(V)$ calculations.) As β decreases (algebraically) the tensile
longitudinal and hoop stress indices at the outer surface decrease,
and the compressive indices at the inner surface increase. These
effects were evident throughout the thickness range.

   Stress-volume integrals are plotted against β for m = 20, η
(compressive tensile strength ratio) = -8 and various t/D values in
Figs. 3a (negative β) and 3b (positive β). It can be seen that if β
lies within the approximate range -10^{-4} to 10^{-4} the effect on $\Sigma(V)$
(and therefore the failure probability) is relatively small. Outside
this range $\Sigma(V)$ varies considerably with β, for example with
$\beta = -5.0 \times 10^{-4}$ $\Sigma(V)$ is 93 per cent smaller than the β = 0 value for
t/D = 0.25. (The β values for the two linear variations in Fig. 1
are -2.9×10^{-4} and -4.3×10^{-4}.)

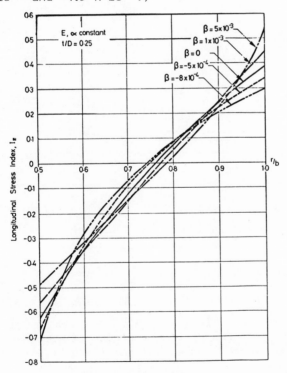

Fig. 2. Longitudinal stress index versus non-dimensional radius for
various β values. Specified surface temperatures($T_i > T_o$).

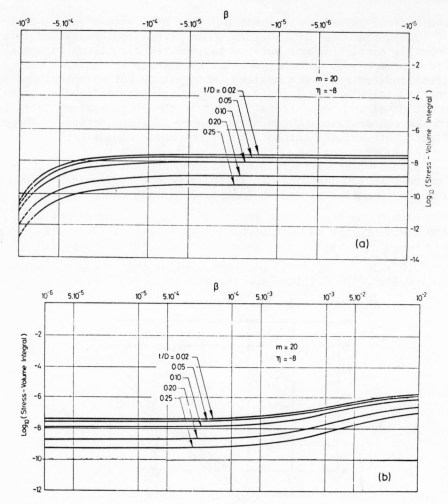

Fig. 3. Log_{10}(stress-volume integral) versus β for various thick-
ness/diameter ratios.   Fixed surface temperatures ($T_i > T_o$)
a) $\beta < 0$, b) $\beta > 0$.

The percentage differences between $\Sigma(V)$ for a given β and that for
$\beta = 0$ decrease with decreasing m, conforming in this respect with
the trends evident in Table 1. (As β approaches 10^{-3} the effective
conductivity at 1000 ^{o}C approaches zero. Smaller values of β are
not therefore considered and calculated I and $\Sigma(V)$ values for β
approaching 10^{-3} must be treated with caution.)

An alternative presentation for m = 20 is given in Fig. 4.
$\Sigma(V)$ decreases with increasing t/D and with decreasing β. From this

figure or Fig. 3, $\Sigma(V)$ for a cylinder with a given t/D, m = 20, and a k - T variation for which β is known can be readily determined. Similar charts for other m values could be readily prepared.

The effects discussed in the foregoing may be regarded as "safe" from the design point of view, since typical property variations are such that conservative estimates of maximum tensile stresses and stress-volume integrals are obtained if material properties are assumed to be independent of temperature.

Specified Surface Temperatures $(T_i < T_o)$

Results for the case of a radially inward heat flow, corresponding in all other respects with those given in Table 1 for radially outward heat flow, are given in Table 3, and again bracketted figures are percentage differences relative to the corresponding value in Table 3 for constant k.

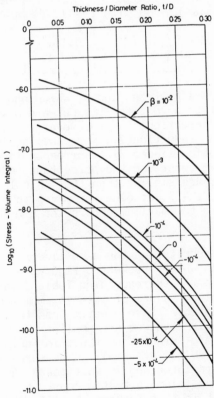

Fig. 4. Log_{10}(stress-volume integral) versus thickness/diameter ratio for various β values. Specified surface temperatures $(T_i > T_o)$.

Table 3. Maximum tensile stress index and stress-volume integrals
          for four k - T variations.
          Specified surface temperatures $(T_i < T_o)$.
          E, α constant. t/D = 0.25.

k - T VARIATIONS	MAXIMUM TENSILE STRESS INDEX	STRESS-VOLUME INTEGRAL, $\Sigma(V)$			
		m = 5 $\times 10^{-2}$	m = 10 $\times 10^{-3}$	m = 15 $\times 10^{-4}$	m = 20 $\times 10^{-6}$
CONSTANT	0.614	0.792	0.350	0.203	1.325
BCRA POLYNOMIAL	0.587 (-4.4%)	0.697 (-12.0%)	0.247 (-29.4%)	0.115 (-43.3%)	0.596 (-55.0%)
LINEAR 1	0.586 (-4.6%)	0.683 (-13.8%)	0.240 (-31.4%)	0.111 (-45.3%)	0.572 (-56.8%)
LINEAR 2	0.568 (-7.5%)	0.612 (-22.7%)	0.185 (-47.1%)	0.073 (-64.0%)	0.324 (-75.5%)

    The maximum tensile stress index, which now occurs at the inner
surface, is greater than that for the outward heat flow condition
but the percentage reductions for the several k variations are
smaller, the maximum reduction(7.4 per cent) again occurring for the
second linear variation(linear 2). Corresponding stress-volume
integrals are also greater and the percentage reductions, though
still considerable, are smaller than the Table 1 values. The reductions increase from 12-23 per cent for m = 5 to 55-75 per cent for
m = 20 and as before the maximum occurs for m = 20 with the second
linear variation.

    Corresponding results for the same linear α - T variation are
given in Table 4, in which bracketted figures are percentage differences relative to the corresponding Table 3 values. It is noteworthy that the Table 4 values of both maximum tensile stress indices
and stress-volume integrals are greater than the Table 2 values and
that whereas the Table 2 values are smaller than the Table 1 values,
the Table 4 values are larger than those in Table 3. The stress
index increases in Table 4 are small. The stress-volume integral
increases however are large and in most cases cancel or substantiall
off-set the decreases relative to the constant k values shown in
Table 3. The result is that for the case of inward heat flow with
specified surface temperatures the combined effects of possible
variations in k and α may lead to an increase in the stress-volume
integral and therefore in the failure probability. The contrasting

effects of typical α - T variations in the cases of inward and outward heat flow is a particularly important outcome of this investigation. A further point of contrast between the Table 4 and Table 2 results is that whereas the percentage changes in the latter are greatest for the linear 2 variation, in the former they are least.

Stress indices and stress-volume integrals were determined using the k - T variations of equation(1), for the same m and t/D values as previously and again with β in the approximate range -10^{-3} to 10^{-2}. The trends in the stress indices were very much as observed for the outward heat flow case (see Fig. 2) in that the maximum tensile indices (which were somewhat greater than for outward heat flow) decreased with decreasing β while the maximum compressive indices increased; here though the former occurred at the inner surface, the latter at the outer. Again these effects occurred throughout the thickness range $(0.02 < t/D < 0.25)$.

Stress-volume integrals for m = 20 and η = -8 are given in Fig. 5. $\Sigma(V)$ decreases with decreasing β, as in Fig. 4, but in this case it increases with increasing t/D, contrary with the trends of Fig. 4 for the outward flow condition. In absolute terms a $\Sigma(V)$ value for the case of inward heat flow is always greater than the corresponding value for outward heat flow; the difference is relatively large for the higher t/D values and becomes smaller as the cylinder becomes thinner, indicating that the direction of heat flow is less important the thinner the cylinder.

Table 4. Maximum tensile stress index and stress-volume integrals for four k - T variations.
Specified surface temperatures $(T_i < T_o)$.
E, α varying. t/D = 0.25.

k - T VARIATIONS	MAXIMUM TENSILE STRESS INDEX	STRESS-VOLUME INTEGRAL, $\Sigma(V)$			
		m = 5 $\times 10^{-2}$	m = 10 $\times 10^{-3}$	m = 15 $\times 10^{-4}$	m =20 $\times 10^{-6}$
CONSTANT	0.624 (+1.6%)	1.010 (+27.5%)	0.493 (+40.9%)	0.312 (+53.7%)	2.214 (+67.1%)
BCRA POLYNOMIAL	0.594 (+1.2%)	0.872 (+25.1%)	0.335 (35.6%)	0.166 (+44.3%)	0.921 (+54.5%)
LINEAR 1	0.593 (+1.2%)	0.853 (+24.9%)	0.324 (+35.0%)	0.159 (+43.2%)	0.875 (+53.0%)
LINEAR 2	0.573 (+0.9%)	0.754 (+23.2%)	0.242 (+30.8%)	0.100 (+37.0%)	0.466 (+43.8%)

As before the changes in $\Sigma(V)$ for β values between -10^{-4} and $+10^{-4}$ were relatively small and the effect of ignoring typical conductivity variations, though uneconomic, is "safe" in design terms.

Specified Heat Flux, Radially Outwards

Table 5 gives principal results for the case a fixed radial heat flux of 50 cals/cm^2/s, radially outwards, with the outer surface temperature held at 0 ^{o}C. As before the t/D ratio was 0.25 and for these calculations E and α were assumed constant. Percentage differences relative to the corresponding constant k value are given in brackets.

The striking feature of these results is that for the k - T variations studied the maximum tensile stress indices(outer surface) and stress-volume integrals are invariably greater than the appropriate values for constant k. This contrasts markedly with the Table 1 results for the case of specified surface temperatures. The increases in the maximum tensile stress index are small; the maximum increase of 7.2 per cent occurs for the linear 2 variation. The increases in $\Sigma(V)$, however, are considerable, ranging from 30-52 per cent for m = 5 to 161-346 per cent for m = 20. As in Table 1 the greatest difference occurs for the linear 2 k variation for each m value, but here, in contrast with the specified surface temperature results in Tables 1 and 3, the better approximations to the BCRA polynomial results are from the linear 2 variation.

Results for the same coverage of the variables with E constant and α linearly temperature-dependent(see above) are given in Table 6 together with percentage differences relative to the Table 5 values.

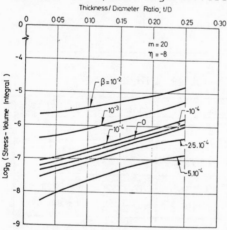
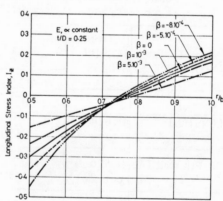

Fig.5. Log_{10}(stress-volume integral) versus thickness/diameter ratio for various β values. Specified surface temperatures($T_i < T_o$).

Fig.6. Longitudinal stress index versus non-dimensional radius for various β values. Specified heat flux, outwards.

Table 5.   Maximum tensile stress index and stress-volume integrals
            for four k - T variations.
            Specified heat flux(Outwards).
            E, α constant.   t/D = 0.25.

k - T VARIATIONS	MAXIMUM TENSILE STRESS INDEX	STRESS-VOLUME INTEGRAL, $\Sigma(V)$			
		m = 5 $\times 10^{-2}$	m = 10 $\times 10^{-4}$	m = 15 $\times 10^{-7}$	m = 20 $\times 10^{-9}$
CONSTANT	0.387	0.224	0.112	0.691	0.467
BCRA POLYNOMIAL	0.412 (+6.5%)	0.334 (+49.1%)	0.238 (+112%)	2.070 (+200%)	1.977 (+323%)
LINEAR 1	0.405 (+4.7%)	0.293 (+30.8%)	0.185 (+65.2%)	1.431 (+107%)	1.217 (+161%)
LINEAR 2	0.415 (+7.2%)	0.340 (+51.8%)	0.245 (+119%)	2.157 (+212%)	2.085 (+346%)

Table 6.   Maximum tensile stress index and stress-volume integrals
            for four k - T variations.
            Specified heat flux (Outwards).
            E, α varying.   t/D = 0.25.

k - T VARIATIONS	MAXIMUM TENSILE STRESS INDEX	STRESS-VOLUME INTEGRAL, $\Sigma(V)$			
		m = 5 $\times 10^{-2}$	m = 10 $\times 10^{-4}$	m = 15 $\times 10^{-7}$	m =20 $\times 10^{-9}$
CONSTANT	0.351 (-9.3%)	0.148 (-33.9%)	0.046 (-58.9%)	0.175 (-74.7%)	0.073 (-84.4%)
BCRA POLYNOMIAL	0.378 (-8.3%)	0.226 (-32.3%)	0.102 (-57.1%)	0.558 (-73.0%)	0.334 (-83.1%)
LINEAR 1	0.371 (-8.4%)	0.200 (-31.7%)	0.081 (-56.2%)	0.397 (-72.3%)	0.213 (-82.5%)
LINEAR 2	0.380 (-8.4%)	0.236 (-30.6%)	0.111 (-54.7%)	0.628 (-70.9%)	0.390 (-81.3%)

Reductions of about 8 per cent occur in the maximum tensile stress indices with the result that these indices become smaller than the constant k value of 0.387 in Table 5. $\Sigma(V)$ values are also reduced again with the nett effect that in the great majority of cases values become smaller than the corresponding constant k value in Table 5. (The only exceptions to this are the polynomial and linear 2 values for m = 5.) It is noteworthy that the reductions do not vary significantly for the different k - T variations. It is also noteworthy that the reductions in Table 6 are considerably greater than those in Table 2 for the specified surface temperature conditions.

Reverting to constant α values, and with E constant, the effec of the general k - T variations of equation(1) were obtained over tl previous β range. Longitudinal stress indices for t/D = 0.25 and a series of β values are shown in Fig. 6 (compare with Fig. 2). It c be seen that as β decreases both the maximum tensile and the maximum compressive stress indices increase. Similar trends were apparent in the hoop stress indices. As a consequence, under these condition $\Sigma(V)$ is greater than the constant k value when k decreases with increasing temperature (i.e. negative β) and vice versa. The resul is that for the commonly encountered ceramics(negative β) neglect o the temperature dependence of k will yield non-conservative failure probability estimates. This is in direct contrast with the conserva tive effect of a similar neglect when the surface temperatures are specified.

Variations of $\Sigma(V)$ with t/D for several β values are shown in Fig. 7. The thicker the cylinder the greater the dependence on β.

Specified Heat Flux, Radially Inwards

Results for inward heat flow with the same heat flux, an inner surface temperature of 0 $^{\circ}$C and E and α assumed constant, are tabulated in Table 7, together with percentage changes relative to the appropriate constant k value. The k - T variations considered give rise to increases in the maximum tensile stress index (inner surface) and stress-volume integral in all cases, in contrast with the Table 3 results for specified surface temperatures. The increases are considerably greater than those given in Table 5 for a radially outward heat flux(maximum of 596 per cent compared with 346 per cent). The closeness of the indices and integrals for the BCRA polynomial and linear 2 variations evident in Table 5 is lost, but again values for the BCRA polynomial variation fall between thos for the two linear variations.

The further effects of temperature-dependence in α are covered in Table 8. Small reductions occur in the maximum tensile stress index and in one case (linear 1 variation) the value is less than the constant k value in Table 7

Table 7.   Maximum tensile stress index and stress-volume integrals
           for four k - T variations.   Specified heat flux(Inwards).
           E, α constant.   t/D = 0.25.

k - T VARIATIONS	MAXIMUM TENSILE STRESS INDEX	STRESS-VOLUME INTEGRAL, $\Sigma(V)$			
		m = 5 $\times 10^{-2}$	m = 10 $\times 10^{-3}$	n_1 = 15 $\times 10^{-4}$	m =20 $\times 10^{-6}$
CONSTANT	0.614	0.792	0.350	0.203	1.325
BCRA POLYNOMIAL	0.668 (+8.8%)	1.280 (+61.6%)	0.864 (+147%)	0.762 (+275%)	7.558 (+470%)
LINEAR 1	0.652 (+6.2%)	1.119 (+41.3%)	0.671 (+91.7%)	0.527 (+160%)	4.644 (+250%)
LINEAR 2	0.674 (+9.8%)	1.354 (+71.0%)	0.958 (+174%)	0.887 (+337%)	9.222 (+596%)

There are significant reductions in the $\Sigma(V)$ values but here, in
contrast with the results for a radially outward heat flux (Tables 5
and 6), the reduced value is never less than the corresponding
constant k value in Table 7. As observed in discussing the Table 6
results, the variations in the reductions for the constant k case and
the different k - T variations are relatively small. The decreases
in Table 8 contrast with the increases shown in Table 4 resulting
from the same α - T variation in the case of specified surface
temperatures.

    The effects of the general form of k - T variations were studied
as before, with both E and α constant. Here also $\Sigma(V)$ values were
practically unaffected for β values numerically smaller than
10^{-4}. For numerically greater β values the effects on $\Sigma(V)$ were
considerable, particularly for the higher t/D ratios. $\Sigma(V)$ is
plotted against t/D in Fig. 8 for sample β values, with m = 20 and
η = -8. $\Sigma(V)$ increases with decreasing β as in the case of a
radially outward heat flux (Fig. 8), but here, in contrast with what
has been established for the other three conditions(Figs. 4,5 and 7),
for β values greater than 10^{-3} $\Sigma(V)$ decreases with t/D, for β = 10^{-3}
$\Sigma(V)$ is practically independent of thickness ratio, and for β values
less than 10^{-3} $\Sigma(V)$ increases with t/D. The occurrence of a
"critical" β for this thermal condition is important; changes in
cylinder thickness will have opposite effects on $\Sigma(V)$ for β values
on either side of the "critical" value. The risk of non-conservative estimates of $\Sigma(V)$ (and therefore of failure probability) in
neglecting negative β values is evident.

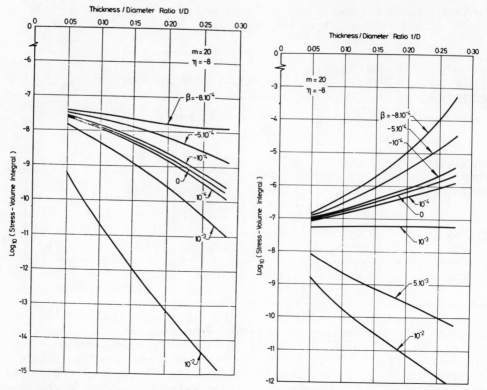

Fig.7. Log$_{10}$(stress-volume integral) versus thickness/diameter ratio for various β values. Specified heat flux, radially outwards.

Fig.8. Log$_{10}$(stress-volume integral) v. thickness/diameter ratio for various β values. Specified heat flux, inwards.

The increased dependence of $\Sigma(V)$ on β for the higher t/D values evident in Fig. 8 has been noted previously in Fig. 7. The lower the Weibull modulus, m, the less dependent is $\Sigma(V)$ on the temperature dependence of α.

CONCLUSIONS

This investigation has shown that generalisions on the effects of the temperature-dependence of the thermal properties of the engineering ceramics on the maximum thermal stress indices and stress-volume integrals developed in hollow cylindrical bodies with radial heat flow must be made with great caution. It has been shown that the thermal conditions assumed in the analysis, e.g. constant temperature difference or constant heat flux, and the direction of heat flow are particularly important, and it is very clear that trends established for one set of thermal conditions cannot be assumed to apply for another.

Table 8.   Maximum tensile stress index and stress-volume integrals
            for four k - T variations.   Specified heat flux (Inwards).
            E, α varying.   t/D = 0.25

k - T VARIATIONS	MAXIMUM TENSILE STRESS INDEX	STRESS-VOLUME INTEGRAL, $\Sigma(V)$			
		m = 5	m = 10	m = 15	m = 20
		$\times 10^{-2}$	$\times 10^{-3}$	$\times 10^{-4}$	$\times 10^{-6}$
CONSTANT	0.569 (-7.3%)	0.644 (-18.7%)	0.196 (-44.0%)	0.078 (-61.6%)	0.347 (-73.8%)
BCRA POLYNOMIAL	0.623 (-6.7%)	1.079 (-15.7%)	0.521 (-39.7%)	0.327 (-57.1%)	2.300 (-69.6%)
LINEAR 1	0.607 (-6.9%)	0.931 (-16.8%)	0.393 (-41.4%)	0.216 (-59.0%)	1.334 (-71.3%)
LINEAR 2	0.630 (-6.5%)	1.150 (-15.1%)	0.586 (-38.8%)	0.388 (-56.3%)	2.885 (-68.7%)

For both outward and inward heat flow with fixed surface
temperatures thermal conductivities which decrease with increasing
temperature result in reductions in the maximum tensile stress index
and stress-volume integral for all the thickness/diameter ratios and
Weibull modulus values considered. The percentage decreases are less
for inward than for outward heat flow but whereas the effect of a
linearly decreasing coefficient of thermal expansion is to further
decrease stress indices and integrals in the case of outward heat
flow with specified surface temperatures, considerable increases
occur with an inward heat flow. Moreover, for a given linearly
dependent coefficient of thermal conductivity, the stress-volume
integral decreases with increasing thickness/diameter ratio for
outward heat flow and increases for inward heat flow.

For a specified radially outward heat flux increases in the
maximum stress indices and stress-volume integrals are caused as a
result of decreases in thermal conductivity with increasing temperature over the range of parameters studied, but these increases can be
practically nullified by the effects of typical variations in the
coefficient of thermal expansion. Maximum stress indices and stress-
volume integrals decrease with increasing thickness for a specified
temperature-dependence of thermal conductivity.

Increases in maximum stress indices and stress-volume integrals
occur due to decreases in conductivity with temperature in the case
of a specified radially-inward heat flux also. These increases are
greater than those for a radially-outward heat flux and they are
significantly reduced as a result of typical variations in the

coefficient of thermal expansion. Variations of stress-volume integral with thickness/diameter ratio can be positive, zero or nega tive, depending on the degree of the temperature-dependence of thermal conductivity.

The trends revealed in this investigation are clearly of impor- tance to the material scientist and designer. However, because of the contrasting features for the two loading conditions, it is particularly important that the correct thermal conditions are established when the effects of temperature-dependence of the material properties are being assessed for a particular design. The thermal conditions considered are the extremes of a range of poss- ible conditions and in a specific intermediate case, further detaile considerations of the effects of temperature-dependence of propertie may be required.

ACKNOWLEDGEMENTS

This work was carried out with the support of the Procurement Executive, Ministry of Defence.

REFERENCES

1. D.J. Godfrey, "A critical review of engineering ceramics relevan to their use in severe thermal environments", Conference on Non- Metallic Materials for the Royal Navy, Manadon, Plymouth,(1975)
2. B.K. Ganguly, K.R. McKinney and D.P.H. Hasselman, "Thermal stres analysis of flat plate with temperature-dependent thermal conduc tivity", J. Amer. Ceram. Soc., 58, (1975) 455.
3. W. Ashcroft, "Mechanical properties of silicon nitride at elevated temperatures", Proc. Brit. Ceram. Soc., 22, (1973) 169.
4. J.E. Restall and C.R. Gostellow, "A limited assessment of selec- ted low-expansion coefficient ceramics with potential for high- temperature applications", Proc. Brit. Ceram. Soc., 22, (1973) 8
5. P. Stanley, H. Fessler and A.D. Sivill, "An engineer's approach to the prediction of failure probability of brittle components". Proc. Brit. Ceram. Soc., 22, (1973) 453.
6. P. Stanley, A.D. Sivill and H. Fessler, "The application and confirmation of a predictive technique for the fracture of brittle components", Proc. 5th International Conf. on Exptl. Stress Anal., Udine, (1974) 2.42.
7. P. Stanley, A.D. Sivill and H. Fessler, "Applications of the fou function Weibull equation in the design of brittle components", Fracture Mechanics of Ceramics, Plenum Press, New York - London, Vol. 3 "Flaws and Testing", (1978) 51.
8. S.P Timoshenko and J.N. Goodier, "Theory of Elasticity", McGraw- Hill Book Company Inc., New York - London, (1970).
9. PAFEC Note 76/1 "Variation of E and α with temperature", University of Nottingham, (1976).

A NONLINEAR CONSTITUTIVE RELATIONSHIP

FOR COMPOSITE PROPELLANTS

Donald L. Martin, Jr.

Propulsion Directorate, US Army Missile Laboratory
US Army Missile Command
Redstone Arsenal, Alabama 35809

INTRODUCTION

The mechanical response and the failure of composite solid
propellants are known to be related to the formation and growth of
vacuoles on the microscopic or macroscopic scale. Failures in composite materials such as composite propellants originate at the
filler particle or binder molecular level. These microscopic
failures, whether adhesive failures between the binder and filler
particles or cohesive failures in the binder, result in vacuole
formation. The cumulative effect on all vacuole formation and
growth is observed on the macroscopic scale as strain dilatation.
The strain dilatation in composite propellants varies with binder
type, binder formulation, filler particles, and presumably filler
particle size distribution. The strain dilatation caused by
vacuole growth and formation results in a nonlinear stress-strain
behavior for these materials.

The experimental investigations of Leeming et al.,[1]
Bornstein,[2] and Martin[3] emphasize the fact that the state-
of-the-art linear viscoelastic theory does not predict propellant
grain stresses and strains under realistic loading conditions. The
most significant portion of the error is due to the constitutive
nonlinearity of the material. Linear thermorheological simple
viscoelastic theory has certain important limitations; however, it
is still the only constitutive theory currently used in most
analyses because no other constitutive equations previously existed
that are as easily incorporated in the analyses.

There has been considerable progress in the development of
nonlinear constitutive theories and methods of approximating

81

specific nonlinear viscoelastic behavior in the last few years.
There will be no attempt in this paper to review the extensive
literature on the general theories of nonlinear viscoelasticity.

BACKGROUND AND THEORY

Farris[4] demonstrated that the mechanical response and the
failure in solid propellants are related to the formation and
growth of vacuoles within the composite material and that this
phenomenon is directly related to the macroscopically observed
strain dilatation. The mechanical properties, particularly Poisson's ratio, are strongly dependent on the dilatational behavior of
these materials. Strain dilatation measurements, even for the uniaxial tensile tests, are cumbersome, tedious, and time consuming.
Therefore, a relationship that would permit the determination of
strain dilatation and Poisson's ratio as a function of strain
utilizing only the stress-strain data would be most desirable.
Insight into the factors which govern the nonlinear response and
failure of composite propellants can be obtained utilizing a model
for dewetting, vacuole formation, and growth of vacuoles which is
based on assumptions regarding the microstructural behavior of the
system. The model can then be used to develop an expression for
dilatation as a function of stress and strain and to compare the
predicted dilatation with experimental measurements of dilatation.
The degree of agreement between the predicted and experimental
values of dilatation gives an indication of the validity of the
model. This approach was used by Farris,[5,6] Fedors and
Landel,[7] and others.[8,9]

The present state-of-the-art of predicting dilatation as a
function of strain is limited because there is very little quantitative information available on the experimentally determined
dilatation under conditions similar to those encountered with solid
propellant rocket motors. Various models describing this phenomenon have appeared in the literature. In developing such a model,
attention is focused on a rigid filler particle contained in an
elastomeric matrix. As the material is strained above some critical value, a vacuole is formed about the filler particle due to
the internal failure of the composite. The source of vacuole
formation may be either adhesive failure of the filler-binder bond
or cohesive failure in the binder near the filler particle.[10,11]
As the strain in the material increases, the vacuoles continue to
increase in size and number. The shape and instantaneous behavior
of these vacuoles seem to follow these assumptions: (1) Vacuoles
form arbitrarily at any magnitude of strain above the critical
value. (2) Each vacuole behaves as an ellipsoid of revolution with
the minor axis determined by the diameter of the enclosed filler
particle. (3) The major axis of the ellipsoidal vacuole increases

linearly with strain at a rate proportional to the size of the filler particle it contains.

Various constitutive theories exist and experimental investigations can be designed to produce specific stress-strain equations. In this section a constitutive relationship for uniaxial tensile data is discussed and the similarity with Farris' equation is noted.[4]

Martin[12] used the constitutive relationship

$$\sigma(\varepsilon,T,t) = \left\{ F(\varepsilon,T,t)e^{-\ell n f(\varepsilon)} \right\} \varepsilon \quad . \tag{1}$$

Equation (1) has been verified experimentally.

Starting with Farris' equation and defining $F = \sigma/\varepsilon$ as the secant modulus, as was used by Martin, the following relationship is obtained:

$$F(\varepsilon) = \frac{\sigma}{\varepsilon} = E_1 \, e^{-\beta\left(\frac{\Delta V}{V_o}\right)1/\varepsilon} \quad , \tag{2}$$

where E_1 is the initial modulus and $F(\varepsilon)$ is meant to indicate F at a given strain. This value will be indicated simply by F in the remainder of this work. After rearrangement one obtains:

$$\frac{F}{E_1} = e^{-\beta(\Delta V/V)1/\varepsilon} \quad . \tag{3}$$

Equation (3) indicates that the nonlinearity of the stress-strain curve (of which the vacuole formation and growth is considered to be the major contributor) should be described by the ratio of F/E_1. Taking the natural logarithm of both sides of Equation (3) results in Equation (4):

$$\beta\left(\frac{\Delta V}{V}\right)\frac{1}{\varepsilon} = \ell n\left(\frac{E_1}{F}\right) \tag{4}$$

Since β is assumed to be independent of strain magnitude, Equation (3) suggests that a plot of $\{\varepsilon\ell n(E_1/F)\}$ versus strain should have a similar shape to the dilatation-strain curve. Figure 1 shows plots of $\{\varepsilon\ell n(E_1/F)\}$ and $(\Delta V/V)$ versus strain. The similarity in the shapes of the two curves indicates that the total nonlinearity, $\{\varepsilon\ell n(E_1/F)\}$, in the materials' behavior may be calculated from an equation similar to that used by Farris.[4] The same reasoning used by Farris results in the expression

$$f(\varepsilon) = S\gamma_{max} \int_{-\infty}^{n} \int_{-\infty}^{n} \frac{e^{-n2/2}}{\sqrt{2\pi}} \, dndn + \gamma_o \varepsilon \quad , \tag{5}$$

where

$f(\varepsilon)$ = total nonlinearity of the materials' behavior

$S = (A - B)\sqrt{2\pi}/\gamma_{max}$

γ_0 = initial slope of the $\{\varepsilon \ln(E_1/F)\}$ versus strain curve

γ_{max} = final slope of the $\{\varepsilon \ln(E_1/F)\}$ versus strain curve

$\overline{\varepsilon}$ = the strain magnitude at the intersection of the initial asymptote of slope γ_0 and the final asymptote of slope γ_{max}

A = the magnitude of $\{\varepsilon \ln(E_1/F)\}$ at $\overline{\varepsilon}$

B = the magnitude of the ordinate at the intersection of the two asymptotes

$n = (\varepsilon - \overline{\varepsilon})/S.$

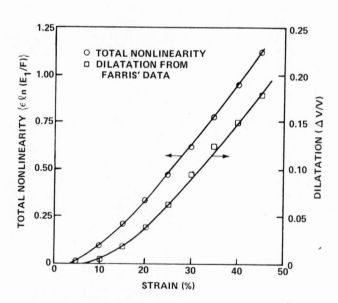

Figure 1. Total nonlinearity and dilatation versus strain for a granular filled elastomer.

Equation (5) then would permit one to estimate the total nonlinearity of the propellant once the parameters have been determined. It would be more useful, however, to obtain a quantity
representing the propellant's nonlinear behavior that could be
readily utilized in the current linear viscoelastic programs to
account for the nonlinear response of the material. With this in
mind, consider Equation (3), which after rearranging becomes

$$\beta \left(\frac{\Delta V}{V} \right) = \varepsilon \ln \left(\frac{E_1}{F} \right) \tag{6}$$

Recalling that β and E_1 are independent of strain and differentiating both sides with respect to strain, one has

$$\beta \ \frac{d\left(\frac{\Delta V}{V_o}\right)}{d\varepsilon} = \ln \frac{E_1}{F} + \left(1 - \frac{E}{F} \right) \tag{7}$$

where E = the instantaneous slope of the stress–strain curve.
Equation (7) now contains two unknowns, β and $d/d\varepsilon(\Delta V/V_o)$. Now,
consider the change of dilatation with respect to strain. Accord-
ing to Farris' data,[5] the quantity $d/d\varepsilon(\Delta V/V_o)$ obtains a
maximum somewhere between the points of maximum stress and rupture
stress and remains constant until failure occurs. Farris proposed
the relationship

$$\frac{d}{d\varepsilon} \left(\frac{\Delta V}{V_o} \right) = c \ V_{fd} \tag{8}$$

where V_{fd} = volume fraction of dewetted solids. Experimental
data were presented for systems that tend to dewet completely.
These data gave excellent agreement with the model represented by
Equation (8). The maximum slope of the dilatation–strain curve is
shown to be directly proportional to the volume fraction of filler
particles in the material. For materials that tend to dewet completely, the proportionality factor is unity ($c = 1$); therefore,

$$\left[\frac{d}{d\varepsilon} \left(\frac{\Delta V}{V_o} \right) \right]_{max} = \phi_f \tag{9}$$

where ϕ_f is the total volume fraction of filler particles in the
system.

With these observations noted, the following simplifying assumptions will be made that will enable the determination of the factor β without having to conduct the dilatation-strain measurements:

(1) The maximum slope of the dilatation-strain curve is obtained at a strain magnitude equal to halfway between the point of maximum stress and breaking stress.

(2) All composite propellants are totally dewetted in the region of the sample where failure occurs.

The assumption that all filler particles in composite propellants are totally dewetted in the region of failure is not too unrealistic for the following reasons. In composite propellants where the bond between the filler particles and the binder is stronger than the cohesive strength of the binder, vacuoles initiate in the binder and propagate to the surface of the filler particle prior to failure. In these materials there is probably a large strain gradient in the region of failure and there would be a high probability of totally dewetted filler particles in this region. One reason that Equation (9) does not appear to be true for all propellants is that the measured dilatation, whether localized or not, is averaged over the total volume of the sample. If the localized band of dewetted filler particles is narrow, the measured dilatation would appear to be much smaller than the actual dilatation in the region of failure. The maximum slope of the dilatation-strain curve is related to the dewetted filler particles, but the dewetted filler particles may be contained in a narrow band around the failure region.

The assumption that the maximum slope of the dilatation-strain curve is obtained at a strain magnitude halfway between the points of maximum stress and breaking stress is approximately correct according to Farris' data. The purpose of this assumption is to fix the point on the stress-strain curve for the calculation of the coefficient β. One reason that failure points are not recommended for this calculation is the variation in ways of selecting failure points. The use of the stress and strain values at $\varepsilon = (\varepsilon_m + \varepsilon_b)/2$ should minimize the variations in methods of selecting failure points as a source of error in the calculation of β.

Substituting Equation (9) into Equation (7) yields the following results after rearranging:

$$\beta = \frac{1}{\phi_f} \left\{ \ln \frac{E_1}{F} + 1 - \frac{E}{F} \right\} \, , \tag{10}$$

to be evaluated at $\varepsilon = (\varepsilon_m + \varepsilon_b)/2$.

With β known, one can then utilize Equation (6) to determine the dilatation as a function of stress and strain. The resulting expression is as follows:

$$\left(\frac{\Delta V}{V_o}\right) = \frac{\varepsilon}{\beta} \, \ln \frac{E_1}{F} \, . \tag{11}$$

This equation was utilized to calculate the apparent volume change for two composite materials where the experimental dilatation-strain data were available. The calculated and experimentally measured dilatation-strain data were in excellent agreement.

To more readily utilize the volume change data obtained in this way for a nonlinear viscoelastic stress analysis, the following method is proposed. It would be more desirable to reflect the propellant nonlinearity in an equivalent Poisson's ratio, which would be allowed to vary with strain magnitude, strain rate, and temperature. The variable Poisson's ratio could then be incorporated into the current linear viscoelastic stress analysis program to reflect the propellant's nonlinear behavior.

To derive an expression of Poisson's ratio as a function of stress and strain, the volume change is approximated by the first strain invariant neglecting higher order terms involving strain.

$$\left(\frac{\Delta V}{V_o}\right) = e_1 + e_2 + e_3 \quad , \tag{12}$$

where e_1, e_2, and e_3 are the strains in the principal directions and are related as follows in the uniaxial condition:

$$e_1 = \varepsilon$$

$$e_2 = e_3 = -\nu e_1 = -\nu \varepsilon \qquad , \tag{13}$$

where ν is Poisson's ratio. Substituting these relationships into Equation (11) results in the following

$$\left(\frac{\Delta V}{V_o}\right) \frac{1}{\varepsilon} = (1 - 2\nu) \qquad . \tag{14}$$

Substituting Equation (14) into equation (6) and rearranging, one obtains the relationship:

$$\nu = \frac{1}{2} \left\{ 1 - \frac{1}{\beta} \, \ln\left(\frac{E_1}{F}\right) \right\} \, . \tag{15}$$

Also, substituting Equation (14) into Farris' equation yields the
following constitutive relationship for the uniaxial nonlinear
behavior:

$$\sigma = E_1 \varepsilon e^{-\beta(1 - 2\nu)}$$

(16)

For the nonlinear viscoelastic treatment, the initial slope
modulus, E_1, is allowed to vary with strain rate and temperature;
the nonlinearity coefficient, β, is allowed to vary with strain
rate and temperature; and the Poisson's ratio, ν, is allowed to
vary with strain rate, temperature, and strain magnitude.

EXPERIMENTAL VERIFICATION

The model used in deriving the expressions for dilatation and
Poisson's ratio as a function of stress and strain was based on
certain assumptions regarding the microstructural behavior of composite materials such as composite propellants. The validity of
the assumptions and the model is indicated by good agreement of
calculated and experimental stress and dilatation data.

Figures 2, 3, and 4 are plots of stress and dilatation versus
strain for a granular filled elastomer. The experimental data were
obtained from Farris' work.[5] The experimental data in these
figures were obtained at 25°C (75°F) and at strain rates of 6.66,
66.6, and 666%/min. The plotted points represent the data read
from Figure 6 in Farris' paper. The solid lines indicate the
values calculated using Equation (16) for stress and Equation (11)
for dilatation. Figures 2, 3, and 4 indicate good agreement
between calculated and experimental values of stress and dilatation
at all strain rates. The dilatation-strain data appear to be
shifted by a constant factor along the strain axis in some cases.
This would indicate a slight error in the dilatation measurements,
which could be attributed to localized dewetting or vacuole formation as discussed previously. Equation (16) is shown to yield
excellent agreement with experimental stress values at all strain
rates. In Figure 4, Equation (11) is shown to also give excellent
agreement with the dilatation experimental data at a strain rate of
666%/min. up to a strain magnitude of 35%. At strain magnitudes in
excess of 35%, the experimental dilatation data fall below the
predicted values. While vacuoles may form arbitrarily at all
strain magnitudes above a critical value, it is reasonable to
assume that the weakest point in the composite material will be
more susceptible to vacuole formations and should contain a larger
population of vacuoles than the remainder of the material. The
formation of vacuoles to relieve the large strain gradient that

would be present at the weakest point in the composite requires a
finite time in which to react. If the composite material is
strained at a rate faster than the material can respond through
microstructural failure and vacuole formation, the dilatation would
appear to be uniform throughout the material until conditions were
such that the rate of dilatation at the weakest point in the material became large enough to promote the formation of a localized
band of a large population of vacuoles. The data presented in
Figure 4 indicate that at a strain rate of 666%/min. near uniform
dilatation is present up to a strain magnitude of approximately
35%. At this strain a localized band of a large population of
vacuoles developed at the weakest point in the composite material,
which then began to have a strong influence on the material's
behavior. As the material is strained in excess of 35%, the volume
increase in the vacuoles within the established localized failure
region is more than the measured average dilatation. This would
explain why the experimental dilatation data presented in Figure 4
became less than the calculated values as the strain magnitude
increased to exceed 35%. At the lower strain rates of 6.66 and
66.6%/min. the localized failure region was established in the
initial portion of the test ($\varepsilon<0.10$), thus explaining why experimental dilatation deviated slightly from the calculated dilatation
for the duration of the test.

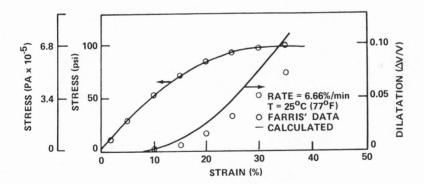

Figure 2.   Dilatation and stress versus strain for
a granular filled elastomer.

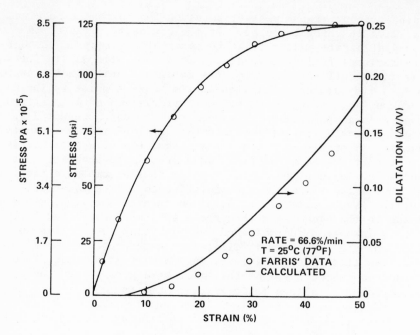

Figure 3.   Dilatation and stress versus strain for
            a granular filled elastomer.

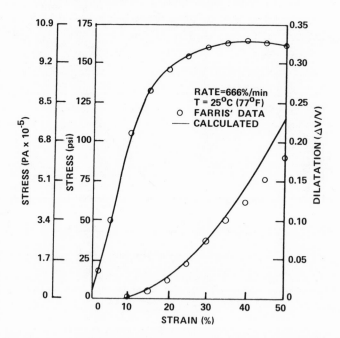

Figure 4.   Dilatation and stress versus strain for
            a granular filled elastomer.

Figures 2, 5, and 6 are plots of stress and dilatation versus strain. The experimental data in these figures were obtained from Figure 3 in Farris' paper.[5] These data indicate the stress and dilatation behavior of a typical granular filled elastomer at a strain rate of 6.66%/min. at 2.5°, 4.4°, −18°C (77°, 40°, and 0°F). The data shown on Figures 2, 5, and 6 indicate excellent agreement between values of stress calculated by Equation (15) and the experimental values at all temperatures. The behavior of the dilatation strain data is similar to that shown on Figures 2, 3, and 4 with the experimental values deviating slightly from the values calculated by Equation (11). The same reasoning as used previously would also explain the deviation of the experimental dilatation from that calculated by Equation (11), as indicated on Figures 2, 5, and 6.

Normally, nominal stress values were corrected to true stress by multiplying the nominal stress value by the extension ratio, λ_1. This correction was originally derived based on the assumption that there is no volume change within the sample for the duration of the test. The relationship considering volume change is derived as follows:

$$\left(\frac{\Delta V}{V_o}\right) = \lambda_1 \lambda_2 \lambda_3 - 1 \qquad , \tag{17}$$

where $\lambda_i = 1 + \varepsilon_i$. Therefore, rearranging Equation (17), one obtains

$$\lambda_2 \lambda_3 = \frac{1 + \left(\dfrac{\Delta V}{V_o}\right)}{\lambda_1} = \frac{A}{A_o} \qquad . \tag{18}$$

Substituting this expression into the equation $\sigma_c = \sigma_o A_o / A$, one obtains the following:

$$\sigma_c = \frac{\sigma_o \lambda_1}{1 + \dfrac{\Delta V}{V_o}} \qquad . \tag{19}$$

Therefore, the true stress σ_c is lower than would be obtained by the normal correction procedures. Equation (19) reduces to the equation normally used as ΔV approaches zero.

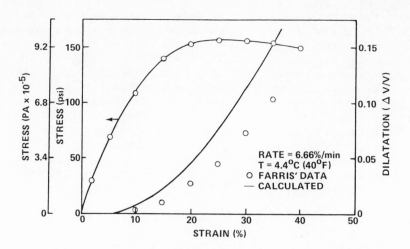

Figure 5.   Dilatation and stress versus strain for
             a granular filled elastomer.

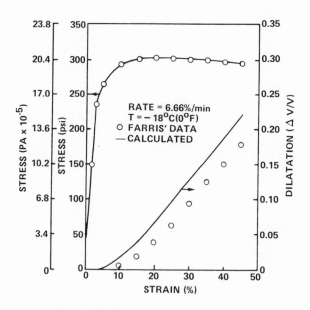

Figure 6.   Dilatation and stress versus strain for
             a granular filled elastomer.

CONCLUSIONS

A nonlinear constitutive relationship is developed for uni-axial stress-strain conditions that is shown to fit experimental data over a wide range of strain rates and temperatures. The total stress-strain nonlinearity of particulate filled composites may be represented by statistical parameters similar to those used by Farris to describe the dilatation-strain behavior. By using Equations (10) and (11), the dilatational behavior of composite propellants may be predicted from the stress-strain data and the total filler content. The relationships derived in this report will enable one to investigate the nonlinear behavior at strain-rates and temperatures where it is very difficult, if not impossible, to obtain measured dilatation data.

The relationship represented by Equation (15) may be utilized, along with experimental stress-strain data obtained at different strain-rates and temperatures, to determine the dependency of an equivalent Poisson's ratio on the strain-rate, temperature, and strain magnitude. The Poisson's ratio thus determined may then be utilized in currently available computer programs to describe transient thermoviscoelastic behavior with the consideration of the materials' nonlinear response

The relation normally used for obtaining corrected stress data by multiplying the nominal values by the extension ratio would yield values of stress and strain at the maximum corrected stress, which were significantly larger than the true stress-strain values would indicate. These errors could conceivably result in the wrong conclusions as to the structural integrity of marginal designs; i.e., one might predict a safe design utilizing stress-strain allowables based on the normal correction method when, in reality, the true stress-strain allowables would indicate the design would be expected to fail. It is therefore recommended that stress-strain allowables for use in future failure analysis be based on true stress as defined by Equation (18).

REFERENCES

1. H. Leeming, Solid Propellant Structural Test Vehicle and System Analysis, LPC Final Report No. 966-F (AFRPL-TR-70-10), Lockheed Propulsion Company, Redlands, California, (March 1970).
2. G. A. Bornstein, Transient Thermoviscoelastic Analysis of a Uniaxial Bar, 8th MBWG, 1, CPIA Publication No. 177, (1968).
3. D. L. Martin, Jr., An Approximate Method of Analysis of Non-linear Transient Thermoviscoelastic Behavior, 8th JANNAF MBWG, 1, CPIA Publication No. 193, (1970).

4. R. J. Farris, The Influence of Vacuole Formation on the Response and Failure of Solid Propellants, Trans. Soc. Rheol., 12, (1968).
5. R. J. Farris, Strain Dilatation in Solid Propellants, 3rd ICRPG MBWG, 1:291 CPIA Publication No. 61U, (October 1964).
6. R. J. Farris, of Granular Filled Elastomers Under High Rates of Strain, J. of Applied Polymer Science, 8:25-35 (1964).
7. R. F. Fedors, and R. F. Landel, Mechanical Behavior of SBR-Glass Bead Composites, JPL Space Program Summary, IV, 37-41.
8. D. M. Elliott, and W. B. Ledbetter, Microstructural Damage and Its Influence on the Dilatation-Strain Properties of Filled Elastomeric Composite Materials, 6th ICRPG MBWG, 1:235-251, CPIA Publication No. 158, (1967).
9. H. C. Jones, and H. A. Yiengst, Dilatometer Studies of Pigment-Rubber Systems, J. of Ind. Eng. Chem., 32:10 (1940).
10. A. E. Oberth, and R. S. Bruenner, Binder-Filler Interaction and Propellant Physical Properties, 4th ICRPG MBWG, 45, (October 1965).
11. J. E. Hilzinger, Microstructural Behavior of Filler Reinforced Elastomeric Binders, 5th ICRPG MBWG, 1:169, CPIA Publication No. 119, (1966).
12. D. L. Martin, Jr., An Approximate Method of Analysis of Nonlinear Transient Thermoviscoelastic Behavior, PhD Dissertation, Department of Mech. Sys. Eng., University of Alabama, (1968).

THERMAL STRESS FRACTURE IN ELASTIC-BRITTLE MATERIALS

A. F. Emery

Department of Mechanical Engineering
University of Washington
Seattle, WA   98195

ABSTRACT

The uncoupled thermoelastic brittle fracture problem is discussed in terms of the types of stress fields produced by surface
heating or cooling and the generic characteristics of the thermally
generated stress intensity factors. Examples of experimental measurements and numerical calculations are given to demonstrate these
general characteristics.

INTRODUCTION

Thermal stresses are usually generated by one of two mechanisms:
the creation of temperature variations within a body because of
imposed surface conditions; temperature distributions caused by the
generation of heat by internal thermal sources, such as joulean dissipation, or by the conversion of mechanical energy to thermal
energy such as that found in stress waves or by fatiguing processes.
The solution of a problem which involves the conversion of mechanical to thermal energy requires the simultaneous solution of the
structural and thermal equations but except for thermally generated
stress waves or strong dissipative effects, as found in plastic
forming, it is rare to have to consider the coupled problem. For
brittle materials and for temperatures caused by surface imposed
conditions such coupled effects are unlikely.

In this paper we discuss the general characteristics of thermal
stresses and the mode I and II stress intensities which are produced
by such stresses in two dimensional structures. No attempt is made
to survey the entire field or even to establish the details of the
calculations or of the solutions. Rather the desire is to impart

to the reader a feeling for the spatial and temporal behavior of the thermal stress fields, particularly in those areas where thermal and mechanical stresses differ most. For detailed solutions the reader is directed to references 1-4.

For a region V with surfaces S, the pertinent equations which model the uncoupled thermal stresses are the first law of thermo-dynamics and conservation of momentum.

a) For the temperature, the use of Fourier's relationship $(q_i = -k \; \partial T/\partial x_i)$ with the first law gives the field equation

$$\frac{\partial \rho c T}{\partial t} = \frac{\partial}{\partial x} (k_x \frac{\partial T}{\partial x}) + \frac{\partial}{\partial y} (k_y \frac{\partial T}{\partial y}) + \frac{\partial}{\partial z} (k_z \frac{\partial T}{\partial z}) + Q \qquad (1a)$$

with an initial condition

$$T(x,y,z,0) = f(x,y,z) \qquad (1b)$$

and boundary conditions of

$$T = g(x,y,z) \text{ on } S_T \text{ for prescribed temperature} \qquad (1c)$$

$$k_n \frac{\partial T}{\partial n} = q(x,y,z) \text{ on } S_q \text{ for prescribed heat flux} \qquad (1d)$$

$$k_n \frac{\partial T}{\partial n} = h(T_\infty - T) \text{ on } S_c \text{ for convective cooling} \qquad (1e)$$

b) The elastic displacements and stresses for static condi-tions are given by the equilibrium equations

$$\frac{\partial \sigma_{ij}}{\partial x_i} + X_j = 0 \qquad (2a)$$

with boundary conditions of the form

$$\left. \begin{array}{l} \sigma_{nn} = t_t(x,y,z) \\ \\ \sigma_{ns} = t_s(x,y,z) \end{array} \right\} \text{on } S_\sigma$$

$$u_i = U_i(x,y,z) \quad \text{on } S_u \qquad (2c)$$

with the constitutive equations for linear elastic mater-ials of the form

$$\varepsilon_{xx} = \alpha_x T + [\sigma_{xx} - \nu(\sigma_{yy} + \sigma_{zz})]/E \tag{2d}$$

$$\varepsilon_{xy} = \sigma_{xy} \left(\frac{2(1+\nu)}{E}\right)$$

In equations 1 and 2 the material properties k, ρc, h, α, ν and E are, in general, functions of temperature, although for the purposes of this paper they will be considered as constant. Furthermore, the temperature fields will be restricted to those for no heat sources, Q = 0, and for convective surface conditions. In general, thermal stresses often involve significant surface radiation, which is highly non-linear, but reasonable estimates of the resulting temperature and stress fields can be made by defining an equivalent heat transfer coefficient, h_r, and incorporating it into equation (1e).

The solution to the problem consists of solving equations (1a-1e) to determine the temperature T(x,y,z,t). The temperature distribution at a specific time, t_1, is then used to compute the corresponding stresses. These stresses are termed quasi-static since no time variation is assumed, but the stresses are presumed to adjust instantaneously to any changes in the temperature field without generating stress waves. This is only true if the characteristic times associated with the stress waves are much shorter than the times associated with the temperature field — a condition which is generally true, even for highly transient thermal shocking. This condition is equivalent to stating that $\partial e/\partial t$ is proportional to $\partial T/\partial t$. Unfortunately $T(x,y,z,t_1)$ does not appear directly in equation 2a. When using the stress formulation (eqs.2) the temperature appears in the compatability equations which ensure that displacements are everywhere continuous and are of the forms (for simply connected regions)

$$(1+\nu)\ \nabla^2\sigma_{xx} + \frac{\partial^2\Sigma}{\partial x^2} + \alpha E\ \left(\frac{1+\nu}{1-\nu}\ \nabla^2 T + \frac{\partial^2 T}{\partial x^2}\right)$$

$$+ \nu\ \left(\frac{1+\nu}{1-\nu}\right)\ \left(\frac{\partial X}{\partial x} + \frac{\partial Y}{\partial y} + \frac{\partial Z}{\partial z}\right) + 2(1+\nu)\ \frac{\partial X}{\partial x} = 0 \tag{3a}$$

$$(1+\nu)\ \nabla^2\sigma_{xz} + \frac{\partial^2\Sigma}{\partial x\partial z} + \alpha E\ \frac{\partial^2 T}{\partial x\partial z} + (1+\nu)\ \left(\frac{\partial X}{\partial z} + \frac{\partial Z}{\partial x}\right) = 0 \tag{3b}$$

The appearance of the temperature in the constitutive equations and in the compatability conditions tends to obscure its role. If we make use of the strain-displacement relations, the consitutive equations (eq. 2d) and the equation of equilibrium (eq. 2a), we may express the problem in terms of the displacements by equations of the form

$$(\lambda + \mu)\,\frac{\partial e}{\partial x} + \mu\nabla^2 u - \beta\frac{\partial T}{\partial x} + X = 0 \tag{4a}$$

where $\beta = \alpha E/(1-2\nu)$ (4b)

Equation (4a) clearly illustrates that the temperature gradient pla
the role of a body force and as such is a useful way of understanding thermal effects provided that the material is elastic. Althoug
in no way does it reduce the mathematical complexity of the problem
Equation (3) shows that temperatures which are linear in the x,y,z
coordinates cause no stresses, although equation (4) shows that suc
a distribution does cause displacements. Likewise, for a body whic
is completely restrained, the constitutive equations (eq. 2d) show
that an equivalent hydrostatic pressure

$$\sigma_{xx} = \sigma_{yy} = \sigma_{zz} = -\beta T \tag{5}$$

must exist at all points in the region and that a corresponding set
of surface tractions must be applied to maintain equilibrium. Sinc
$\partial T/\partial x$ is proportional to the heat flux, it is often more helpful to
consider that the displacements are caused by the heat flow rather
than by the temperature per se, and since $\nabla^2 T$ is proportional to the
net heat flow to a region, the stresses are best visualized as a
result of unbalanced heat flow or since $\nabla^2 T$ is also proportional to
$\partial T/\partial t$ as the result of time varying temperatures. The visualizatiou
are even more appropriate when two dimensional multiply-connected
regions are considered. Then in addition to the usual compatability
equations (eq. 3) we also require that around each contour C_α
(except C_0) the following relationships hold.

$$\int_{C_\alpha} q_n\,ds = 0 \tag{5a}$$

$$\int_{C_\alpha} (y\frac{\partial T}{\partial n} + x\frac{\partial T}{\partial s})ds = \int_{C_\alpha} (y\frac{\partial T}{\partial s} + x\frac{\partial T}{\partial n})ds = 0 \tag{5b}$$

for a stress free state. The first condition is simply that there
be no net heat flowing into or out of the contour. The second is
not as tractable, but it can be shown that if the temperature distribution in the neighborhood of the contour is steady and no local
heat sources are present, then equations (5b) will be satisfied.

We are thus led to the observation that linearly varying
temperatures in rectangular cartesian coordinates and two dimensional steady state temperatures in the absence of heat sources,
with no net heat flow out of any multiple contours will not give
rise to any stresses. (Because linear varying temperatures do not

give rise to stresses in unrestrained regions, many authors auto-
matically discard this portion of the temperature field in calculating the stresses. Care should be taken in following this approach
since it should be remembered that the displacements are dependent
upon the entire temperature expression and that for regions with displacement boundary conditions the stresses will be dependent upon
the linear portion of the temperature solution.)

Another way of examining the problem is to cast it as a two
part problem such that

$$\sigma = \sigma^T + \sigma^I$$

$$u = u^T + u^I$$

(6)

where

σ^T, u^T represent any convenient solution to the thermal problem
without regard to the boundary conditions or body forces
associated with the problem.

σ^I, u^I represent an isothermal problem with boundary conditions
such that $\sigma^T + \sigma^I$ and $u^T + u^I$ satisfy the desired
boundary conditions on σ and u.

Two frequent applications of this method are

a)   Let the displacements be zero. This requires surface
     tractions of the amount $\sigma_n = -\beta T$, $\sigma_s = 0$ and body forces
     $X = \beta\, \partial T/\partial x$ etc. The isothermal solution must correspond
     to the imposition of opposite body forces and such surface
     tractions as to ensure the satisfaction of the boundary
     conditions. This method is commonly used in finite element
     analyses although it is usually done in one step in typical
     computer codes. Viewing the solution in this fashion often
     helps to understand some of the numerical errors of such
     computations. For example if a finite element code is
     capable of an accuracy to within 10%, then since the first
     step of the two part solution generates stresses in the
     amount βT, one should expect a final accuracy of the order
     of $0.1\ \beta T$. For two-dimensional steady-state problems for
     which the stresses are zero, errors of this order would
     likely exist.

b)   Let the stresses be zero and the region adopt the calcu-
     lated displacements. The second part of the solution
     would then impose the correct displacements.

These two approaches, in conjunction with the body force analogy, are probably the best visualization of a thermal stress problem. The next section presents an example of each approach and illustrate characteristics which are typical of thermal stress problems.

APPLICATION OF THE SUPERPOSITION METHOD

The following examples are taken from typical two dimensional problems. Similar solutions exist for three dimensional problems, but the visualization of the solution and the complexity of the application of the methods diminishes their pedagogical value.

Example 1

Consider a two dimensional anaular region partially filled with a liquid (5) as illustrated in figure 1. Under the liquid, the high heat transfer coefficient causes the surface temperature to be nearly equal to the fluid temperature. Above the fluid level, the poor vapor heat transfer coefficient causes the inner surface temperature to be closer to the exterior temperature. In steady-state, there is a net flow of heat through the contours $r = r_i$ and $r = r_o$, and since equation 5a and possibly eq. 5b are not satisfied, a non-zero state of stress is likely to exist. Furthermore, at the liquid vapor interface, there is a very large circumferential temperature gradient $\partial T/\partial \theta$ which leads one to expect a locally strong thermal stress field. However, the solution for this problem with symmetry about the y axis reveals that the stresses respond only to the average temperature difference

$$\int_0^\pi [T(r_i,\theta) - T(r_o,\theta)]d\theta \tag{7a}$$

and to the difference of the first moment of the temperature

$$\int_0^\pi y[T(r_i,\theta) - T(r_o,\theta)]d\theta \tag{7b}$$

Since one can choose a smooth variation of $T(r_i,\theta)$ and $T(r_o,\theta)$, which have differences of the average temperature and first moment equal to those of the true temperature field, one should not be surprised that the resulting stress distribution, as shown on figure 2, is smooth.

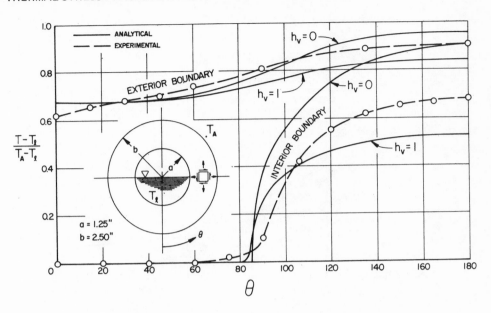

Fig. 1.   Analytical and experimental temperature profiles for
          several values of the inner surface heat transfer coeffi-
          cient h (from ref. 5)

Now let us apply the second superposition method by considering an annulus with a vertical cut at $\theta = \pi$. For this now simply
connected region, with a steady state distribution, there are no
stresses, but the thermal deflections cause the cut surfaces to
open and to rotate. Closing the cut requires a force and a moment,
neither of which can be expected to introduce an unusual stress
pattern. (Generally speaking the cut surfaces, cannot be matched by
just closing and rotating the surfaces, but St. Venant's principle
assures us that the effect of such local imperfections in the
surface joining will be confined to the joint area. In this case,
the closure is exact.)

Thus, although the direct solution gives use to an unexpectedly
smooth stress field, the visualization of the stresses as the sum of
two stress fields provides a deeper understanding of the situation.

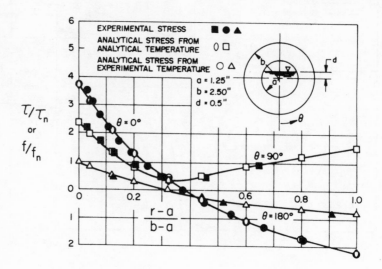

Fig. 2.   Maximum Shear Stress Distribution
          [τ_n is the value at the inner surface at $\theta = 180°$]
          (from ref. 5)

Example 2

Most transient thermal stress fields can be visualized as the sudden change of a surface temperature and the slow penetration of the thermal disturbance into the material. Consider a simple rectangular plate, figure 3, initially at the uniform temperature T_i. Let the surface be exposed to an ambient temperature of 0 through a spatially and temporally constant heat transfer coefficient h. The temperature field is illustrated on figure 3 for several values of the non-dimensional time $\kappa t/W^2$. If we imagine that the stress fields are:

a) σ^T - stresses generated by assuming that the ends of the plate are restrained from movement in the y direction but movement in the x direction is free. The only non-zero stress is given by

$$\sigma^T_{yy} = \frac{-\alpha E T(x)}{1-\nu}$$

(8a)

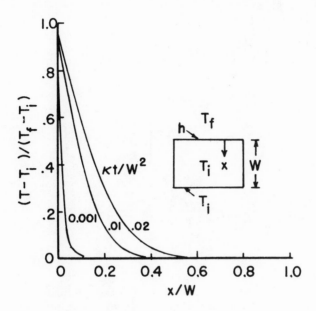

Fig. 3. Temperature distribution in a plate cooled on one edge and maintained at the initial temperature at the other edge for Bi = 100.

which generates a boundary force and moment at the restrained ends of

$$F^T = \text{Force} = \frac{-\alpha E}{1-\nu} \int_0^W T(x)\,dx$$

$$M^T = \text{Moment} = \frac{-\alpha E}{1-\nu} \int_0^W (x - \frac{W}{2})\, T(x)\,dx$$

b) σ^I - stresses associated with the application of a force and a moment which are equal but opposite to those generated by σ^T. At distances W away from the ends, this stress is given by

$$\sigma^I = \frac{-F^T}{W} - \frac{12\,(x-\frac{W}{2})}{W^2}\,M^T \tag{8b}$$

The total stress is thus

$$\sigma = \sigma^T + \sigma^I = \frac{E}{1-\nu}\left\{-T(x) + \frac{1}{W}\int_0^W T(x)\,dx\right.$$

$$\left. + \frac{12\,(x-w/2)}{w^2}\int_0^W (x - \frac{W}{2})T(x)\,dx\right\} \tag{8c}$$

A graph of $\sigma(x,t)$ for several values of time is shown on figure 4. This graph demonstrates clearly the most obvious and unexpected feature of thermal stresses, namely the generation of stresses far from the region in which the thermal effects are to be found. This feature of action at a distance not only confuses the analyst but places severe requirements upon any analysis which uses numerical techniques since the nodal point density (e.g. the element mesh) must be different in different regions to adequately investigate the problem. That is we usually need a very fine thermal mesh near the surface, but we can tolerate a very coarse thermal mesh deeper in the bar where the temperature changes have not penetrated, while a fairly fine mesh is needed everywhere to calculate the stresses.

A non-dimensionalization of the problem reveals that the important parameters are

$$\frac{x}{W} \ , \ \frac{H}{W} \ , \ \frac{hW}{k} \ , \ \frac{\kappa t}{W^2} \ , \ \frac{x}{\sqrt{\kappa t}} \ , \ \frac{h\sqrt{\kappa t}}{k}$$

The latter 4 groupings represent

$\frac{hW}{k}$ the Biot number measures the importance of the surface heat transfer as compared to internal conduction heat transfer. High values of the Biot number correspond to specifying the surface temperature, while low values correspond to insulated surfaces.

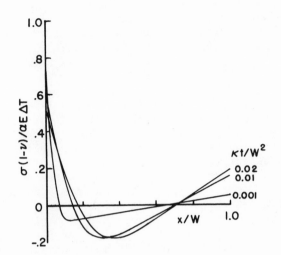

Fig. 4.   Stress distribution in a plate cooled on one edge and
maintained at the initial temperature at the other edge
for Bi = 100.

$\dfrac{\kappa t}{W^2}$ The Fourier modulus represents the ratio of the rate of
energy storage to the rate of thermal conduction

$\dfrac{x}{\sqrt{\kappa t}}$ is a measure of the depth of penetration of the thermal
effect

$\dfrac{h\sqrt{\kappa t}}{k}$ is a measure of the ratio of surface heat flux to the
heat conducted out of the region of thermal penetration.

To compare two thermal stress fields, all 6 of these groupings must
be equal, although after the pulse has penetrated a reasonable
distance the effect of the last group is often small and it can

be neglected.   Reference 6 discusses the effect of this term for a
solution of a ceramic quench test.   The stress pattern given by
equation 8c is characterized by a high surface tensile stress, a
central compressive zone, and a far side tensile stress.   Initially
the surface tension is high, the zero stress occurs near the cooled
surface and the compressive stress is low.   With time, the surface
tension diminishes, the zero point moves inward, and the compression stresses increase.   Finally the zero point becomes relatively
stable and all stresses gradually decrease to their steady state
value of zero.   Figure 5 illustrates the movement of the edge of
the thermal effect (defined as the point at which the temperature
change is 0.1%) and the movement of the zero stress point.   Clearly,
the thermal penetration is rapid and a very strong function of the
Biot modulus, while surprisingly, the zero point of the stress moves
very slowly and is apparently quite insensitive to the surface
thermal conditions.   Figure 6 shows the maximum surface tensile
stress and the time at which it occurs for two cases: a) a
symmetrically cooled plate of thickness 2W; b) a plate of thickness
W which is cooled on one side and maintained at the original temperature at the other surface.   A third case in which the surface
x = W is insulated has a different temperature distribution than
does case b, but during the time required for the peak stresses to
develop ($\kappa t/W^2$ < 0.05) the temperature pulse has not reached the
surface at x_T = W and consequently the stresses are equal.   Because
the moment M^T is zero for the symmetrical case, the stresses are
significantly higher than for the asymmetrical case.   However the
asymmetrical stresses develop much faster, leading to the conclusion
that the time of thermal stress failure is strongly problem dependent.   Figures 7 and 8 illustrate the temporal behavior of the
surface tensions and the maximum compression.   For all Biot numbers
the maximum compression and the far side tensile stresses are roughl
equal for a considerable time interval.

FRACTURE OR FAILURE

     The failure of ceramics which are exposed to severe thermal
stresses can be regarded either as a fracture mechanics problem
associated with well defined internal or surface flaws or as a
statistical failure problem.   The former presumes that the interaction of the thermal stresses and a flaw produce a singular state
of stress at the tips of cracks of the form.

$$\sigma = \frac{K}{r^n} f(\theta).$$

where the order of the singularity n depends upon the opening angle
of the crack (0 < n < 1) and K is called the stress intensity factor
Stress intensity factors are classified as mode I, II or III for
stresses which lead to crack opening, crack shearing or out-of-plane

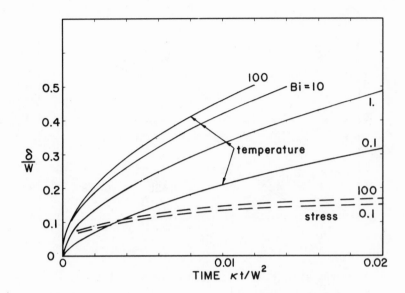

Fig. 5.   Thermal penetration depth and the location of the point
of zero stress for a plate cooled on one edge and maintained at the initial temperature of the other edge for
Bi = 100.

(torsional twisting) shear deformations respectively.   Using
Griffith's theory of fracture, that the crack will extend whenever
the elastic energy G contained within the body is greater than that
needed to create new crack surfaces, the fracture criterion can be
expressed as (7)

$$K < K_c \quad \text{no fracture}$$

(10)

$$K = K_c \quad \text{fracture}$$

where K_c is a material constant.   The stress intensity factor K is
a function of the state of stress, the shape and orientation of the
crack and the body and the boundary conditions.   Sih (8) has shown
that steady state temperatures can induce such a singular state of
stress and thus can lead to crack growth.   Emery (9) demonstrated

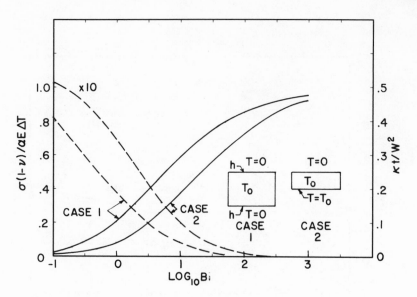

Fig. 6.   Maximum stress at the cooled edge and time of occurrence
           for a plate initially at T_o with respect to the fluid
           temperature. (——— stress,   ----- time).

that the temporal character of the temperature field was contained
in the time variation of K and that n was a constant for all time.

The statistical view point is represented by use of the Weibul
or weakest link theory which assumes that failure will occur whenev
the value of σ_w defined by

$$\left(\frac{\sigma_w - \sigma_o}{\sigma_m}\right)^m = \frac{1}{V}\int_V \left(\frac{\sigma - \sigma_o}{\sigma_m}\right)^m dV \tag{11}$$

exceeds a critical value. For structures with singular states of
stress, the integral tends to emphasize those regions near the
singular points and consequently the statistical and the fracture
mechanics view points should lead to comparable results. The
Weibull stress can be regarded as a measure of the overall tendency

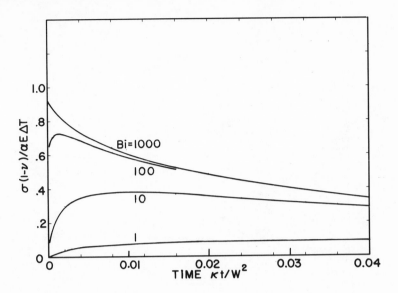

Fig. 7.   Stress at the cooled edge of a plate whose other edge is
           maintained at the original temperature.

of a body to fail because of a distribution of flaws and of stresses.
If σ_w is to be applied to a fracture problem with a defined flaw,
the experimental determination of σ_w should be done on a comparable
specimen.

Batdorf and Heinisch (10) and Batdorf and Crose (11) have shown that
for both volume and surface cracks that the two approaches are
mathematically the same in the limit of small regions of singular
stress and $\sigma/\sigma_m - 1 \ll 1$. We may then approach the determination
of the tendency for thermal failure by either method. Since the
statistical approach requires that we know the Weibull modulus m
and the reference stress σ_m, it is easier from a computational point
of view to use the fracture mechanics approach and to determine the
behavior of K.

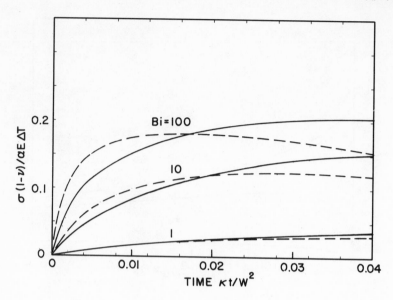

Fig. 8.   Stress of the un-cooled edge of the plate (———)
The maximum compressive stress (— —) in the plate.

NUMERICAL EVALUATION OF K

The stress intensity factor can be determined by:

a.   Calculating the stresses near the crack tip and fitting
with a curve. This method requires a very fine mesh
near the crack tip which frequently leads to high computing costs.

b.   Using the relationship

$$K = \frac{EG}{1-\nu^2}; \quad G = \frac{\partial U}{\partial b}$$

where G is the strain energy release rate and b is the
crack length. The procedure is to calculate the strain
energy for a number of different crack lengths and to
determine G through differentiation of the resulting strain
energy versus crack length relationship.

c.   The best method is to again consider the problem as a two part problem.

I.   Assume that the crack surfaces are welded shut and calculate the stresses σ_c on the plane of the crack.

II. Treat the isothermal problem of applying stresses $-\sigma_c$ to the crack face and determine the resulting value of K.   The stress intensity is then simply that generated by the crack opening stresses of part II.   Of course one must still evaluate K and although sometimes curve fitting is done, it is most usual to use:

(1)   Special formulations - Super finite elements (12,13)
Here the basis function for the finite element is expanded to include the singular term and the stress intensity factor plays the role of a Lagrangian multiplier which represents one of the degrees of freedom of the solution.

(2)   Influence Functions   (14,15)
Consider a crack whose face is loaded by the stress σ_0 over the interval 0 to ζ and let the stress intensity factor be $K(\zeta)$.   Then for any stress $\sigma_c(\zeta)$, the superposition principle gives

$$K = \int_o^b K(\zeta) \frac{d\sigma_c}{d\zeta} \, d\zeta \tag{12}$$

Once the function $K(\zeta)$ is known, the stress intensity factor for any stress distribution $\sigma_c (\zeta)$ $0 \leq \zeta \leq b$ can be easily evaluated.

(3)   Crack Face Stress Fitting (16,17,18).   In this method, one determines the stress intensity for a specific crack and structural geometry in the form

$$K = \alpha_0 K^0 + \alpha_1 K^1 + \alpha_2 K^2 \, --- \, \alpha_n K^n \tag{13a}$$

where K^i is the stress intensity factor generated by a stress of the form $\sigma^o = (\zeta/b)^i$ where
ζ is the distance along the crack surface from
the tip. Given the stresses σ_c from method cI,
we write

$$\sigma_c = \alpha_o + \alpha_1 \left(\frac{\zeta}{b}\right)^1 \cdots + \alpha_n \left(\frac{\zeta}{b}\right)^n, \qquad (13b)$$

determine the values of α_i through a least
squares fitting procedure and apply eq (13a)
to determine K.

Of the different methods, a and b can be applied to any problem,
elastic, plastic or temperature dependent material properties.
Because method c relies upon superposition, it cannot be applied to
plastic or non-linear problems. Method cI requires a special finite
element program. Methods cII and cIII require a dictionary of influence functions K^i, which can be computed by other techniques or
can even be obtained through suitable simplifications and approximations to existing solutions. (Reference 19 discusses a particularly simple approach to evaluating these influence functions.)

EXAMPLES OF THERMAL FRACTURE CALCULATIONS

The following examples of two dimensional fracture problems are
given to illustrate typical temporal and spatial characteristics of
the stress intensity factors for unrestrained bodies. In all of the
examples, except for the annulus problem, the steady state stresses
are zero-leading to zero values of K. For restrained boundary conditions, as in the case of the annulus, steady state non-zero values
of K will exist.

A. Two Dimensional Mode I Fracture

1. Edge Crack in a Suddenly Cooled Plate
    Returning to the problem shown on figure 3, the stress in-
    tensity factors were calculated as a function of time for
    several different Biot numbers for edge cracks on both the
    cooled face and the opposite face (15). Typical values
    are shown on figure 9. For cracks on the cooled face,
    the stress intensity factors start at 0 and quickly rise
    to a high value as the tensile stress zone increases in
    size. Even though the tensile stresses are diminishing
    with time, the movement of the zero point inward with the
    consequent reduction in the compressive stresses and in-
    crease in the tensile zone size increases the force and
    the moment on the crack fact with a consequent increase
    in K. For later times the nearly stationary position of

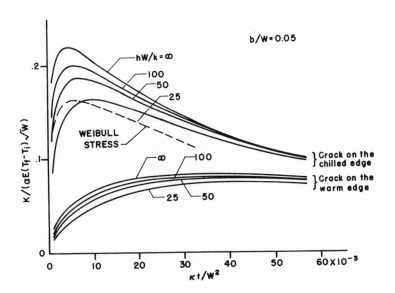

Fig. 9.   Computed stress intensity factors for a crack ($b/w = 0.05$) in a plate and the Weibull stress for an uncracked plate.

the zero point and the continually reducing tensile stresses results in a slow decay of K. For cracks on the uncooled face, both the tensile stress and the tensile stress zone increase slowly with time and then the stresses slowly reduce to zero so that the behavior of K is much less abrupt.

The behavior of the Weibull stress σ_w is also displayed, and even though it does not show quite as much temporal variation, the trend is the same as that of K.

Figure 10 compares the computed stress intensity factor and values determined from photoelastic studies (20).

2.   Edge Crack in the Annulus
A crack was placed in the annulus of figure 1 at $\theta = 270°$. Because of the circumferential variation of the temperature profile, the crack is subjected to strong opening

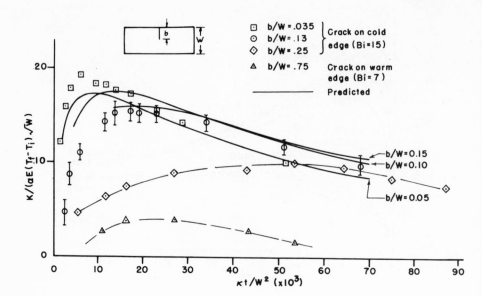

Fig. 10.   Measured and computed stress intensity factors for edge
           cracks in a plate with Bi = 15 (from ref 15)

forces.   Stress intensity factors computed by means of
method cII gave values within 12% of those measured.

3.   Edge Discontinuties
     To generalize the problem, consider the situation of an
     edge discontinuity as illustrated in figure 11 and 12
     (20).   Here we have either a fillet or a notch at the
     edge of the plate.   For the fillet, immediately upon
     cooling, the temperature penetration is so slight that
     the stress is equal to that found on the bar.   However
     as soon as the thermal pulse penetrates into the depth,
     the edge begins to simulate a plate, with a fillet, under
     tension and bending and the fillet acts as a stress raiser
     The ratio of the stresses at the root of the fillet to
     stresses in a plane plate is in good agreement with values
     reported by Isida (20).   When the fillet is insulated so
     that there is no sudden temperature change, the magnifi-
     cation is substantially delayed and remains less than
     that for the uninsulated case.

         When the crack is replaced by a notch of opening
     angle 2α, the order of the stress singularity differs from

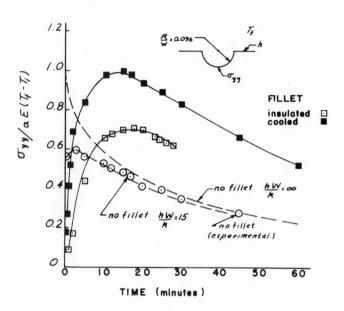

Fig. 11.   Stress at the fillet root and plate edge (from ref 20)

$\sqrt{r}$. Figure 12 illustrates the variation of $\overline{K}$ with time
for several angles where the generalized factor $\overline{K}$ is
defined by $\sigma_{yy} = \overline{K}/(2\pi r)^n$ near the tip of the crack.
The experimentally determined values of n agreed with the
theoretical to within 3% at 90% confidence. The figure
also shows values of the computed factors for the 90°
notch, based upon a finite element program with a super
element to account for the stress singularity. The
slight differences are attributed to inaccuracies in the
computed temperature profile and uncertainties in the
thermal and material properties and in the thermal boundary conditions.

All of the examples presented thus far illustrate the basic
characteristics of mode I stress intensity factors; namely:
a fairly quick rise to a maximum and then a slow decay towards

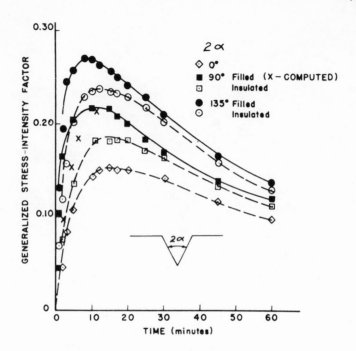

Fig. 12. Generalized stress intensity factors for notches and edge cracks in an edge-cooled plate from ref 15.

zero with time. At early times the greater the surface heat transfer, the higher the value of K. At later times the faster thermal penetration associated with the higher surface heat transfer leads to a faster thermal and stress equilibrium and a faster decay of K with time. However, since fracture, if possible, would occur at or before the time of peak K, the subsequent behavior is of lesser importance. On the other hand it is critical to recognize that there is an interaction between time and crack depth which may lead to unusual crack behavior as demonstrated in figure 13. Here we have K versus crack depth for several values of time (6). Because of the shape of the curve we see that if the critical stress intensity is reached at given time for an initial flaw, it will extend until its length is such that any further extension will decrease K because the tip has penetrated too far into the compressive zone. Any further extension must take place in concert with the movement of the zero stress point and the simultaneously reduction in both the tensile and the compressive stresses.

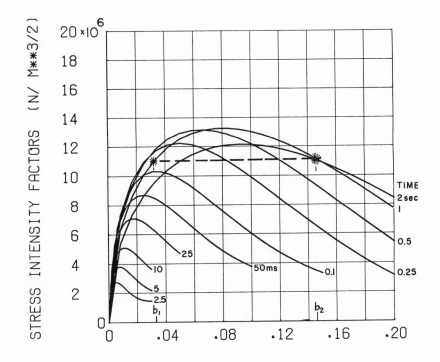

Fig. 13. Stress intensity factor for different initial crack
lengths. The dashed line represents a controlled crack
extension (from ref 6.)

As a result each crack tends to undergo a controlled extension to a maximum depth. Any subsequent thermal shocking
will not extend the crack further. This behavior is markedly
different from a mechanical loading for which most cracks
would continue to extend until catostrophic failure occurs.
Evans and Weiderhorn (21) give a good discussion of crack
extension as it pertains to typical ceramics.

B.   Mode II Stress Intensity Factors

Mode I stress intensity factors exist whenever there is a
temperature variation over the surface of the crack, i.e. when
the heat flux is parallel to the crack. On the other hand,
whenever the flow disrupts the flow of heat, mode II, or sliding, deformations exist. Although the computation of K_{II} is,
in principle, just as straightforward as the computation of
K_I, in practice it is much more difficult. Consider the crack
shown on figure 3 in the case where heat is flowing vertically,
not horizontally. Because the crack interrupts the heat flux,

the flux is infinite at the crack tip, the temperature is of the form $T = K_T \sqrt{r} \ f(\theta)$ leading to severe problems in the accurate assessment of the tip temperature field. For brittle materials, the state of stress is well represented by $K/\sqrt{r}$, (5, 7, 8) but because of convection and radiation across the flaw volume, the flux is never quite as singular as $K_T/\sqrt{r}$ and the stress intensities should be lower. Kassir (22,23) and Sekins (24,25) discuss the steady state behavior of such crack for a variety of crack inclinations to the heat flux and for a range of several parameters. Figure 14 illustrates the variation of K_{II} for the case where $T = 1$ on the upper left edge, $T = -1$ on the lower left edge and all other surfaces are insulated. The curve which represents the calculation of K_{II} without consideration of the singularity in the temperature, $K_T \sqrt{r}$, differs by a negligible amount until long after the peak value of K_{II} has passed.

In comparison to the value of K_I for the edge cooled plat K_{II} is 1/5 to 1/10 of K_I. Measurements made in the annulus wi the liquid half filling the pipe and a crack at the liquid lev gave $K_{II}/K_I \sim 1/7$, which is in agreement with the above estimates.

CONCLUSIONS

The assessment of the possibility of the thermal fracture of brittle materials depends upon an accurate evaluation of the therma stresses and the determination of the resulting stress intensity factors. The stress intensity factors can be calculated in a varie of ways ranging from the very precise to approximate, but only for a limited number of geometries. However the main difficulty is in determining the thermal stress field because of its unusual charact and its dependence upon boundary conditions at points far from the region of thermal activity.

Examination of a number of examples suggests that the best visualization of the thermal stresses and any associated fracture can be made by considering the problem to be the combination of thermal and isothermal problems or by considering that the prime effect of the temperature is in the generation of thermal strains and that the thermal stresses are simply the result of the region trying to accommodate these strains.

For unrestrained bodies, the tendency of the stress intensity factor to rise to a maximum and then to decay to zero restricts' the crack extension and does not permit the use of a proof-test concept. It also implies that repeated thermal loadings may yield a uniform distribution of flaws even when the initial distribution was highly non-uniform.

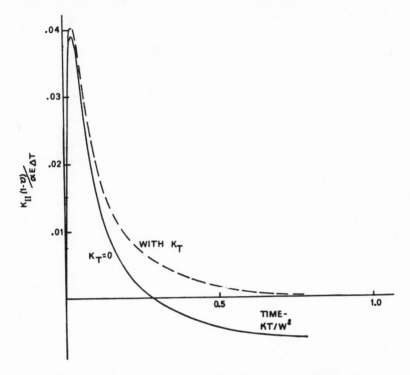

Fig. 14. Mode II stress intensity factor for a temperature
suddenly applied to the cracked edge of a plate.
—— Singularity in the temperature considered.
-- Singularity in the temperature not considered.

ACKNOWLEDGEMENTS

A portion of this study was supported by NASA grant
NGL 48-002-004

NOMENCLATURE

b	crack depth
c	specific heat capacity
C	contour
e	dilation
E	modulus of elasticity
h	surface heat transfer coefficient
k	thermal conductivity
K	stress intensity factor
q	heat flux
Q	volumetric heat source
r	radius

S   surface
t   time, traction
T   temperature
u   displacements
V   volume
x   cartesian coordinate
X   body force
α   linear coefficient of thermal expansion
ε   strain component
Σ   sum of normal stresses
λ´,μ   Lame constants
θ   angle
σ   stress component

subscripts
i   initial
f   fluid
n   normal

REFERENCES

1. B. A. Boley and J. H. Weiner,"Theory of Thermal Stresses"
       John Wiley and Sons, New York, 1960
2. W. Nowacki, "Thermoelasticity" Pergamon Press, London 1962
3. D. J. Johns, "Thermal Stress Analysis" Pergamon Press,
       London, 1965
4. Journal of Thermal Stresses, Hemisphere Publ. Washington DC
5. A. F. Emery, A. S. Kobayashi and C. F. Barrett, Thermal
       Stresses in Partially Filled Annuli, SESA J,6:602, 1966
6. A. F. Emery and A. S. Kobayashi, The Transient Stress
       Intensity Factor for Edge and Corner Cracks in Quench
       Test Specimens, to appear Am. Ceramics J.
7. Experimental Techniques in Fracture Mechanics, vol 1,
       A. S. Kobayashi, ed. Soc Exp Stress Anal. Westport, 1973
8. G. C. Sih, On the Singular Character of Thermal Stresses
       Near a Crack Tip, Trans ASME, J. Appl Mech, 29:587, 1962
9. A. F. Emery, P. K. Neighbors, A. S. Kobayashi and W. J. Love
       Stress Intensity Factors in Edge Cracked Plates Subjected
       to Transient Thermal Singularities, Trans ASME, J. Press
       Ves Tech, 99;100,1977
10. S. B. Batdorf and H. L. Heinisch, Jr., Fracture Statistics
       of Brittle Materials with Surface Cracks, Rept. UCLA-
       ENG-7703, Univ. of Calif. at Los Angeles; 1977
11. S. B. Batdorf and J. G. Crose, A Statistical Theory for the
       Fracture of Brittle Structures Subjected to Nonuniform
       Polyaxial Stresses, Trans ASME, J Appl Mech, 41:459, 1964
12. A. F. Emery, F. J. Cupps and P. K. Neighbors, The Use of
       Singularity Programming in Finite Element Calculations

of Elastic Stress Intensity Factors, Plane and Axi-
symmetric--Applied to Thermal Stress Fracture,
"Computational Fracture Mechanics" E. F. Rybicki and
S. E. Benzley, eds., ASME, New York, 1975

13. S. E. Benzley and Z. E. Beizinger, CHILES-A Finite Element
Computer Program that Calculates the Intensities of
Linear Elastic Singularities, Sandia Lab Rept SLA-73-
0894, 1973

14. A. F. Emery, Stress Intensity Factors for Thermal Stresses
in Thick Hollow Cylinders, Trans ASME, J. Basic Engin,
88:45, 1966

15. A. F. Emery, G. Walker and J. A. Williams, A Green's
Function for Stress Intensity Factors of Edge Cracks
and its Application to Thermal Stresses, Trans ASME,
J. Basic Engin, 91:618, 1969

16. A. S. Kobayashi, A. F. Emery, W. J. Love and N. Polvanich,
Surface Flaws in Thermally Shocked Hollow Cylinders,
Proc. 3rd Int. Conf. in Str. Mech. in Reactor Tech.,
3:11, 1975

17. A. S. Kobayashi, A. F. Emery and W. J. Love, Surface Flaw
in a Pressurized and Thermally Shocked Hollow Cylinder,
Int. J. Press Ves and Piping, 5:103, 1977

18. G. G. Chell, The Stress Intensity Factors for Part Through
Thickness Embedded and Surface Flaws Subject to a
Stress Gradient, Engin Fract Mech, 8:331,1976

19. E. F. Rybicki and M. F. Kanninen, A Finite Element
Calculation of Stress Intensity Factors by a Modified
Crack Closure Integral, Engin Fract Mech, 9:931, 1977

20. A. F. Emery, J. A. Williams and J. Avery, Thermal Stress
Concentration Caused by Structural Discontinuities,
SESA J, 9:558, 1969

21. A. G. Evans and S. M. Wiederhorn, Proof Testing of Ceramic
Materials- An Analytical Basis for Failure Prediction,
Int. J. Fract., 10:379, 1974

22. M. K. Kassir, Stress Intensity Factors for an Insulated
Half-Plane Crack, Trans ASME, J. Appl Mech. 43:107, 1976

23. M. K. Kassir, Thermal Crack Propagation, Trans ASME,
J. Basic Engin, 93:643, 1971

24. H. Sekine, Thermal Stress Singularities at Tips of a Crack
in a Semi-Infinite Medium under Uniform Heat Flow,
Engin Fract Mech, 7:713, 1975

25. H. Sekine, Thermal Stresses near Tips of an Insulated Line
Crack in a Semi-Infinite Medium under Uniform Heat Flux,
Engin Fract Mech, 9:499, 1977

ON CALCULATING THERMALLY INDUCED STRESS SINGULARITIES

Morris Stern

Department of Aerospace Engineering and Engineering
Mechanics
The University of Texas at Austin
Austin, Texas 78712

INTRODUCTION

A problem of increasing interest and importance is the
numerical calculation of stress intensity factors for cracked or
otherwise imperfect structures due to thermal as well as mechanical loading. The finite element method has become one of the
most popular and effective computational approaches to stress
analysis, and a large number of fairly successful schemes have
been introduced to treat the stress singularities of interest in
fracture mechanics as outlined in [1]. One of the simplest yet
most effective methods for treating such problems is the introduction of special "singular elements" to model the fields in the
immediate neighborhood of the exceptional point without requiring
an overly refined (and expensive) mesh.

In the following paragraphs a particularly versatile family
of singular finite elements will be described with particular
reference to their use in calculating stress intensity factors in
quasistatic uncoupled thermoelastic analyses of cracked structures.

THE SINGULARITIES

Consider, for example, a notched plane region of thermally
isotropic material, and suppose the edges of the notch to be
insulated as indicated in Fig. 1. In the absence of heat sources
a steady temperature field $T(r,\theta)$ in the neighborhood of the
notch tip will satisfy Laplaces equation

$$\nabla^2 T = 0 \quad ; \quad r > 0 \quad , \quad |\theta| < \alpha \qquad (1)$$

and have vanishing normal derivative on the notch flanks:

$$\frac{dT}{dn} = 0 \quad , \quad \theta = \pm \alpha \qquad (2)$$

Separation of variables and imposition of the boundary conditions yields eigensolutions of the form

$$T(r,\theta) = r^\lambda [A \sin \lambda\theta + B \cos \lambda\theta] \qquad (3)$$

provided

$$A \cos \lambda\alpha \pm B \sin \lambda\alpha = 0 \qquad (4)$$

A reasonable requirement for a physically meaningful temperature field is that its gradient be square integrable, which translates into the requirement $\lambda > 0$. The smallest eigen-pair satisfying this restriction is

$$\lambda = \pi/2\alpha \quad , \quad T = kr^{\pi/2\alpha} \sin \pi/2\alpha \; \theta \qquad (5)$$

and for $\alpha > \pi/2$ this particular temperature field has an unbounded gradient (like $r^{-(1-\pi/2\alpha)}$ with "intensity" $k\pi/2\alpha$) at $r = 0$.

In the particular case of an insulated slit or crack along $\theta = \pm \pi$ we have $\lambda = 1/2$ and the temperature field in the neighborhood of the crack tip $(r > 0, |\theta| < \pi)$ is

$$T(r, \theta) = 2k \sqrt{r} \sin \theta/2 + \text{remainder} \qquad (6)$$

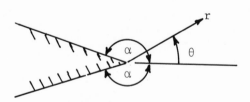

Fig. 1. Insulated traction free notch

where the remainder goes to zero faster than $\sqrt{r}$. The dominant term in the temperature gradient has components

$$Q_r = \frac{\partial T}{\partial r} = k/\sqrt{r} \; \sin \theta/2$$

$$Q_\theta = \frac{1}{r} \frac{\partial T}{\partial \theta} = k/\sqrt{r} \; \cos \theta/2$$

(7)

We may call k the thermal gradient intensity factor in analogy to stress intensity factors described below.

Returning to the original case, the associated plane elasticity problem with traction free conditions on the notch edges (for $\alpha > \pi/2$) admits a solution with unbounded stresses at the notch tip. Again, physical (or mathematical) considerations dictate that the stresses also be square integrable. A separation of variables eigensolution for the displacement components is of the form

$$u(r,\theta) = r^\eta f(\theta;\eta)$$

with η the smallest positive solution of

$$\sin 2\eta\alpha = \pm \; \eta \; \sin 2\alpha$$

(see, for example [2]). For α between $\pi/2$ and π the smallest positive solution of

$$\frac{\sin 2\eta\alpha}{2\eta\alpha} = - \frac{\sin 2\alpha}{2\alpha}$$

yields an eigenvalue η ranging between 1 and 1/2, and the stresses in this case become unbounded near the notch tip like $r^{-(1-\eta)}$.

Again, for the particular case of a traction free crack along $\theta = \pm \pi$ the smallest positive eigenvalue is $\eta = 1/2$ and the corresponding eigensolution for displacement components may be witten (see [3]) in the form

$$u_r = \frac{\sqrt{2r}}{8\mu} \left\{ K_1 [(2\kappa-1) \cos \theta/2 - \cos 3\theta/2] \right.$$

$$\left. + K_2 [-(2\kappa-1) \sin \theta/2 + 3 \sin 3\theta/2] \right\}$$

$$u_\theta = \frac{\sqrt{2r}}{8\mu} \left\{ K_1 [-(2\kappa+1) \sin \theta/2 + \sin 3\theta/2] \right. \tag{8}$$

$$\left. + K_2 [-(2\kappa+1) \cos \theta/2 + 3 \cos 3\theta/2] \right\}$$

where μ is the shear modulus of the material, κ is a dimensionless elastic constant whose value ranges between 1 and 3, and K_1, K_2 are parameters of the eigensolution called the opening and shear mode stress intensity factors respectively. These names stem from the observation that on the crack extension ($\theta = 0$) the dominant terms in the normal and shear stress are

$$\sigma = \frac{K_I}{\sqrt{2r}} \quad , \quad \tau = \frac{K_{II}}{\sqrt{2r}} \tag{9}$$

This singular temperature and displacement field behavior is typical in a large class of problems associated with boundary discontinuities such as occur at notches, cracks, and reentrant corners, or at an abrupt change in material properties on the boundary, or in the nature of the prescribed boundary conditions. Frequently, as in the example above, one can perform simple analyses of the heat transfer and elastostatic eigenvalue problems to determine that the primary field (temperature or displacement in the example) varies as r^λ with λ a (known) positive number less than unity.

In a direct finite element formulation using the usual polynomial basis elements, a relatively fine grid is needed in the neighborhood of the singularity, especially if one wants to infer a reliable estimate of the stress intensity factors from the finite element solution. We show here that a single versatile family of singular finite element triangles can be used to permit fairly coarse grids in the neighborhood of simultaneously occurring stress and temperature gradient singularities which produce fairly accurate estimates of the singularity intensities.

SINGULAR TRIANGLE ELEMENT FAMILY

We outline briefly here the development of one member of the family of singular elements. A more detailed treatment may be found in [4]. The particular features characterizing the family may be summarized as follows:

i) Linear fields are contained in the element shape functions; this insures that homogeneous strain and uniform temperature gradient fields are modelled exactly.

ii) The element shape functions also contain Kr^λ where r is distance along a ray emanating from the singular vertex and $0 < \lambda < 1$, thus the gradient becomes singular like $r^{-(1-\lambda)}$ at the origin. Furthermore the number and placement of nodes in the element is independent of λ but is chosen so that

iii) Members of the singular family are compatible with standard polynomial elements and each other in the sense that interpolation fields defined by nodal values are continuous across element boundaries.

It is convenient to introduce a set of natural coordinates on the triangle which are closely related in structure to polar coordinates in the plane, hence the name triangular polar coordinates. The linear coordinate transformation (see Fig. 2)

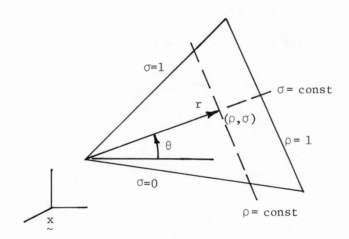

Fig. 2. - Triangular polar coordinates

$$\underset{\sim}{x} = \underset{\sim}{x}_1 + \rho[(\underset{\sim}{x}_2 - \underset{\sim}{x}_1) + \sigma(\underset{\sim}{x}_3 - \underset{\sim}{x}_2)] \tag{10}$$

defines a "radial" coordinate ρ and an "angular" coordinate σ in terms of the cartesian coordinates of the triangle vertices. In particular, along a σ = constant line ρ is proportional to distance from the origin node, and the ρ = constant lines are parallel to the side opposite the singular vertex. The triangle interior is described by $0 < \rho < 1$, $0 < \sigma < 1$, and the Jacobian of the coordinate transformation is $J = 2A\rho$ where A is the area of the triangle, so that a differential area element becomes

$$dA = 2A \, \rho d\rho d\sigma \tag{11}$$

Note that the coordinate transformation is itself singular (i.e., not invertible) at the origin where $\rho = 0$ but σ is arbitrary.

To avoid extraneous details we limit the discussion here to the six node element compatible with standard quadratic elements. Other cases are treated in more generality in [4].

The shape functions on the element are required to approximate fields of the form

$$\phi(r,\theta) = p(\theta)r^{\lambda} + \{\phi_o + q(r,\theta)\} \tag{12}$$

where $0 < \lambda < 1$ and $\lim_{r \to 0} r^{-\lambda} q(r,\theta) = 0$. The local finite element approximation is taken in the form

$$\phi^h(\rho,\sigma) = P(\sigma)\rho^{\lambda} + Q(\rho,\sigma) \tag{13}$$

where P and Q are polynomials in the indicated arguments.

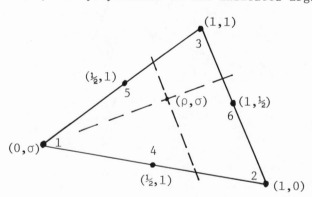

Fig. 3. Nodal locations and numbering

The requirement that Q contain linear fields ties down three degrees of freedom, and for compatibility with quadratic elements on the side $\rho = 1$ we require that P also be quadratic which adds three mode degrees of freedom. Thus, in terms of nodal values the local approximation on the element shown in Fig. 3 becomes

$$\phi^h(\rho,\sigma) = \phi_1 + \frac{\rho}{\beta-1} \{(1-\sigma)[\beta\phi_2 - 2\phi_4 + (2-\beta)\phi_1]$$

$$+ \sigma[\beta\phi_3 - 2\phi_5 + (2-\beta)\phi_1]\}$$

$$+ \rho^\lambda \left\{ \frac{1-\sigma}{\beta-1} [2\phi_4-\phi_2-\phi_1] + \frac{\sigma}{\beta-1} [2\phi_5-\phi_3-\phi_1] \right.$$

$$\left. + \sigma(1-\sigma)[4\phi_6 - 2\phi_2 - 2\phi_3] \right\}$$

$$= \sum_{a=1}^{6} [N_a(\rho,\sigma) + M_a(\sigma)\rho^\lambda]\phi_a \qquad (14)$$

where the shape functions $N_a(\rho,\sigma)$ are linear in ρ and $\rho\sigma$ while the $M_a(\sigma)$ are no worse than quadratic, and we have written β for $2^{1-\lambda}$. In particular, a comparison of Eq. $(14)_1$ and (12) shows that if the edge $\sigma = 0$ falls along the line $\theta = \theta_o$, then the coefficient $p(\theta_o)$ of the r^λ term is approximated by

$$p(\theta_o) \approx (\frac{\rho}{r})^\lambda P(0) = \frac{1}{(\beta-1)L^\lambda} \{2\phi_4 - (\phi_1 + \phi_2)\} \qquad (15)$$

where L is the length of the side connecting nodes 1 and 2 (i.e. the $\sigma = 0$ side). We will use Eq. (15) later to estimate the stress intensity factor in a particular example.

This element has been incorporated in the element library of TEXGAP-2D, a finite element code for the solution of plane or axisymmetric elastostatic and heat transfer problems. The following example illustrates the quality of the results which can be obtained with a relatively coarse mesh, especially around the singular point.

EXAMPLE

The example problem has been treated earlier using an integral equation approach by Konishi and Atsumi [5]. The problem

consists of an infinite elastic strip with a uniform heat flow
across its width disturbed by an insulating crack in the center
of the strip parallel to its edges. Because of symmetry only one
quarter of the strip need be modelled as indicated in Fig. 4,
where dimensions, boundary data, and a typical finite element mesh
are shown. The four elements adjacent to the crack tip are the
singular elements described in the preceeding section; the remaining 34 elements are standard quadratic isoparametric quadra-
laterals.

If the initial (unstressed) temperature of the strip is T_o
then symmetry dictates that the axial displacement along the extended crack line as well as the line through the center of the
crack vanish. To fix the rigid body displacement of the strip
we also find it convenient to fix the transverse displacement of
the crack tip. Along the edges on which the temperature is not
prescribed the natural boundary condition is vanishing heat flux,
and of course the traction components are zero where the corresponding displacement components are not specified.

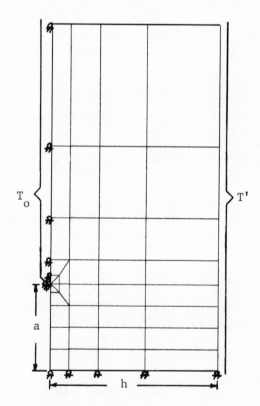

Fig. 4. - Finite element mesh on one quarter of a cracked strip

The constitutive equation for linear isotropic thermoelastic materials is of the form

$$\varepsilon_{ij} = 1/2(u_{i,j} + u_{j,i}) = 2\mu\sigma_{ij} + \delta_{ij}[\lambda\sigma_{kk} + \alpha(T - T_o)]$$

where the ε_{ij} are components of the infinitesimal strain tensor defined as the symmetric part of the displacement gradient, σ_{ij} are components of the stress tensor, λ and μ are elastic constants and α is the coefficient of thermal expansion. The stresses in the neighborhood of the crack tip behave like $r^{-1/2}$ whereas the temperature increase is of the order $r^{1/2}$. Thus the dominant part of the local displacement field around the crack tip is still given by Eq. (8) with the contribution due to nonuniform temperature not showing up until terms of the order of $r^{3/2}$ become significant.

From Eq. (8) we conclude that the shear displacement discontinuity across the crack is of the form

$$[[v]] = u_r\Big|_{\theta=-\pi} - u_r\Big|_{\theta=\pi} = \frac{\sqrt{2}\,(\kappa+1)}{2\mu}\, K_2\,\sqrt{r} + \dots \qquad (16)$$

while symmetry dictates that $[[u]] = 0$ and consequently $K_1 = 0$. Alternatively, in front of the crack the transverse displacement is of the form

$$u = -u_\theta\Big|_{\theta=0} = \frac{\sqrt{2}\,(\kappa-1)}{4\mu}\, K_2\sqrt{r} \qquad (17)$$

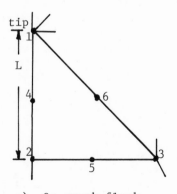

a) On crack flank

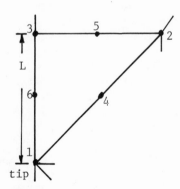

b) On crack extension

Fig. 5. - Singular elements at crack tip

while again K_1 vanishes since the boundary condition requires $v = 0$.

The singular element on the crack face is shown again in Fig. 5a with node 1 at the crack tip and the side $\sigma = 0$ (nodes 1, 4 and 2) along the crack. According to Eq. (15) the finite element approximation to the shear displacement discontinuity across the crack is obtained from nodal point values as

$$[[v]]^h = \frac{2}{(\sqrt{2}-1)\sqrt{L}} \; [2v_4 - (v_1 + v_2)] \; \sqrt{r} \tag{18}$$

Comparison of Eq. (16) and (18) enables us to infer the stress intensity factor from the nodal displacements:

$$K_2^h = \frac{2(\sqrt{2}+2)\mu}{(1+\kappa)\sqrt{L}} \; [2v_4 - (v_1 + v_2)] \tag{19}$$

In a similar manner we infer a value for the stress intensity factor from displacement data of the singular element bordering the crack extension indicated in Fig. 5b: –

$$K_2^h = \frac{2(\sqrt{2}+2)\mu}{(\kappa-1)\sqrt{L}} \; [2u_6 - (u_1+u_3)] \tag{20}$$

The stress intensity factor for an infinite plate $(h/a{\to}\infty)$ was given by Florence and Goodier [6] as

$$K_2^\infty = \frac{\alpha^* \mu q a^{3/2}}{1+\kappa} \tag{21}$$

where $q = (T'-T_o)/h$ is the undisturbed temperature gradient component normal to the crack and $\alpha^* = (1+\nu)\alpha$ for plane strain with ν Poisson's ratio.

A more accurate estimate of the stress intensity factor using the same mesh can be obtained by combining the use of sungular elements with the extended singularity integral representation for stress intensity factors described in [7] and [8]. Without furnishing details here, we record the following path independent representation for the shear mode stress intensity factor:

$$K_2 = \int_\Gamma (\underset{\sim}{t}\cdot\underset{\sim}{\hat{u}} - \underset{\sim}{\hat{t}}\cdot\underset{\sim}{u})ds + \int_A \alpha^* T\hat{\theta} \; da \tag{22}$$

In Eq. (22) Γ is a contour enclosing the crack tip and terminating on the crack flanks, A is the region interior to Γ , and $\underset{\sim}{t}$, $\underset{\sim}{u}$, T are respectively the traction and displacement on Γ and the temperature field in A . These are obtained from the finite element solution of the original problem.

Finally $\hat{t}$, $\hat{u}$ and $\hat{\sigma}$ are special singular kernel functions which are given explicitly in [8].

For the extremely coarse 38 element grid shown in Fig. 4 (a/h = .5, L/h = .05), the finite element solution gave the values listed below for non-dimensionalized stress intensity factor. The $\theta = 0$ value is within 2 percent of the singularity integral value but the calculation on the crack flank $\theta = -\pi$ is nearly 10 percent high. This might be due to the fact that the temperature is constant on $\theta = 0$ but grows like $r^{1/2}$ on $\theta = -\pi$. Results for a more refined grid with 6 singular elements surrounded by 100 standard quadratic elements (not shown) are also tabulated. The value obtained by the singularity integral varied less than .1 percent over a wide range of contours and furthermore, even the value obtained using the 36 element grid is within 1 percent.

K_2/K_2^∞ from:	Singularity Integral	Singular Element Crack Extension	Crack Flank
38 element grid	.795	.781	.872
106 element grid	.802	.810	.843

Konishi and Atsumi [5] calculate a value of approximately .6 for this case which is significantly different from the results produced above. While it does not establish which result is more correct, the comparison of K_2/K_2^∞ for a range of h/a in Fig. 6 would seem to indicate that the values obtained using singular elements are more plausible.

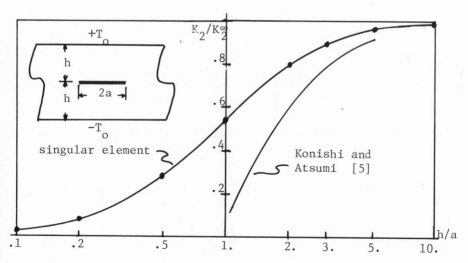

Fig. 6. - Effect of finite strip width

CONCLUSIONS

Special finite elements whose derivative fields contain
singularities of the power type can be·used, in conjunction with
an asymptotic analysis of the singularity fields expected in the
actual problem, to effectively model such situations in linear
thermoelastic analyses by the finite element method. The problem
treated above is an example of such use. The method would probably show to considerably more advantage in transient problems
where the savings in coarse modelling of the singularity neighborhood is reflected at each step in the solution.

The element family is easily extended to three dimensional
wedges with the singularity along one edge, as detailed in [4].
Finally, in two dimensional problems, additional accuracy for
a given element mesh can be obtained by combining the use of singular elements with the extended contour integral representation
for stress intensity factors described in [7].

REFERENCES

1. R.H. Gallagher, A Review of Finite Element Techniques in
    Fracture Mechanics, in: "Proceedings of the First Internation-
    al Conference on Numerical Methods in Fracture Mechanics,"
    Swanwea (1978).

2. A.H. England, On Stress Singularities in Linear Elasticity,
    Int. J. Engng. Sci., 9:571 (1971).

3. P.C. Paris and G.C. Sih, Stress Analysis of Cracks in: "Frac-
    ture Toughness Testing and its Application," ASTM STP 381
    (1965).

4. M. Stern, Families of Consistent Conforming Elements with
    Singular Derivative Fields, Int. J. Num. Meth. Engng.,
    14:409 (1979).

5. Y. Konishi and A. Atsumi, The Linear Thermoelastic Problem of
    Uniform Heat Flow Disturbed by a Two-dimensional Crack in a
    Strip, Int. J. Engng. Sci., 11:1 (1973).

6. A.L. Florence and J.N. Goodier, Thermal Stresses Due to Dis-
    turbance of Uniform Heat Flow by an Insulated Ovaloid Hole,
    J. Appl. Mech., 27:635 (1960) (A.S.M.E. Trans., 82, Ser. E).

7. M. Stern, The Numerical Calculation of Thermally Induced
    Stress Intensity Factors, J. Elast., 9:91 (1979).

8. M. Stern, E.B. Becker and R.S. Dunham, A Countour Integral
    Computation of Mixed-mode Stress Intensity Factors, Int. J.
    Fracture, 12:359 (1976).

THERMOMECHANICAL PARAMETERS DUE TO FIRE

IN UNSTEADY HEAT-CONDUCTION ANALYSIS

Dietrich Hartmann

Associate Professor
University of Dortmund
Institute of Structural Mechanics

The concept of structural safety is reconsidered. In order
to get a computable standard value for safety, the term reliability
is introduced. A structure can be regarded as "reliable" if the
given loading is not greater than its structural resistance. It
is evident that the total analysis directly depends on the accuracy
with which the load intensity is determined. The loading "fire",
in comparison to conventional loads such as dead and live loads,
etc., is considerably more complicated because of its physical
complexity. This paper deals with the loading "fire" and demonstrates how fire can be embedded in a thermomechanical analysis.

Contrary to presently used approaches to structural safety in
case of fire which are more or less experimentally oriented, a
rational concept of analysis is needed which takes into consideration all physical laws of fire development, fire expansion, heat
transfer and heat conduction. Since this problem is basically a
nonlinear, unsteady-state, field problem, its complexity requires
a computer oriented solution. It is, therefore, mandatory that
a standardized thermomechanical software-system be developed in
the future. To achieve this objective the thermal loading caused
by fire must be assessed as realistically as possible. The
present paper is concerned with the parameters that are absolutely
necessary for a realistic analysis of structures subjected to
"catastrophic" type of fire loading.

The determination of temperature distributions
$\theta = \theta(x,y,z,t)$ in the structure is governed by a partial, quasi-
harmonic, differential equation of Fourier type. In the case of
a three-dimensional continuum this equation can be written as

$$\frac{\partial}{\partial x}\left(\lambda_x \frac{\partial \theta}{\partial x}\right) + \frac{\partial}{\partial y}\left(\lambda_y \frac{\partial \theta}{\partial y}\right) + \frac{\partial}{\partial z}\left(\lambda_z \frac{\partial \theta}{\partial z}\right) = \rho c \frac{\partial \theta}{\partial t} \tag{1}$$

where λ_x, λ_y, λ_z are the thermal conductivities in the direction of the co-ordinates x,y,z and ρc is the heat capacity. The above equation describes the heat conduction of temperature transferred from the fire room. The process of heat transfer produced by the last layer between the fire room and the exterior of the structure is governed by the boundary condition

$$\lambda_x \frac{\partial \theta}{\partial x} \ell_x + \lambda_y \frac{\partial \theta}{\partial y} \ell_y + \lambda_z \frac{\partial \theta}{\partial z} \ell_z + \alpha(\theta - T_G) = 0 \tag{2}$$

where ℓ_x, ℓ_y, ℓ_z are the direction cosines of the surface normals. The term α is the heat transfer coefficient and T_G is the gas temperature of the surrounding space. Solution of the differential equation (Eq. (1)) requires knowledge of the initial temperature field θ at the time $\bar{t}$. Thus

$$\theta(x, y, z, \bar{t}) = \bar{\theta} \tag{3}$$

In addition to Eqs. (1), (2) and (3) the solution for θ must satisfy the boundary condition that the given temperature θ on part $\tilde{A}$ of the structure is prescribed:

$$\theta(x, y, z, t) = \tilde{\theta}; \ x, y, z \in \tilde{A} \tag{4}$$

Consequently, an unsteady-state boundary value problem is obtained which can be solved by the finite element method. Discretization in space leads to a matrix differential equation[1]

$$[H]\ \{\theta\} + [P]\ \frac{\partial\{\theta\}}{\partial t} + \{F\} = 0 \tag{5}$$

where [H] is the conductivity matrix, [P] is the heat capacity matrix and {F} is the generalized thermal load vector. Furthermore, $\{\theta(t)\}$ is the vector of the node temperatures. If the time-dependent process described by Eq. (5) is divided into finite elements in the time dimension and the Galerkin method is applied to a linear Lagrangean interpolation function for the temperature θ, the matrix differential equation (5) can be approximated by a simple matrix relation[2]

$$\left[\frac{2}{3}\ [H] + \frac{1}{\Delta t}\ [P]\right]\{\theta\}_i + \left[\frac{1}{3}\ [H] - \frac{1}{\Delta t}\ [P]\right]\{\theta\}_{i-1}$$

$$+ \frac{1}{3}\ \{F\}_{i-1} + \frac{2}{3}\ \{F\}_i = 0 \tag{6}$$

At present, there are numerous (well tested) finite-element-
program—systems which enable the computation of the discrete node
temperature vector $\{\theta\}_i$ from the known vector $\{\theta\}_{i-1}$. In the
first step of the process i is equal to 1 and $\{\theta\}_o$ corresponds to
the initial boundary condition, Eq. (3).

In the present work the ICES—STRUDL III-program system (version 4.4) was applied for determination of the temperature distribution in plane and spatial steel-frames composed from I-profiles.
The version 4.4 was established by the German firms MAN and MBB.
The I-shaped cross-sections of the frames were idealized and
modelled by one- and two-dimensional finite elements. The heat
conduction and heat transfer phenomena related to flanges can be
modeled by one-dimensional elements ("bars"), while those for the
webs are described by two-dimensional isoparametric elements having
8 nodes per element. Figure 1 schematically illustrates the sub-
division of the continuum into finite elements.

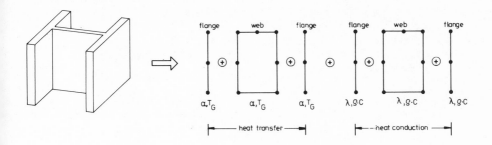

Fig. 1.   Finite element idealization

As Figure 1 shows, the total problem is composed of
individual phenomena. Furthermore, the same figure outlines the
set of parameters which represent the necessary input data for solution of the present thermo-analysis problem. In the case of
steel-frame structures, the thermomechanical parameters α,
$\lambda = \lambda_x = \lambda_y = \lambda_z$, ρc and T_G are used as input parameters in the
program. Most of them are nonlinear. It can be shown that
errors of about 130% and even more can appear in the temperature
distribution θ, if the nonlinearity is neglected. Hence, the
accuracy of the $\alpha, \lambda, \rho c$ and T_G input data must be very high.

The parameters $\alpha, \lambda, \rho c$ can be easily found experimentally
because it is possible to create conditions comparable with those
of the fire room. Determination of the fire room temperature T_G,
however, is too complicated and needs the input of a computer since
T_G depends on a number of factors influencing the values of the
T_G-function. For practical purposes only the most important
factors can be considered. In the present investigation of

steel-frames the T_G-distribution is calculated independently of the heat analysis program. (A pertinent user-oriented FEM-program should be developed in the future.)

Assuming that the _fire_ _room_ is secluded by walls, slabs, doors, etc., and only communicates with its exterior by certain _openings_, the rate of internal heat generation or _fire_ _load_ _q_, can be considered as one of the most important factors. The load q can be determined from

$$q = \frac{\sum_\nu m_\nu H_\nu}{\sum_i F_i} \quad ; \qquad \begin{array}{l} \nu = 1,2,\ldots,M \\[4pt] i = 1,2,\ldots,N \end{array} \qquad (7)$$

where m_ν is the weight of the νth part of the combustible material in the fire room in kg. The term H_ν is the calorific value in Mcal/kg and F_i is the area of the ith part of the boundary surface in m^2. The load q provides the heating of the fire-room-gas by combustion of the burnable material. It is desirable to determine average values of the load q for general classes of buildings such as hotels, office buildings, because data m_ν and H_ν are known for special types of buildings (i.e., q = 33 Mcal/m^2 for office-buildings). The openings are characterized by the opening factor χ_A defined as

$$\chi_A = \sum_{j=1}^{n} A_j \quad \frac{\sum_j h_j}{n} \quad \sum_i F_i \quad ; \qquad \begin{array}{l} j = 1,2,3,\ldots,n \\[4pt] i = 1,2,3,\ldots,N \end{array} \qquad (8)$$

where A_j is the jth opening in the fire room in m^2 and h_j its height in m. The quantity χ_A is the second parameter of importance. If the value χ_A do not exceed a given limit—in most practical cases this can be assumed—the fire development proceeds in such a way that the air exchange equalizes turbulences. Hence, the temperature distribution T_G, which is normally a function of the co-ordinates x,y,z and the time t, becomes a function of one variable t. Thus

$$T_G = T_G(x,y,z,t) \implies \tilde{T}_G(t) \qquad (9)$$

Now, the approximated distribution of $\tilde{T}_G$ depending on the time t can be found by the thermal equilibrium equation

$$I_G = I_L \qquad (10)$$

in which the thermal energy I_G generated by combustion is equated with the thermal energy I_L lost by heat radiation I_R by heat transfer I_T and by convection I_C. Therefore

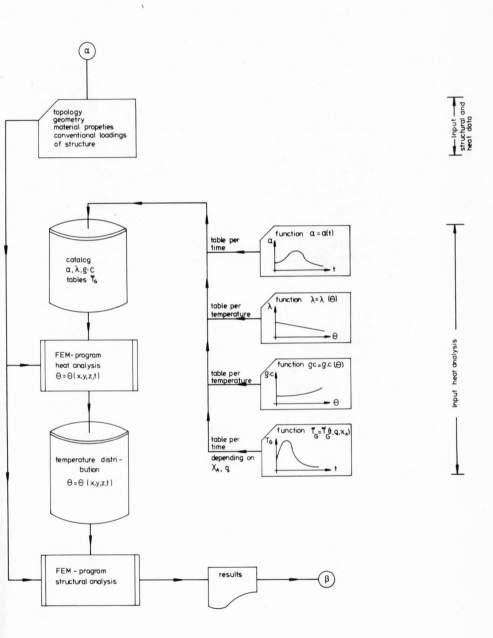

Fig. 2.   General flow-chart

$$I_L = I_R + I_T + I_C \tag{11}$$

The left hand-side, I_G, of Eq. (10) is directly proportional to the opening factor X_A. Evaluation of the energy relation, Eq. (10) leads to a differential equation for T_G and its derivative $\frac{d\tilde{T}_G}{dt}$. This equation can be solved by means of the finite difference method for various values X_A and q (separated from the heat analysis). A set of tables for the temperature T_G per unit time depending on the factors X_A and q are obtained[4] and can be stored on a computer disc along with the corresponding tables for $\alpha, \lambda, \rho c$. Figure 2 demonstrates storage of the input data necessary for the heat analysis.[5] Furthermore, the total coherence between structural data, heat analysis data heat analysis and structural analysis is flow-charted to facilitate the reader's comprehension.

REFERENCES

1. O. C. Zienkiewicz and C. J. Parekh, Transient field problems: two-dimensional and three-dimensional analysis by isoparametric finite elements, Int. Journ. Num. Meth. 2:61-71 (1970).
2. J. Pittr and W. Kohler, Calculation of transient temperature in space and time dimensions, Int. Journ. Num. Meth. 8: 625-631 (1974).
3. M.A.N. - STRUDL III, Benutzerhandbuch für statische Berechnungen publicated by M.A.N. Nürnberg, Abt. Organisation u. Datenverarbeitung, 384-394, Update Januar 1978.
4. O. Petterson, S.-E. Magnusson, and J. Thor, Brandschutztechnische Bemessung von Stahlkonstruktionen, internal research publication, 1977.
5. D. Hartmann, Brandverhalten von Gesamttragkonstruktionen, Teil 1, Temperaturverteilung, Forschungsbericht des Bundesministeriums für Raumordnung, Bauwesen und Städtebau, 1978.

THERMOELASTIC BUCKLING OF PLATES IN A CYLINDRICAL

GEOMETRY AGAINST AN ELASTIC BACK SUPPORT

L. D. Simmons

Multnomah School of Engineering
University of Portland
Portland, Oregon 97203

R. W. Wierman

Westinghouse Hanford Company
Richland, Washington 99352

INTRODUCTION

A plate which is fixed at its edges to a strong edge
support structure will develop large compressive stresses
when heated from ambient temperature more rapidly than the
support structure. Determining the response of the plate
to this situation requires stability analysis to ascertain
whether the plate might buckle, or whether the constrained
thermal expansion will lead to compressive stresses exceeding the yield point because it did not buckle. A special
case is considered here, both analytically and experimentally, in which the plate is curved slightly into a
cylindrical shape and the convex face of the plate is
against a supporting surface. This case is more complex
because the buckling mode will be a harmonic rather than
the fundamental mode which is usually encountered.

BUCKLING AND ITS IMPORTANCE

Thermal expansion of a plate which is firmly fixed at
its edges causes a buildup of compressive stress in the
plate. This can lead to one of two consequences: The
compressive yield stress can be reached causing permanent
deformation; or the plate can buckle before reaching the
yield point. Buckling, which in effect limits stress

because further expansion just increases the curvature
of the buckled plate, can be considered beneficial in ma:
cases.   This is particularly true when it is important t
avoid yielding of the plate or when stresses approaching
the yield value could lead to failure of edge support
structure.

     Figure 1 illustrates the special case of a plate
with slight curvature forming a segment of a cylinder an
with support on the back surface of the plate.   In this
case, failure to buckle could lead to compressive yieldi:
of the plate or large enough forces to cause failure of
the I-beam embedments or the concrete supporting them.
.Buckling, on the other hand, could limit the stress and
support forces.   In this case, if the heat source was on
the concave side of the plate, buckling would also cause
a gap between the plate and concrete which would reduce
heat flow to the concrete.   The overall effect on plate,
edge supports, and concrete would be quite different if
buckling occurred.

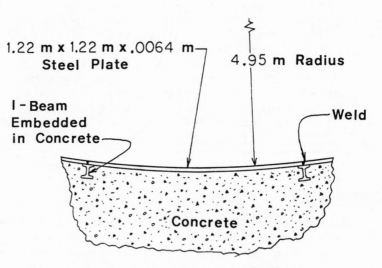

Fig. 1.   Edge view of a slightly curved thin plate with
           back support.

Experience has shown that engineers commonly make
two intuitive errors regarding the effects of heating on
the configuration in Figure 1. First, it is easy to conclude that expansion of the plate would simply force it
against the concrete and the concrete support would prevent buckling. It is true that the initial curvature
would promote buckling toward the concrete, and that buckling in the fundamental mode is prevented by the concrete.
However, as the stress continues to rise above the critical
level for fundamental mode buckling, it can reach the level
necessary for some harmonic mode which permits the plate to
buckle away from the concrete. The second common error is
to assume a strong edge support structure is rigid. This
is a common simplifying assumption in stress analysis, but
is completely invalid in analysis of stresses caused by
constrained thermal expansion. It will be demonstrated
later that even the slight deflection under stress of a
very strong edge support has significant stress relieving
effect on the thermal stress in the plate.

THEORY

In analyzing whether buckling will play a significant
stress-limiting role in a particular case, it is first
necessary to predict whether buckling is likely to occur
before yield stress is reached. In the case considered
here, only harmonic buckling is possible. Unfortunately
the literature on buckling focuses on fundamental mode
buckling because that is appropriate in most cases.
Reference 1 (pp.474-488) gives the theoretical basis for
analyzing plates resting on an elastic foundation and for
analysis of buckling, but the combination of buckling of
plates resting on an elastic foundation is not developed.
The theoretical approach which will be developed here has
limited validity because of the many simplifying assumptions which were necessary. However it will serve to
illustrate the important parameters in predicting buckling
and will yield approximate predictions of critical stress
for buckling in various modes.

Critical Stress for Harmonic Modes

If the curvature of the plate in Figure 1 is slight
then the critical buckling stress will be approximately
that for a flat plate of the same dimensions. The
critical stress required to initiate buckling is determined by the energy method[2]. Either assuming the work
done by the forces deflecting the plate exceeds the
energy of bending with no stretching at buckling, or

assuming the edges rigidly fixed and some stretching in the middle at buckling, leads to the same equation for t critical forces in the plane of the plate (x,y plane).

$$1/2 \iint \left[N_x \left(\frac{\partial w}{\partial x}\right)^2 + N_y \left(\frac{\partial w}{\partial y}\right)^2 + 2N_{xy}\frac{\partial w}{\partial x}\frac{\partial w}{\partial y}\right] dx \; dy$$

$$+\frac{D}{2} \iint \left\{ \left(\frac{\partial^2 w}{\partial x^2} + \frac{\partial^2 w}{\partial y^2}\right)^2 - 2(1-\nu)\left[\frac{\partial^2 w}{\partial x^2}\frac{\partial^2 w}{\partial y^2} - \left(\frac{\partial^2 w}{\partial x\partial y}\right)^2\right]\right\} dx \; dy = \tag{1}$$

N_x, N_y, and N_{xy} are the critical forces per unit length, w (x,y) is the deflection of the plate from its un-buckled shape, ν is Poisson's ratio and

$$D = \frac{Eh^3}{12(1-\nu^2)}$$

where E is elastic modulus and h is plate thickness. Choosing a deflection equation for a rectangular plate compatible with the built-in edge condition[3]

$$w = \frac{w_o}{4} \left(1 - \cos\frac{m\pi x}{a}\right)\left(1 - \cos\frac{n\pi y}{b}\right) \tag{2}$$

where w_o is maximum deflection, a and b are plate length in the x and y directions respectively, and in the x-direction

m = 2 for fundamental,

m = 4 for first harmonic,

m = 6 for second harmonic, etc.

with corresponding values of n for fundamental and harmonics in the y-direction. If (2) is substituted into (1) and the integrations carried out with N_x, N_y, D and ν taken as constants and $N_{xy} = 0$, the result is an equation for the critical forces

$$\frac{1}{2}\left[N_x\frac{3w_o^2}{64} m^2 \frac{b}{a}\pi^2 + N_y\frac{3w_o^2}{64} n^2 \frac{a}{b}\pi^2\right]$$

$$+ \frac{D}{2}\frac{\pi^4}{32} w_o^2 \left(\frac{3}{2}\frac{bm^4}{a^3} + \frac{3}{2}\frac{an^4}{b^3} + \frac{m^2n^2}{ab}\right) = 0 \tag{3}$$

The shear component N_{xy} is taken as zero because the forces are due to thermal expansion and are assumed to be strictly compressive. Uniformity of N_x and N_y would depend on uniformity of the edge restraint (each would be constant along an edge only if the edge restraint was rigid or deflected uniformly, so this assumption is an approximation). As will be shown in the next section, for rigid edge restraints the compressive stress produced by restrained thermal expansion is independent of length of the plate, so it will be assumed that, even for elastic rather than rigid restraint, $N_x \simeq N_y$. With this assumption and with the stress $\sigma = N/h$, (3) can be solved for critical stress.

$$\sigma_{cr} = -\frac{2}{3} \pi^2 \frac{D}{h} \frac{(\frac{3}{2} b \frac{m^4}{a^3} + \frac{3}{2} \frac{an^4}{b^3} + \frac{m^2 n^2}{ab})}{(m^2 \frac{b}{a} + n^2 \frac{a}{b})} \qquad (4)$$

If $m = n = 2$, equations (3) and (4) reduce to the corresponding equations given by reference 3 for fundamental mode buckling.

Figure 2 illustrates the plate shape for fundamental and first and second harmonic modes. It should be emphasized that the derivation of (4) involved several simplifying assumptions and therefore it will yield approximate results. However, with stiff edge supports and slight curvature (large cylindrical radius) the predicted values of σ_{cr} should be reasonably close to true values. Note that, for any particular mode, σ_{cr} is proportional to the square of plate thickness and is inversely proportional to the square of the plate length and width dimensions. Thick plates would reach a larger stress before buckling and large plates require less stress to buckle as would be expected.

Stress Produced by Constrained Thermal Expansion

The strain in the x-direction in the plane of the plate would be[4]

$$\varepsilon_x = \frac{1}{E} [\sigma_x - \nu(\sigma_y + \sigma_z)] + \alpha \Delta T \qquad (5)$$

where α is the coefficient of thermal expansion and ΔT the change in temperature. For a thin plate $\sigma_z \cong 0$ and, if the edge supports are approximately rigid, $\sigma_y = \sigma_x = -\sigma$ (compressive stress due to restrained expansion). If we take the center of the plate as reference, then outward displacement of the edge relative to the center is

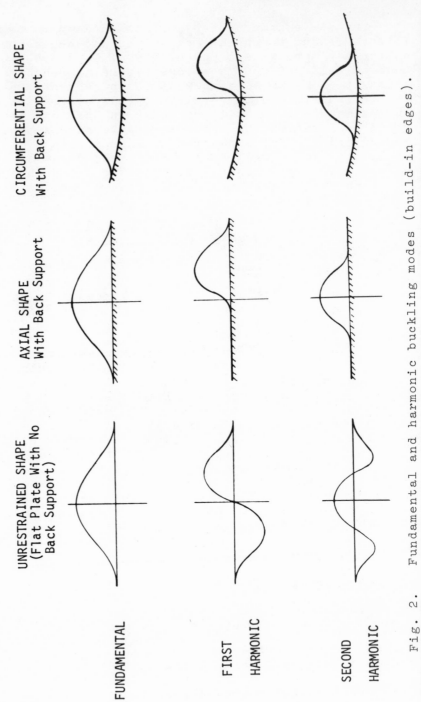

Fig. 2.   Fundamental and harmonic buckling modes (build-in edges).

$\delta_x = \varepsilon_x \frac{a}{2}$ or from (5)

$$\delta_x = -\frac{a}{2E}\,\sigma(1-\nu) + \alpha\,\Delta T\,\frac{a}{2} \qquad (6)$$

Now if the edge supports are rigid and do not deflect, $\delta_x = 0$ and the compressive stress produced by thermal expansion would be

$$\sigma = \frac{\alpha\,\Delta T\,E}{(1-\nu)} \qquad (7)$$

which depends only on temperature. On the other hand if the edge supports are elastic, as most real structures would be, the edge deflection of the plate would be given by (6), and the compressive stress produced by thermal expansion would be

$$\sigma = \frac{\alpha\Delta T\,E}{(1-\nu)} - \frac{2E}{a(1-\nu)}\,\delta_x \qquad (8)$$

The second term on the right side (due to elastic deflection of the edge support) will have a large effect in reducing stress in the plate. In the case of elastic deflection of the edge support, determining the relationship between stress and temperature change would require solving (6) simultaneously with the equation for displacement of the edge support structure under the (assumed uniform) load from the plate. This would involve fairly straightforward structural analysis of the support structure if it remained isothermal. Such analysis would be useful in estimating the relationship between σ and ΔT which could be used to estimate the T at which critical stress would be reached. However, in reality the support structure would be heated by either the same heat source which heats the plate or by conduction from the plate. The displacement of the support structure is the sum of that due to the load from the plate and that due to thermal expansion of the structure itself. An accurate structural and thermal analysis would likely require finite element analysis primarily because the heat transfer analysis would be too complex for analytical solution.

Stress versus Temperature Diagram to Predict Buckling

Figure 3 shows the relationships between σ and T for a 1.22m x 1.22m x .0064m steel plate with fairly stiff edge support structure. Such a graph could be used to estimate at what temperature buckling is likely to occur. Three straight lines are shown (assuming α, ν, and

E as constants) for stress as function of plate temperature. If the edge supports were truly rigid, the stres
would grow very rapidly as the plate was heated and at a
temperature would be much greater than would occur if th
edge support structure could deflect elastically. The
most realistic σ-T curve would be that including elastic
deformation and thermal expansion of the edge structure.
Predicting whether buckling or yielding would occur firs
as the plate was heated (and the corresponding temperature) would require knowing or estimating this curve an
anticipating which buckling mode is most likely to occur

EXPERIMENTS AND RESULTS

      Two experiments were performed at Hanford Engineeri
Development Laboratory[5] to determine the effects of rapi
heating on plates in a cylindrical geometry like Figure
The first (designated FT-3) used radiant heating of the
steel plate so that the plate could be observed and
photographed, as well as deflection measurements made
during the heating. The second test (designated LT-1)
used a spill of high temperature liquid sodium to heat
the plate. This gave extremely rapid rise in plate
temperature but precluded visual observation and extensi
deflection measurements.

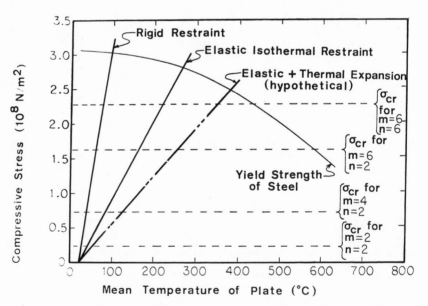

Fig. 3.   Stress versus temperature diagram for restrain
          ed thermal expansion.

Test FT-3

Figure 4 shows a cross-section of the test set-up to use radiant heaters to achieve rapid temperature rise in a steel plate.   A very stiff edge support structure was provided around the edges of the test article with 2.5cm steel reinforcing bar tying the sides together.   The heaters were tungsten-filament quartz tube lamps (about 49 kw) with ceramic embedded coiled filament heaters (about 20 kw) filling in any spaces between them. Viewing ports were provided for still and movie photography, and displacement was also measured at five points using linear potentiometers.   Figure 3 was drawn for the test article in Figure 4.

Because of the thermal inertia of the ceramic, the ceramic embedded heaters rose in temperature somewhat slowly, so it was necessary to preheat them for 15 minutes

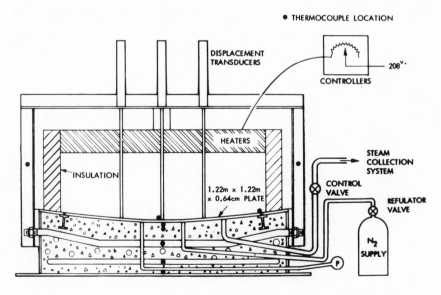

Fig. 4.   FT-3 test configuration (cross-section view).

at the beginning of the test. Response of the quartz tu
heaters was very rapid, however, and they were able to
heat the large steel plate quite rapidly as shown in
Figure 5. The plate was also reasonably uniformly heate
The concrete surface beneath the plate followed the plat
temperature fairly closely until buckling occurred. Aft
buckling the concrete surface temperature was determined
by heating from the plate across a fairly large gap and
cooling by the still cold concrete beneath the surface,
and it rose more slowly.

    The observed sequence of significant events in the
response of the plate to the rapid heating was as follow

    (1)   Gentle buckling at about 7½ minutes when the
           plate temperature reached about 325°C with
           second harmonic shape in both axial and
           circumferential directions (m=6,n=6).

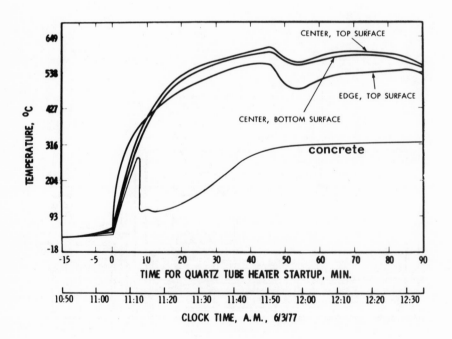

Fig. 5.   Temperature history for FT-3 test.

(2)   Violent jump and loud noise accompanying a tran-
       sition to fundamental mode in the axial direction
       and second harmonic mode in the circumferential
       direction (m = 6, n = 2) at 40 minutes when the
       temperature reached about 580°C.

Figure 3 gives an approximate explanation of the observed
events. As the temperature rose to 325°C, the stress in
the plate increased and reached the critical level for
second harmonic/second harmonic buckling, and buckling
was initiated. Further heating resulted in increased
curvature of the buckled plate with compressive stress
remaining approximately constant. At a little over 400°C
the yield stress was reached and compressive yielding
commenced. Yielding relieved the stress so that, as
heating continued, the stress followed approximately the
yield strength line, causing stress to drop below the
level necessary to maintain the second harmonic/second
harmonic configuration. At about 580°C, because of insufficient stress to maintain its shape, the plate popped
violently into the fundamental/second harmonic shape.

   Post-test examination of the plate after it had
cooled confirmed that there was considerable residual
tension in the plate due to compressive yielding, and

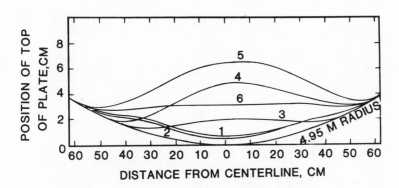

Fig. 6.   Plate geometry at various times during FT-3
           test (along circumferential centerline). (1)
           pretest, 21°C; (2) start quartz tube heaters,
           49°C; (3) after buckling, 8 min, 354°C; (4)
           before loud pops, 39 min, 577°C; (5) after pops,
           43 min, 582°C; (6) post-test, 21°C.

after the plate was cut free of the edge supports it was
found to have residual bow due to yielding in bending at
some time during the test. Figure 6 gives the shape of
the plate at various times during the test. Note that,
because of slight residual tensile stress due to welding
the plate to the embedments, the plate was not initially
against the concrete back support. Heating, and expansion
moved it toward the support, but it buckled before reaching the concrete.

Test LT-1

      The second test made use of a large test article in
the form of a basin with a number of steel plate and
structural components incorporated in it. The purpose of
the test was to determine the response of all these components to a very rapid rise in temperature achieved by
spilling liquid sodium into the test article. Figure 7
shows a view of the test article looking down into it
from above. The cylindrical curved plate in this case
was 0.61m x 1.22m x 0.64cm and was along the right wall
of Figure 7, between the floor and the horizontal shelf.

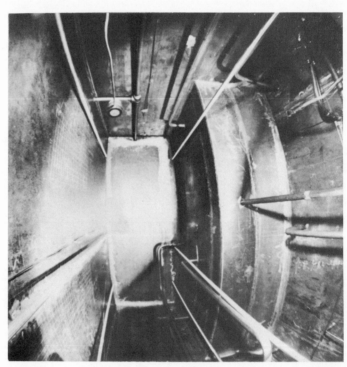

Fig. 7.   LT-1 test article (looking down from top).

The gap between the plate and concrete was measured by a
gap transducer behind the center of the plate. When 593°C
sodium was spilled into the test article, the temperature
of the plate rose very rapidly to about 490°C as shown
in Figure 8. In this test, with the plate being half as
large in the axial direction as the plate in the FT-3
test, the critical stress for second harmonic/second
harmonic buckling would have been 6.55×10^8 N/m^2 which
is considerably above the yield stress. However, as
shown in Figure 9 the plate did buckle less than a minute
after the spill. In Figure 9, as in Figure 6, the initial
gap was due to residual stress in the plate from welding
it to embedments at its edges. From the fact that the
plate had buckled, and from the post-test shape of the
plate, it was concluded that it had buckled directly to
the fundamental/second harmonic mode in this case.

SUMMARY AND CONCLUSIONS

     The approximate relation for critical stress to
initiate harmonic mode buckling (equation 4) and the
stress-temperature diagram for a particular plate configuration (Figure 3 being an example) were developed to
permit predicting buckling of a cylindrical plate with a
back support. The experimental results give some hope
that the approach may be successful. The results of the
FT-3 test, especially, seem to be explained reasonably
well by the theoretical relationships. However, because
of the great difficulty in measuring stress or strain at
high temperature (especially when temperature is changing
rapidly) stress measurement was not attempted in either
of the tests. As a result the confirmation of the theoretical results must be considered primarily qualitative.
Also nothing in the development given here would allow
predicting which mode of buckling would be most likely to
occur. It is clear intuitively that the second harmonic
configuration should be most likely in the circumferential direction (Figure 2); because the fundamental is
prevented by the back support, and the second harmonic is
the first available shape with the symmetry required by
the symmetric edge conditions. In the axial direction,
however, it is not at all clear why the fundamental
(away from the back support) should not be the most
likely mode, since it requires the lowest critical stress.
That apparently was the case in the LT-1 test, but in the
FT-3 test the stress initially passed the level required
for the fundamental mode and buckled into the second
harmonic shape. Clearly more analysis of this question
is needed. It may be that mode was influenced by

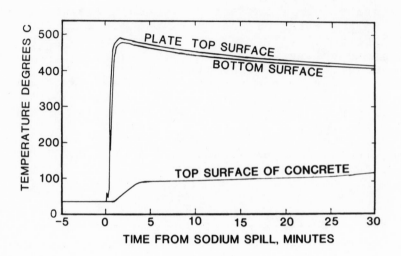

Fig. 8.   Temperature history for LT-1 test.

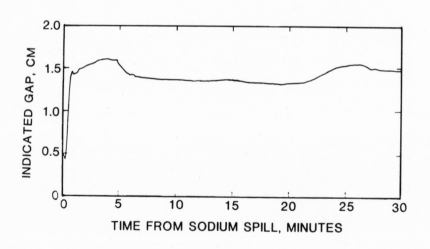

Fig. 9.   Indicated gap between plate and concrete (cent
            of plate) for LT-1 test.

initial stress asymmetries due to initial welding of the plate. If this is the case more tests should be performed to either eliminate residual stresses due to welding or determine their effects by planned revisions in the welding sequence.

In both tests reported here the plates did buckle, but the buckling stress was quite high. Subsequent continued rise in temperature caused the yield strength of the material to drop, and, since the stress in the buckled plate was high, yield·strength dropped below the actual stress and yielding occurred. Buckling, therefore, did not serve to avoid yielding. It did limit the force exerted on the edge support structure, however, and it did reduce the heat transfer to the concrete back support. Development and understanding of a diagram like Figure 3 for a particular situation could permit choosing plate dimension and thickness to achieve desired results regarding buckling and yielding. For example, substantially larger plates in either test would have permitted buckling at a much lower stress, and subsequent yielding could probably have been avoided.

ACKNOWLEDGEMENTS

The analytical and experimental work reported here was done at Hanford Engineering Development Laboratory as a part of projects sponsored by the U.S. Department of Energy.

REFERENCES

1. W. Nowacki, "Thermoelasticity", Addison-Wesley (1962).
2. S. Timoshenko and J.M. Gere, "Theory of Elastic Stability," 2nd ed., McGraw-Hill, New York (1961), Chapter 9.
3. S. Timoshenko, "Theory of Plates and Shells," McGraw-Hill, New York (1940), pp. 320-321.
4. Crandall, Dahl, and Lardner, "An Introduction to the Mechanics of Solids," 2nd ed., McGraw-Hill, New York (1972), p.289.
5. R.W. Wierman, L.D. Muhlestein, and L.D. Simmons, Experimental Evaluation of Cell Liners , in: "Coolant Boundary Integrity Considerations in Breeder Reactor Design", American Society of Mechanical Engineers, New York (1978).

STRESSES DUE TO THERMAL TRAPPING IN SEMI-ABSORBING MATERIALS
SUBJECTED TO INTENSE RADIATION

J. P. Singh, J. R. Thomas, Jr., and D. P. H. Hasselman

Departments of Materials and Mechanical Engineering
Virginia Polytechnic Institute and State University
Blacksburg, Virginia 24061

ABSTRACT

Analytical results are presented for the thermal stresses resulting from "thermal trapping" in semi-absorbing materials subjected to symmetric and assymmetric radiation heating and convective cooling with finite heat transfer coefficient, h. The transient stresses during heat--up in both cases were found to be an
inverse function of the heat transfer coefficient, h, and increase
monotonically with the optical thickness, μa. In contrast, the
steady state stresses were independent of h and exhibited a maximum
at $\mu a \simeq 1.3$ for symmetric heating and at $\mu a = 2$ for assymmetric
heating with zero stresses at $\mu a = 0$ and ∞. For the symmetrically
heated plate, the transition from the transient to the steady-state
condition involved a reversal in the sign of thermal stress at any
position. For the assymmetrically heated plate the steady state
maximum tensile thermal stresses occur at the position of the highest temperature, ie. the front surface.

INTRODUCTION

Many engineering structures such as aerospace vehicles and
concentrating solar collectors are subjected to high intensity radiant heat fluxes. In order to withstand the high temperature involved, such structures must be constructed of materials with high
melting point and good mechanical properties. Unfortunately, such
materials tend to be brittle and are susceptible to catastrophic
failure under the influence of transient and steady state thermal
stresses resulting from non-linear temperature distribution[1,2].
Therefore, for the selection of the optimum material and reliable

157

engineering design, it is necessary to analyze the thermal stresses arising from radiation heating and understand the variables which control their magnitude.

The analysis of thermal stress failure of an opaque material subjected to black body radiation was performed by Hasselman[3] some time ago. Subsequently, the effect of the spectral dependence of absorption coefficient was taken into account to include materials completely transparent below and opaque above a given wavelength[4] For many materials, however, the incident radiation may be absorbed throughout the thickness and depending upon the absorption properties, it may result in so called "thermal trapping" effect in which the interior temperature of the body may exceed the ambient temperature. Thermal stresses in a flat plate due to this type of heat transfer were analyzed recently by Hasselman et. al.[5] for symmetric radiation heating with convective cooling for two limiting values o heat transfer coefficient, h, equal to zero and infinity. Subsequenly two follow-up studies by Thomas et. al.[6] and Singh et. al.[7] included the case of symmetric and assymmetric radiation heating and convective cooling with finite values of h.

The purpose of this paper is to review and report the main results for stresses in the latter two studies. For further detail the reader is referred to the original papers.[6,7]

ANALYSIS

The analysis of thermal stresses in the flat plate first required obtaining the temperature distribution, which was subsequently utilized to calculate the thermal stresses. In this analysis the reflectivity of the material was assumed to be low and hence the effect of multiple internal reflections was neglected. It was also assumed that the maximum value of the stress in the plate will be reached before the temperature becomes sufficiently high that the effect of re-emission must be taken into account[8]. The validity of this latter assumption can easily be substantiated with numerical examples[5,6]. Finally it was assumed that the emissivity and absorptivity of the plate are independent of wavelength. However, the spectral dependence of these properties can easily be incorporated in the analysis, if needed.

The general analytical approach along with the heating and cooling conditions are as follows:

A. Symmetric Radiative Heating and Convective Cooling

The rate of internal heat generation (g''') within the plate subjected to symmetric normally incident radiation q_o is given by:

$$g''' = 2\mu\epsilon q_o e^{-\mu a} \cosh(\mu x) \tag{1}$$

where μ is the absorption coefficient, ϵ is the emissivity ($\epsilon=1-r$, where r is the reflectivity), a is the half thickness of the plate and x is the through-the-thickness coordinate with $x=0$ at the center of the plate.

The temperature solutions were obtained by solving the differential equation[9]:

$$\frac{\partial^2 T}{\partial x^2} + g'''(x)/k = (1/\kappa)\frac{\partial T}{\partial t} \tag{2}$$

where k is the thermal conductivity, t is the time and κ is the thermal diffusivity.

The initial and boundary conditions were:

$$T(x,0) = T_o; \frac{\partial T}{\partial x}(0,t) = 0 \tag{3}$$

$$\frac{\partial T}{\partial x}(a,t) = -\frac{h}{k}\{T(a,t) - T_o\} \tag{4}$$

$$\frac{\partial T}{\partial x}(-a,t) = \frac{h}{k}\{T(-a,t) - T_o\} \tag{5}$$

where h is the convective heat transfer coefficient.

The solution of temperature T, thus obtained from differential equation (2), was utilized to obtain the expressions for thermal stress from[10]:

$$\sigma_{y,z} = \frac{\alpha E}{1-\nu}[- T + \frac{1}{2a}\int_{-a}^{a} Tdx + \frac{3x}{2a^3}\int_{-a}^{a} Txdx] \tag{6}$$

which yields:

$$\sigma_{y,z} = \frac{-\alpha E}{1-\nu}\Sigma_{n=o}^{\infty} B_n(t)\cos(\lambda_n x) + \frac{\alpha E}{a(1-\nu)}\Sigma_{n=o}^{\infty}\frac{B_n(t)}{\lambda_n}\sin(\lambda_n a) \tag{7}$$

where λ_n are the roots of the transcendental equation

$$\lambda_n \tan(\lambda_n a) = h/k$$

and

$$B_n(t) = \frac{G_n}{\kappa\lambda_n^2}(1 - e^{-\kappa\lambda_n^2 t})$$

with $G_n = \dfrac{2\kappa\mu\varepsilon q_o e^{-\mu a}}{N_n k}$ $[\dfrac{\mu \cos (\lambda_n a)(e^{\mu a}-e^{-\mu a}) + \lambda_n \sin \lambda_n a(e^{\mu a}+e^{-\mu a})}{\mu^2 + \lambda_n^2}]$

and $N_n = a + \dfrac{h}{\lambda_n^2 k} \cos (\lambda_n a)$

B.   Assymmetric radiation Heating and Convective Cooling

In this case, the radiation q_o is normally incident to only one face of the plate at $x = -a$ and the plate is cooled by convection at $x = a$. The intensity of internal heat generation (g''') within the plate can be shown to be:

$$g'''(x) = \mu\varepsilon q_o e^{-\mu(a+x)} \tag{8}$$

The initial and boundary conditions are:

$$T(x,0) = T_o; \frac{\partial T}{\partial x}(-a,t) = 0 \tag{9}$$

$$-k \frac{\partial T}{\partial x}(a,t) = h[T(a,t) - T_o] \tag{10}$$

The temperature solutions were obtained by solving the differential equation (2) subjected to the conditions (9) and (10). Similar to case A, the temperature solution, thus obtained, was utilized to find out the expression for thermal stress. The thermal stress in this case is given by:

$$\sigma_{y,z} = \frac{-\alpha E}{1-\nu} \Sigma_{n=0}^{\infty} B_n(t) \cos\{\lambda_n (x+a)\}$$

$$+ \frac{\alpha E}{2a(1-\nu)} \Sigma_{n=0}^{\infty} \frac{B_n(t)}{\lambda_n} \sin (2\lambda_n a)$$

$$+ \frac{3x\alpha E}{2a(1-\nu)} \Sigma_{n=0}^{\infty} B_n(t) [\frac{\sin(2\lambda_n a)}{a\lambda_n} + \frac{1}{a^2\lambda_n^2} \cos(2\lambda_n a) - \frac{1}{a^2\lambda_n^2}] \tag{11}$$

where λ_n are the roots of the transcendental equation

$$\lambda_n \tan (2\lambda_n a) = h/k$$

and $B_n(t) = G_n[1 - e^{-\kappa\lambda_n^2 t}]/\kappa\lambda_n^2$

$$\text{with } G_n = \frac{\kappa q_o \epsilon \mu}{N_n k} \left\{ \frac{\mu + e^{-2\mu a}[\lambda_n \sin(2\lambda_n a) - \mu \cos(2\lambda_n a)]}{\mu^2 + \lambda_n^2} \right\}$$

$$\text{and } N_n = \frac{1}{2\lambda_n}[\sin(2\lambda_n a)\cos(2\lambda_n a) + 2\lambda_n a]$$

NUMERICAL RESULTS AND DISCUSSION

For the convenience of data presentation, the results will be presented in non-dimensional form as follows:

The non-dimensional time:

$$t^* = \kappa t/a^2 \tag{12}$$

the non-dimensional temperature:

$$T^* = (T - T_o) k/\epsilon q_o a \tag{13}$$

and the non-dimensional stress:

$$\sigma^* = \sigma(1-\nu)k/\alpha E \epsilon q_o a \tag{14}$$

For simplicity, the quantities t^*, T^* and σ^* hereafter will be referred to as time, temperature and stress, respectively.

Figure 1 shows the thermal stresses in the surface and the center of a symmetrically heated plate as a function of time for a range of values for h/k with an arbitrarily chosen value of the optical thickness $\mu a = 3$. Stresses reach the steady state value after an initial transient. For a given optical thickness μa, the value of the steady state stress is independent of h. This is so because the steady state temperature profile within the plate is independent of the heat transfer coefficient h, even though the magnitude of the temperature for a given profile is an inverse function of the heat transfer coefficient. The values of the maximum transient stress, however, decrease with increasing h in contrast to many literature solutions for transient thermal stresses in convection heat transfer environment where the thermal stresses increase with increasing value of h.

Another noteworthy point in figure 1 is the change in the sign of the thermal stresses in the plate with time for finite values of h/k. At small values of time the surface stress is compressive whereas the stress in the center is tensile. However, at longer times the stresses reverse their sign and become tensile and compressive in the surface and the center, respectively. This

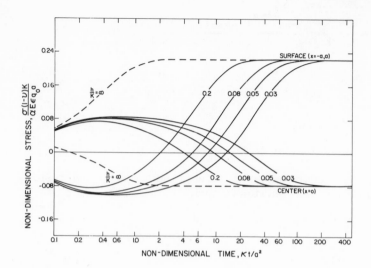

Fig. 1. Maximum thermal stress as a function of time in a partially
absorbing flat plate symmetrically heated by thermal radiation and cooled by convection for various values of h/k
(cm.$^{-1}$) with optical thickness $\mu a=3$ and $a=1$ cm.

phenomenon is a result of the non-uniform heat generation inside
the plate and the associated "thermal trapping" effect. Due to a
larger portion of the incident radiation being absorbed in the
surface, initially the surface is at higher temperature than the
interior. During this time period, the heat loss from the surface
by convection is very small because of the low surface temperature
in combination with finite value of h and the heat flows primarily
from the surface to the interior. This results in a temperature
distribution which causes a compressive stress in the surface and
tensile stress in the center. However, after a long period of
time ($t \rightarrow \infty$) the surface gets hot with an appreciable amount of heat
loss by convection which eventually will be equal to the incident
radiative heat flux at steady state. Under this condition, due to
the internal heat generation, the interior of the plate will be at
higher temperature than the surface resulting in a tensile stress
in the surface and a compressive stress in the center of the plate.
However, as $h/k \rightarrow \infty$, no stress reversal in the surface takes place
because the surface temperature remains constant (equal to the ambient temperature) at all times resulting in tensile surface stress
for $t > 0$.

For the symmetrically heated plate, figures 2 and 3 show the
maximum values of transient and steady state stress as a function
of optical thickness, μa. It is interesting to note that the
maximum value of the transient stress increase with optical thickness whereas the steady state stress exhibits a maximum at $\mu a \approx 1.3$

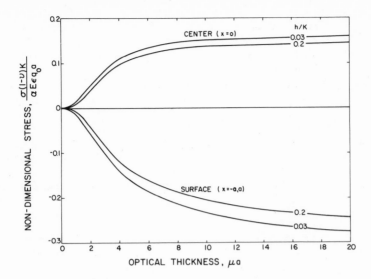

Fig. 2. Magnitude of maximum transient thermal stresses as a func-
        tion of optical thickness μa in partially absorbing flat
        plate symmetrically heated by thermal radiation and cooled
        by convection for various values of h/k(cm.[-1]) with a=1 cm.

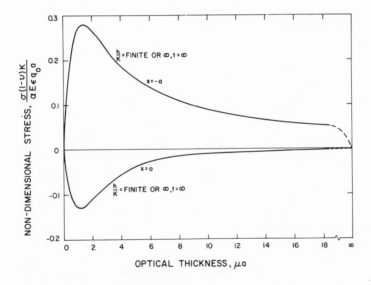

Fig. 3. Magnitude of maximum steady state thermal stresses as a
        function of optical thickness μa in a partially absorbing
        flat plate symmetrically heated by thermal radiation and
        cooled by convection for various values of h/k(cm.[-1]) with
        a=1 cm.

and approaches zero at $\mu a=0$ and ∞. For $\mu a=0$, no heat is absorbed, resulting in zero stress. However, for $\mu a \to \infty$, although the heat is transmitted through the plate, the steady state temperature profile is linear resulting in zero stress at $t=\infty$. At $\mu a \approx 1.3$, the non-uniformity in the internal heat generation and the associated thermal trapping effects are maximum resulting in the maximum stress.

For an assymmetrically heated plate, figure 4 shows the values of maximum tensile thermal stress as a function of time for various values of h/k with an arbitrarily chosen value of $\mu a=3$. Similar to figure 1, the stress reaches a constant steady state value independent of h and the value of maximum transient stress is an inverse function of h.

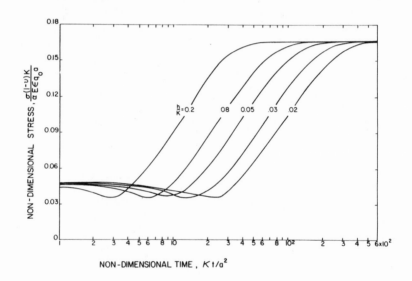

NON-DIMENSIONAL TIME, Kt/a^2

Fig. 4. Maximum tensile thermal stress as a function of time in a partially absorbing flat plate assymemetrically heated in front by thermal radiation and cooled at the rear surface by convection for various values of h/k(cm.$^{-1}$) with $\mu a=3$ and a=1 cm.

Figure 5 shows the spatial distribution of the steady state tensile stress ($t \to \infty$) in the assymmetrically heated plate. It is important to note that the maximum tensile stress occurs in the front face (x = -a) of the plate which is obviously at the highest temperature. This initially unexpected result was shown to be due to a concave downward temperature distribution[11] as the direct result of an assymmetrically non-uniform internal heat generation. This temperature distribution contrasts with the usual concave upward temperature distribution found in many cases where the heat

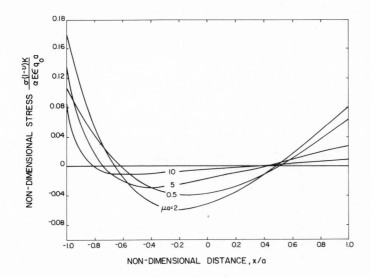

Fig. 5. Spatial distribution of steady state thermal stresses in a
        partially absorbing flat plate assymmetrically heated in
        front by thermal radiation and cooled at the rear surface
        by convection for various values of μa with h/k(cm.$^{-1}$)
        finite and a = 1 cm.

is exchanged only at the outside boundaries of the structure.   For
this latter case, the maximum tensile stresses frequently occur at
or near the position of the lowest temperature.

    Figure 6 summarizes the values of maximum tensile transient
and steady state thermal stress as a function of μa in the assymmetrically heated plate.   Similar to the observation made for
symmetric heating case (Figures 2 and 3) the steady state stress
exhibits a maximum, which occurs at μa=2 because of the maximum
thermal trapping effect.   However, the maximum transient stress for
finite value of h is a monotonically increasing function of μa and
will fall between the stress plots for h=0 and h=∞ (Figure 6).
From Figures (4) and (6), it is obvious that for 10.7 < μa < ∞,
the maximum transient stress is greater than the corresponding
maximum steady state stress and for 0 < μa < 10.7, the converse is
true.

    A comparison of the results for thermal stresses in the symmet-
rically and assymmetrically heated flat plates indicates that the
nature of the stress profiles in both cases are qualitatively similar
(Figures 1-6). However, the magnitude of the maximum stress in case
of symmetric heating is higher than that in case of assymmetric
heating.   The lower value of stress in case of assymmetric heating
is due to the fact that the non-uniform thermal expansion can be

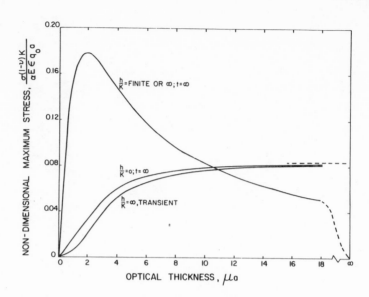

Fig. 6. Magnitude of maximum tensile thermal stresses as a function
        of μa in a partially absorbing flat plate assymmetrically
        heated in front by thermal radiation and cooled at the rear
        surface by convection with h/k(cm.$^{-1}$) finite and a = 1 cm.

partly accommodated by bending which is not the case in symmetric
heating.

     The above discussions have the following implications: First,
it is important to note that although the steady state stress is
zero at μa=0 and ∞, the magnitude of stress could be significant
for the intermediate values of μa. Therefore, it is preferable to
have a transparent material to avoid thermal stress fialure. However, this criterion for material selection may not be compatible
with other performance criteria. For example, for high efficiency
of the solar collectors, it is desirable to have high values of the
absorption coefficient and emissivity to keep the amount of transmitted and reflected radiation to a minimum. Clearly a compromise
between high collector efficiency and high thermal stress resistance
will have to be made.

     Finally, the observation of the maximum tensile thermal stress
in the hottest part (front face) of the assymmetrically heated plate
could be very critical for failure by the stress corrosion mechanism
Stress corrosion, which could be promoted by environmental conditions, is generally a thermally activated process and is expected
to increase with stress and temperature. Therefore, in addition
to having low thermal stress, it is important to maintain low temperature levels in the plate to ensure high thermal stress resistance and long life. Thus, care must be taken in selection of

material and heat transfer conditions in order to keep the critical regimes at low temperature. It is quite obvious then, that the incidence of thermal stress failure in a plate subjected to radiation heating can be minimized by proper selection of material in combination with the optimum choice of the optical thickness and heat transfer conditions.

ACKNOWLEDGEMENTS

The present study was conducted as part of a larger research program on the thermo-mechanical and thermal behavior of high temperature structural materials supported by the Office of Naval Research under contract N00014-78-C-0431.

REFERENCES

1. W. D. Kingery, Factors Affecting Thermal Stress Resistance of Ceramic Materials, J. Amer. Ceram. Soc. 38:3 (1955).
2. D.P.H. Hasselman, Unified Theory of Thermal Shock Fracture Initiation and Crack Propagation in Brittle Ceramics, J. Amer. Ceram. Soc. 52:600 (1969).
3. D.P.H. Hasselman, Thermal Shock by Radiation Heating, J. Amer. Ceram. Soc. 46:229 (1963).
4. D.P.H. Hasselman, Theory of Thermal Shock Resistance of Semi-transparent Ceramics Under Radiation Heating, J. Amer. Ceram. Soc. 49:103 (1966).
5. D.P.H. Hasselman, J. R. Thomas, M. P. Kamat and K. Satyamurthy, Thermal Stress Analysis of Partially Absorbing Brittle Ceramics Subjected to Radiation Heating, J. Amer. Ceram. Soc. (in press).
6. J. R. Thomas, Jr., J. P. Singh and D.P.H. Hasselman, Analysis of Thermal Stress Resistance of Partially Absorbing Ceramic Plate Subjected to Assymmetric Radiation, I: Convective Cooling at Rear Surface, J. Amer. Ceram. Soc. (in review).
7. J. P. Singh, J. R. Thomas, Jr., and D.P.H. Hasselman, Thermal Stresses in Partially Absorbing Flat Plate Symmetrically Heated by Thermal Radiation and Cooled by Convection, J. Thermal Stress (in review).
8. R. Viskanta and E. E. Anderson, Heat Transfer in Semitransparent Solids, in: "Advances in Heat Transfer", F. Irvine, Jr., and James P. Hartnett, eds., Academic Press, NY, 11:317-441 (1975).
9. H. S. Carslaw and J. C. Jaeger, "Conduction of Heat in Solids", 2nd Ed., Oxford at Clarendon Press (1960).
10. B. A. Boley and J. H. Weiner, "Theory of Thermal Stresses", John Wiley and Sons, New York (1960).

11. K. Satyamurthy, D.P.H. Hasselman and J. P. Singh, Effect of
       Nature of Concavity of Temperature Distribution on Posi-
       tion and Sign of Maximum Thermal Stress, J. Appl. Mech.
       (in review).

INSTABILITY OF PARALLEL THERMAL CRACKS

AND ITS CONSEQUENCES FOR HOT-DRY ROCK GEOTHERMAL ENERGY

Zdeněk P. Bažant

Professor of Civil Engineering
Northwestern University
Evanston, Illinois 60201

ABSTRACT

Review of recent work on instabilities of crack systems and applications to the hot-dry rock geothermal energy scheme is presented. The basic variational formulation of the crack stability problem is outlined and the critical states of a system of parallel equidistant cooling cracks propagating into a halfspace are explained and analyzed. The solution, which shows that at a certain critical crack length-to-spacing ratio every other crack suddenly jumps ahead at constant temperature while the remaining cracks stop growing and subsequently close, determines the crack width and is of importance for heat withdrawal from hot rock by circulation of water in cooling cracks. Some typical numerical results obtained by finite elements are presented and the effect of the temperature drop profile on the critical crack length is discussed. Finally, some other applications, such as parallel cooling cracks or drying shrinkage cracks in reinforced solids, such as concrete, are pointed out.

INTRODUCTION

In the fracture problem, the boundaries of the elastic solid are not fixed because the crack surfaces represent boundaries too. Therefore, even under assumptions of small strain and linear elasticity, the uniqueness theorem of thermoelasticity[1] does not apply when the crack length is not given. Consequently, one must expect that various configurations of thermal cracks are possible for the same evolution of the temperature field within the given solid.

When uniqueness is lacking, one must investigate stability and may expect that some evolution paths of the system of thermal cracks may be unstable and therefore impossible.

A system of parallel equidistant cooling cracks that spreads into an elastic halfspace in the normal direction is the simplest yet practically important example. According to linear fracture mechanics,[2] solutions of various crack spacing and also solutions of equally long cracks or various periodically varying lengths are possible, the number of possible solutions being infinite. In practice, for given initial spacing only one solution can take place because of stability conditions, as has been recently dis-covered.[3,4] As the cooling penetration depth grows the cracks first extend while remaining equally long. But at a certain crack length-to-spacing ratio they may reach a critical state beyond which equally long cracks are unstable. It appears that at the critical state every other crack stops growing while the remaining cracks suddenly jump ahead at no change in the tempera-ture fields.[3,4] Subsequently the leading cracks resume a stable growth but later another critical state may be reached and a similar phenomenon observed.*

The problem has important implications for the creation of cracks in hot-dry rock mass for the purpose of harvesting geo-thermal energy.[6-8] It has also many other applications: cracks in solidifying lava beds,[9] drying shrinkage cracks in mud flats,[10] in permafrost and in concrete. Several papers have recently been devoted to analyzing the crack stability in the context of these problems, especially the hot-dry rock geothermal energy scheme. The purpose of this brief paper is to present a review and summary of the theoretical development made in several preceding papers on this subject[3,4,10-12] and to discuss some consequences for the hot-dry rock geothermal energy scheme.

CONDITIONS OF STABILITY FOR A GENERAL CRACK SYSTEM

The work, W, that needs to be done to produce cracks of length a_i may be expressed as:

$$W = U(a_1, a_2, \ldots, a_n; D) + \sum_{i=1}^{N} \int_0^{a_i} 2\gamma_i \, da_i' \tag{1}$$

where U represents the strain energy of the elastic body, $2\gamma_i$ is the specific energy of crack extension of the ith crack (material property), and D is a loading parameter which in our problem

---

*From a somewhat different point of view, instability of cooling cracks was analyzed by Hasselman.[5]

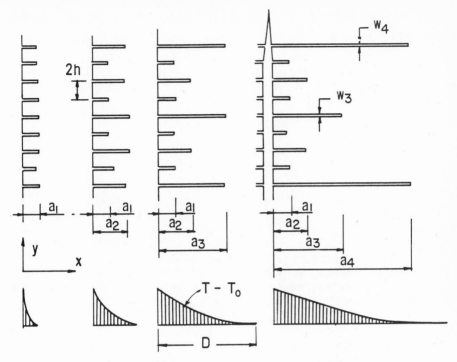

Fig. 1. Parallel Crack System in a Halfspace.

represents the penetration   depth of cooling or drying. The
equilibrium state of cracks is characterized by zero value of the
first variation of work, i.e.,

$$\delta W = \sum_{i=1}^{m} \left(\frac{\partial U}{\partial a_i} + 2\gamma_i\right)\delta a_i + \sum_{j=m+1}^{k} \frac{\partial U}{\partial a_j} \delta a_j = 0 \tag{2}$$

where $i = 1,\ldots,m$ are the cracks which grow ($\delta a_i > 0$), $j = m+1$,
$\ldots,k$, are those which close ($\delta a_j < 0$), and $i = k+1,\ldots,N$ are
eventual further cracks which remain stationary ($\delta a_i = 0$). Since
Eq. (2) must be satisfied for any admissible δa_i, we must have for
equilibrium crack extensions

$$\text{for } \delta a_i > 0: \ -\frac{\partial U}{\partial a_i} = 2\gamma_i, \ K_i = K_{c_i}; \text{ for } \delta a_i < 0: \frac{\partial U}{\partial a_i} = 0, \ K_i = 0.$$
$$\tag{3}$$

The first condition is the well-known Griffith fracture criterion,[2]
and K_i is the stress intensity factor. For plane strain,
$\partial U/\partial a_i = -K_i^2/E'$, where $E' = E(1 - v^2)$, E = Young's modulus,

ν = Poisson ratio,[2] and $K_{c_i} = (2\gamma_i E')^{\frac{1}{2}}$ = critical stress intensity factor. The case $K_i > K_{c_i}$ or $-\partial U/\partial a_i > 2\gamma_i$ cannot happen because Eq. (2) yields $\delta W < 0$ for $\delta a_i > 0$, which is an unstable situation. The case $K_i < 0$ or $-\partial U/\partial a_i < 0$ yields $\delta W < 0$ for $\delta a_i < 0$, which is also unstable. So it follows that only the following crack length variations are admissible:

$$
\begin{aligned}
&\text{for } K_i = K_{c_i}: \ \delta a_i \geq 0; \\
&\text{for } 0 < K_i < K_{c_i}: \ \delta a_i = 0; \\
&\text{for } K_i = 0: \ \ \delta a_i \leq 0.
\end{aligned}
\tag{4}
$$

Let us now examine the possibility of a critical state (bifurcation point) in which there exists an adjacent equilibrium state of crack lengths $a_i + \delta a_i$ at the same loading parameter D. The equilibrium state is characterized by the condition $W_i = \partial W/\partial a_i = 0$, and this condition would have to hold both for a_j and $a_j + \delta a_j$ at the same D. Hence, W_i for these two states must be the same, i.e. $\delta W_i = \Sigma_j \ (\partial W_i/\partial a_j)\delta a_j = 0$ at constant D. Noting that $\partial W_i/\partial a_j = \partial^2 W_i/\partial a_i \partial a_j$, we thus obtain the critical state condition:[3, 4, 11]

$$
\sum_{j=1}^{n} W_{ij}\delta a_j = 0
\tag{5}
$$

in which

$$
W_{ij} = \frac{\partial^2 W}{\partial a_i \partial a_j} = \frac{\partial^2 U}{\partial a_i \partial a_j} + 2 \frac{\partial \gamma_i}{\partial a_i} \delta_{ij} H(\delta a_i)
\tag{6}
$$

according to Eq. (1); $j = 1,\ldots,n$ runs through given set of crack lengths which are assumed to vary independently; H is the Heaviside step function, and for a homogeneous material, in which $\gamma_i = \gamma = $ constant, we have $W_{ij} = \partial^2 U/\partial a_i \partial a_j = \partial G_i/\partial a_j$. Since $\partial U/\partial a_i = -K_i^2/E'$, Eq. (5) yields for a homogeneous material ($\partial \gamma_i/\partial a_i = 0$) the expression $W_{ij} = W_{ji} = K_i \ \partial K_i/\partial a_j$.

Equations (5)-(6) may be also obtained directly from the condition that an equilibrium crack system ($\delta W = 0$) is stable if the second variation $\delta^2 W$ is always positive, and unstable if it can be negative, i.e.[3,4,11]

$$
\delta^2 W = \frac{1}{2} \sum_{i=1}^{n} \sum_{j=1}^{n} W_{ij}\delta a_i \delta a_j
\begin{cases}
> 0 \text{ stable} & \text{(all admissible } \delta a_i) \\
= 0 \text{ critical} & \text{(some admissible } \delta a_i) \\
< 0 \text{ unstable} & \text{(some admissible } \delta a_i)
\end{cases}
\tag{7}
$$

Evidently $2\delta^2 W = \Sigma\Sigma \ \delta a_i(W_{ij}\delta a_j)$ which equals zero if Eq. (5) holds, and so Eq. (5) leads to the critical state according to Eq. (7).

If matrix W_{ij} is positive definite, stability is assured. Note, however, that if it is not positive definite, the crack system may still be stable if those δa_i which make $\delta^2 W$ negative are not admissible according to Eq. (4). This last restriction constitutes an important difference from buckling problems. In fact, there exist cases where W_{ij} is not positive definite yet the crack system is stable.[3]

In the cracked halfspace (Fig. 1) there are infinitely many cracks, but assuming a periodic pattern of crack lengths we may limit ourselves to just a few independent crack lengths. We denote their number as N (N = 2 in Fig. 1), and we then consider the body to be limited by the symmetry lines.

If Eq. (5) is fulfilled for some non-zero δa_i that are admissible according to Eq. (4) we have vanishing $\delta^2 W$ and the state is a critical one, so the crack lengths can change at no change of loading. Since Eq. (6) is a system of homogeneous linear equations, the critical state can be obtained only if some principal minor, $\det_n (W_{ij})$, of (N × N) matrix W_{ij} vanishes (n ≤ N). The quadratic form in Eq. (7) is positive definite if each principal minor $\det_n (W_{ij})$ is positive. So,[3,4,11]

$$\det_n (W_{ij}) \begin{cases} > 0 & \text{stable (all n, any admissible } \delta a_i) \\ = 0 & \text{critical (some n, admissible } \delta a_i) \\ < 0 & \text{unstable (some n, admissible } \delta a_i) \end{cases} \qquad (8)$$

Note that we must check the principal minors of all sizes n ≤ N. The eigenvector that corresponds to $\det_n (W_{ij}) = 0$ may be determined from Eq. (5) and must satisfy Eq. (4) to obtain instability.

If one considers two independent crack lengths a_1 and a_2 (Fig. 1), one possibility is that only a_2 extends while a_1 remains constant. Then, we need to check $\det_n (W_{ij}) = W_{22} = \partial^2 W / \partial a_2^2$, and Eq. (7) yields $\delta^2 W = \frac{1}{2} W_{22} (\delta a_2)^2$. Since for a homogeneous material $W_{22} = \partial^2 U / \partial a_2^2 = -\partial G_2 / \partial a_2$ where $G_2 = K_2^2 / E'$, and also $K_2 \neq 0$, it follows[3,4,11]

$$\frac{\partial K_2}{\partial a_2} \begin{cases} < 0 & \text{(stable)} \\ = 0 & \text{(critical)} \\ > 0 & \text{(unstable)} \end{cases} \qquad (9)$$

at D = const., $\delta a_2 > 0$ ($K_2 = K_c$) and $\delta a_1 = 0$. This condition is indeed found to govern in our problem.

Further, it is necessary to check the (2 × 2) determinant condition. The critical state is now given by $\det_2 (W_{ij}) = W_{11} W_{22} - W_{12} W_{21} = 0$, and the eigenvector ($\delta a_1$, δa_2) is given by $W_{11} \delta a_1 + W_{12} \delta a_2 = 0$ and $W_{21} \delta a_1 + W_{22} \delta a_2 = 0$. Thus, if $K_1 > 0$ and $K_2 > 0$, we have $\delta a_2 / \delta a_1 = -W_{11} / W_{21} = -W_{12} / W_{22}$, and because $W_{12} = W_{21}$

(Eq. 6) we get $W_{12} = W_{21} = \pm\sqrt{W_{11}W_{22}}$. Numerical calculations indicate that $\partial K_1/\partial K_2$ is always negative, i.e., $W_{12} > 0$, and so $W_{12} = W_{21} = \sqrt{W_{11}W_{22}}$. Thus we get the critical state $W_{11}W_{22} - W_{12}^2 = 0$ at $\delta a_2/\delta a_1 = \sqrt{W_{11}/W_{22}}$. But this would be possible only if $\delta a_1 \neq 0$ and $\delta a_2 \neq 0$ were admissible increments according to Eq. (4). Obviously, if $K_1 = K_2 = K_c$, the signs of W_{11} and W_{22} are both positive (Eq. 9), and so $\delta a_2/\delta a_1$ is obtained as negative.[4,11] This, however, violates the admissibility condition in Eq. (4) because $K_1 = K_2 = K_c$. Therefore, violation of the (2×2) determinant condition does not indicate instability of our system of parallel cooling cracks provided the cracks are equally long.

For the case $K_1 = 0$ and $K_2 = K_c$, it is of interest to test the (2×2) determinant condition again. Because $\partial U/\partial a_i = -K_i^2/E'$, we have $\det_2(W_{ij}) = 4K_1K_2(K_{1,1}K_{2,2} - K_{1,2}K_{2,1})/E'^2$ where $K_{i,j} = \partial K_i/\partial a_j$. Thus, at $K_1 = 0$ we must always have[11] $\det_2(W_{ij}) = 0$. To obtain the corresponding eigenvector, we may note that, according to Eq. (6), $K_2K_{2,1} = K_1K_{1,2} = 0$ if $K_1 = 0$. Thus, we see that

$$\det_2(W_{ij}) = 0; \quad \partial K_2/\partial a_1 = 0 \quad (\text{at } K_1 = 0). \tag{10}$$

The eigenvector corresponding to the (2×2) determinant condition satisfies Eq. (5), i.e., $W_{22}\delta a_2 + W_{21}\delta a_1 = 0$. Hence, $K_2K_{2,2}\delta a_2 = -K_2K_{2,1}\delta a_1$, and assuming that $\partial K_2/\partial a_2 \neq 0$ and substituting Eq. (10) we conclude that the eigenvector $(\delta a_1, \delta a_2)$ which corresponds to Eq. (10) is:[11]

$$\delta a_1 < 0, \quad \delta a_2 = 0 \quad (\text{at } K_1 = 0 \text{ and constant D}). \tag{11}$$

This is admissible according to Eq. (4). So, here we have a critical state that is characterized by vanishing of the (2×2) determinant.

The following picture emerges.[3,11] The cracks, which may be assumed to start at equal spacing b (b = 2h), extend at first at equal length as the penetration depth of cooling D increases. The (2×2) determinant vanishes first, but this does not represent instability as already explained. Subsequently, the derivative $\partial K_2/\partial a_2$ vanishes, and this represents a critical state of crack arrest and bifurcation of equilibrium path. Subsequent crack extension at $a_1 = a_2$ represents an unstable, impossible path because $\partial K_2/\partial a_2 > 0$. Cracks a_2 at critical state jump ahead at constant a_1 and D (Fig. 2) and then gradually extend with increasing D. At the same time, cracks a_1 stop growing and K_1 gradually diminishes at increasing a_1 and D until K_1 becomes zero; this is the second critical state (Fig. 2), which corresponds to vanishing (2×2) determinant. In this state cracks a_1 suddenly close over a finite length at constant a_2 and D (Fig. 2), and cracks a_2 grow thereafter at increasing D. Because cracks a_1 are closed, this is equivalent to the initial situation of a single crack length, but with doubled

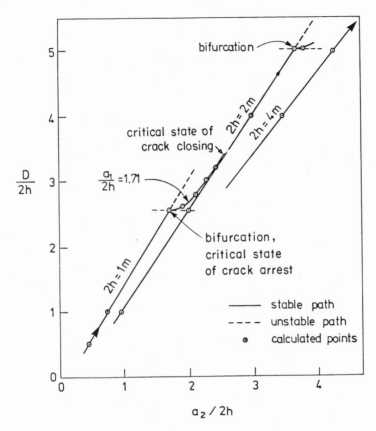

Fig. 2. Evolution of Crack Lengths a_i as a Function of Cooling
Penetration Depth D (Error Function Profile) at a Given
Initial Crack Spacing 2h (after Ref. 11).

spacing. The process then gets repeated at doubled spacing, quad-
rupled spacing, etc.

APPLICATIONS

     Practical cases were solved by the finite element method.
Assuming that heat is transported solely by conduction in the
solid, the temperature drop profile is given by the error function. The resulting equilibrium crack length for this profile
and for typical properties of westerly granite are plotted as
curve 1 in Fig. 3, and the post-critical crack growth for the
error function profile has been shown in Fig. 2. The dashed
portion of the curve represents unstable, unreachable states,
and the critical states of crack arrest are marked by circles.
The postcritical decrease of the stress intensity factor K_2 of
the arrested cracks is shown in Fig. 4 where

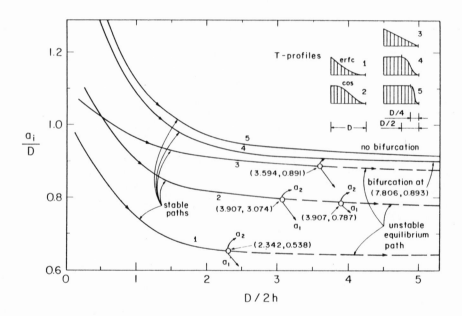

Fig. 3.   Finite Element Results on the Equilibrium Crack Lengths
          and Critical States for Various Temperature Drop Profiles
          in the Halfspace (after Ref. 12).

$\kappa_2 = K_2(1- \nu^2)/(E\alpha\Delta T\sqrt{2h})$ = nondimensional stress intensity factor,
ΔT = temperature drop, ν = Poisson ratio, α = thermal dilatation
coefficient.   The plots of the stress intensity factor variation
with a_2 at constant a_1, from which the critical states in Fig. 3
were obtained, are shown in Fig. 5.

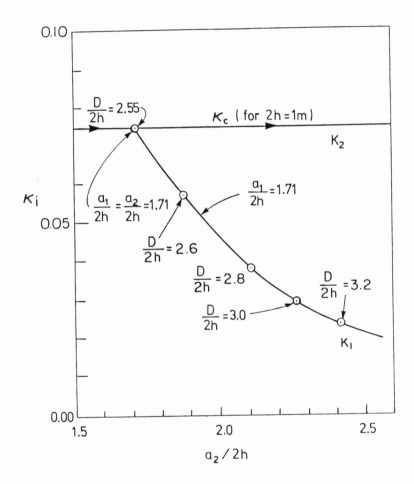

Fig. 4. Post-critical Variation of Nondimensional Stress Intensity Factors κ_i (for Error Function Profile) (after Ref. 11).

Consider now the withdrawal of heat from hot dry rock. We imagine that we already have a large vertical primary crack, produced for example by hydraulic fracturing. The halfspace surface represents one wall of this primary crack, and due to circulation of water in the primary crack, cooling cracks are produced in the halfspace. For the creation of cracks that would serve as effective additional circulation channels for cooling water, we are

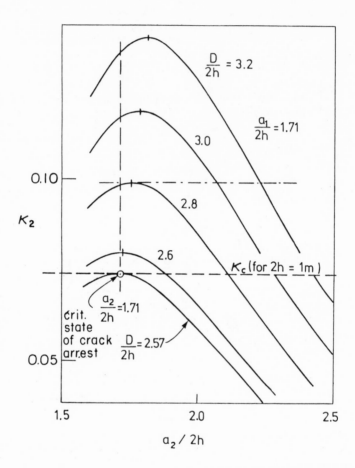

Fig. 5. Variation of Nondimensional Stress Intensity Factor
with One Crack Length, the Other Crack Length Being
Constant (after Ref. 11).

interested in obtaining a large crack spacing. This is because
at given temperature the thermal crack width w is roughly proportional to crack spacing b (= 2h), and the rate of viscous flow
through the cracks is proportional to b^3. The location of the
critical point on curve 1 in Fig. 3 is relatively favorable for

the hot-dry rock geothermal energy scheme; it indicates that the
crack spacing would double approximately when the length of the
cracks equals the double of their spacing.

If the flow of water through the cracks becomes more intense,
a significant amount of heat will be withdrawn by the water, and
this will cause the average temperature profile to become steeper
at the front. Precise calculation of the temperature field is
difficult, and for this reason several characteristic profiles
with progressively steeper fronts, as shown in Fig. 3, were
selected for computations of crack growth and instability. The
results for these five profiles are shown by the five curves in
Fig. 3.

The numerical results for various profiles may be approximately
described by the formula[12]:

$$\left(\frac{a_2}{2h}\right)_{crit} \approx \left(\frac{0.29}{0.48 - x_c/D}\right)^2 \quad \text{if } \frac{x_c}{D} < 0.48; \text{ else } \infty \tag{12}$$

where a_2 = crack length; $2h$ = crack spacing, D = cooling penetration depth, and x_c = distance from the halfspace surface to the
centroid of the profile of the temperature drop.

The following general trend may be observed in Fig. 3: The
steeper the front of the temperature profile, the larger is the
crack length-to-spacing ratio at which every other crack gets
arrested due to instability. This last result is not particularly
favorable to the withdrawal of heat from hot dry rock by the
assumed water circulation in the cooling cracks. The early doubling of crack spacing (and width) for the error function profile,
which corresponds to pure heat conduction in rock, causes early
doubling of crack spacing (and width). This leads to increased
water flow in the cooling cracks, which in turn produces a temperature drop profile with a steeper front. This, however, inhibits
further doubling of crack spacing (and increase of the width of
cracks) and thus impedes the water circulation in the cooling
cracks. Obviously, other mechanisms must be also at work to obtain
efficient heat withdrawal from the rock.

Calculations have also been done taking the anisotropic properties of granite into account.[12] However, the resulting picture
is not significantly different. More elaborate estimates of the
water flow and crack spacing may be found in Refs. 11 and 12.

Another application which has been studied numerically is
the thermal cracks and the drying shrinkage cracks in reinforced
solids, such as reinforced concrete.[13] It is found that even
very light reinforcement close to the surface has a very large

effect on suppressing the instability of crack arrest, while even
a very heavy reinforcement cannot prevent crack instability deeper
in the solid if the solid is unreinforced. So, the effect of reinforcement on crack stability, and thus crack spacing, is essentially local.

A detailed analysis of the water circulation in a large hydraulic fracture whose width responds to water temperature and
pressure was presented in Ref. 14.

ACKNOWLEDGMENT

Support under Los Alamos Scientific Laboratory contract No.
N68-85-8539G-1 with Northwestern University and under previous
National Science Foundation Grants AER75-00187 and ENG 75-14848-
A01 is gratefully acknowledged. The writer is also thankful for
partial support under a Guggenheim Fellowship awarded to him for
1978-79. Finally, it is acknowledged that during the initial
phase of this project in 1975-76 S. Nemat-Nasser served as co-
principal investigator and co-director. Valuable discussions with
him and L. M. Keer at that time are deeply appreciated.

REFERENCES

1. B. A. Boley and J. H. Weiner, "Theory of Thermal Stresses,"
    John Wiley, New York (1960).
2. J. F. Knott, "Fundamentals of Fracture Mechanics," Butter-
    worth, London (1973).
3. Z. P. Bažant and H. Ohtsubo, Stability and Spacing of Cooling
    or Shrinkage Cracks, "Advances in Civil Engineering through
    Engineering Mechanics," Preprints of 2nd Annual Eng. Mech.
    Div. Specialty Conference held at North Carolina State
    Univ., Raleigh in May 1977, published by Am. Society of
    Civil Engineers.
4. Z. P. Bažant and H. Ohtsubo, Stability Conditions for Prop-
    agation of a System of Cracks in a Brittle Solid, Mechan-
    ics Research Communications, Vol. 4, pp. 353-366 (1977).
5. D. P. H. Hasselman, Unified Theory of Thermal Shock, Fracture
    Initiation and Crack Propagation in Brittle Ceramics,
    Journal of the American Ceramic Society, Vol. 52,
    pp. 600-604 (1969).
6. R. D. McFarland, Geothermal Reservoir Models — Crack Plane
    Model, Report LA-5947-MS, Los Alamos Scientific Laboratory,
    Los Alamos, N.M., (April 1975).
7. M. Smith, H. Potter, D. Brown and R. L. Aamodt, Induction and
    Growth of Fractures in Hot Rock, in "Geothermal Energy,"
    edited by P. Kruger and C. Otte, pp. 251-268, Stanford
    University Press, Stanford, California (1973).

8. M. C. Smith, R. L. Aamodt, R. M. Potter, and D. W. Brown,
    Proceedings of the 2nd Geothermal Energy Symposium, San
    Fransisco, California (1975).

9. C. R. B. Lister, On the Penetration of Water into Hot Rock,
    Geophysics Journal of the Royal Astronomical Society,
    Vol. 39, pp. 465-509 (1974).

10. A. H. Lachenbruch, Depth and Spacing of Tension Cracks,
    Journal of Geophysical Research, Vol. 66, p. 4273 (1961).

11. Z. P. Bažant, H. Ohtsubo, and K. Aoh, Stability and Post-
    Critical Growth of a System of Cooling or Shrinkage
    Cracks, Intern. Journal of Fracture, Vol. 15, pp. 443-456
    (1979).

12. Z. P. Bažant and A. B. Wahab, Instability and Spacing of
    Cooling or Shrinkage Cracks, Journal of the Engng. Mech.
    Division, Proc. ASCE, Vol. 105, pp. 873-889 (1979).

13. Z. P. Bažant and A. B. Wahab, Stability of Parallel Cracks
    in Solids Reinforced by Bars, Intern. Journal of Solids
    and Structures, Vol. 16, pp. 97-105 (1980).

14. Z. P. Bazant and H. Ohtsubo, Geothermal Heat Extraction by
    Water Circulation through a Large Crack in Dry Hot Rock
    Mass, Intern. J. of Numerical Methods in Geomechanics,
    Vol. 2, pp. 317-327 (1978).

APPENDIX

As has been mentioned, inequalities (4) represent the most significant difference from typical buckling problems. Fig. 6, however,
shows an example of a buckling problem that is analogous to ours.
The potential energy V of this structure may be expressed as a
function of midspan deflections a_1 and a_2 ; $V = V(a_1, a_2)$. The
first buckling mode is characterized by $a_2 = -a_1$. If we, however,

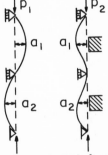

Figure 6

Analogous Problem

install rigid blocks as shown, so as to prevent buckling to the
left side, then the first admissible buckling mode is the second
buckling mode of the unconstraint system ($a_2 = a_1$), corresponding
to the second eigenvalue. The quadratic form for potential energy
$V(a_1, a_2)$ ceases to be positive definite already before the lowest
buckling load P is reached.

THERMAL STRESSES IN COAL CONVERSION PRESSURE VESSELS

BUILT OF LAYERED CONSTRUCTION

Theodore R. Tauchert

Department of Engineering Mechanics
University of Kentucky
Lexington, KY   40506

SUMMARY

Radially varying temperature and thermal stress distributions
in layered pressure vessels designed for use in coal conversion
processes are investigated. A least-squares residual method which
incorporates Lagrange multipliers for satisfaction of initial,
boundary and interface conditions is employed in the case of
transient heat conduction. Assuming orthotropic elastic behavior,
a general solution is formulated for the stresses induced by combined temperature, pressures and initial interferences between
layers. The effects of heat-transfer resistance at layer contact
surfaces are illustrated through several numerical examples.
Representation of a layered vessel as a homogeneous cylinder having
an effective thermal conductivity, for the purpose of predicting
thermoelastic response, is also explored.

INTRODUCTION

An increase in coal usage, including coal gasification and
liquefaction, represents one measure for alleviating the severe
energy problems which this country faces during the coming years.
It is anticipated that as commercial feasibility of coal conversion
processes is established there will be a demand for large diameter
thick-walled reactor vessels capable of operating at high temperatures and pressures. For coal gasification the vessels probably
will be 8 - 11 m (25-35 ft.) in diameter, 30 - 90 m (100-300 ft.)
high, and will operate at pressures of 7 - 10 MPa (1000-1500 psi)
and temperatures in the 800 - 2000 K (1000-3000°F) range. The
vessels for coal liquefaction will be of somewhat smaller diameter,
but will be exposed to pressures of 14 - 21 MPa (2000-3000 psi) and

temperatures in the neighborhood of 650 - 750 K (700-900°F). Based on such combinations of size, pressure and temperature the vessels will necessarily be constructed by a technology other than single-wall steel. One of the most feasible classes of technology is layered steel fabrication*. Rules for layered vessels were approve recently by the American Society of Mechanical Engineers, and are included in the Winter 1978 Addenda[1] to Section VIII, Divisions 1 and 2 of the ASME Boiler and Pressure Vessel Code.[2] For the most part the rules governing design and analysis of layered vessels are basically the same as those for single-wall vessels. However with regard to thermal stress analysis the Addenda state explicitly that:

> "When calculating thermal stresses in layered
> pressure vessels, consideration shall be given to
> heat-transfer resistance at layer contact surfaces."

As this statement implies, imperfect thermal contact between the individual layers of the vessel can lead to a significant increase in thermal stress. This is especially true for transient tempera-ture conditions, such as during start-up or shut-down of a reactor vessel.

Characterization of the heat-transfer resistance between metal surfaces has not been fully developed, although it is known that the thermal resistance is a function of contact pressure, surface roughness, and to a lesser extent temperature. Measurements of the transfer resistance of layered structures have been reported by Weills and Ryder[3], and Maruyama, et al.[4] Also, the Nooter Corporation** is presently investigating the thermal resistance for layered cylindrical pressure vessels. Typical values for steel laminates are given in Table 1.

Only a few papers dealing with thermal stresses in layered pressure vessels have appeared in the literature. Suto et al.[5] calculated the thermal stress distributions in the zone of a nozzle in a spiral sheet vessel caused by a transient temperature distribution. Maruyama et al.[4] used their experimentally deter-mined values of thermal resistance to compute the transient thermal stresses in the neighborhood of weldments in a two-layer vessel. And Mancuso and Rauschenplat[6], basing their analysis upon the

---

* An analysis of alternate designs for coal conversion pressure
   vessels is being conducted at the University of Kentucky, with
   funds provided by the Department of Energy.
** Results of the study being conducted by the Nooter Corporation
   (St. Louis) will be published in the near future.

Table 1. Measured Values of Thermal Resistance Between Steel Layers[4]

Surface Finish	Contact Stress (MPa)	Thermal Resistance (10^{-3} $m^2 K/W$)
Machined (max. 10 μ)	0	0.97
	6	0.30
	25	0.17
As Rolled (max. 30 μ)	0	1.00
	6	0.40
	25	0.34
Machined (max. 100 μ)	0	3.80

values of contact resistance given in Ref. 3, examined the stresses at the circumferential weld between the courses of a shrunk-fit layered vessel during start-up and shut-down operating transients. In each of the above mentioned studies the stresses were analyzed by finite element methods.

The purpose of the present paper is to investigate the effects of heat-transfer resistance upon thermal stresses in the walls of layered cylindrical vessels. A least-squares residual method is employed to obtain an approximate analytical solution in the case of temperature transients. Based upon the approximate temperature distribution the thermal stresses can be computed without difficulty. With slight modification the approach can be applied to initially nonhomogeneous and temperature sensitive materials.

Also considered in this study is the possibility of accounting for the effects of heat-transfer resistance through the use of effective thermal conductivities.

CALCULATION OF TEMPERATURE DISTRIBUTION

Consider a long n-layer cylindrical vessel which contains a gas at temperature $T_G(t)$, and which is exposed to an external ambient temperature $T_A(t)$. It is presumed that T_G and T_A are spacewise constant, in which case the temperature within the cylindrical wall varies in the radial direction only. For the typical j-th layer, bounded by inner radius $r^{(j)}$ and outer radius $r^{(j+1)}$, the temperature $T^{(j)}(r,t)$ is governed by the Fourier heat conduction equation

$$k^{(j)} \left(\frac{\partial^2 T^{(j)}}{\partial r^2} + \frac{1}{r} \frac{\partial T^{(j)}}{\partial r} \right) - \eta^{(j)} \frac{\partial T^{(j)}}{\partial t} = 0 \quad (j = 1,2,\ldots,n) \qquad (1)$$

in which $k^{(j)}$ and $\eta^{(j)}$ denote the radial coefficient of heat conduction and the heat capacity for layer j, respectively.

In terms of a prescribed initial temperature distribution $f^{(j)}(r)$ for the j-th layer, the initial conditions are

$$T^{(j)}(r,0) = f^{(j)}(r) \tag{2}$$

Assuming that convection occurs on the cylindrical surfaces of the vessel through boundary conductances h_G and h_A, the respective boundary conditions are

$$r = r^{(1)}: \quad -k^{(1)} \frac{\partial T^{(1)}}{\partial r} = h_G[T_G(t) - T^{(1)}] \tag{3}$$

$$r = r^{(n+1)}: \quad k^{(n)} \frac{\partial T^{(n)}}{\partial r} = h_A[T_A(t) - T^{(n)}] \tag{4}$$

If there is imperfect thermal contact at radius $r^{(j)}$ between layers (j-1) and (j), the appropriate interface conditions become[8]

$$r = r^{(j)}: \quad k^{(j-1)} \frac{\partial T^{(j-1)}}{\partial r} = k^{(j)} \frac{\partial T^{(j)}}{\partial r} \tag{5}$$

$$k^{(j-1)} \frac{\partial T^{(j-1)}}{\partial r} = \frac{1}{R^{(j)}} [T^{(j)} - T^{(j-1)}] \tag{6}$$

for j = 2,3,...n. Condition (5) expresses continuity of heat flux at the interface. Condition (6) states that the difference between t two surface temperatures is proportional to the heat flux, the constant of proportionality being the thermal contact resistance $R^{(j)}$.

Stationary Temperature

In the case of time-independent internal and external ambient temperatures T_G and T_A, the general solution for the stationary temperature $T^{(j)}(r)$ is[8]

$$T^{(j)} = C_1^{(j)} \ln r + C_2^{(j)} \tag{7}$$

The 2n arbitrary constants $C_1^{(j)}$, $C_2^{(j)}$ are found by applying the two boundary conditions (3)-(4) and the 2(n-1) interface conditions (5)-(6).

Transient Temperature

Obtaining an exact solution for the transient temperature distribution $T^{(j)}(r,t)$ in an n-layer cylinder involves the determination of n sets of eigenvalues and eigenfunctions (say $\beta_\ell^{(j)}$, $X_\ell^{(j)}(r)$ with $\ell = 1,2,\ldots,\infty$ and $j = 1,2,\ldots,n$); in order to evaluate the eigenvalues it is generally necessary to numerically solve complicated transcendental equations. Even in those cases where the exact temperature distribution can be found, the mathematical form of the solution usually is so complex that the corresponding thermoelastic problem cannot be solved in closed form. The approach which is followed here is to employ a least-squares residual method[7] to obtain an approximate analytic solution to the heat conduction problem. The approximate solution for the temperature is of a sufficiently simple form that the thermal stresses can be found in a straightforward manner.

For the present problem we seek an approximate solution $\hat{T}^{(j)}(r,t)$ to equation (1) of the form

$$\hat{T}^{(j)} = \sum_i \sum_\ell C_{i\ell}^{(j)} r^i e^{-\beta_\ell t} \tag{8}$$

where $C_{i\ell}^{(j)}$ denote an arbitrary number of undetermined coefficients. A sufficiently large number of different functions (r^i and $e^{-\beta_\ell t}$) must be chosen in order that the approximate solution yields an acceptable degree of accuracy. While only a few powers of r may be needed, inclusion of a relatively large number of different coefficients β_ℓ may be required if the solutions are to be valid for both large and small times t. As illustrated in the numerical examples which follow, a suitable set of values β_ℓ often can be found by selecting the characteristic roots of the exact solution to a related heat conduction problem (see Appendix).

It is presumed that the initial temperature distribution $f^{(j)}(r)$ in (2) can be represented by the truncated power series

$$f^{(j)}(r) = \sum_i A_i^{(j)} r^i \tag{9}$$

and that the internal and external ambient temperatures can be represented as

$$T_G(t) = \sum_\ell T_{G\ell} e^{-\beta_\ell t} \tag{10}$$

$$T_A(t) = \sum_\ell T_{A\ell} e^{-\beta_\ell t} \tag{11}$$

where nonzero steady-state temperatures can be handled by taking $\beta_0 \equiv 0$, in which case $T_G(\infty) = T_{G0}$ and $T_A(\infty) = T_{A0}$.

The initial, boundary and interface conditions impose restrictions upon the undetermined coefficients $C_{i\ell}(j)$ in the approximate solution (8). For example, by substituting (8) and (9) into initial condition (2) it is found that the coefficients $C_{i\ell}(j)$ must satisfy the equation

$$b_i^{(j)} = \sum_\ell C_{i\ell}^{(j)} - A_i^{(j)} = 0 \tag{12}$$

for all values of i and j. Likewise the boundary conditions (3) and (4) impose the restrictions

$$c_\ell = \sum_i C_{i\ell}^{(1)} \{-i \, k^{(1)} [r^{(1)}]^{i-1} + h_G [r^{(1)}]^i \} - h_G T_{G\ell} = 0 \tag{13}$$

$$d_\ell = \sum_i C_{i\ell}^{(n)} \{i \, k^{(n)} [r^{(n+1)}]^{i-1} + h_A [r^{(n+1)}]^i \} - h_A T_{A\ell} = 0 \tag{14}$$

for all values of ℓ, whereas the interface conditions (5) and (6) require

$$e_\ell^{(j)} = \sum_i \{C_{i\ell}^{(j-1)} i \, k^{(j-1)} [r^{(j)}]^{i-1} - C_{i\ell}^{(j)} i \, k^{(j)} [r^{(j)}]^{i-1} \} = 0 \tag{15}$$

$$f_\ell^{(j)} = \sum_i \{C_{i\ell}^{(j-1)} \left[i \, R^{(j)} k^{(j-1)} [r^{(j)}]^{i-1} + [r^{(j)}]^i \right] - C_{i\ell}^{(j)} [r^{(j)}]^i \} = 0 \tag{16}$$

for j = 2,...,n and all values of ℓ. The restrictions (12)-(16) can be used to eliminate certain of the coefficients $C_{i\ell}(j)$ from the assumed solution (8). However the algebra involved in the elimination process often becomes extremely involved, and so instead we employ the method of Lagrange multipliers.

At this point we introduce a functional E defined as

$$E = \int_{t=0}^{t_0} 2\pi L \sum_{j=1}^n \int_{r^{(j)}}^{r^{(j+1)}} [\mathcal{R}^{(j)} (\hat{T})]^2 r \, dr \, dt + \sum_{j=1}^n (\sum_i \lambda_i^{(j)} b_i^{(j)})$$

$$+ \sum_{\ell} \mu_\ell c_\ell + \sum_{\ell} \nu_\ell d_\ell + \sum_{j=2}^{n} (\sum_{\ell} \xi_\ell^{(j)} e_\ell^{(j)} + \sum_{\ell} \psi_\ell^{(j)} f_\ell^{(j)}) \qquad (17)$$

in which $\mathcal{R}^{(j)}(\hat{T})$ denotes the residual for layer j obtained by substituting the approximate solution (8) into the heat conduction equation (1). Also $\lambda_i(j)$, $\mu_\ell, \ldots, \psi_\ell(j)$ are the Lagrange multipliers for the constraint conditions (12)-(16), respectively. Thus the squares of the residuals of the differential equation for each layer have been integrated over the volume of the cylinder (assuming a length L) and over the time interval $0 < t < t_0$, where t_0 is an arbitrarily chosen upper limit. Clearly t_0 should be at least as large as the largest time of interest in the particular problem.[7] For an approximate solution which satisfies the initial, boundary and interface conditions and has a least-squared error, we require that

$$\frac{\partial E}{\partial c_{i\ell}}(j) = \frac{\partial E}{\partial \lambda_i}(j) = \frac{\partial E}{\partial \mu_\ell} = \cdots = \frac{\partial E}{\partial \psi_\ell}(j) = 0 \qquad (18)$$

In the case of a cylinder with material properties which are independent of temperature, equations (18) represent a system of $N+M$ linear algebraic equations in the N coefficients $c_{i\ell}^{(j)}$ and M Lagrange multipliers.

CALCULATION OF THERMAL STRESSES

For the stress calculations we suppose that the n-layer cylindrical vessel is subject to uniform internal and external pressures p_G and p_A, in addition to the radial temperature distribution analyzed earlier. We shall also allow for the possibility of an initial radial interference between each layer of the vessel, as would result from longitudinal weld shrinkage or fabrication by shrink fitting.

Because of the axisymmetry, the radial displacement component $u^{(j)}$ is a function of r only and the circumferential displacement is zero. Therefore the relevant strain-displacement relations for the jth-layer are

$$\varepsilon_{rr}^{(j)} = \frac{du^{(j)}}{dr}, \quad \varepsilon_{\theta\theta}^{(j)} = \frac{u^{(j)}}{r}, \quad \varepsilon_{zz}^{(j)} = \varepsilon \qquad (19)$$

in which the axial strain is assumed to have a constant value ε which will be determined as part of the solution.

Taking the material to be cylindrically orthotropic, the non-zero stress components are given by

$$\sigma_{rr}^{(j)} = A_{11}^{(j)} \frac{du^{(j)}}{dr} + A_{12}^{(j)} \frac{u^{(j)}}{r} + A_{13}^{(j)} \varepsilon - \beta_1^{(j)} \overline{T}^{(j)}$$

$$\sigma_{\theta\theta}^{(j)} = A_{12}^{(j)} \frac{du^{(j)}}{dr} + A_{22}^{(j)} \frac{u^{(j)}}{r} + A_{32}^{(j)} \varepsilon - \beta_2^{(j)} \overline{T}^{(j)} \qquad (20)$$

$$\sigma_{zz}^{(j)} = A_{13}^{(j)} \frac{du^{(j)}}{dr} + A_{32}^{(j)} \frac{u^{(j)}}{r} + A_{33}^{(j)} \varepsilon - \beta_3^{(j)} \overline{T}^{(j)}$$

in which $A_{mq}^{(j)}$ and $\beta_m^{(j)}$ denote the elastic stiffnesses and stress-temperature coefficients for the jth-layer. Here $\overline{T}^{(j)} = T^{(j)} - T_0$ denotes the temperature rise from the stress-free temperature T_0. The stresses (20) must satisfy the equation of equilibrium

$$\frac{d\,\sigma_{rr}^{(j)}}{dr} + \frac{\sigma_{rr}^{(j)} - \sigma_{\theta\theta}^{(j)}}{r} = 0 \qquad (21)$$

or in terms of the displacement $u^{(j)}$

$$\frac{d^2 u^{(j)}}{dr^2} + \frac{1}{r}\frac{du^{(j)}}{dr} - \frac{A_{22}^{(j)}}{A_{11}^{(j)}} \frac{u^{(j)}}{r^2} = \frac{A_{32}^{(j)} - A_{13}^{(j)}}{A_{11}^{(j)}} \frac{\varepsilon}{r}$$

$$+ \frac{B_1^{(j)}}{A_{11}^{(j)}} \frac{d\overline{T}^{(j)}}{dr} + \frac{\beta_1^{(j)} - \beta_2^{(j)}}{A_{11}^{(j)}} \frac{\overline{T}^{(j)}}{r} \qquad (22)$$

Using the method of variation of parameters, the general solution to equation (22) is found to be[9,10]

$$u^{(j)} = D_1^{(j)} r^{a^{(j)}} + D_2^{(j)} r^{-a^{(j)}} + g^{(j)} \varepsilon\, r + \chi^{(j)}(r) \qquad (23)$$

where $D_1^{(j)}$ and $D_2^{(j)}$ are arbitrary constants, and*

$$a^{(j)} = \sqrt{\frac{A_{22}^{(j)}}{A_{11}^{(j)}}}, \qquad g^{(j)} = \frac{A_{32}^{(j)} - A_{13}^{(j)}}{A_{11}^{(j)} - A_{22}^{(j)}} \qquad (24)$$

---

*Note that $g^{(j)} \equiv 0$ in the case of an isotropic material.

and also

$$\chi^{(j)}(r) = h_1^{(j)} r^{a^{(j)}} \int r^{-a^{(j)}} \bar{T}^{(j)} dr + h_2^{(j)} r^{-a^{(j)}} \int r^{a^{(j)}} \bar{T}^{(j)} dr \quad (25)$$

in which

$$h_1^{(j)} = \frac{a^{(j)} \beta_1^{(j)} - \beta_2^{(j)}}{2a^{(j)} A_{11}^{(j)}}, \qquad h_2^{(j)} = \frac{a^{(j)} \beta_1^{(j)} + \beta_2^{(j)}}{2a^{(j)} A_{11}^{(j)}} \quad (26)$$

Substitution of (23) into (20) yields the thermal stresses

$$\sigma_{rr}^{(j)} = \ell_1^{(j)} r^{a^{(j)}-1} D_1^{(j)} + m_1^{(j)} r^{-a^{(j)}-1} D_2^{(j)} + q_1^{(j)} \varepsilon + \phi_1^{(j)}(r)$$

$$\sigma_{\theta\theta}^{(j)} = \ell_2^{(j)} r^{a^{(j)}-1} D_1^{(j)} + m_2^{(j)} r^{-a^{(j)}-1} D_2^{(j)} + q_2^{(j)} \varepsilon + \phi_2^{(j)}(r)$$

$$\sigma_{zz}^{(j)} = \ell_3^{(j)} r^{a^{(j)}-1} D_1^{(j)} + m_3^{(j)} r^{-a^{(j)}-1} D_2^{(j)} + q_3^{(j)} \varepsilon + \phi_3^{(j)}(r)$$

$$(27)$$

where

$$\ell_\alpha^{(j)} = a^{(j)} A_{1\alpha}^{(j)} + A_{\alpha 2}^{(j)}$$

$$m_\alpha^{(j)} = -a^{(j)} A_{1\alpha}^{(j)} + A_{\alpha 2}^{(j)}$$

$$q_\alpha^{(j)} = (A_{1\alpha}^{(j)} + A_{\alpha 2}^{(j)}) g^{(j)} + A_{\alpha 3}^{(j)}$$

$$\phi_\alpha^{(j)}(r) = A_{1\alpha}^{(j)} \frac{d\chi^{(j)}}{dr} + A_{\alpha 2}^{(j)} \frac{\chi^{(j)}}{r} - \beta_\alpha^{(j)} \bar{T}^{(j)}$$

$$(28)$$

The $2n+1$ arbitrary constants in the solution, namely $D_1^{(j)}$, $D_2^{(j)}$ and ε, are next found by applying two boundary conditions, $2(n-1)$ interface conditions, and an equation of force equilibrium in the axial direction. The stress boundary conditions are

$$r = r^{(1)}: \qquad \sigma_{rr}^{(1)} = -P_G \quad (29)$$

$$r = r^{(n+1)}: \qquad \sigma_{rr}^{(n)} = -P_A \quad (30)$$

whereas the interface conditions are taken to be

$$r = r^{(j)}: \quad \sigma_{rr}^{(j-1)} = \sigma_{rr}^{(j)} \tag{31}$$

$$u^{(j-1)} = u^{(j)} - \delta^{(j)} \tag{32}$$

Equation (31) expresses continuity of radial stress between adjacent layers, while (32) permits specification of an arbitrary interference at each interface. That is, $\delta^{(j)}$ represents the difference between the initial outer radius of layer (j-1) and the initial inner radius of layer (j).

Finally, for a vessel with closed ends, the longitudinal stresses $\sigma_{zz}^{(j)}$ must satisfy the equilibrium condition

$$2\pi \sum_{j=1}^{n} \int_{r^{(j)}}^{r^{(j+1)}} \sigma_{zz}^{(j)} r \, dr = (p_G - p_A)\pi [r^{(1)}]^2 \tag{33}$$

Thus, once the layer temperature distributions $T^{(j)}$ have been found, the coefficients $D_1^{(j)}$, $D_2^{(j)}$ and ε can be obtained by applying the conditions (29)-(33). In the case of the steady-state temperature (7), the functions $\chi^{(j)}(r)$ and $\phi_\alpha^{(j)}(r)$ appearing in the expressions for the displacement (23) and stresses (27) become

$$\chi^{(j)} = C_1^{(j)} [h_1^{(j)} r^{a^{(j)}} \int r^{-a^{(j)}} \ell nr \, dr + h_2^{(j)} r^{-a^{(j)}} \int r^{a^{(j)}} \ell nr \, dr]$$

$$+ \bar{C}_2^{(j)} [h_1^{(j)} r^{a^{(j)}} \int r^{-a^{(j)}} dr + h_2^{(j)} r^{-a^{(j)}} \int r^{a^{(j)}} dr] \tag{34}$$

$$\phi_\alpha^{(j)} = C_1^{(j)} [h_1^{(j)} \ell_\alpha^{(j)} r^{a^{(j)}-1} \int r^{-a^{(j)}} \ell nr \, dr$$

$$+ h_2^{(j)} m_\alpha^{(j)} r^{-a^{(j)}-1} \int r^{a^{(j)}} \ell nr \, dr + (\frac{A_{1\alpha}^{(j)}}{A_{11}^{(j)}} \beta_1^{(j)} - \beta_\alpha^{(j)})\ell n \, r]$$

$$+ \bar{C}_2^{(j)} [h_1^{(j)} \ell_\alpha^{(j)} r^{a^{(j)}-1} \int r^{-a^{(j)}} dr$$

$$+ h_2^{(j)} m_\alpha^{(j)} r^{-a^{(j)}-1} \int r^{a^{(j)}} dr + (\frac{A_{1\alpha}^{(j)}}{A_{11}^{(j)}} \beta_1^{(j)} - \beta_\alpha^{(j)})] \tag{35}$$

in which we take $\bar{C}_2{}^{(j)} = C_2{}^{(j)} - T_0$ to account for the fact that the stresses depend upon the change in temperature from the stress-free temperature T_0.

For the approximate transient temperature (8) the corresponding functions are

$$\chi^{(j)} = \sum_i \sum_\ell \bar{C}_{i\ell}{}^{(j)} [h_1{}^{(j)} r^{a^{(j)}} \int r^{-a^{(j)}+i} dr$$

$$+ h_2{}^{(j)} r^{-a^{(j)}} \int r^{a^{(j)}+i} dr] e^{-\beta_\ell t} \qquad (36)$$

$$\phi_\alpha{}^{(j)} = \sum_i \sum_\ell \bar{C}_{i\ell}{}^{(j)} [h_1{}^{(j)} \ell_\alpha{}^{(j)} r^{a^{(j)}-1} \int r^{-a^{(j)}+i} dr$$

$$+ h_2{}^{(j)} m_\alpha{}^{(j)} r^{-a^{(j)}-1} \int r^{a^{(j)}+i} dr + (\beta_1{}^{(j)} \frac{A_{1\alpha}{}^{(j)}}{A_{11}{}^{(j)}} - \beta_\alpha{}^{(j)}) r^i] e^{-\beta_\ell t} \qquad (37)$$

where $\bar{C}_{i0}{}^{(j)} = C_{i0}{}^{(j)} - T_0$ (when $\beta_\ell = 0$) and $\bar{C}_{i\ell}{}^{(j)} = C_{i\ell}{}^{(j)}$ otherwise ($\beta_\ell \neq 0$).

NUMERICAL EXAMPLES

We shall now examine the effects of heat-transfer resistance upon the radial distributions of temperature and thermal stress in layered vessels. Numerical results will be presented for cylinders constructed of various numbers of equal thickness layers. For this example the inner and outer radii of the vessel are taken to be 5.0 and 5.2 m, respectively. Each layer is assumed to be A 543 steel, with isotropic thermal and elastic constants given by

$$k^{(j)} = 46.3 \text{ W/mK}, \quad \eta^{(j)} = 4.27 \times 10^6 \quad \text{J/m}^3\text{K}$$

$$A_{11}{}^{(j)} = A_{22}{}^{(j)} = A_{33}{}^{(j)} = 253 \text{ GPa}, \quad A_{12}{}^{(j)} = A_{13}{}^{(j)} = A_{32}{}^{(j)} = 103 \text{ GPa},$$

$$\beta_1{}^{(j)} = \beta_2{}^{(j)} = \beta_3{}^{(j)} = 5.78 \times 10^6 \text{ Pa/K}$$

As a first case, stationary ambient temperatures $T_G = T_{G0}$ and $T_A = 0$ are prescribed on the cylindrical surfaces of the vessel, and convection of heat into the cylinder is presumed to occur through boundary conductances $h_G = h_A = 288$ W/m^{2}K. Figs. 1(a)

and 2(a) show computed values of the steady-state temperatures
(solid lines) in 6 and 12-layer cylinders, respectively. The
corresponding circumferential thermal stresses $\sigma_{\theta\theta}/\overline{T}_{G0}$ are given
in Figs. 1(b) and 2(b). (Here $\overline{T}_{G0} = T_{G0} - T_0$ denotes the difference between the absolute gas temperature and the stress-free
temperature for the cylinder; for this example $T_0 = 0$.) Results
are presented for three different values of thermal resistance:
$R^{(j)} = 0$ (the limiting case of ideal thermal contact); $R^{(j)} =$
0.34×10^{-3} m^{2}K/W (a typical value for rolled steel plates when the
contact stress equals or exceeds 25 MPa (see Table 1 or Ref. 4);
and $R^{(j)} = 3.80 \times 10^{-3}$ m^{2}K/W (a relatively high value, appropriate
for rough machined surfaces under zero contact stress). The thermal
resistance is assumed to be the same at each interface of a vessel.
For the cases considered it is seen that as the value of the resistance increases, the total temperature difference across the wall
of the cylinder increases; likewise the magnitudes of the thermal
stresses increase.

Also shown in the figures (as dashed curves) are distributions
of temperature and stress calculated using reduced values of thermal
conductivity $\overline{k}$, and zero thermal resistance. The aim here is to
determine whether the temperatures and the resulting thermal
stresses in multilayer vessels can be adequately predicted by
idealizing the vessel as a homogeneous cylinder having an apparent
or "effective" value of thermal conductivity. Effective conductivity $\overline{k}$ is defined here to be that value for which the difference
in steady-state temperature across the thickness of the homogeneous
cylinder is equal to the temperature difference across the layered
vessel.* As shown in Figs. 1 and 2 the radial distributions of the
steady-state temperature and circumferential stress calculated for
the "equivalent" homogeneous cylinders provide good approximations
to the average distributions of these quantities throughout the
corresponding layered vessels. Later we shall investigate how
adequately the same idealization will predict the average transient
thermoelastic response.

The effect of the number of layers N and the contact resistances $R^{(j)}$ upon the maximum absolute value of the circumferential
stress in layered vessels is illustrated in Fig. 3. These results
are based upon the same geometry and thermal boundary conditions
described earlier. Note that an increase in N and/or an increase

---

* To obtain the effective conductivity $\overline{k}$ for a vessel having a
  given number of layers n and thermal resistances $R^{(j)}$, it is convenient to first construct a plot of the change in steady-state
  temperature across the thickness of a vessel versus the vessel's
  thermal conductivity. Once the steady-state temperature has been
  calculated for a particular layered vessel, the corresponding
  effective conductivity can be read from this plot.

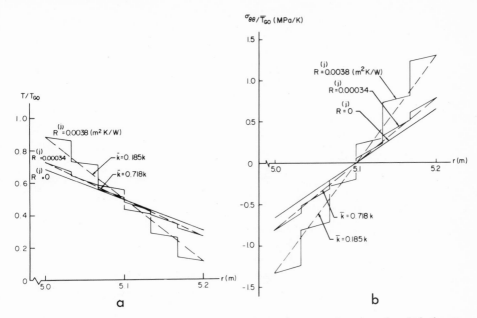

Fig. 1.   Stationary response of a 6-layer steel vessel with heat-transfer resistances $R^{(j)}$.   (a) temperature; (b) circumferential stress.

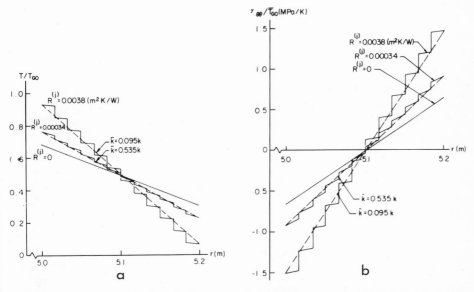

Fig. 2.   Stationary response of a 12-layer steel vessel with heat-transfer resistances $R^{(j)}$.   (a) temperature; (b) circumferential stress.

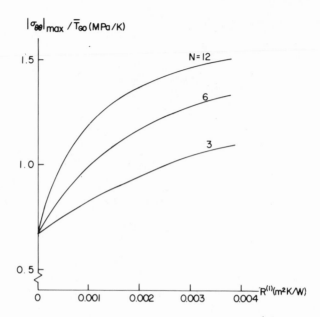

Fig. 3. Effect of heat-transfer resistances $R^{(j)}$ upon the maximum circumferential stress in 3, 6 and 12-layer vessels.

in the values $R^{(j)}$ result in an increase in the maximum stress. For each case investigated the maximum circumferential stress is compressive, and occurs at the inner surface of the cylinder.

We next consider the transient response of a vessel consisting of 6 layers of A 543 steel, with constant heat-transfer resistances $R^{(j)} = 0.34 \times 10^{-3}$ m²K/W at the interfaces. A rapid rise in the gas temperature within the vessel is presumed to occur, while the outside ambient temperature remains constant. In particular, we take

$$T_G(t) = T_{G0}(1 - e^{-0.025\,t}), \quad T_A(t) = 0$$

where time t is expressed in seconds. Temperature distributions at three selected times, calculated using the least-squares residual method described earlier, are shown as solid curves in Fig. 4(a). For this problem the approximate solution (8) for the temperature within the typical j-th layer was based upon 3 powers of r (r^0, r^2, r^4) and 5 values of β_ℓ ($0, 0.555 \times 10^{-3}, 0.387 \times 10^{-2}$,

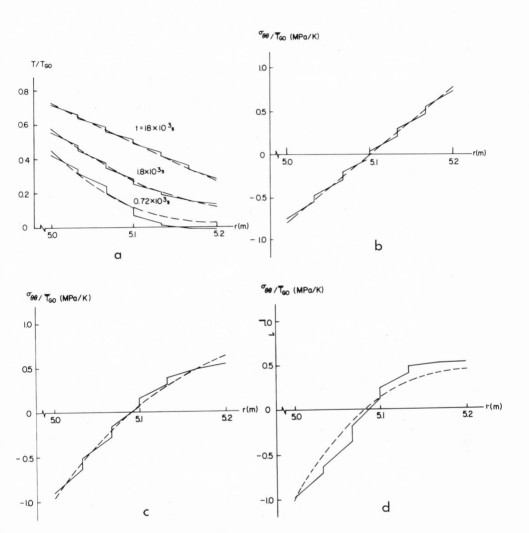

Fig. 4. Transient response of a 6-layer steel vessel with $R^{(j)}$ = 0.34 x 10^{-3} m^2K/W. (a) temperature distributions; (b)-(d) circumferential stress distributions for t = 18 x 10^3s, 1.8 x 10^3s and 0.72 x 10^3s, respectively.

0.0120, 0.0250)*. The upper limit of integration t_0 in the functional E (equation (17)) was taken as 18×10^3s (5 hr.). It can
be seen that the temperature profile corresponding to this elapsed
time differs slightly from the exact steady-state temperature distribution for the 6-layer cylinder plotted in Fig. 1(a).

In calculating the temperatures at times $t = 1.8 \times 10^3$s (0.5
hr.) and $t = 0.72 \times 10^3$s (0.2 hrs.) the value of t_0 was taken to
be 3.6×10^3s. The fact that the solution for $t = 0.72 \times 10^3$s
predicts a small but finite negative temperature at the outer surface of the cylinder indicates that the approximation is not particularly accurate for this small value of time. As mentioned earlier
it is often necessary to include a very large number of different
coefficients β_ℓ in order to obtain an accurate short-time solution.

Also shown in Fig. 4(a) (dashed curves) are the exact transient temperature solutions (see Appendix) for a homogeneous cylinder subject to an instantaneous rise T_{G0} in gas temperature.
In representing the layered vessel as a homogeneous cylinder, the
effective thermal conductivity was taken to be $\bar{k} = 0.718 k$ (see
Fig. 1(a)). It is observed that the homogeneous representation
yields temperature distributions which agree well with the average
temperatures computed for the layered vessel, particularly in the
cases of elapsed times $t = 18 \times 10^3$s and $t = 1.8 \times 10^3$s. The idealization also leads to thermal stress values which compare favorably
with results for the layered cylinder, as shown in Fig. 4(b)-(d).

As a last example, calculated temperatures and stresses are
reported for two different designs of a 9.45 m (31 ft.) diameter
reactor vessel for use in the coal gasification process HYGAS**.
The prescribed operating pressure and temperature are $p_G = 8.50$ MPa
(1233 psi) and $T_G = 1228$ K (1750°F), while ambient conditions outside the vessel are selected arbitrarily as $p_A = 0$ and $T_A = 294$ K
(70°F); the corresponding boundary conductances are taken to be
$h_G = 288$ W/m²K and $h_A = 11.4$ W/m²K. Since Division 2, Section
VIII of the ASME Boiler and Pressure Vessel Code[2] does not permit
a design temperature in excess of 755 K (900°F), a refractory liner
is provided in each design.

The first design considered is a "concentric-layered-arc"

---

* The values β_ℓ used in the approximate solution were taken equal
  to the lowest five characteristic roots of the exact solution
  (see Appendix) for the temperature in a single-layer vessel
  having thermal conductivity $k = 46.3$ W/mK.
** HYGAS refers to a high-Btu gasification process developed by the
   Institute of Gas Technology.

vessel, consisting of a corrosion-resistant fluid-tight shell
(SA 387 steel in this case), surrounded by concentric layers of
steel (A 543) having high strength and toughness. This vessel can
be fabricated by pressing shell arcs around the partially completed
vessel, and butt welding each new arc to the previously attached
one.

Table 2 indicates the general layout of the concentric-layered-
arc vessel, and shows the stationary temperature and stress levels,
calculated as outlined earlier. In this analysis the heat-transfer
resistance between each steel layer is taken to be $R(j) = 0.34$ x
10^{-3} m^2K/W. Values of the thermal and mechanical properties for
the refractories and steels were selected from Refs. 2 and 11.

A preliminary stress analysis for this design indicated that
unless expansions joints were provided in the refractory liner,
uneven thermal expansion of the refractory and steel layers would
result in crushing of the liner. The stresses reported in Table
2 therefore correspond to a design in which the two layers of refractory material contain joints which permit unrestricted circumferential movement. Thermal stresses throughout the steel wall
then range from 49 MPa compression in layer 3 to 50 MPa tension
in layer 18. At the outer surface of the vessel the temperature
induced circumferential stress constitutes over 20% of the total
stress. If heat-transfer resistances between the steel layers
were ignored the thermal stress at this location would represent
only about 10% of the total stress.

The second design considered is a "shrunk-fit" shell, consisting of cylindrical layers of steel (SA 516-70 in this case) successively shrunk-fit together. This vessel can be fabricated by
heating each additional steel layer to approximately 755 K (900°F),
and slipping it over the previously assembled layers. Usually a
slight interference Δ is introduced between the inner radius of
the added layer and the outer radius of the partially completed
shrunk-fit cylinder. Assuming thin-walled layers, the corresponding
initial radial interferences (prior to any of the shrink fits) can
be shown to be $\delta(1) = \Delta$, $\delta(2) = \Delta/2$, $\delta(3) = \Delta/3,...$; for the present design the shrink-fit interference is presumed to be $\Delta = 0.8$ mm
(1/32 in.). The heat-transfer resistances are again taken to be
$R(j) = 0.34$ x 10^{-3} m^2K/W.

Table 3 shows the general layout and calculated steady-state
temperatures and circumferential stresses for this design. The
total temperature variation through the 5 steel layers is 24 K,
and the resulting thermal stresses vary from 42 MPa compression in
layer 3 to 40 MPa tension in layer 7. On the outside surface of
the cylinder the stresses caused by the temperature and the shrink
fitting represent 30% and 20%, respectively, of the stress produced
by the internal pressure.

Table 2. Temperature and Stresses for the
Concentric-Layered-Arc Pressure Vessel Design

Layer No.	Material	Radius r(m)	Temp Change $\overline{T}$(k)	Circumferential Stress, $\sigma_{\theta\theta}$ (MPa)		
				Due to P_G	Due to $\overline{T}$	Total
1	Insulating Refractory	4.725 4.814	922 770	0 0	0 0	0 0
2	Dense Alumina Refractory	4.814 4.992	770 276	0 0	0 0	0 0
3	SA 387 Steel	4.992 5.023	276 274	196 194	−49 −40	147 154
4	A 543 Steel	5.023 5.036	273 272	194 194	−37 −34	158 160
. . .						
18	A 543 Steel	5.201 5.214	248 247	188 187	48 50	235 237

Table 3. Temperature and Stresses for the
Shrunk-Fit Pressure Vessel Design

Layer No.	Material	Radius r(m)	Temp. Change $\overline{T}$(K)	Circumferential Stress, $\sigma_{\theta\theta}$ (MPa)			
				Due to $\delta(j)$	Due to P_G	Due to $\overline{T}$	Total
1	Insulating Refractory	4.725 4.814	922 768	0 0	0 0	0 0	0 0
2	Dense Alumina Refractory	4.814 4.992	768 269	0 0	0 0	0 0	0 0
3	SA 516-70 Steel	4.992 5.056	269 265	−43 −42	137 135	−42 −27	52 66
. . .							
7	SA 516-70	5.248 5.312	249 245	26 26	130 128	27 40	183 194

DISCUSSION

The heat conduction and thermal stress formulations described
in this paper provide a suitable means for calculating radial variations of temperature and stress in layered cylinders having prescribed heat-transfer resistances. From the numerical results
presented it can be concluded that these resistances generally
cannot be ignored in the computation of thermal stresses. For the
cases investigated it appears also that representation of the
layered vessel as a homogeneous cylinder having an effective thermal
conductivity $\bar{k}$ leads to a relatively accurate prediction of the
average steady-state and transient responses.

The present study, however, represents only a preliminary
effort toward the general three-dimensional problem of thermoelastic behavior of layered pressure vessels. Consideration has
been confined here to constant axisymmetric heat-transfer resistances; in general the resistances are functions of contact stress
and temperature, and may vary in both the circumferential and axial
directions. Moreover the assumption has been made that ideal
mechanical contact (continuity of radial stress and displacement)
is maintained over each interface; in practice interlayer gaps
exist, the size of which depends upon the contact stress. Analysis
of an actual pressure vessel is often further complicated by the
need to consider complex geometries (including various head configurations, welded joints, penetrations and other structural
discontinuities), three-dimensional loadings, and more general
material behaviors (e.g., temperature-dependent properties). In
such instances analysis by finite element or other numerical
techniques generally will be required. Further study on a number
of these problems is underway.

Acknowledgment

This work has been supported by the U.S. Department of Energy,
Contract No. DE-AS05-77ET05378. The author wishes to express his
thanks to Dr. M. A. Kalam for his assistance with the calculations
described in the Appendix.

References

1. ASME Boiler and Pressure Vessel Code, Section VIII, "Pressure ·
        Vessels: Division 1, and Division 2 - Alternative Rules,
        Winter 1978 Addenda" (1978).
2. ASME Boiler and Pressure Vessel Code, Section VIII, "Pressure
        Vessels: Division 1, and Division 2 - Alternative Rules"
        (1977).
3. N. D. Weills and E. A. Ryder, "Thermal Resistance Measurements
        of Joints Formed Between Stationary Metal Surfaces", J. Heat
        Transfer, TRANS. ASME, Vol. 71 (1949), p. 259.

4.   T. Maruyama, H. Togawa and S. Kimura, "Analysis of Transient
     Thermal Stress and Thermal Fatigue Strength on Multiwall
     Vessel and Nonintegral Reinforcing Structure", Third Intl.
     Conf. on Pressure Vessel Technology, Tokyo, April 1977.

5.   K. Suto, T. Hada and H. Kawano, "Study of Elevated-Temperature
     Strength of Coilayer Vessel with Nozzles", Technical Review,
     Mitsubishi Heavy Industries, Ltd., June 1972.

6.   R. A. Mancuso and H. C. Rauschenplat, "Thermal Transient
     Analysis in Layered Pressure Vessels", ASME Pressure Vessel
     and Piping Conf., San Francisco, 79-PVP-13 (1979).

7.   T. R. Tauchert, "A Residual Method With Lagrange Multipliers
     for Transient Heat Conduction Problems", J. Heat Transfer,
     TRANS. ASME, Vol. 99 (1977), p. 495.

8.   H. S. Carslaw and J. C. Jaeger, Conduction of Heat in Solids,
     2nd ed., Oxford Univ. Press, Oxford (1959).

9.   B. I. Birger, "Thermal Stresses in an Anisotropic Cylinder",
     Izvestiya Vuz, Aviatsionnaya Tekhnika, Vol. 14 (1971), p. 24.

10. T. R. Tauchert and N. N. Hsu, "Shrinkage Stresses in Wood Logs
     Considered as Layered, Cylindrically Orthotropic Materials",
     Wood Sci. Technol., Vol. 11 (1977), p. 51.

11. The Handbook of Castable Refractories, Harbison-Walker Refrac-
     tories, Pittsburgh (1977).

12. G. E. Myers, Analytical Methods in Conduction Heat Transfer,
     McGraw-Hill, New York (1971).

Appendix. Transient Temperature in a Homogeneous Cylinder with Convection Boundary Conditions

Consider a long homogeneous cylinder $a < r < b$ of thermal diffusivity $\kappa \equiv k/\eta$. Initially the temperature is zero throughout the
cylinder. For times $t > 0$ the inner surface $r = a$ is exposed to
convection into a gas at temperature T_G through a boundary conductance h_G; the outer surface $r = b$ is subject to convection into
air at temperature T_A through conductance h_A. Using the method
of separation of variables, with expansion of the function of the
radial coordinate in a series of Bessel functions [8,12] the temperature distribution is found to be

$$T(r,t) = C_1 \ell n r + C_2 + \sum_{n=1}^{\infty} A_n R_0(\alpha_n r) e^{-\kappa \alpha_n^2 t} \qquad (A.1)$$

The eigenfunctions $R_0(\alpha_n r)$ in (A.1) are given by

$$R_0(\alpha_n r) = J_0(\alpha_n r) + b_n Y_0(\alpha_n r) \qquad (A.2)$$

where J_0 and Y_0 are Bessel functions of order zero of the first and second kind, respectively. It can be shown that the corresponding eigenvalues α_n represent the roots of the transcendental equation

$$[k\alpha J_1(\alpha a) + h_G J_0(\alpha a)][-k\alpha Y_1(\alpha b) + h_A Y_0(\alpha b)]$$

$$-[-k\alpha J_1(\alpha b) + h_A J_0(\alpha b)][k\alpha Y_1(\alpha a) + h_G Y_0(\alpha a)] = 0 \qquad (A.3)$$

Constants appearing in the general solution (A.1) include

$$C_1 = \frac{T_G - T_A}{\ln\left(\frac{a}{b}\right) - \dfrac{\kappa}{ah_G} - \dfrac{\kappa}{bh_A}} \;, \quad C_2 = \frac{T_A\left(\ln a - \dfrac{\kappa}{ah_G}\right) - T_G\left(\ln b + \dfrac{\kappa}{bh_A}\right)}{\ln\left(\frac{a}{b}\right) - \dfrac{\kappa}{ah_G} - \dfrac{\kappa}{bh_A}} \qquad (A.4)$$

and

$$A_n = \frac{-2}{\alpha_n}\left\{C_1\left[b \ln b\, R_1(\alpha_n b) - a \ln a\, R_1(\alpha_n a) + \frac{1}{\alpha_n} R_0(\alpha_n b)\right.\right.$$

$$\left. - \frac{1}{\alpha_n} R_0(\alpha_n a)\right] + C_2\left[b\, R_1(\alpha_n b) - a\, R_1(\alpha_n a)\right]\right\}$$

$$/\{b^2[R_1^2(\alpha_n b) + R_0^2(\alpha_n b)] - a^2[R_1^2(\alpha_n a) + R_0^2(\alpha_n a)]\} \qquad (A.5)$$

$$b_n = -\frac{h_G\, J_0(\alpha_n a) + \kappa\alpha J_1(\alpha_n a)}{h_G\, Y_0(\alpha_n a) + \kappa\alpha Y_1(\alpha_n a)} \qquad (A.6)$$

where orthogonality[12] of $R_0(\alpha_n r)$ has been used in evaluating the coefficients A_n in (A.5).

Analysis of the thermal stresses induced in a homogeneous cylinder by the temperature variation (A.1) can be carried out by substituting (A.1) into the equations for stress given earlier.

Table 2. Temperature and Stresses for the
Concentric-Layered-Arc Pressure Vessel Design

Layer No.	Material	Radius r(m)	Temp Change $\overline{T}$(k)	Circumferential Stress, $\sigma_{\theta\theta}$ (MPa) Due to p_G	Due to $\overline{T}$	Total
1	Insulating Refractory	4.725 4.814	922 770	0 0	0 0	0 0
2	Dense Alumina Refractory	4.814 4.992	770 276	0 0	0 0	0 0
3	SA 387 Steel	4.992 5.023	276 274	196 194	−49 −40	147 154
4	A 543 Steel	5.023 5.036	273 272	194 194	−37 −34	158 160
. . .						
18	A 543 Steel	5.201 5.214	248 247	188 187	48 50	235 237

Table 3. Temperature and Stresses for the Shrunk-Fit Pressure Vessel Design

Layer No.	Material	Radius r(m)	Temp. Change $\bar{T}$(K)	Circumferential Stress, $\sigma_{\theta\theta}$ (MPa)			
				Due to $\delta(j)$	Due to P_G	Due to $\bar{T}$	Total
1	Insulating Refractory	4.725 4.814	922 768	0 0	0 0	0 0	0 0
2	Dense Alumina Refractory	4.814 4.992	768 269	0 0	0 0	0 0	0 0
3	SA 516–70 Steel	4.992 5.056	269 265	−43 −42	137 135	−42 −27	52 66
. . .							
7	SA 516–70 Steel	5.248 5.312	249 245	26 26	130 128	27 40	183 194

AN APPROACH TO LIFE PREDICTION OF DOMESTIC GAS FURNACE CLAM

SHELL TYPE HEAT EXCHANGERS

Brian Leis, Allen Hopper, Nu Ghadiali, Carl Jaske
and Gene Hulbert

Battelle's Columbus Laboratories
505 King Avenue
Columbus, Ohio 43201

INTRODUCTION

During normal operations, the heat exchanger shells of domestic gas furnaces experience alternately rising and falling temperatures. These cyclic temperatures in turn induce cyclic strain
patterns in the shell. The fatigue life of the shell is primarily
dependent upon the temperature range of the heating cycle, rate of
heating and cooling, the cyclic strain amplitudes, the cyclic
straining rates, the material of which the shell is made, and the
ambient environment.

The determination of the thermal strain distributions in a
heat exchanger for a given heating cycle and the prediction of the
fatigue life of the heat exchanger for that cycle are very difficult and complex problems. Strain ranges are influenced by the
cyclic temperature distributions, the shell geometry, material and
manufacturing imperfections, and the way in which the heat exchanger is restrained at its edges at installation.

This paper outlines an approach to crack initiation life
prediction of heat exchangers under coupled thermal-mechanical
cycling. A key to the approach is the introduction of a damage
assessment technique that reflects the biaxial thermal-mechanical
loading experienced by the heat exchanger.

An application of the prediction technique is made to a heat
exchanger whose response to a temperature cycle has been determined
experimentally in terms of measured strains and analytically in
terms of stresses. The analytical approach utilized thin-shell
finite element stress analysis of an exchanger subjected to a

measured temperature response. The predicted and experimentally observed life are compared.

CRACK INITIATION LIFE PREDICTION UNDER COUPLED THERMAL-MECHANICAL CYCLING

The life prediction process in any structural application seeks to transform applied loads into some measure of damage which can be integrated and compared with some appropriate failure criterion. The term applied loads is used here in a generic sense. It infers the action of factors such as mechanical force or displacement, chemical (aggressive) environment, temperature, etc., which serve to degrade the structure's utility. For the present paper it infers the action of coupled thermal-mechanical loading.

The degradation process under thermal-mechanical loading might involve burn through or gross yielding, or crack initiation and propagation. Typically because burn through and gross yielding can be adequately guarded against during the design phase, thermal-fatigue crack initiation and propagation is the predominating degradation-mechanism under coupled thermal-mechanical loading. In general applications, both initiation and growth may be of concern. This paper focuses on the crack initiation stage because design against crack initiation is most desirable in gas furnace heat exchanger applications. Also, the visual appearance of cracking was used as a failure criterion in actual cyclic testing of heat exchangers.

General Approach to Life Prediction

Given that the scope of the paper is restricted to crack initiation life prediction under thermal-mechanical loading, analysis of the degradation process follows as outlined in Figure 1. Evident in this figure are the two transformations that must be made in mapping far-fielding thermal-mechanical loads into damage at critical areas in a structure.

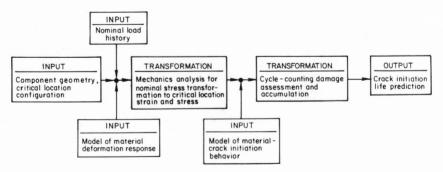

Fig. 1. Essential features in the direct fatigue-analysis procedure

The first transformation entails the determination of critical areas in the structure. This determination requires the results of appropriate stress analysis which are applied at each reversal of the loading in the context of the deformation theory of plasticity. Such a calculation gives rise to mechanical stresses and strains for each cycle of the loading. If a thermal component is involved in the loading, it must be embodied in the stress analysis so that the computed mechanical stresses and strains reflect the influence of the thermal cycle.

The second transformation assesses the damage done in terms of the ranges of these mechanical stresses and strains using material fatigue resistance data that reflect the damage process in the structure. For example, for thermal-mechanical cycling the material's data used should match the component's thermal-mechanical cycle. Alternately, data developed under other conditions could be used if the difference between the damage rate process under those conditions could be analytically matched to that in the structure. This analytical match is provided by a "damage parameter", equal values of which mean equal damage rates. For the case at hand, a damage parameter will be introduced to match uniaxial isothermal fatigue resistance data with the biaxial thermal-mechanical loading within the heat exchanger. This damage parameter will be detailed after the input data necessary to make the life prediction are discussed.

Material's Reference Data Base

As identified in Figure 1, two basic forms of materials data are required to make a crack initiation fatigue life prediction. The first of these pertains to deformation response whereas the second describes the material's fatigue resistance. Because of complications in experimentation, these reference data are developed using uniaxial samples tested under mechanical strain control with either isothermal or non-isothermal conditions. Because the heat exchanger analyzed as the illustrative example in the next section is made of AISI 1010 mild steel, the present section deals with this material. Furthermore, because the necessary reference data base available in the literature was inadequate, the data to be presented have been developed specifically for the purpose of the heat exchanger analysis. Finally, because of the ease in developing isothermal versus thermal-mechanical data, both types of data have been developed with the emphasis on the isothermal case, and ways of relating these data have been examined.

Deformation response. The relationship between stable stress amplitude, $\Delta\sigma/2$, and strain amplitude $\Delta\varepsilon/2$, the control parameter for the tests reported, is shown in Figure 2. It is evident that the cyclic stress-strain response for isothermal tests was considerably different than that for thermal-mechanical tests. In applications where inelastic stresses or strains are calculated appropriate,

stress-strain relationships should be used. In general, these relationships should account for cyclic hardening and/or cyclic soften
ing as a function of strain level, strain rate, temperature level,
and number of strain reversals. However, development and applicatio
of such relationships on a reversal-by-reversal basis throughout the
loading history is often quite complex and costly. For use in
fatigue analysis, cyclic stress-strain curves for half-life (approxi
mately stable) conditions are usually reasonable approximations to
use in the calculation of stress or strain. When thermally cycled
heat exchangers are being analyzed, the thermal-mechanical stress-
strain curves such as those shown in Figure 2, are appropriate.

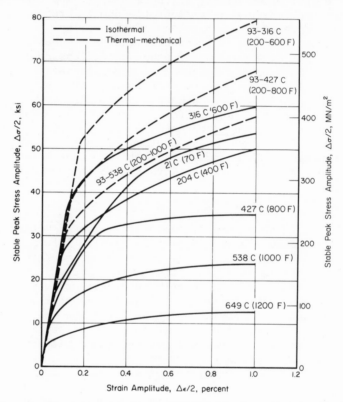

Fig. 2. Cyclic stress-strain response for AISI 1010 steel.

Note that the experiments used to develop stress-strain data
usually employ uniaxial smooth specimens while most practical applications involve multiaxial states of stress and strain. In order to
employ these data in multiaxial situations, the following equivalenc
criteria are introduced.[1] In two dimensions, the equivalent stress
range, $\overline{\Delta\sigma}$, is related to the principal stress ranges, $\Delta\sigma_1$ and $\Delta\sigma_2$, b

$$\overline{\Delta\sigma} = \left[\Delta\sigma_1^2 - \Delta\sigma_1\Delta\sigma_2 + \Delta\sigma_2^2\right]^{1/2} \quad . \tag{1}$$

Similarly, equivalent strain range, $\overline{\Delta\varepsilon}$, is related to the principal strain ranges, $\Delta\varepsilon_1$, $\Delta\varepsilon_2$ and $\Delta\varepsilon_3$, by

$$\overline{\Delta\varepsilon} = \frac{1}{\sqrt{2(1+\nu)}}\left[(\Delta\varepsilon_1-\Delta\varepsilon_2)^2 + (\Delta\varepsilon_2-\Delta\varepsilon_3)^2 + (\Delta\varepsilon_3-\Delta\varepsilon_1)^2\right]^{1/2} \qquad (2)$$

where ν is Poisson's ratio. Mendelson and Manson[2] have shown that stress-strain data such as that shown in Figure 2 can be related by an equation of the form

$$\overline{\Delta\varepsilon_T} = \overline{\Delta\varepsilon_p} + \frac{2(1+\nu)\overline{\Delta\sigma}}{3E} \qquad (3)$$

where $\overline{\Delta\varepsilon_T}$ is the total equivalent strain range, $\overline{\Delta\varepsilon_p}$ is the plastic equivalent strain range, and E is the elastic modulus. As true for most materials, it was found that the plastic strains $\overline{\Delta\varepsilon_p}$ could be related to stress by a power function of the form

$$\overline{\Delta\sigma} = K(\overline{\Delta\varepsilon_p})^n \quad , \qquad (4)$$

where K is the strength coefficient and n is the strain hardening exponent. The values of these constants for the data shown in Figure 2 are listed in Table 1. The regime of approximately elastic cyclic stress-strain response is defined by the elastic limit stress, σ_0, and strain, ε_0, values.

Table 1. Constants for Thermal-Mechanical Cyclic Stress-Strain Amplitude ($\Delta\sigma$ Versus $\Delta\varepsilon_p$) Curves for AISI 1010 Steel

Limits of Temperature Cycle, °C	E,[a] GPa (ksi)	ε_0, percent	σ_0, MPa (ksi)	n	K, MPa (ksi)	C
93-316	200 (29,000)	0.176	352 (51.0)	0.270	3999 (580)	0.0173
93-371[b]	197 (28,500)	0.154	303 (44.0)	0.239	2861 (415)	0.0126
93-427	193 (28,000)	0.136	262 (38.0)	0.207	2275 (330)	0.0102
93-482[b]	190 (27,500)	0.125	236 (34.3)	0.179	1724 (250)	0.0079
93-538	186 (27,000)	0.115	214 (31.0)	0.151	1372 (199)	0.0064
93-593[c]	184 (26,700)	0.105	193 (28.0)	0.151	1138 (165)	0.0054
93-649[c]	182 (26,400)	0.0958	174 (25.3)	0.151	931 (135)	0.0044

(a) Determined from average tensile values at mean temperature of cycle.
(b) Values of σ_0 and K were interpolated on a log-linear basis with reference to maximum temperature and values of n were interpolated on a linear basis with reference to maximum temperature.
(c) Values of σ_0 and K were extrapolated on a log-linear basis with reference to maximum temperature. Values of n were not extrapolated as there is not basis for doing so.

Fatigue resistance and damage parameter. The isothermal data
were reported in[3] whereas the thermal-mechanical data were in[4]. Wi
regard to the isothermal data, tensile properties, cyclic stress
strain behavior, and thermal-expansion behavior were considered in
addition to fatigue life. Fatigue curves were developed at 21, 204
316, 427, 538, and 649°C, for lives between 10^3 and 10^6 cycles to
failure. These data were obtained from constant-amplitude, fully
reversed, strain cycling tests of axially loaded specimens. Therma
mechanical fatigue behavior was investigated for constant-amplitude
fully reversed strain cycling of uniaxially loaded specimens at thr
ranges of temperature — (1) 93-316°C, (2) 93-427°C, and (3) 93-538°
The experiments were conducted either with maximum strain in phase
with maximum temperature or with it out of phase with maximum temperature, at strain rates that compare reasonably with that which
occurred in the heat exchangers.

Detailed comparisons have been made between the measured fatig
lives for isothermal conditions and for the thermal-mechanical conditions. No acceptable correlation could be found between the results for the two types of tests when they are compared on the basi
of inelastic strain ranges, as evident in Figure 3. Further, it ha
been demonstrated that using isothermal tests to predict the therma
mechanical fatigue life at the same strain level is very nonconserva
tive since for the same strain range the isothermal fatigue life is
significantly longer than the thermal-mechanical fatigue life.

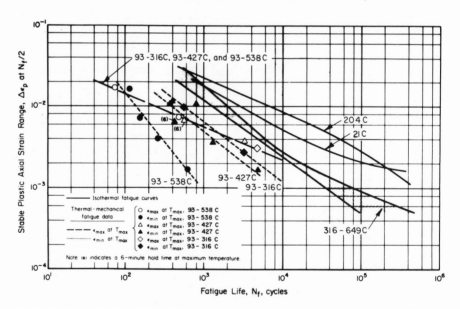

Fig. 3.   Comparison of thermal-mechanical and isothermal fatigue
          behavior of AISI 1010 steel based on plastic strain range.

It is found that a reasonably consistent and conservative correlation between the two types of tests could be obtained when they are compared on the basis of the cyclically stable stress amplitude. The stable stress amplitude is that cyclic stress induced by the applied cyclic strain at a number of cycles equal to one half the fatigue life. This improved correlation, when compared to that for plastic strain, may be artificial in that at low lives fatigue resistance is less sensitive to stress than it is to inelastic strain. Also, the degree of correlation depends on the definition of an appropriate effective test temperature for the thermal-mechanical tests.

In establishing the correlation, the mean equivalent test temperature, T_e, defined by the relations

$$T_e = T_{max} \qquad\qquad T_{max} < 316°C \qquad ,$$

and

$$T_e = \frac{T_{max} + 316}{2} \qquad 316°C < T_{max} \leq 649°C \qquad (5)$$

has been employed. Here T_{max} is the maximum temperature reached in the thermal cycle.

For purposes of comparison, the isothermal and thermal-mechanical fatigue lives, N_f, are plotted against cyclically stable stress amplitude, $\Delta\sigma/2$, in Figure 4. Examination of this figure shows that, on the basis of stable stress amplitude, the isothermal fatigue curves serve as a lower bound for the fatigue life of AISI 1010 steel under thermal-mechanical cycling. Close inspection of this figure reveals that the average fatigue life for a thermal-mechanical cycle, at a given cyclic stress range and effective temperature, differs from the corresponding isothermal fatigue life by a factor of about 3.3.

The curves in Figure 4 are easily represented in equation form since the fatigue life data lie along approximately straight lines on the log-log plot. Thus,

$$b \log N_f = \log \frac{\Delta\sigma}{2} - \log A$$

or

$$N_f = \left(\frac{\Delta\sigma}{2}/A\right)^{1/b} \qquad , \qquad\qquad (6)$$

where A and b are constants that vary with temperature, $\Delta\sigma/2$ is the stable stress amplitude and N_f is the fatigue life. Values for A and b determined from the isothermal curves in Figure 4 are listed in Table 2.

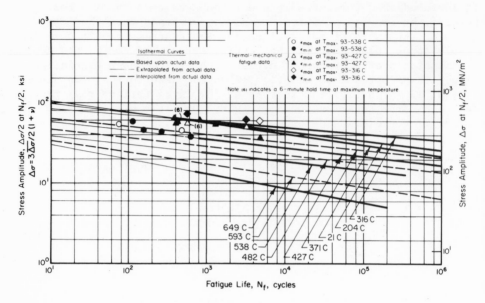

Fig. 4.   Comparison of thermal-mechanical and isothermal fatigue
          behavior of AISI 1010 steel based on stress amplitude

Values for the constants A and b at intermediate temperatures
can be determined from Table 2 by logarithmic and linear interpolation, respectively, between adjacent table temperatures.   It
should be kept in mind that the temperature to be used here is the
test temperature for isothermal tests and the effective temperature
(given by Equation (5)) for thermal-mechanical tests.

To show how Table 2 is used, assume T_e = 149°C.   Then from Table
2:

$$\log A_{149} = \log 979 - \frac{149-21}{204-21} \left[\log 979 - \log 862\right]$$

or

$$A_{149} = 895.6 \text{ MPa (129.9 ksi)}$$

and

$$b_{149} = -0.135 + \frac{230}{330} (0.135 - 0.115) = -0.121 \qquad .$$

Table 2.   Constants for Isothermal, Stable Stress Amplitude/Fatigue
           Fracture Life Relationship for AISI 1010 Steel

Temperature, °C	A, MPa (ksi)	b
21	979 [142]	-0.135
204	862 [125]	-0.115
316	645 [93.5]	-0.073
371[a]	571 [82.8]	-0.090
427	506 [73.4]	-0.106
482[b]	412 [59.8]	-0.106
538	337 [48.9]	-0.106
593[c]	323 [46.9]	-0.142
649	312 [45.3]	-0.180

(a) Curve obtained by taking the loga-
     rithmic average of A and the arith-
     metic average of b for the 316°C
     and 427°C curves.
(b) Curve obtained by taking the loga-
     rithmic average of A and the arith-
     metic average of b for the 427 and
     538°C curves.
(c) Curve obtained by taking the loga-
     rithmic average of A and the arith-
     metic average of b for the 538 and
     649°C curves.

      In the context of a life prediction, the ordinate on each of
Figures 3 and 4 serves as a damage parameter. In the case of Figure
3, which uses plastic strain range as an ordinate, the multiaxial
equivalence of the uniaxial value is numerically equal to the uniaxial value. As such, the ordinate on Figure 3 serves equally well
in both uniaxial and multiaxial problems. With reference to Figure
4, note that the ordinate is expressed as a uniaxial value along with
the relationship necessary to obtain the equivalent multiaxial
stress. Because the stress-based approach provides the best and
most consistently conservative (from design viewpoint) correlation
of the currently available parameters, including more complex forms
such as products of stress and strain, equivalent stress is adopted
for use as a damage parameter.

Estimating Thermal-Mechanical Fatigue Resistance

      As noted in the introduction, details of the stress-strain field
in the heat exchanger have been evaluated experimentally in terms of
strain and analytically in terms of stresses. Because of this,
techniques to estimate fatigue resistance of the heat exchanger
independently in terms of both stresses and strains are needed.

Predicting fatigue life from calculated stresses. Thermal
elastic stress analysis were used to compute principal stresses.
Valid applications of these results would, therefore, be restricted
to situations where the stresses are elastic. Under these circumstances, fatigue life can be determined directly from Figure 4 for a
stable stress amplitude calculated by using an equivalent stress
range computed from Equation (1).

In situations where the equivalent stress range exceeds twice
the proportional limit stress, σ_0, listed in Table 1, use of Figure
would predict a fatigue life considerably shorter than expected. To
account for this overconservatism, it is necessary to introduce the
material stress-strain behavior to estimate a more realistic value o
stress amplitude. With reference to Figure 5, such a value can be
obtained by entering the appropriate inelastic stress-strain curve
at a strain equal to $\Delta\sigma/2E$. To use the data shown in Figure 2,
strain must be converted to percent. A note of caution: this procedure assumes that the computed elastic strain does not differ
significantly from that under inelastic action (the strain equivalence principal). Thus, it should be employed only when this assumption is reasonable. Life can now be predicted using the data in
Figure 4.

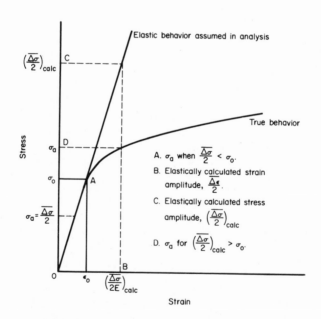

Fig. 5.   Illustration of the calculations of σ_a when the stress
         amplitude exceeds the elastic limit.

Predicting fatigue life from measured strains. The strain state in thin shells, such as heat exchangers, is three dimensional. In the present program, principal strains (ε_1 and ε_2) were measured on the surface of the shell so that the strain through the thickness (ε_3) had to be calculated. When the principal strain directions are not known, it is possible to determine the principal strains using a strain-gage rosette[5].

In order to calculate fatigue life from experimentally determined strains, equivalent strain must first be computed from the two measured principal strains and the through thickness principal strain determined from material deformation behavior. After determining the equivalent strain, equivalent stress used in life prediction can be easily obtained from the material equivalent stress–strain curve such as that shown in Figure 2.

When the principal strain components are elastic, the through thickness principal strain is directly related to the in-plane principal strains by Hooke's Law. Utilizing the usual assumption that for thin shells, $\sigma_3 = 0$, and the relationship between principal strains, Equation (2) becomes

$$\overline{\Delta\varepsilon} = \frac{1}{\sqrt{2(1-\nu^2)}} \left[(1-\nu)^2(\Delta\varepsilon_1-\Delta\varepsilon_2)^2 + (\nu\ \Delta\varepsilon_1+\Delta\varepsilon_2)^2\right.$$
$$\left. + (\Delta\varepsilon_1+\nu\ \Delta\varepsilon_2)^2\right]^{1/2} \quad . \tag{7}$$

This gives the effective strain range directly in terms of the in-plane principal surface strains measured by the strain gages. For predicting the fatigue life of a heat exchanger, it is clear that the strains must be measured at the two times that give the maximum value of $\overline{\Delta\varepsilon}$. When the prinicipal strains are in phase (or directly out-of-phase) with each other as shown schematically in Figure 6, it is not difficult to determine the maximum strain ranges. However, if there is some arbitrary phase relationship between ε_1 and ε_2, it is necessary to compute the strains at a number of points in the heating cycle and determine the maximum value of $\overline{\Delta\varepsilon}$ from plots of ε_1 and ε_2.

Continuing within the context of elastic strains, the general material stress-strain response given by Equation (3) simplifies to

$$\overline{\Delta\sigma} = \frac{3E\ \overline{\Delta\varepsilon}}{2(1+\nu)} \quad , \tag{8}$$

where $\overline{\Delta\varepsilon}$ is the measured equivalent strain range (in inch/inch). The value of $\overline{\Delta\sigma}$ given by Equation (8) is then used in conjunction with Figure 4 to predict life. Note that Figure 4 is expressed in terms of uniaxial stress amplitude, $\Delta\sigma/2$, whereas Equation (8) is expressed in terms of effective stress range $\overline{\Delta\sigma}$. Calculation of fatigue life

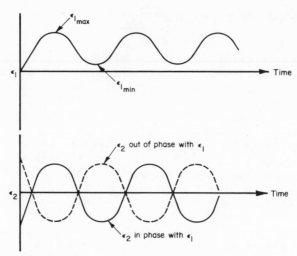

Fig. 6. Schematic of relationships between principal strains.

thus requires the use of the relationship between $\Delta\sigma$ and $\overline{\Delta\sigma}$ which is given on the ordinate of Figure 4.

When the equivalent strain range, as calculated under elastic assumptions (Equation 7), exceeds the proportional strain range $(2\varepsilon_0)$, the principal strains are no longer explicitly related by Hooke's Law. The relationship between the principal strains can, however, be determined using an iterative technique given by Mendelson and Manson[2].

Essentially, the technique consists of determining elastic and plastic components of the total equivalent strain which satisfy Equation (3):

$$\overline{\Delta\varepsilon_T} = \overline{\Delta\varepsilon_p} + \frac{2}{3}(1+\nu)\frac{\overline{\Delta\sigma}}{E}$$

Note that in the equation $\overline{\Delta\sigma}$ is a nonlinear function of $\overline{\Delta\varepsilon_p}$. A form of Equation (3) which is more convenient for computation is obtained by substituting Equation (4) into Equation (3) to eliminate $\overline{\Delta\sigma}$ which results in

$$\overline{\Delta\varepsilon_T} = \overline{\Delta\varepsilon_p} + \frac{2(1+\nu)K}{3E}(\overline{\Delta\varepsilon_p})^n \qquad (9)$$

This last relation provides a form convenient for an iterative solution when it is rewritten in the form

$$\left(\overline{\Delta\epsilon_p}\right)_i = \overline{\Delta\epsilon_T} - C\left[\left(\overline{\Delta\epsilon_p}\right)_{i-1}\right]^n \quad , \quad i = 1,\ldots,j \qquad , \qquad (10)$$

where the constant C is given by

$$C = \frac{2(1+\nu)K}{3E} \quad . \qquad (11)$$

A suitable initial value of the equivalent plastic strain is obtained by assuming that the equivalent plastic strain is equal to the equivalent total strain. The values of C and n used in Equation (10) are also given in Table 1.

The application of the iterative technique results in a relationship between the equivalent total strain $\overline{\Delta\epsilon_T}$ and equivalent plastic strain $\overline{\Delta\epsilon_p}$, which can then be used in conjunction with Figure 3 to determine a strain-based prediction of fatigue life. Alternately, the equivalent plastic strain $\overline{\Delta\epsilon_p}$ can be used with Equation (4) to calculate the corresponding equivalent stress range to provide a stress-based prediction of life using Figure 4.

As an example of the procedures described above, consider the case where $T_{max} = 427°C$, ϵ_1 and ϵ_2 are in phase, and the measured values are

$$\epsilon_{1_{max}} = -100 \times 10^{-6}$$
$$\epsilon_{1_{min}} = -3000 \times 10^{-6}$$
$$\epsilon_{2_{max}} = 200 \times 10^{-6}$$
$$\epsilon_{2_{min}} = -2000 \times 10^{-6} \quad .$$

The plastic strain ranges $\Delta\epsilon_{1p}$, $\Delta\epsilon_{2p}$ are then equal to 2900 x 10^{-6} and 1800 x 10^{-6}, respectively. These values in Equation (7) yield

$$\overline{\Delta\epsilon} = \frac{\sqrt{2}}{3}(1100^2 + 6500^2 + 7600^2) \times 10^{-6} = 4743 \times 10^{-6} \quad .$$

This value of strain is then taken to be approximately the total strain range, $\overline{\Delta\epsilon_T}$, in Equation (10). The true value of the plastic strain range, $\overline{\Delta\epsilon_p}$, is then the solution of the equation

$$0.004743 = \overline{\Delta\epsilon_p} + 0.0102\,(\overline{\Delta\epsilon_p})^{0.207} \quad ,$$

which is satisfied by $\overline{\Delta\epsilon_p} = 0.00194$. The strain-based life is then found from Figure 3 to be approximately 2500 cycles. The equivalent stress range is found from Equation (4) as

$$\overline{\Delta\sigma} = 2275\,(0.00194)^{0.207} = 625 \text{ MPa (90.6 ksi)}$$

which corresponds to a stable stress amplitude $\Delta\sigma$ = 717 MPa (104 ksi). The stress-based life is then seen to be $\sim$300 cycles from the isothermal data of Figure 4. Also,

$$N_f = \left[\frac{\left(\frac{717}{2}\right)}{571}\right]^{-11.2} = 183 \text{ cycles} \quad ,$$

from Equation (6). The values of A = 571 MPa (82.8 ksi) and 1/b = -11.2 were obtained from Table 2 and correspond to an effective temperature, T_e, of 371°C. The order of magnitude difference in the strain-and stress-based life predictions is not too surprising in light of Figures 3 and 4. Figure 3 shows little sensitivity in the life for a small change in strain amplitude whereas Figure 4 shows that the life is very sensitive to small changes in the stress amplitude. Also the strain-based prediction uses the actual thermal-mechanical fatigue resistance from Figure 3, whereas the stress-based prediction uses the isothermal curve at 371°C in Figure 4. Close inspection of Figure 4 shows that the thermal-mechanical data for T_{max} = 427°C (triangular symbols) fall about a factor of 10 in cycles above the 371°C isothermal curve near $\Delta\sigma$ = 717 MPa (104 ksi).

Flowchart of fatigue — life estimation procedures. In order to simplify the implementation of the discussed procedures, the flow-chart of Figure 7 was prepared. It shows the various paths that can be taken and refers to specific equation numbers in the text at each point. The entire process can be easily computerized for making repetitive calculations at many points throughout a heat exchanger. Such calculations provide useful quantitative values for assessing weak spots in the unit. The accuracy of the actual life prediction is subject to errors introduced by scatter in material behavior, errors in measurement and calculations, manufacturing and installa-tion variations, and possible environmental degradation. Since these items reflect areas of ignorance, they are usually treated by apply-ing a safety factor to the predictions in the design process. This factor can be applied to either the stress or life, realizing that a safety factor of 2.0 on stress is equal to one of 38.4 on cyclic life for a fatigue curve with slope 1/b = -9 (see Table 2).

Note that the flowchart in Figure 7 represents in detail the approach to fatigue life prediction shown in Figure 1, as applied in an application to thermal-mechanical fatigue analysis. With refer-ence to Figure 1, the analysis to this point pertains to damage assessment given stress and strain at critical locations in the heat exchanger. The next section discusses how these stresses and strains are obtained and details the use of the approach in an application to an actual heat exchanger.

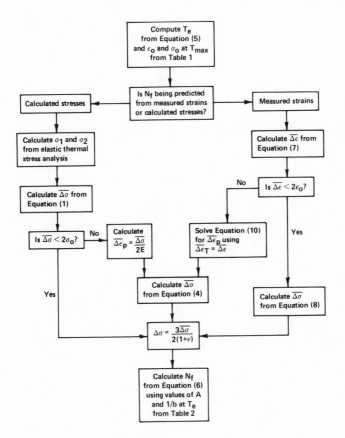

Fig. 7. Flowchart for calculating N_f.

APPLICATION OF THE APPROACH

The life prediction approach presented uses either measured strain values, calculated strains, or calculated stresses. In the application which follows, the approach is illustrated by its application in the context of strains which were measured, along with temperature profiles, the latter for use in thermal stress analyses. This allowed a comparison of life to first observed cracking with predictions made from calculated and measured strains.

A clam-shell heat exchanger used in some domestic gas furnaces
was instrumented with strain gages and thermocouples as shown in
Figure 8. The thermocouples were spot welded to the shell in a two-
inch grid pattern.

Fig. 8. Heat exchanger showing thermocouple and strain-gage
         placement.

Gas was burned inside the lower portion of the shell using a
standard commercial burner, with exhaust gasses exiting the top. The
test performed on this exchanger was to burn gas for 180 seconds at
110 percent of the rated 43,960 J/sec (150,000 BTU/hr) input and
then to run the furnace fan for 215 seconds with the burner off.
There was no period of combined burner and fan operation because
other work showed that omitting such a period does not significantly
alter the thermal cycle. This heating/cooling cycle was repeated

until a crack was observed in the heat exchanger. The presence of a crack was based upon periodic visual inspections of the exchanger shell. Two exchangers were cycled to failure. One unit cracked between about 600 and 1200 cycles and the other cracked between about 1400 and 1700 cycles. Thus, the observed spread in experimental cyclic life was about 600 to 1700 cycles.

During the cyclic test, temperatures and strains were measured at various times during the 395 second cycle. As an example of the type of data obtained, Figure 9 shows the cyclic response of the strain gage nearest the area that cracked along with the temperature at that location for a somewhat shorter cycle of 280 seconds. Figure 10 shows the temperature profile that was measured after 240 seconds of heating. The peak temperatures for this heating period were about 28°C (50°F) higher than those measured at 180 seconds (the heating period of the cyclic tests). The strains and stresses at this time (240 seconds) are the extreme values and were slightly (<10 percent) larger than those at 180 seconds.

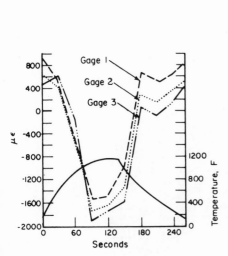

Fig. 9. Recorded strains and temperature.

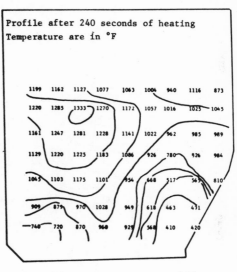

Fig. 10 Temperature profile used in elastic analysis.

To calculate thermal stresses the computer code SNAP[6], a shell finite element code, was employed. The loading was the appropriate measured thermal distribution referenced to an assumed stress free temperature of 27°C. Because the clam shell heat exchanger is so geometrically complex, only a partial model could be accommodated by the computer. This partial model, with interpolated temperatures, is shown in Figure 11. The boundary conditions for this model consisted

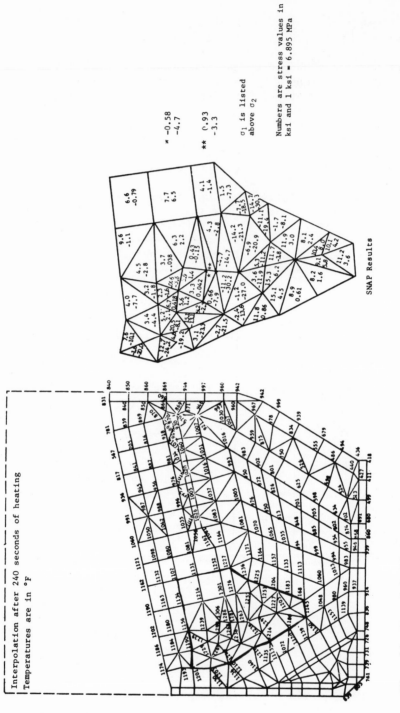

Fig. 11. Partial finite-element model with interpolated temperature model.

Fig. 12. Calculated principle stresses in the critical location of the model after about 240 seconds of heating.

of (a) symmetry conditions along the edges joining the two halves, (b) clamping conditions on the edges to prevent in-plane motion, and (c) no restraint on the upper boundary of the finite element model. The principle stresses calculated in the critical region of the shell are shown in Figure 12.

Predicted cyclic life of the heat exchanger was calculated in two ways — (1) from experimentally measured strains after 240 seconds of heating and (2) from elastically calculated stresses after 240 seconds of heating. The procedures described earlier (see Figure 7) were followed in making these calculations of cyclic life. From measured strain data, a cyclic life of about 1500 cycles was predicted, which is in good agreement with the experimental results of 600 to 1700 cycles. It should be pointed out that the equivalent temperature procedure was experimentally correlated up to maximum temperatures of only 538°C, whereas the peak temperatures in this case were near 650°C. Thus, its use in this case was an extrapolation that has not been strictly correlated at these high temperatures. A life of 60,000 cycles was predicted from the calculated stresses. This is longer than observed experimentally, because the stresses calculated assuming very simplified boundary conditions in modeling this geometrically complex exchanger configuration do not reflect those in the actual heat exchanger. Also, significant inelastic action did take place under the imposed experimental conditions, action which is inadequately accounted for using the confined plasticity assumptions inherent in the approach of Figure 7. An elastic-plastic finite element analysis would have been required to account this factor more fully. However, the elastic analysis did identify a region of high stress that was quite close to where failure occurred. Therefore, it could be used to give qualitative indications of potential trouble spots even though the quantitative life predictions are overly optimistic.

ACKNOWLEDGEMENT

The authors acknowledge the American Gas Association, Inc. for their support of this work.

REFERENCES

1. Manson, S. S., "Thermal Stress and Low-Cycle Fatigue", McGraw-
      Hill Book Company (1966).
2. Mendelson, A., and Manson, S. S., "Practical Solution of Plastic
      Deformation Problems in Elastic-Plastic Range", NASA Technica
      Report T.R. R-28 (1959).
3. Jaske, C. E., "Low-Cycle Fatigue of AISI 1010 Steel at Tempera-
      tures Up to 1200 F (649 C)", J. of Pressure Vessel Tech., 99,
      (3), (1977) pp 423-443.
4. Jaske, C. E., "Thermal-Mechanical, Low-Cycle Fatigue of AISI 101
      Steel", Thermal Fatigue of Materials and Components, ASTM STI
      612, American Society for Testing and Materials, Philadelphia
      (1979) pp 170-198.
5. Perry, C. C., and Leisner, H. R., "The Strain Gage Primer",
      Second Edition, McGraw-Hill Book Company (1962).
6. Whetstone, W. D., "SNAP — Structural Network Analysis Program",
      (December 1970) Lockheed Missiles and Space Company, LMSC-HRI
      D 162812.

Table 1. Constants for Thermal-Mechanical Cyclic Stress-Strain Amplitude ($\Delta\sigma$ Versus $\Delta\epsilon p$) Curves For Aisi 1010 Steel

Limits of Temperature Cycle, °C	E, (a) GPa (ksi)	ϵ_0, percent	σ_0, MPa (ksi)	n	K, MPa (ksi)	C
93-316	200 (29,000)	0.176	352 (51.0)	0.270	3999 (580)	0.0173
93-371(b)	197 (28,500)	0.154	303 (44.0)	0.239	2861 (415)	0.0126
93-427	193 (28,000)	0.136	262 (38.0)	0.207	2275 (330)	0.0102
93-482(b)	190 (27,500)	0.125	236 (34.3)	0.179	1724 (250)	0.0079
93-538	186 (27,000)	0.115	214 (31.0)	0.151	1372 (199)	0.0064
93-593(c)	184 (26,700)	0.105	193 (28.0)	0.151	1138 (165)	0.0054
93-649(c)	182 (26,400)	0.0958	174 (25.3)	0.151	931 (135)	0.0044

(a) Determined from average tensile values at mean temperature of cycle.
(b) Values of σ_0 and K were interpolated on a log-linear basis with reference to maximum temperature and values of n were interpolated on a linear basis with reference to maximum temperature.
(c) Values of σ_0 and K were extrapolated on a log-linear basis with reference to maximum temperature. Values of n were not extrapolated as there is not basis for doing so.

Table 2. Constants for Isothermal, Stable Stress
Amplitude/Fatigue Fracture Life Relationship for Aisi 1010 Steel.

Temperature, °C	A, MPa (ksi)	b
21	979 [142]	-0.135
204	862 [125]	-0.115
316	645 [93.5]	-0.073
371(a)	571 [82.8]	-0.090
427	506 [73.4]	-0.106
482(b)	412 [59.8]	-0.106
538	337 [48.9]	-0.106
593(c)	323 [46.9]	-0.142
649	312 [45.3]	-0.180

(a)   Curve obtained by taking the logarithmic average of A and the arith-
metic average of b for the 316°C
and 427°C curves.
(b)   Curve obtained by taking the logarithmic average of A and the arith-
metic average of b for the 427 and
538°C curves.
(c)   Curve obtained by taking the logarithmic average of A and the arith-
metic average of b for the 538 and
649°C curves.

STATISTICAL FRACTURE ANALYSIS OF BRITTLE MATERIALS IN THERMALLY

STRESSED COMPONENTS

G. G. Trantina

Corporate Research and Development
General Electric Company
Schenectady, New York 12301

ABSTRACT

Thermal stress fracture in brittle materials is treated with a Weibull statistical analysis technique. The probability of failure and size effect is predicted by combining a risk analysis with finite element heat transfer and stress analysis. In a thermal transient the maximum probability of failure can occur at times greater than the time of maximum thermal stress. In many situations the thermal stress in a structure increases with size while, due to the size effect, the strength of the structure decreases. Thermal shock tests of silicon carbide and alumina demonstrate the scatter in the fracture strength. The average fracture strength of alumina is well-predicted, but the strength of silicon carbide disks is apparently affected by a nonhomogeneous flaw population. Applications of the statistical analysis technique to thermal stress situations in gas turbine vanes and combustors are reviewed.

INTRODUCTION

Structural ceramic materials such as silicon carbide and silicon nitride are being considered for gas turbine power plants. Much of this effort has been expended on automotive and stationary gas turbine engines while some effort has been devoted to aircraft gas turbine applications[1]. In these applications ceramic materials offer the potential of high temperature capability and low cost. However, unique design techniques are required to account for the brittle nature of these materials and the statistical distribution of defects that control the strength. A number of investigators[2-6] have explored various aspects of statistical design of ceramic

229

components.

Fracture induced by transient or steady state thermal stresses represents one of the most important concerns with structural ceramic materials in gas turbines. Hasselman[7] and others have extensively studied the fracture behavior of ceramic materials in thermal stress environments. A complete consideration of the statistical fracture of brittle materials in thermally stressed components involves finite element analysis, verification with thermal stress fracture experiments, and application to components. The objectives of this paper are to describe (1) a thermal stress fracture analysis where finite element analysis is used to predict size effects and probability of failure, (2) thermal shock fracture of silicon carbide and alumina disks and (3) the application of the analysis technique to various gas turbine components.

THERMAL STRESS FRACTURE ANALYSIS

Ceramic materials are brittle, and if the strength of many geometrically identical specimens were to be measured by thermal shock, it would be discovered that considerable scatter exists in the observed strength. The reason for this scatter is to be found in the scatter in the size of the small material flaws that control the strength of these materials. The scatter in observed strength of ceramics has significant implications in thermal stress fracture analysis. The scatter implies a statistical nature in the structural strength. An expression of this statistical nature is found through Weibull's weakest link concept. The important point that will come out of the analysis is that there is a size effect in ceramic structures: small stressed volumes appear to have higher strength than larger stressed volumes.

To account for the effects of size and stress distribution in thermal stress fracture analysis of brittle materials, the failure mechanism of these materials must, first of all, be characterized. The strength of ceramic materials depends on the stress required to propagate small inherent flaws which are distributed throughout the ceramic material. The fracture stress, σ_f, is related to the flaw size, a, for situations where no subcritical crack growth occurs, by the relation

$$\sigma_f = K_c/Y\sqrt{a} \tag{1}$$

where K_c is the value of the stress intensity factor required to propagate a crack and Y is a geometric factor which accounts for the crack shape, orientation, and location. Because of a range of flaw sizes, there is a corresponding variation in strength since K_c is a constant material property. This strength variation can be characterized by a weakest link model due to Weibull[8] where

$$P = 1 - e^{-R} \tag{2}$$

where P is the probability of failure and R is the risk of rupture. R is defined for the two-parameter distribution (σ_o, m) by the integral of the stress, σ, over the volume, V,

$$R = \int (\sigma/\sigma_o)^m \, dV = kV(\sigma_{max}/\sigma_o)^m \tag{3}$$

where σ_o is a normalizing constant, m is the Weibull modulus, k is the load factor, and σ_{max} is the maximum stress in a specimen or a structure. These relationships can be used to correlate the fracture behavior of specimens and structures. The Weibull modulus is a measure of the scatter of the strength distribution, a small m value indicating a large amount of scatter. For example, for a group of 20 specimens with m = 7, the strength of the weakest specimen would be about 1/2 of the strength of the strongest specimen. The Weibull modulus for a material can be determined from the scatter in a set of strength measurements or from the size effect. The load factor, k, is a measure of the uniformity of the stress distribution and is defined from equation (3) as

$$k = \int (\sigma/\sigma_{max})^m \, dV/V \tag{4}$$

For uniform tension, k = 1. Thus, the unique material parameter that is required is the Weibull modulus and the key feature that must be calculated is the load factor. The advantage of using the Weibull statistical approach is its versatility and simplicity. Other variations and approaches have been proposed but extensive verification is necessary before applying these theories.

' A number of aspects of the risk analysis should be considered further:

1. Sometimes, only surface flaws control the fracture of ceramic materials. In these cases, a surface area statistical formulation may provide a more satisfactory result than the volume formulation for materials that fail as the result of surface imperfections. A surface area formulation is the same as the volume formulation presented except surface area is substituted for volume.

2. A two-parameter Weibull distribution function has been used. A three-parameter distribution function can be used where a threshold stress is subtracted from the stress. However, most estimates of the threshold stress are relatively small compared to the mean strength, and, unless

there are physical grounds for an upper limit on the flaw
size, the threshold stress should be assumed to be zero.

3. The maximum principal tensile stress in the structure was
    used since most failure criteria are nearly equivalent to
    this criterion in the first stress quadrant. Only tensile
    stresses are considered since cracks will not tend to
    propagate under compressive loading and, in any case, for
    thermal stresses, the compressive stresses are usually not
    larger than the tensile stresses.

Size Effect and Probability of Failure

The size effect that occurs in ceramic materials due to the
higher probability of encountering a larger critical flaw with increasing volume of stressed material can be illustrated by using
equation (3). For a mean fracture strength, σ_f, and a probability
of failure of 0.5 for a structure with a maximum stress, σ_{max},

$$\frac{\sigma_{max}}{\sigma_f} = \left[\frac{k_s V_s}{k_f V_f}\right]^{-\frac{1}{m}} \tag{5}$$

where σ_f/σ_{max} is similar to a factor of safety and s and f refer to
the structure and fracture specimen, respectively. This size effect is illustrated in Figure 1 where the strength of the structure
(σ_{max}) relative to the fracture strength decreases with increasing
effective volume (area) of the structure relative to the effective
volume (area) of the specimen. For m = 7, a typical value for a
ceramic material, the relative strength is reduced by a factor of
2 when the relative effective volume (area) is increased by a
factor of 100.

Another effect of size must be considered for thermal stress
situations. For constant material properties and thermal shock
conditions (constant h - heat transfer coefficient and constant
ΔT - temperature difference between environmental temperature and
initial temperature of specimen or structure) the thermal stress
increases with size for many geometries. For example, for a disk
with the flat faces insulated, the thermal stress (σ/E α ΔT) where
E is the elastic modulus and α is the coefficient of thermal expansion, increases with the Biot modulus (bh/k) where b is the radius
of the disk and k is the thermal conductivity of the material[9].
Thus, for constant material properties and thermal shock conditions,
the thermal stress increases with the surface area of the faces of
the disk - $(bh/k)^2$ as illustrated in Figure 1. For Biot modulus
values of 1 and 10, the thermal stress increases by a factor of 3
as the area is increased by a factor of 100. For Biot modulus
values of 10 and 100, the thermal stress increases by a factor of
1.6 as the area is increased by a factor of 100.

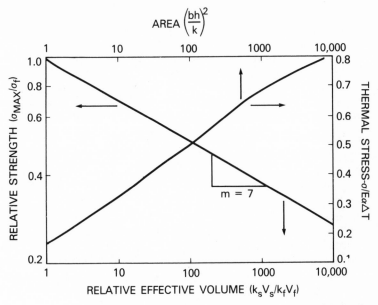

Figure 1.   Strength and thermal stress as a function of size.

The probability of failure (P) due to thermal stresses is influenced significantly by the scatter in the strength distribution.
In order to calculate P, σ_o must be known. By using the approximation that σ_f occurs at P = 0.5, an expression relating σ_f and σ_o
is obtained. For m greater than 3, this introduces a maximum error
of only about 1% and simplifies the relationship for computations.
The probability of failure of a structure is then given by equation
(2) with the risk of rupture expressed as

$$R = \left(\frac{k_s V_s}{k_f V_f}\right)\left(\frac{\sigma_{max}}{\sigma_f}\right)^m \ell n2 \qquad\qquad (6)$$

Finite Element Analysis

In order to determine the precise probability of failure of a
structure subjected to general thermal and mechanical loadings, the
integral $\int \sigma^m \, dV$ or the load factor must be evaluated. Since the
steady state and transient thermal stresses are often calculated by
means of finite element computer programs, it is desirable to incorporate the statistical analysis into the programs so that the stress

and probability calculations can be performed simultaneously and as a function of time.

For a finite element model of a structure, the risk of rupture becomes an integration performed over the element volume ΔV_n and a summation over all the elements in the structure. For a constant strain type element, the integration is easily performed for each element having the value $\sigma^m \Delta V_n$ where the stress is evaluated at the center of the element. However, for a higher order element, the integration has to be performed within the element by some numerical integration scheme. The obvious choice is to use the Gaussian integration rule which already is used in most finite element computer programs, thereby minimizing the necessary changes. Hence once the stresses at the Gauss points have been found, the maximum principal tensile stress at each point has to be raised to the power m and multiplied by the three weight functions and by the Jacobian at that point. The value for the volume integral is obtained by summing over all Gauss points in an element. The risk of rupture is finally obtained by summing the results from all elements Then, at each time step in the transient heat transfer and stress analysis, the probability of failure can be calculated.

This probabilistic finite element analysis was used to calculate the temperatures, stresses, and probability of failure of a silicon carbide (SiC) disk insulated at the flat faces and subjected to a thermal shock by cooling the rim. The experimental results of these thermal shock tests are described in the next section. The 2.8 in. (7.1 cm) diameter disks are heated in a furnace to a uniform temperature, T_o, and then transferred to an ambient fluidized bed. The material properties and heat transfer conditions are summarized in Table 1. The Biot modulus is 0.2. The results of the transient analysis are in good agreement with Reference 9 and are summarized in Figure 2. The temperature at the center relative to the initial temperature (1000°C) decreases with time. The circumferential stress at the rim where the tensile stress is largest reaches a maximum at about 25 sec. The effective volume (m = 7) reaches a maximum - the most uniform stress distribution - at about 50 sec. However, the probability of failure would be a maximum at about 25 sec. since the probability of failure is proportional to $k_s V_s \sigma_{max}^m$ (equation 6) and m is typically around 7. For smaller values of m and larger values of the Biot modulus the time of maximum probability of failure is more than twice the time of maximum stress[10,11].

THERMAL STRESS FRACTURE EXPERIMENTS

Disks of SiC and alumina were thermally shocked with a system that includes a 1700°C furnace with an inside dimension of 8 in. × 8 in. × 6 in. (20.3 cm × 20.3 cm × 15.2 cm) and an ambient temperature fluidized bed. A double door on the bottom of the furnace can be quickly opened with air cylinders allowing a transfer mechanism

Table 1. Material Properties and Heat Transfer Conditions

C - specific heat		$0.3 \dfrac{\text{BTU}}{\text{lb. °F}}$	$0.3 \dfrac{\text{cal.}}{\text{g °K}}$
ρ - density		$0.112 \dfrac{\text{lb.}}{\text{in.}^3}$	3.1 g/cm^3
k - thermal conductivity		$300 \dfrac{\text{BTU-in.}}{\text{hr. ft.}^2 \text{ °F}}$	43 W/m °K
E - elastic modulus		$51 \times 10^6 \text{ psi}$	352 GPa
ν - Poisson's ratio		0.15	0.15
α - coefficient of thermal expansion		$2.7/\text{°F}$	$4.8/\text{°C}$
h - heat transfer coefficient		$60 \dfrac{\text{BTU}}{\text{hr. ft.}^2 \text{ °F}}$	$340 \text{ W/m}^2 \text{ °K}$

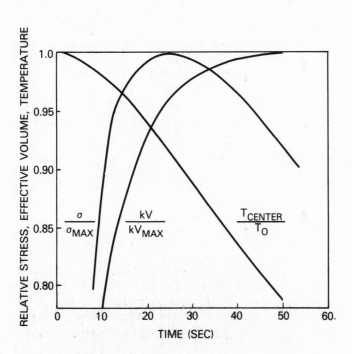

Figure 2. Temperature, stress and effective volume during thermal shock of disk.

to move a specimen or structure from the furnace to the fluidized
bed in less than 1 sec. By adjusting the temperature of the furnace, a thermal shock of varying severity can be applied to a
specimen or structure. The quenching medium is aluminum oxide
fluidized with air. The flat faces of the disk are insulated with
disks of rigid, fibrous zirconia board. A thermocouple is mounted
in contact with the center of a face of the ceramic disk.

Techniques for using fluidized beds for thermal shock testing
have been developed[12]. The heat transfer coefficient of the
fluidized bed is determined by quenching a copper sphere with a
thermocouple located at the center of the sphere. By recording the
temperature decay during the quench the heat transfer coefficient
can be calculated. The computed h is 60 BTU/hr. ft.^{2} °F (340
W/m^2 °K). The measured temperature decrease with time of the
center of the SiC disk agreed well with calculated temperatures
from the transient finite element analysis described in the
previous section.

Ten disks of self-bonded silicon carbide (NC435 - Norton
Company)[13] were tested in thermal shock. The disks (2.8 in. diameter - 0.2 in. thick) were subjected to progressively more severe
thermal shocks by successively increasing the initial temperature
in increments of about 30°C. The failure temperature difference
between the initial temperature and the room temperature bath was
taken as the average of the survival temperature difference and the
successive one at which failure was observed. The ten failed disks
are shown in Figure 3a along with their failure temperature difference (ΔT). The increased degree of cracking and crack branching
with increased ΔT is attributed to an increased amount of stored
energy at fracture. The average fracture strength computed by using
the finite element stress analysis results ($\sigma = 0.056$ E α ΔT) was
10.3 ksi (71 MPa) and the Weibull modulus from the scatter in
fracture strength was about 4. Based on bend strength tests of bars
of two sizes and spin tests of bars and disks (maximum stress at
center of disk) of the same material[14], the strength in thermal
shock should have been more than twice the computed value. The
previous results indicated a consistent value of the Weibull modulus
of about 7 from scatter in the strength values and from the size
effect based on effective area[14]. Also, the fracture mirrors indicated surface flaws but the fracture origin could not be detected[14]
In the thermal shock specimens the fracture origins were obvious
and were subsurface as shown in three examples in Figure 4. Thus,
the lower strength, the lower Weibull modulus, and the nature of
the subsurface failures indicates that the material contained a
nonhomogeneous flaw population with severe subsurface flaws near
the rim of the disks.

Additional thermal shock tests of magnesia-doped alumina[14]
disks (0.4 in. thick and 2.6 in. diameter) were performed. The

Figure 3a.   Thermally shocked silicon carbide disks with ΔT (°C) required for fracture.

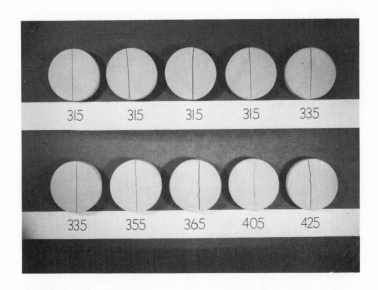

Figure 3b.   Thermally shocked alumina disks with ΔT (°C) required for fracture.

Figure 4.   Fracture surfaces near the rim of three silicon carbide
             disks.

ten failed disks are shown in Figure 3b along with their failure
temperature difference. The average computed[9] fracture strength
(σ = 0.10 E α ΔT) was 14.8 ksi (102 MPa) and the Weibull modulus
was about 8. Using m = 8 and the average fracture strength of 37.1
ksi (256 MPa) for 35 bend specimens[14], the predicted thermal shock
strength of the disks based on the relative effective volumes is
15.4 ksi (106 MPa).

APPLICATIONS TO ENGINE COMPONENTS

       Thermal stress fracture induced by rapid changes in gas
temperature represents one of the most important concerns when
dealing with structural ceramic materials in gas turbines. The
statistical failure analysis has been applied to various engine
components. In this section applications to gas turbine vanes and
combustors will be reviewed.

Vanes

       A three-dimensional steady state and transient thermal analysis
was made for a small (1 in. - 25 mm) ceramic vane. The boundary
conditions were the heat sinks at the inner and outer support lugs
and the gas heat transfer on the airfoil and the platforms (Figure
5). The calculated thermal gradients were approximately the same

for various silicon carbides and silicon nitrides although their
thermophysical properties were somewhat different[1]. For example,
the thermal conductivity varied by a factor of about 3.5. The
thermal gradients were about the same because of the small size of
the ceramic vane. Three dimensional finite element stress analysis
(Figure 5) was conducted by using the temperature conditions at 1
sec. into a severe decrease in gas temperature - the time with the
largest thermal gradients. The thermal stresses were, of course,
proportional to the elastic modulus and the coefficient of thermal
expansion. For the silicon carbide and silicon nitride materials
that were considered, the thermal stresses were low, the stress
distributions were nonuniform (similar to 3-point bending), and
thus the probability of failure was low. The maximum thermal stress
for reaction-bonded silicon nitride was 6.4 ksi.

The critical thermal stress often occurs in the trailing edge
of a gas turbine vane. Two types of specimens have been used to
simulate this thermal shock condition[5]. A ceramic wedge and a
tapered ceramic disk subjected to repeated thermal shock have been
analyzed. Boron-doped sintered silicon carbide wedge and tapered

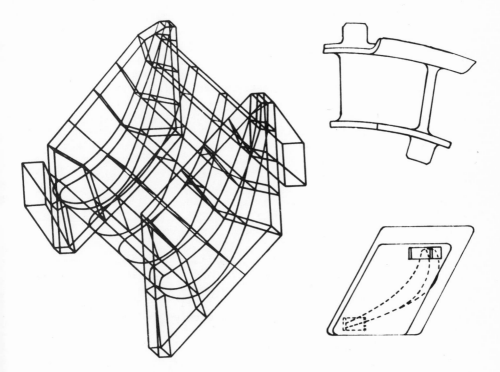

Figure 5.   Ceramic gas turbine vane and three-dimensional finite
            element mesh.

disk specimens have been tested. In the thermal shock test of a
ceramic wedge, the specimen is heated in a furnace and then lowered
into a cold air stream. Surface temperatures of a wedge were measured as a function of time so that effective heat transfer coefficients could be determined from these calibration temperature
measurements. Finite element heat transfer and stress analysis
were performed followed by a statistical risk analysis. The computed maximum stress was 38 ksi (260 MPa) and the predicted probability of failure was 1 in 25. Twelve specimens were tested and
none failed. Similar analysis was performed for the thermal shock
of the tapered disk which is alternately subjected to heating and
cooling shocks in fluidized beds where the heat transfer coefficien
is known from calibration tests. The computed maximum stress was
37 ksi (255 MPa) and the predicted probability of failure was 4 in
10. The higher predicted probability of failure compared to the
wedge specimen is attributed to a more uniform stress distribution.
Thirteen specimens were tested and 4 failed. These predictions
compared reasonably well with the results of the thermal shock
experiments.

Combustors

     Ceramic materials in the gas turbine combustion system offer
the potential of increased operating temperatures and corrosion
resistance. However, the brittle characteristics of ceramics must
be accommodated by unique designs such as hybrid ceramic-metal combustion systems[15-17]. Ceramic materials were used as a low stress,
high temperature liner resiliently mounted within a metal shell
that withstands differential pressure. Conventional air cooled
metal features were retained in the head end where fuel and all air
are introduced. In Figure 6 three types of ceramic combustor liner
are shown. A 3.5 in. (89 mm) diameter stave-type combustor liner
and conical transition piece is shown in Figure 6a along with the
metal shell. In Figures 6b and 6c, 6 in. (152 mm) diameter ceramic
tube and ring combustor liners are shown with their air-cooled
metal head end. The stave and ring combustors are made of Silcomp[TM]
silicon/silicon carbide ceramic composite material[17] and the tube
combustor is made of REFEL silicon carbide[18].

     Thermal stress and probability of failure calculations for
ceramic combustor designs were performed along with material property measurements and laboratory tests of ceramic combustor
liners[16]. The stress levels are relatively low ($\sim$ 6 ksi - 41 MPa)
for an assumed axial and radial temperature gradient in the tube,
ring, and stave 6 in. (152 mm) diameter combustors. The ring design seems to have the greatest potential for scaling to larger
sizes while the stave design is nearly as adequate. However, as
expected, the probability of failure increased for all three
designs because of an increase in volume of stressed material. A
hot streak temperature distribution resulted in stresses in the

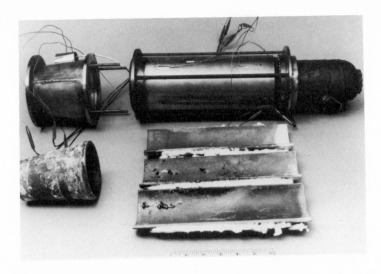

Figure 6a. Ceramic stave combustor liner, ceramic conical
            transition piece and metal shell after testing.

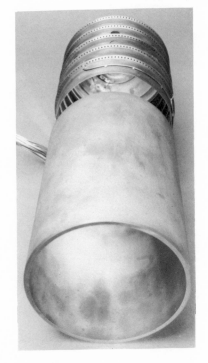

Figure 6b. Ceramic tube   Figure 6c. Ceramic rings
   combustor liner and metal   combustor liner and metal
   head end.   head end.

stave and tube which were three times the stresses in the ring
design.

Bend specimens were cut from one end of a 6 in. (152 mm) diameter, 10 in. (254 mm) long REFEL silicon carbide tube to assure a
representative flaw distribution[16]. Strength, strength scatter,
and strength degradation were evaluated at 2200°F (1200°C) and were
incorporated in the stress analysis. A low probability of failure
was predicted. The REFEL tube was tested in the hybrid ceramic-
metal combustion system at gas temperatures in excess of 2660°F
(1460°C), 16 atm pressure, and 6.9 lb/s (3.1 kg/s) mass flow while
burning No. 2 fuel oil. Successful ceramic performance was demonstrated at 2000°F (1090°C) steady state and at 2000°F (1110°C) per
minute shutdown cooling rate. The stave combustor was successfully
tested to a firing temperature of over 3000°F (1650°C) and through
a series of transients from a ceramic temperature of 2150°C (1175°C)
to 400°F (205°C)[15]. Also, the ring combustor has been extensively
tested in the combustion environment.

CONCLUSIONS

Thermal stress fracture in brittle materials must be treated
with statistical analysis techniques. The probability of failure
and size effect can be predicted by combining a risk analysis with
finite element heat transfer and stress analysis. The Weibull
weakest link theory provides an adequate basis for the risk analysis.
In many situations, the thermal stress in a structure increases
with size while, due to the size effect, the strength of the structure decreases. In a thermal transient the maximum probability of
failure can occur at times greater than the time at which the maximum stress occurs. Thermal shock tests of silicon carbide and
alumina disks demonstrated the scatter in the fracture strength and
a technique of computing the thermal stress probability of failure.
The average fracture strength of the alumina disks was well-
predicted, but the strength of the silicon carbide disks was apparently affected by a nonhomogeneous flaw population. Applications
of the statistical analysis technique to thermal stress situations
in gas turbine vanes and combustors were reviewed.

ACKNOWLEDGMENTS

The author would like to express his appreciation to Jack
Craig for performing the thermal shock tests, to Sally Fabian for
handling some of the computer calculations, to Curt Johnson for
providing the fractography, and to Horst deLorenzi for useful discussions and computing assistance.

REFERENCES

1. A. Brooks and A. I. Bellin, "Benefits of Ceramics to Gas Turbines," to be published in proceedings of AGARD Specialists Meeting on Ceramics for Turbine Engine Applications, Cologne, Germany (October 1979).
2. W. H. Dukes, "Brittle Materials: A Design Challenge," Mech. Eng. (November 1975), pp. 42-47.
3. E. M. Lenoe, "Probability-Based Design and Analysis - The Reliability Problem," Ceramics for High Performance Applications, Brook Hill, Chestnut Hill, Mass. (1974), pp. 123-145.
4. D. G. S. Davies, "The Statistical Approach to Engineering Design in Ceramics," Proc. Brit. Ceram. Soc., Vol. 22 (1973), pp. 429-452.
5. G. G. Trantina and H. G. deLorenzi, "Design Methodology for Ceramic Structures," ASME J. Eng. Power, Vol. 99, No. 4 (1977), pp. 559-566.
6. A. Paluszny and W. Wu, "Probabilistic Aspects of Designing with Ceramics," ASME J. Eng. Power, Vol. 44, No. 4 (1977), pp. 617-630.
7. D. P. H. Hasselman, "Unified Theory of Thermal Shock Fracture Initiation and Crack Propagation in Brittle Ceramics," J. Am. Ceram. Soc., Vol. 52, No. 11 (1969), pp. 600-604.
8. W. Weibull, "A Statistical Theory of the Strength of Materials," Proceedings of the Royal Swedish Institute of Engineering Research, No. 151 (1939).
9. J. C. Jaeger, "Thermal Stresses in Circular Cylinders," Phil. Mag., Vol. 36 (1945), pp. 418-428.
10. S. S. Manson and R. W. Smith, "Theory of Thermal Shock Resistance of Brittle Materials Based on Weibull's Statistical Theory of Strength," J. Am. Ceram. Soc., Vol. 38, No. 1 (1955), pp. 18-27.
11. F. L. Wilson, "Statistical Analysis of Thermal Shock Disks (Flat Faces Insulated)," J. Appl. Phy., Vol. 39, No. 3 (1968), pp. 1403-1407.
12. E. Glenny, et al., "A Technique for Thermal Shock and Thermal Fatigue Testing Based on the Use of Fluidized Solids," J. Inst. Metals, Vol. 87 (1958), pp. 294-302.
13. G. G. Trantina, "Fracture of a Self-Bonded Silicon Carbide," Am. Ceram. Soc. Bull., Vol. 57, No. 4 (1978), pp. 440-443.
14. G. G. Trantina and C. A. Johnson, "Spin Testing of Ceramic Materials," Fracture Mechanics of Ceramics, Vol. 3, edited by R. C. Bradt, D. P. H. Hasselman, and F. F. Lange, Plenum Press, New York (1978), pp. 177-188.
15. C. M. Grondahl and B. W. Gerhold, "A Hybrid Ceramic-Metal Combustion System," ASME Paper No. 76-GT-22 (1976).

16. G. G. Trantina and C. Grondahl, "Demonstration of Ceramic Design Methodology for a Ceramic Combustor Liner," ASME J. Eng. Power, Vol. 101, No. 3 (1979), pp. 320-325.

17. R. L. Mehan, W. B. Hillig, and C. R. Morelock, "Si/SiC Ceramic Composites: Properties and Applications," to be published in Proceedings of Conference on Composites and Advanced Materials, Am. Ceram. Soc. (January 1978).

18. C. W. Forrest, P. Kennedy, and J. V. Shennan, "The Fabrication and Properties of Self-Bonded Silicon Carbide Bodies," Special Ceramics 5, British Ceramic Research Assoc. (1970), pp. 99-123.

THERMALLY-INDUCED STRESSES IN INSULATING CYLINDER LINERS FOR

INTERNAL COMBUSTION ENGINES[1]

William J. Craft[2]
David E. Klett[2]

School of Engineering
North Carolina A & T State University
Greensboro, N. C.   27411

INTRODUCTION

The survivability of a ceramic cylinder liner for an internal combustion engine was investigated. The study is deemed to be important because the use of such liners in fuel-injected engines may increase fuel efficiency by reducing quench zone thickness and heat loss through the cylinder wall and by permitting effective secondary power extraction. Four ceramic liner candidates having potential for surviving this extreme environment were examined, viz. two densities of both silicon nitride and silicon carbide.

A finite element thermal code was used to generate temperature distributions in the cylinder wall. These results were applied with the internal pressure distributions as input to a finite element stress analysis computer code to determine the ceramic liner stresses. Results for each of the four candidates under the least favorable conditions were well below the tensile and compressive ultimates. This was due largely to the constrained thermal expansion of the cylinder. Each ceramic liner candidate reduced the heat loss through the cylinder wall by approximately a factor of eight and resulted in maximum wall temperatures on the order of 1100 F.

---

[1] Supported by General Motors Research Division, Warren, Michigan
[2] Associate Professor of Mechanical Engineering,
   North Carolina A & T State University

Although the study indicated the survivability of ceramic liners, economic and manufacturing feasibility remain to be investigated. Further analytical studies are needed to focus on combustion chamber and piston crown environments, and on piston seal and lubrication techniques.

In choosing ceramic liner candidates, one must be exceedingly careful if realistic predictors are desired. For example, an estimate too low for Youngs Modulus would result in too optimistic a thermal shock resistance as would also be the case in a low estimate of the coefficient of thermal expansion for the same material. A low Youngs Modulus, as would result from lower density material, would be accompanied by a lower conductivity leading to larger thermal gradients which would aggravate thermal stresses and possible surface melt conditions. On the other hand, general thermal growth could lead to more interference between the liner and block housing giving a compressive hoop stress and providing protection from fracture by this compressive preload of the liner. Hence the analysis was conducted by means of literature values of material properties, where possible, in the generation of predicted stresses. These stresses were next compared to conservative values of fracture strength for each candidate material.

Since the liner problem was one of combined thermal and mechanical stress, among other things, it was not apparent at the outset whether the anticipated environment would favor high or low density materials each with its peculiar characteristics. Other factors in material selection were raw material cost and cost of fabrication. Obviously the less expensive forms of any candidate would be commercially preferable, i.e., moderate density, lower purity requirements, etc. For these reasons and because environmental stability in a high temperature combustible atmosphere appeared acceptable in silicon nitride and silicon carbide, they were chosen as the liner candidates. Two densities of each candidate, about 100% and 80 to 90% of theoretical density, were each chosen to represent reaction sintered and hot pressed manufacturing techniques.

The adhesive assumed for the analysis was a high compliance Dow Corning 96-083 silicone adhesive.[1]   Although this adhesive was specified, any low-modulus, high temperature adhesive or potting compound to locate and secure the ceramic sleeve within a conventional engine block might be suitable. The adhesive would also act to save the cost of precision machining of the tapered engine bore and outer ceramic liner. It also served to provide the dominant liner insulation material. All material and thermal properties used in the finite element analyses are given in Table 1. This information was the result of a general literature search on materials.[2,5]   Even with the two relatively common ceramic

Table 1
Material Properties

	DENSITY ρ g/cc	CONDUCTIVITY κ BTU/sec-in-°F	SPECIFIC HEAT C_p BTU/lbm F	YOUNGS MODULUS E psi	POISSONS RATIO ν	SHEAR MODULUS G psi	LINEAR COEFFICIENT OF EXPANSION α °F^{-1}	TEMPERATURE AT WHICH PROPERTY IS QUOTED T °F
CAST IRON	7.60	6.75×10^{-4}	0.11	1.50×10^{7}	0.30	5.77×10^{6}	6.0×10^{-6}	(Cast Iron Properties Assumed temperature independent)
SILICON NITRIDE (HIGH DENSITY)	3.18	2.31×10^{-4}	0.27	4.60×10^{7} 4.38×10^{7} 4.22×10^{7} 4.15×10^{7} 4.04×10^{7} 3.92×10^{7}	0.23 0.23 0.23 0.23 0.23 0.23	1.87×10^{7} 1.78×10^{7} 1.72×10^{7} 1.69×10^{7} 1.64×10^{7} 1.59×10^{7}	1.1×10^{-6} 1.3×10^{-6} 1.5×10^{-6} 1.8×10^{-6} 2.0×10^{-6} 2.3×10^{-6}	0 400 800 1200 1600 2000
SILICON NITRIDE (MODERATE DENSITY)	2.80	8.56×10^{-5}	0.27	1.30×10^{7} 1.24×10^{7} 1.19×10^{7} 1.17×10^{7} 1.14×10^{7} 1.11×10^{7}	0.23 0.23 0.23 0.23 0.23 0.23	5.30×10^{6} 5.04×10^{6} 4.84×10^{6} 4.76×10^{6} 4.63×10^{6} 4.51×10^{6}	1.1×10^{-6} 1.2×10^{-6} 1.5×10^{-6} 1.8×10^{-6} 2.23×10^{-6} 2.32×10^{-6}	0 400 800 1200 1600 2000
SILICON CARBIDE (HIGH DENSITY)	3.21	6.95×10^{-4}	0.27	6.75×10^{7} 6.60×10^{7} 6.58×10^{7} 6.54×10^{7} 6.49×10^{7} 6.42×10^{7}	0.17 0.17 0.17 0.17 0.17 0.17	2.88×10^{7} 2.82×10^{7} 2.81×10^{7} 2.79×10^{7} 2.77×10^{7} 2.74×10^{7}	1.6×10^{-6} 1.8×10^{-6} 2.5×10^{-6} 3.0×10^{-6} 3.2×10^{-6} 3.5×10^{-6}	0 400 800 1200 1600 2000
SILICON CARBIDE (MODERATE DENSITY)	2.60	3.47×10^{-4}	0.27	3.00×10^{7} 2.95×10^{7} 2.90×10^{7} 2.89×10^{7} 2.88×10^{7} 2.04×10^{7}	0.17 0.17 0.17 0.17 0.17 0.17	1.28×10^{7} 1.26×10^{7} 1.24×10^{7} 1.24×10^{7} 1.23×10^{7} 8.70×10^{6}	1.6×10^{-6} 1.8×10^{-6} 2.5×10^{-6} 3.0×10^{-6} 3.2×10^{-6} 3.5×10^{-6}	0 400 800 1200 1600 2000
ADHESIVE	1.11	2.24×10^{-6}	0.31	4.0×10^{3} 2.0×10^{3} 1.7×10^{3} 1.4×10^{3} 1.0×10^{3} 0.4×10^{3}	0.36 0.37 0.39 0.45 0.49 0.50	1.47×10^{3} 7.30×10^{2} 6.12×10^{2} 4.83×10^{2} 3.36×10^{2} 1.34×10^{2}	0.80×10^{-4} 1.05×10^{-4} 1.43×10^{-4} 1.89×10^{-4} 1.98×10^{-4} 2.00×10^{-4}	0 400 800 1200 1600 2000

candidates, much data, particularly at elevated temperatures, was
deficient, thus requiring some conjecture. Furthermore, as with
many brittle materials, available literature showed very much
scatter in results from batch to batch, as a function of manu-
facturer and manufacturing processing. As a base line, cast iron
values are also included in Table 1.

In order to build credible estimates of survival into the
study, conservative values of fracture stress were researched.
Again little high temperature information could be found on the
candidates. Further, conversion between modulus of rupture and
uniaxial tension values could not be made with great confidence due
to the unknown Weibull parameters linking such testing methods.
Figure 1 depicts the assumed tensile values of strength. No compressive mode failure was assumed probable.

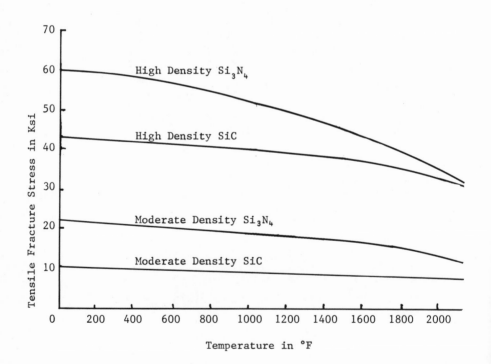

Figure 1:   Tensile Uniaxial Fracture Strength of Each Ceramic vs. T

Model Geometry

The geometry of the model chosen for the investigation is illustrated in Figure 2. It consists of a cylinder with a 4.00 inch bore of sufficient length to accommodate a 5.00 inch stroke. A cooling water jacket surrounds the cylinder at an outer radius of 3.15 inches giving a total cylinder wall thickness of 1.15 inch. The cylinder wall itself consists of an iron casting with a tapered bore with inner radius at the top of 2.45 inches and inner radius at the bottom of 2.25 inches. A tapered ceramic sleeve with constant inner radius of 2.00 inches and outer radii of 2.40 inches at the top and 2.20 inches at the bottom is bonded in the cast iron cylinder with a silicone adhesive layer 0.05 inch thick. The cylinder is topped by a cast iron cylinder head with a hemispherical combustion chamber. A 50 mil gasket material with the properties of the silicone is assumed between the block and head. It is expected that the head, valves, ports, etc. would also be insulated, this study however is confined to the liner.

Axisymmetric finite element computer codes were used in the thermal and stress analysis. The nodal point grid representing a section of the plane of symmetry of the cylinder and a portion of the cylinder head is shown in Figure 3. The grid consisted of 15 divisions in the radial direction, typically arranged from the inner to the outer surface as 7 ceramic elements, 2 adhesive material elements and 6 cast iron elements. In the axial direction, 38 divisions were used, beginning in the cylinder head area and extending 0.30 inches below the ceramic liner.

Thermal Analysis and Methods

The termal analysis of the ceramic liner was done using the finite element thermal code HTCON extensively modified to run on our former computer, a CDC 3300 machine, with a very small core limit of 103K bytes. [6,7] Although this code could generate non-steady temperature distributions and it allows time dependent boundary conditions, the complexity of the rate processes occuring within the cylinder of an IC engine made the full usage of these features appear intractable for this problem, especially in view of the limited speed of the available computer facilities. Thus, rather than attempt a full simulation by HTCON of the actual thermal processes occuring during every few degrees of crankshaft rotation for the entire four stroke cycle for sufficient cycles to achieve cyclic steady state, a steady state approach was adopted for input to the thermal code.

To make this simplified method approach a reasonable thermal model, a separate program was written to calculate time-averaged gas temperatures at various axial locations in the cylinder based on

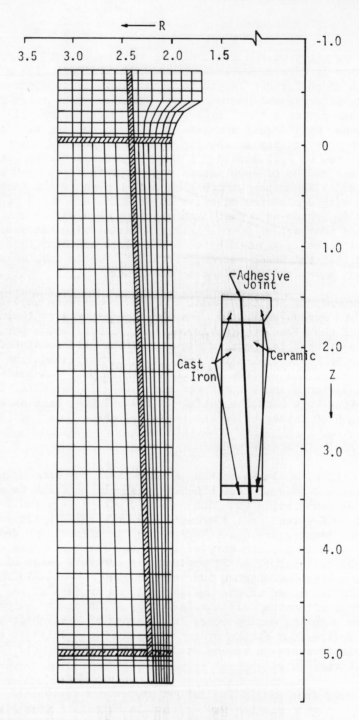

Figure 3. Cylinder Section of Finite Element Grid

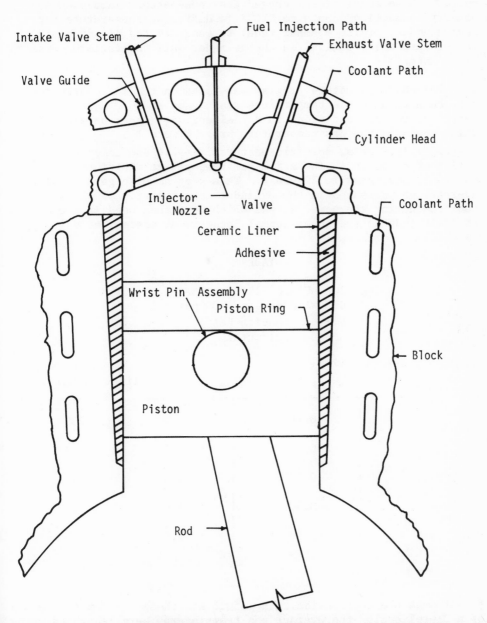

Figure 2. Section of Piston and Cylinder Assembly of Ceramic-
           Lined Engine

the thermodynamic processes for each of the four strokes in the
cycle. These local time-averaged gas temperatures were used as
boundary condition inputs to HTCON (a different temperature for
each boundary finite element face) to generate the steady state
temperature distributions in the cylinder wall for various ceramic
liner materials.

The gas temperature program was based on the ideal Otto cycle
but the maximum cycle temperature was chosen as 5030°R to more
nearly represent conditions in a real engine. Other parametric
values used in the gas temperature program were T_{intake} = 600°R,
P_{intake} = 14.7 psia, specific heat ratio = 1.4 and compression
ratios of 8 and 16.* The axial gas temperature distributions
generated are shown in Figure 4. This program also gave local
average gas pressures in the cylinder as a function of piston
position during the power stroke which were then used in the stress
analysis code as the inner normal boundary stresses, Table 2.

TABLE 2

PRESSURE DROP DURING PISTON MOVEMENT IN POWER
STROKE WITH P_{max} = 1700 psi
16:1 Compression Ratio
(Inches)

Pressure Condition	From Z =	to Z =	Pressure (psia) (liner normal pressur
1	−0.40	0	1700
2	−0.40	0.5	489
3	−0.40	1.5	169
4	−0.40	3.5	61
5	−0.40	5.0	38

* Although the compression ratio of 16 was chosen as representative
of a Diesel cycle, the average gas temperatures were generated using
the Otto cycle analysis since it was felt that the effect of higher
temperatures at TDC would be approximately offset by lower temperatures at BDC, each case produced similar inner wall temperatures
and overall temperature profiles.

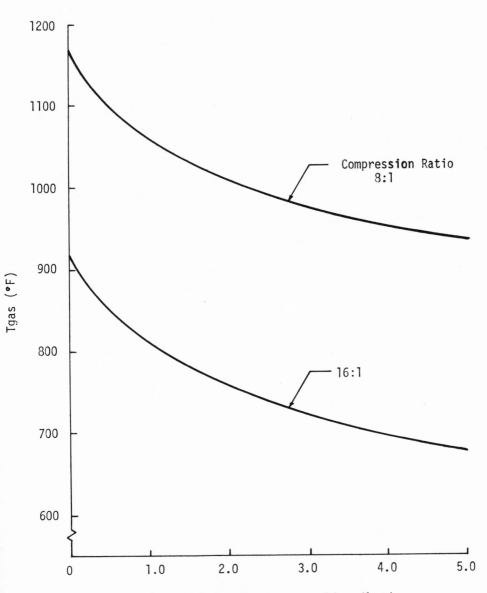

Figure 4. Axial Gas Temperature Distributions
Z(Inches from Top of Cylinder)

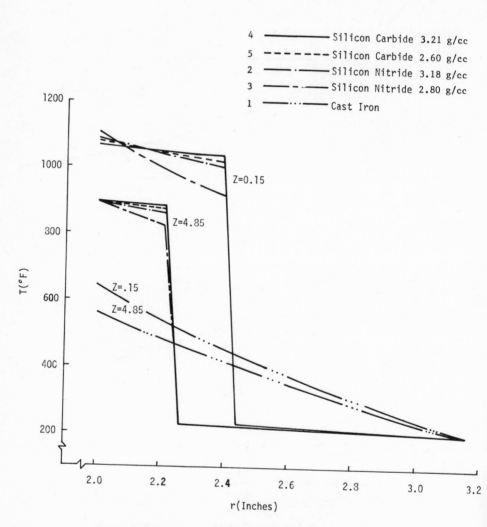

Figure 5. Radial Temperature Distributions at two Axial Locations
           (Z=.15 and Z=4.85")

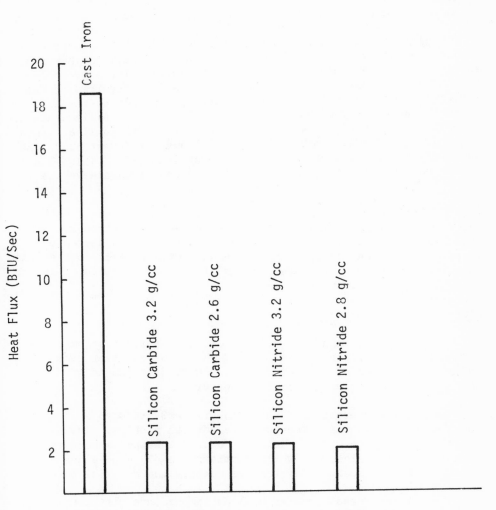

Figure 6. Total Heat Flux Through Cylinder Wall

The other thermal boundary conditions employed were that of a constant temperature of 200 F at the outer cylinder surface and upper cylinder head surface and adiabatic surfaces at Z = 0 and Z = 5 inches. Computer runs were obtained for each of the four ceramic liner cases and for a solid cast iron cylinder for comparison purposes. The output distribution, in the form of temperatures at all nodes was collected for use in the stress analysis code.

Typical radial temperature distributions in the cylinder wall derived by this method for two different axial locations (Z = .15 and Z = 4.85) are shown in Figure 5. The interesting aspect of these distributions is that the steepest gradient occurred in the adhesive bond layer which was to be expected in view of the low thermal conductivity of most adhesives. Low density silicon nitride (2.8g/cc) offered the steepest gradient in the ceramic liner yielding the highest wall temperature (1100°F) and the lowest bond line temperature (918°F) near the top of the cylinder (Z = .15). The corresponding wall temperature for a solid cast iron sleeve was 645°F.

A separate program was written to calculate the total heat flux through the cylinder wall for the various cases. The results are shown in Figure 6 which indicates about an eight-fold decrease in heat loss for each ceramic liner candidate with respect to the solid cast iron cylinder. Of course action of the adhesive was dominant in producing this insulation, Figure 5.

Stress Analysis Description

The structural analysis of the ceramic liner was accomplished by means of the finite element method where the computer code, AMG054, for bodies of revolution, was modified to run in grid sizes 16 x 40 or 20 x 75 for our small computer. [8]

The finite element grid shown in Figure 3 produced a 16 x 39 nodal point array considered adequate for the problem. The element distribution chosen was non-uniform to give a better aspect ratio in areas of greatest interest while permitting a dense covering near points of high stresses near free boundaries and material interfaces.

The cylinder and liner as well as temperature distributions, surface tractions and all other loading and boundary conditions were assumed to be axi-symmetric. This should provide a realistic model for a single cylinder except for piston interference and dynamic side loading during operation. Since the cast iron block is of high stiffness in comparison to the liner, any extension or diminution of the cast iron required for the fabrication of multiple cylinder geometries should not have changed the results significantly.

A complete statement of the stress code needs is:

      (1)   Geometry definition
      (2)   Boundary condition definition
      (3)   Surface shear/pressures
      (4)   Material property definition
      (5)   Temperature distribution

Other major assumptions for the stress analysis were:

1.   The geometry definition grid was as depicted in
     Figure 2 – unchanged for all runs.

2.   The displacement boundary conditions were that the
     axial displacement along the bottom row was con-
     strained to be equal to 0. No constraint existed
     for any radial displacements.

3.   The surface shear stresses at the inner wall of
     the liner were assumed negligible. The inner
     surface normal tractions (σ_r) were assumed given
     by Table 3 as the piston moved downward from TDC.

4.   The mechanical properties of each liner material
     were given by Table 1. Note in all cases
     isotropy was assumed while material properties
     were allowed to vary with temperature, except
     in the comparison case of cast iron for which a
     low density casting was assumed.[9]

The temperature distributions resulted in thermal expansion at
the boundaries between dissimilar materials leading to interference
stresses while temperature differentials or thermal gradients within
each material also led to thermal shock stresses due to variable
expansion and accompanying stiffness constraints of that material.
Thermal shock stresses are a major factor in failure of axisymmetric geometrics with large radial thermal gradients and in cases
of lesser gradients where the Youngs Modulus, Coefficient of
Expansion product was large. In cases of axisymmetric geometries
of joined materials of different thermal expansions, interference
stresses can dominate, even in bodies at a uniform temperature,
particularly if that temperature is displaced sufficiently from
that of the stress-free state.

Discussion and Results

The structural analysis input parameters have been fully
defined for code, AM6054. Computer runs were made for each material
including a full cast iron wall and for each of five pressure

distributions from the combustion chamber to the piston top representing 5 pressures of the power stroke, Table 2. Thus a total of
20 problems were investigated for the ceramic liners and 5
additional problems for the equivalent cast iron block.

In each case the hoop stress reached the most critical value
and was used as the basis of comparison for failure analysis. It
was found that each of the five conditions could be depicted by a
fairly small stress envelope for each of the four ceramics and for
the cast iron wall. These results, shown in Figure 7, demonstrate
that each ceramic was in compression - generally a desirable condition for these brittle materials.

There is little literature data available on compressive
failure of ceramics but it is common to assume compressive strength
at between 8 times the tensile strength following Griffiths Theory,[1]
or three times the tensile strength following some biaxial four-
quadrant experimental data. [11,12] In either case, a comparison between Figure 7 and Figure 1 indicates that the working
stresses reached only a fraction of the compressive ultimate.
Tensile mode stresses were small and would probably only be
encountered in warmup or in erratic conditions of operation. The
size effect due to Weibull volume flaw theory was not taken into
account because, (1) no reliable data was thought to exist for
these materials under the conditions of this problem and (2) after
the analyses, it was determined that there were no tensile-mode
stresses which were significant.

An interesting feature of the flat temperature distribution
achieved throughout the ceramics (Figure 4) was that the main
source of stress was due to the interference of the ceramic and
adhesive rather than thermal shock usually associated with uneven
temperatures in a material. The adhesive attempted to grow
considerably during heating due to a high coefficient of expansion,
but acted in a nearly hydrostatic way due to its highly constraining geometry. Mechanically induced stresses, as from the internal
pressures, were shown to be small in comparison to this interference stress. Of course, this same radial interference
produced relatively large tensile hoop stresses in the cast iron.
These values were, however, smaller than fracture data for cast
iron. [9]

In effect then, all materials were well below critical
tensile-mode values and all liners were below 1100°F at the inner
surface where no melt problem should occur. It remained to compare
radial displacements to an equivalent cast iron wall to find out
what differences existed in the possible piston design for a
ceramic lined engine.

(Hoop Stress in Ksi)

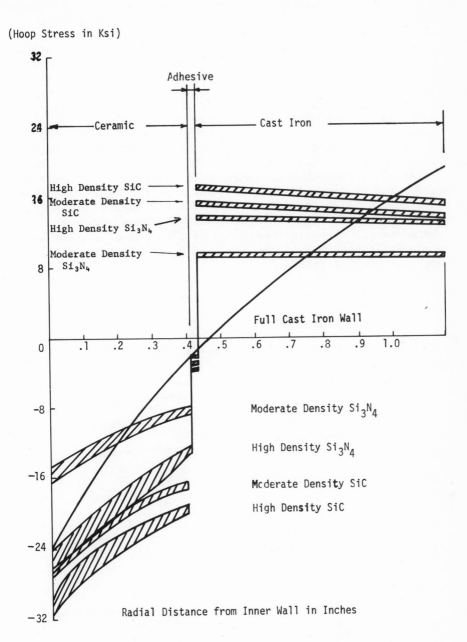

Figure 7. Variation Envelope of Hoop Stress σ_θ Through the
Cylinder Wall for One Cycle at Z=0.1" From Top of Head

Several interesting phenomena circumvented the anticipated outward radial displacement field at the liner's inner surface (r = 2.0" and 0" ≤ Z ≤ 5"). Obviously, a radial displacement investigation was important in order to determine piston friction and compression seal and material compatibility during operation and to ascertain the possibility of seizure. The higher temperatures of the ceramic liner and its lower stiffness in resisting internal combustion pressures would have led one to believe that larger radial growth would occur while the resisting effect of the higher expansion adhesive would counter such displacements.

Figure 8 revealed this radial growth information for each of the four candidate materials along with a basic cast iron block. All materials showed an increased growth nearer the combustion chamber. The displacement of cast iron appeared to be sandwiched between that of high density SiC and moderate density SiC while the results are less for each Si_3N_4 candidate. These graphs are generated at the pressure conditions of 1700 psi at TDC but they were nearly independent of the pressure distribution as it changed through the power stroke.

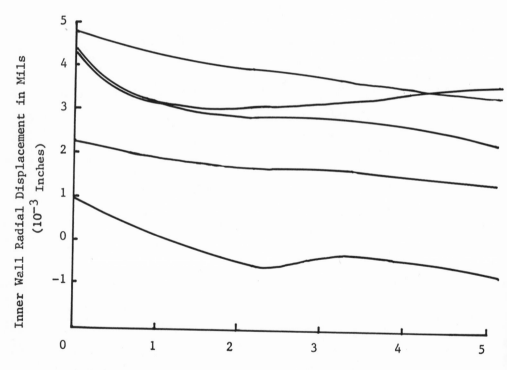

Figure 8: Radial Displacements for 0 < Z < 5 at the Inner Wall

Of some concern in the analysis of the radial displacement at the cylinder wall was the fact that the usual aluminum alloy pistons would likely be unsuitable because there would surely be a melt problem with this high wall temperature environment. Even without melt, it was likely that the high thermal expansion of the piston coupled with its resulting higher steady-state (mean time-averaged) temperature would produce excessive interference and seizure.

Future investigation should be directed toward an alloy or material change for the piston with the possibility of an insulating ceramic piston cap which may itself require a thin coating or plate of say a refractory metal over the ceramic cap to fight potential erosion associated with the injection of fuel and the combustion process and the ensuing fatiguing thermal cycling. Another difficulty to be addressed is piston ring seal or other seal design associated with a ceramic wall. It may be possible to place rings further down on the piston skirt so that a cooler temperature regime would be encountered or even to discontinue the liner, in this case, in the vicinity of the ring seal area by reverting to cast iron and conventional rings. Of course, much less heat loss occurs in the lower region of the liner.

One natural and needed extension of the liner idea for more retention of exhaust heat would be the application of a low conductivity liner to a combustion chamber where potentially the greatest heat loss occurs to the cooling system. This problem was not investigated in detail, however, because the ceramic liner survival was considered more critical and complex than the ceramic-lined combustion chamber.

Conclusions

What has been demonstrated by this study is that under the assumed conditions of structural and thermal environment of a fuel-injected ICE – Internal Combustion Engine: (1) survival of a ceramic liner can be obtained and (2) insulation value of a liner is such that a great reduction of heat loss can be expected. This heat loss reduction if extended to an insulated combustion chamber as well may remove the need for all or part of the conventional cooling system as well as allow secondary energy extraction. Cooling system reduction and the lower mass of the ceramic liner and chamber may both reduce the complexity of engine design, increase efficiency and help reduce weight. In addition to this study, other recent investigations of insulated liner engines have taken a major step toward feasibility while questions of cost trade-offs are still to be answered. Much of the work to date has seen agency support for speciality military diesel engines [13-15]. In some of these studies, turbocompounding in conjunction with the adiabatic design may lead to 50% thermodynamic efficiencies [14] compared to 30% to 35% for current diesels.

REFERENCES

1. Dow Corning 96-083 Silicone Adhesive (New Product Information)
    Dow Corning Corp. (Encapsulants & Sealants Dept.), Midland,
    Michigan 48640.
2. Craft, William J., 'Creation of a Ceramics Handbook,' Final
    Report NASA Grant NGR-34-012-013, March 1, 1976.
3. McLean, A.F., Fisher, E.A. and Bratton, R.J., 'Brittle Materials
    Design, High Temperature Gas Turbine, Interum and Final Report,'
    AMMRC-CTR-72-19, 73-32, and 74-26.
4. Lange, F. F., Dense Si_3N_4 and SiC: Some Critical Properties for
    Gas Turbine Application, ASME Pater, 72-GT-56, March, 1972.
5. Lynch, J.F., Ruderer, C.G., and Duckworth, W. H., "Engineering
    Properties of Ceramics: Data Book to Guide Materials Selection
    for Structural Applications," AD 803 765, Batelle Memorial
    Institute, Columbus, Ohio, June, 1966.
6. Wilson, E.L., and Nickel, R.E., "Application of the Finite
    Element Method to Heat Conduction Analysis," Nuclear Engineer-
    ing and Design 4, 3, pp. 276-286, October, 1966.
7. Chaloupka, A. B., "A Computer Program for the Analysis of Two
    Dimensional Heat Conduction Using the Finite Element Technique,"
    M. S. Thesis, U. S. Naval Post Graduate School, June, 1969,
    AD690450.
8. Brisbane, J. J., "Heat Conduction and Stress Analysis of Solid
    Propellant Rocket Motor Nozzeles," Rohm and Haas Co., Redstone
    Research Laboratories, Huntsville, Alabama, Technical Report
    S-198, February, 1969.
9. Materials Selector 73, Reinhold Publishing Co., Inc., Stamford,
    Connecticut, Vol. 76, No. 4.
10. Griffith, A. A., Philosophical Transactions of the Royal Society
    of London, A221, 163, 1921.
11. Ely, Richard E., 'Strength Results for Ceramic Materials Under
    Multi-axial Stresses,' U. S. Army Missile Command, Report
    No. RR-TR-68-1, AD-670126, April, 1968.
12. Adams, Marc and Sines, George, 'Determination of Biaxial Com-
    pressive Strength of a Sintered Alumina Ceramic,' Journal of
    the American Ceramic Society, Vol. 59, No. 7-8, July - August,
    1976.
13. Stang, J., and Johnson, K., "Advanced Ceramics for Diesel
    Engines," U. S. Army Tank-Automotive Development Center,
    Technical Report No. 12131, January, 1976.
14. Kamo, R., and Bryzik, W., "Adiabatic Turbocompound Engine,
    Performance Prediction," SAE paper 780068, Detroit, Michigan,
    February, 1978.
15. Kamo, R., Wood, M., and Geary, W., "Ceramics for Adiabatic
    Diesel Engines," CIMTECH 4th, St. Vincent, Italy, June, 1979.

THERMAL CREEP OF COKE-OVEN JAMB FRAMES

E. N. Kuznetsov and A. T. Hopper

Applied Solid Mechanics Section
Transportation and Structures Department
Battelle Columbus Laboratories
Columbus, Ohio 43201

INTRODUCTION

The jamb frame is one of the major structural elements of a coke-oven end-closure system. The structural behavior of the jamb interacting with other end-closure elements determines the sealing performance of the system as a whole. One of the main causes of poor sealing is the accumulation of reversible and/or irreversible jamb distortions. In particular, the jamb frame often deforms in its plane so that the posts deflect towards each other--a deformation called hourglassing.

The objective of this study is to reveal a possible mechanism accounting for hourglassing. At normal operating temperatures and stresses in the jamb, plasticity is not a likely major contributor to the jamb distortion. Therefore, an analysis has been formulated which addresses creep at elevated temperatures as a potential cause of jamb hourglassing. The analysis is intended not only to determine the jamb in-plane deformation resulting from a given temperature pattern, but to establish such patterns as are likely to cause jamb hourglassing.

Problem Statement and Working Assumptions

In what follows, the jamb is considered as a rectangular frame subjected to elevated temperatures and thermal gradients in its plane. Typical cross sections of actual cast iron jamb frames are shown in Figure 1. The available data on the mechanical properties of cast iron[1] suggest that within the range of stress not exceeding ±10 ksi, the modulus of elasticity of cast iron is practically

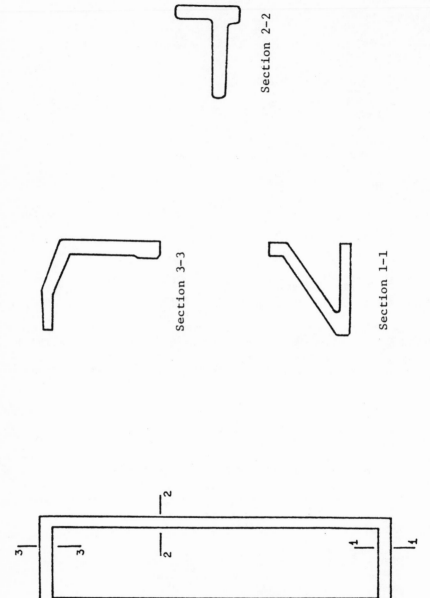

Figure 1. Typical Jamb Cross Sections

the same in tension as it is in compression. Also, this modulus
remains essentially unchanged under temperatures up to 1,000 F.
However, at this temperature, and stress levels approaching 10 ksi,
creep becomes pronounced.

For practical purposes, creep is often taken to obey a relation of the form[2,3,4]:

$$\varepsilon_c = \sigma^k \cdot \Omega(t) \quad , \tag{1}$$

where ε_c is the strain component due to creep, σ is the total
stress, $\Omega(t)$ is some explicit function of time t, and k is a material parameter. Both Ω and k depend on temperature. For steady
creep, a linear approximation for $\Omega(t)$ is used

$$\Omega(t) = at \quad . \tag{2}$$

To determine the parameters a and k, the data (1, Figure 38)
on the steady creep rate can be employed. At any given temperature, the creep rate (percent per hour) depends on the stress
level:

$$\dot{\varepsilon}_c = c(\sigma) \tag{3}$$

Having values c_1 and c_2 for two different stress levels, σ_1
and σ_2, permits Equations (1), (2), and (3) to be combined to
obtain

$$k = \frac{\ln c_1 - \ln c_2}{\ln \sigma_1 - \ln \sigma_2} ; \quad a = \frac{c_1}{\sigma_1^k} \quad . \tag{4}$$

The next step is to transform Equation (1) into a relation
between the bending moment M and curvature κ in a bar. If y is the
distance from the neutral axis to a given fiber, the strain is
related to the curvature by

$$\varepsilon = \kappa y \tag{5}$$

Equation (1) yields

$$\sigma = \left(\frac{\kappa_c}{\Omega}\right)^{1/k} |y|^{1/k-1} y \tag{6}$$

where κ_c denotes the curvature due to creep.

The bending moment is given by

$$M = \int_A \sigma y \, dA = \left(\frac{\kappa_c}{\Omega}\right)^{1/k} \int |y|^{(k+1)/k} \, dA = \left(\frac{\kappa_c}{\Omega}\right)^{1/k} \cdot J \qquad (7)$$

where A is the bar cross sectional area, and J is its generalized moment of inertia. For a b by h rectangular cross section

$$J = 2b \int_0^{h/2} y^{(k+1)/k} \, dy = 2^{-(k+1)/k} \frac{k}{2k+1} \, bh^{(2k+1)/k}$$

$$(8)$$

$$= 2b \frac{k}{2k+1} \left(\frac{h}{2}\right)^{\frac{2k+1}{k}} .$$

The bending moment--curvature relation may now be written as

$$\Omega_c = \left(\frac{M}{J}\right)^k \Omega = a \left(\frac{M}{J}\right)^k t \qquad (9)$$

and the total curvature including elastic and creep components reads

$$\kappa = \kappa_e + \kappa_c = \frac{M}{EI} + a \left(\frac{M}{J}\right)^k t . \qquad (10)$$

Basic Equations and Solution Technique

Equation (10) is basic to the subsequent jamb frame analysis which can be conveniently performed by the force method. To this end, the jamb frame is cut at the middle of the upper cross bar (Figure 2a) and two unknown forces--the bending moment M and the axial force N--are introduced. Now, mutual displacements--angular, θ, and linear, δ--at the cut need to be determined. Each displacement consists of two components, one resulting from the unknown forces M and N and the other caused by external actions. Both of the displacements can be expressed in terms of curvature taken as a function of the longitudinal coordinate, s (one half of each displacement is taken because of symmetry):

$$\theta = \int_0^{s_3} \kappa(s) ds \qquad (11)$$

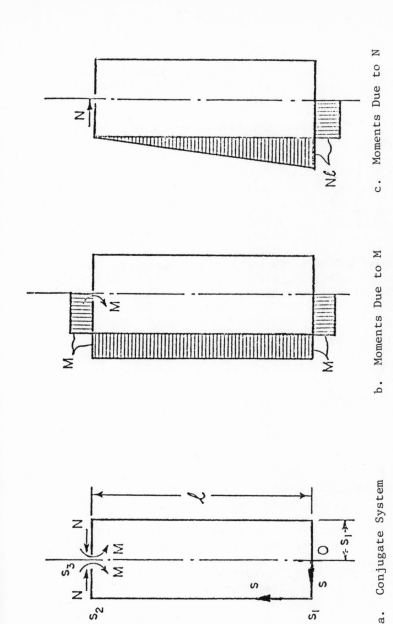

a. Conjugate System

b. Moments Due to M

c. Moments Due to N

Figure 2. Conjugate System and Bending Moment Diagrams

$$\delta = \theta_1 \ell + \delta_v = \ell \int_0^{s_1} \kappa(s)ds + \int_{s_1}^{s_2} ds \int_{s_1}^{s} \kappa(s)ds \quad , \tag{12}$$

where ℓ denotes the jamb length. Thus, the problem is reduced to the evaluation of curvatures resulting from each of the above reasons--i.e., unknown forces and external actions.

The effect of the basic unknowns is given by Equation (10), where the bending moment is taken as

$$M(s) = M + \begin{cases} N(s_2 - s_1) & 0 \le s < s_1 \\ N(s_2 - s) & s_1 \le s < s_2 \\ 0 & s_2 \le s \le s_3 \end{cases} \tag{13}$$

Here, the second addend shows the bending moment produced by N. The diagrams of the bending moments produced by M and N are shown in Figures 2b and 2c, respectively. Substituting Equation (13) into (10) yields

$$(s) = \begin{cases} \dfrac{M + N\ell}{EI_1} + \dfrac{a_1 t}{J_1^{k_1}} (M + N\ell)^{k_1} & 0 \le s < s_1 \\[3mm] \dfrac{M + N(s_2 - s)}{EI_2} + \dfrac{a_2 t}{J_2^{k_1}} [M + N(s_2 - s)]^{k_2} & s_1 \le s < s_2 \\[3mm] \dfrac{M}{EI_3} + \dfrac{a_3 t}{J_3^{k_3}} M^{k_3} & s_2 \le s \le s_3 \end{cases} \tag{14}$$

where the parameters a and k characterize creep and depend on the respective temperatures of the three frame segments.

The curvature produced by external actions results from thermal gradients, $\Delta T(s)$, varying along the frame axis

$$\kappa_t(s) = \alpha \, \frac{\Delta T(s)}{h(s)} \tag{15}$$

where α is the coefficient of thermal expansion, and $h(s)$ is the depth of the jamb cross section (which may be different for the upper and lower cross bars and for the vertical bar).

The resolving system of equations of the force method can be written schematically in the form

$$\theta(M,N,t) = \theta_t$$

$$\theta(M,N,t) = \delta_t \tag{16}$$

Here all the terms are to be evaluated by substituting the preceding expressions for the curvatures, $\kappa(s)$ and $\kappa_t(s)$, into Equations (11) and (12). The evolving non-linear system of Equation (16) was solved on the computer by means of the self-correcting finite increment method[5]. Let $M(t)$ and $N(t)$ represent an approximate solution for the moment of time t. Then, increment time by a reasonably small amount, τ. As a result, both of the unknowns acquire increments, m and n, respectively, which can be found from the linearized system of equations

$$\frac{\partial\theta}{\partial M}\,m + \frac{\partial\theta}{\partial N}\,n = \theta_t - \frac{\partial\theta}{\partial t}\,\tau - \theta(M,N,\tau) \;\bigg|\; = \theta_t - \theta(M,N,t+\tau)$$

$$\frac{\partial\delta}{\partial M}\,m + \frac{\partial\delta}{\partial N}\,n = \delta_t - \frac{\partial\delta}{\partial t}\,\tau - \delta(M,N,\tau) \;\bigg|\; = \delta_t - \delta(M,N,t+\tau) \;\;. \tag{17}$$

The last term on the right hand side of each equation represents a correction which keeps error from accumulating as successive steps. This increases the accuracy of the solution and permits larger increments of the independent variable t to be used. Within each time increment, the forces M and N can be assumed constant, which makes both θ and δ depend linearly on t. Taking advantage of this fact, the last two terms on the right hand sides can be combined as shown to the right of the vertical line in Equations (17).

Solving the system of Equations (17) for a current step allows the determination of the basic unknowns

$$M(t + \tau) = M(t) + m$$

$$N(t + \tau) = N(t) + n \tag{18}$$

and passage to the next time step. As soon as the basic unknowns at a fixed moment of time are determined, the entire pattern of the stress state and displacements for this moment can be established.

Discussion of Numerical Results

A computer program utilizing the above approach was written and used for a parametric study. The program provides a history of the stress and strain states of a jamb frame subjected to

prescribed temperatures and temperature differences measured in the
plane of the jamb which can account for hourglassing. Specifically,
the deflection curve of the jamb post as well as the values of the
bending moment and of the thrust in the upper cross piece are calculated at various times. The output arrangement permits the evolution of the stress and strain states to be followed from the onset
of thermal action through the whole period of exposure to a given
temperature regime. Also computed are the residual deflections
and forces which would be present after the removal of the temperature action.

The quantitative results are mainly determined by the creep
rates. To avoid extrapolation, a relatively wide range of stresses
(from 1000 to 10,000 psi) is taken as a base. For this reason,
the creep rate (that corresponding to 10,000 psi) is rather high.
However, the working stress, as a rule, is much lower than 10,000
psi and gives rise to a more moderate creep rate.

Under some of the temperature and thermal gradient combinations studied, the jamb frame develops pronounced hourglassing.
Normally in a jamb frame the upper and lower cross pieces are more
rigid than the posts. Also, the temperatures of the frame inner
faces (i.e., of those exposed to hot gases and to direct radiation
from passing coke) are usually higher. Figure 3a shows a typical
pattern of temperatures (the first number) and thermal gradients
(in parenthesis) in the jamb frame. Under this pattern, two somewhat different mechanisms for hourglassing were revealed.

If outward bowing (reverse hourglassing) of the jamb is possible to some extent, the frame initially acquires the shape shown

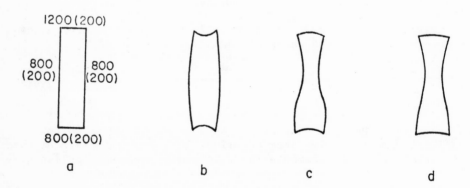

Figure 3. Jamb Frame Distortion

a) Temperature and Thermal Gradient Pattern
b, c, d) Successive Deformed Configurations

in Figure 3b. Here the stiffer cross pieces override the resistance of the vertical posts and their trend to hourglass. Later,
the upper cross bar, which is exposed to a higher temperature, will
relax and gradually give in. As a result, the frame successively
assumes the configurations shown in Figures 3c and d. The corresponding numerical data are presented in the following table
(deflections are positive if directed inward).

Time Hours	Post Deflection (in mils) at Equally Spaced Locations x/ℓ				Forces in the Upper Crossbar	
	0.2	0.4	0.6	0.8	Bending Moment in-kips	Axial Force lb
1	-32	-57	-49	-31	164	14
5	-32	-42	-36	-20	159	46
10	-29	-35	-25	-11	155	72
50	-21	-11	10	20	141	157
100	-16	2	28	36	138	202
500	-6	32	72	75	119	309
1000	-1	45	91	91	111	355
5000	9	73	132	128	96	456
10,000	13	84	149	142	89	497

If outward bowing is precluded by the brickwork, a compression stress will develop in the outer fiber of the post. This
stress relaxes gradually and then upon reducing all the thermal
gradients, the post will acquire some inward deflection rather
than return into its initial rectilinear configuration.

In reality, both the described mechanisms can combine and
give rise to appreciable hourglassing.

ACKNOWLEDGEMENT

The authors would like to acknowledge the American Iron and
Steel Association and the United States Environmental Protection
Agency who cosponsored this work.

REFERENCES

1. Iron Castings Handbook (1971).
2. Y. N. Rabotnov, Creep in Structural Elements (In
    Russian), Moscow (1966).
3. W. Flugge, Viscoelasticity, Blaisdell, Waltham, MA,
    (1967).
4. F. Carofalo, Fundamentals of Creep and Creep Rupture,
    McMillan Co., New York (1965).
5. J. T. Oden, Finite Elements of Nonlinear Continua,
    McGraw-Hill, New York (1972).

A STUDY OF PART THROUGH CRACKS IN A REACTOR BELTLINE SUBJECTED TO THERMAL SHOCK

W. H. Peters

ESM Dept.
Virginia Polytechnic Institute
    & State Univ.
Blacksburg, VA 24061

J. G. Blauel

Fraunhofer Institut
    für Werkstoffmechanik
D-7800 Freiburg
Federal Republic of
    Germany

INTRODUCTION

A problem which has received considerable attention in the past few years is the analysis of a cracked reactor beltline region which is subjected to a thermal shock loading condition. This attention has resulted in many techniques for estimating the Stress Intensity Factor (K_I) distributions along the front of part through cracks for this problem. The purpose of the study presented herein is to utilize some of these more recent developments for a linear elastic analysis of hypothetical elliptical surface cracks subjected to a time-varying non-uniform stress distribution which results from a thermal shock loading. The problem is further complicated because the initial crack is assumed to be in a weld region with the possibility of extending into a base material region with different material properties. A technique which is based on linear elastic fracture mechanics is chosen to predict K_I distributions for the geometry and loading employed. In order to gain confidence in the solution, it is compared with other available analytical and numerical solutions from the literature. It is then used to predict crack extension and subsequent arrest with consideration given to the varying material parameters present.

GEOMETRY, MATERIAL PROPERTIES AND LOADING CONDITIONS

The vessel geometry is shown in Figure 1. The sketch shows a region from the beltline of a cylindrical reactor vessel with gross measurements of

Inner Radius (R_i) = 2047 mm
Wall Thickness (W) = 192 mm.

The vessel contains an inner cladding of thickness 7 mm. For the
fracture mechanics analysis, the presence of this cladding was ignored (because of its very ductile nature), however its effect was
included in the thermal analysis.

The initial crack was assumed to exist in a "safe flaw detection" region as defined by ultrasonic testing capabilities. The
crack was also assumed to initiate in a weld region with a width
of 40 to 60 mm. If one choses symmetry of the crack position in
this region, then a variety of crack geometries which can exist
may be defined. As seen in Figure 1, the crack may have aspect
ratios of a/b = 1, a/b < 1, and a/b → 0 where a is the crack depth
and b is the half surface length. The initial aspect ratio of the
crack will be a/b ⩾ 0.33 thus confining the initial crack to the
weld region.

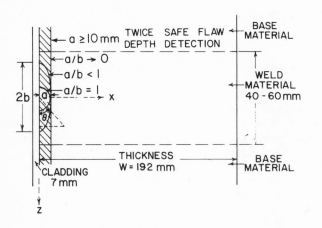

Figure 1   Vessel Geometry and Problem Notation

The predic-
tion of crack
initiation and
crack arrest can
be made by com-
paring the Stress
Intensity Factor,
K_I, at a point
along the crack
front with the
initiation frac-
ture toughness
K_{Ic} and the ar-
rest fracture
toughness K_{Ia},
respectively, at
the same point.
The values of
K_{Ic} as a function of temperature for the base material and weld
material considered in this study were determined from WOL specimens and are shown in Figure 2. The values of arrest fracture
toughness K_{Ia} as a function of temperature are also shown in this
figure. The K_{Ia} values were established by shifting the K_{Ic} (T)-
curve according to the procedure outlined in the ASME Boiler and
Pressure Vessel Code-Section XI.

As was mentioned previously, the primary loading condition
used in this study was that of thermal shock. The introduction of
cold coolant into a warm vessel causes a temperature gradient
through the wall thickness which can result in high tensile hoop

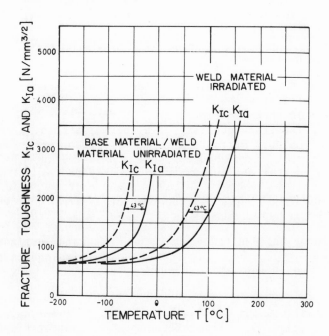

Figure 2   Toughness Values as a Function of Temperature for the
           Problem Considered

and axial stresses on the inner.surface of the vessel. The hoop
stress which is only investigated in this study is noted as σ_ϕ
and decreases through the vessel wall thickness (x-direction in
Figure 1). The temperature gradient and resulting stress distribution obtained from a standard finite element procedure are given
in detail in a report by Peters and Blauel (1979). The contributions of a stress resulting from relatively small internal pressure load and residual stresses due to the cladding attachment procedure were also included in the calculations. The resulting hoop
stress distribution will be discussed in more detail in the following section.

An important observation should be made at this point. The
effect of the sudden cooling of the inner vessel wall results in a
high tensile stress which will tend to increase the K_I value for a
crack located in this region. At the same time, this cooling
causes the fracture toughness (K_{Ic} or K_{Ia} – see Figure 2) to

decrease in value. It is the ratio of stress intensity factor to fracture toughness, K_I/K_{Ic} or K_I/K_{Ia}, which determines crack extension or arrest, respectively. After establishing a technique for estimating K_I, these ratios will be used to predict growth for cracks of varying geometry.

TECHNIQUE FOR ESTIMATING K_I

A typical hoop stress distribution for a selected time t_2 after the start of the shock for which crack initiation is expected (see Fig. 11) is shown in Figure 3. The total stress ($\sigma_{\phi,tot}$) which acts on the crack consists of a thermal stress, a pressure stress and a residual stress. This total stress decreases linearly for small crack sizes such as those considered in this study. It can be noted that the total stress is composed of a linear part (σ_{Linear}) a uniform part ($\sigma_{Uniform}$). Based on this observation and the principle of superposition the total stress intensity factor ($K_{I,tot}$) can be written as

$$K_{I,tot} = K_{IL} + K_{IU}$$

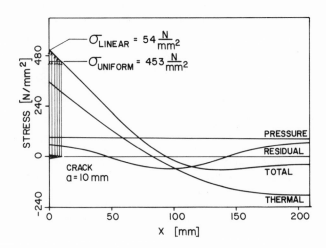

Figure 3 Hoop Stress (σ_ϕ) Distribution for the Selected Time t_2 after the Start of Cooling.

where K_{IL} is the stress intensity factor which results from the linear part of the load (σ_{Linear}) and K_{Iu} is the stress intensity factor which results from the uniform part of the load ($\sigma_{Uniform}$). Kobayashi, Enetanya and Shah (1974) utilized an alternating numerical technique developed originally by Kantorovich and Krylow (1964) (and further refined by Smith, Emery and Kobayashi (1967)) to esti-

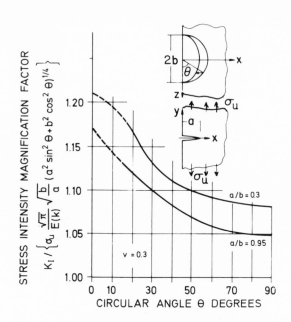

Figure 4 K_I Estimation for Uniform Stress (σ_U)
           Loading.   After Kobayashi et al (1974)

mate the stress intensity factor for various aspect ratio cracks in a semi-infinite body subjected to both uniform remote loading and linearly varying crack face pressure. Henceforth, this procedure will be noted as the Kobayashi-Technique in this paper.

The results of the Kobayashi-Technique are shown in Figure 4 and Figure 5 and have been adapted to the notation of the present study. The K_I values are presented in a normalized form

(E(k) is the elliptic integral of the first kind and is defined by Kobayashi et al (1974)) as a function of the circular angle θ (see Figure 1 also) and can easily be interpolated to make estimates for the aspect ratio of cracks which would exist in the weld region in the present study.

As was mentioned previously, the estimation of K_I using the Kobayashi-Technique for both the linear and uniform stress loading was made with the simplifying assumption of a semi-infinite body containing an elliptical surface crack. Since in the present study the crack depth to wall thickness ratio (a/W) for all calculations was $\leqslant 0.25$, it was thought that the effect of the outer vessel wall on the crack tip solution would be negligible. For the geometry considered in the present study, Kobayashi et al (1977) have shown the curvature correction factor would be approximately equal to one. In a later study of Kobayashi et al (1979), these curvature correction factors have been updated, but they remain unchanged for crack sizes and geometry considered in the present study.

In order to gain confidence in applying the Kobayashi-Technique,

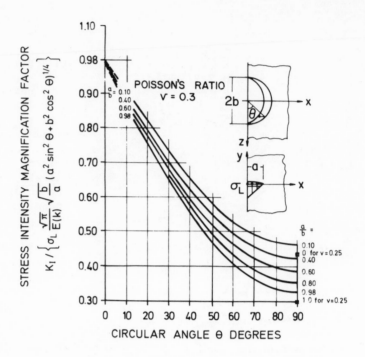

Figure 5 K_I Estimation for Linear Stress (σ_L) Loading.   After
Kobayashi et al (1974)

three comparisons were made with recent solutions of thermal shock
problems.   In Figure 6, the finite element results of Schmitt and
Keim (1978) are shown together with the results of the Kobayashi-
Technique.   The geometry analyzed was a cylindrical vessel with
inner radius R_i = 1635 mm and wall thickness, W = 160 mm.   The
crack was a semi-elliptic surface crack of aspect ratio a/b = 0.33
and depth a = 40 mm.   The loading used in the analysis was pure
thermal shock (no internal pressure loading or residual stress)
and two methods (J-integral and displacement extrapolation method)
were used to predict K_I values along the crack front.   The K_I
values calculated by the Kobayashi-Technique using the geometry
and loading parameters of Schmitt and Keim (1978) are shown in
Figure 6 (note that ϕ is defined according to Schmitt's notation).
There is good agreement but the Kobayashi-Technique predicts values
which are slightly conservative when compared with the finite-element results.

A second comparison was made with the finite element results
of Ayres (1975).   Ayres predicted the K_I values along the front of

a semi-elliptic crack of geometry a = 25.4, b = 42.3 mm and a/W = 0.2 subjected to a loading condition of thermal shock. The vessel geometry was identical to that used in the Heavy Section Steel Technology Program (HSST) at Oak Ridge National Laboratory (see Cheverton et al (1978)). The K_I estimates predicted by the Kobayashi-Technique yielded values which were 10% conservative when compared to the finite element results of Ayres as can be seen in Figure 7.

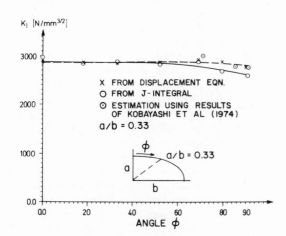

Figure 6   Comparison of the Technique of Kobayashi et al (1974) with the Finite Element Results of Schmitt and Keim (1978).

In order to predict crack initiation and arrest for the thermal shock experiments conducted in the HSST program at Oak Ridge National Laboratory, Merkle (1974) developed a semi-analytical technique to estimate K_I values along the front of semi-elliptic surface cracks subjected to a linear varying load such as that which occurs during thermal shock. This procedure was successfully used by Cheverton, Iskander and Bolt (1978) to predict crack behavior in the Oak Ridge experiments.

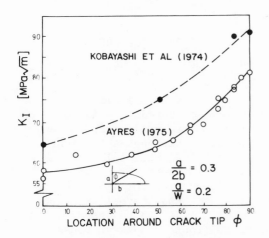

Figure 7   Comparison of the Technique of Kobayashi et al (1974) with the Finite Element Results of Ayres (1975).

periments. Merkle's procedure includes correction factors to account for the interaction of the crack with both the inner and

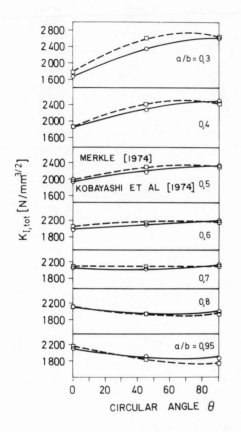

outer vessel wall. A comparison is made in Figure 8 between $K_{I,tot}$ values calculated by the Kobayashi-Technique and Merkle's semi-analytical procedure using the thermal loading and geometry conditions in the present study. For these calculations, the crack depth, a, was assumed to be 10 mm, the aspect ratio was allowed to vary between a/b = 0.30 (low aspect semi-ellipse) and a/b = 0.95 (approximately semicircular), and the loading was that for the time t_2 as shown in Figure 3. As can be seen in Figure 8, the agreement is good except for cracks of a/b ≤ 0.4 at values of θ = 45°. In this region, Merkle's procedure estimates $K_{I,tot}$ to be 10% higher than the values obtained from the Kobayashi-Technique.

STRESS INTENSITY AND CRACK BEHAVIOR PREDICTIONS

Several observations can be made from Figure 8 which allow simplification of the procedure which will be used to predict crack behavior. Although this figure is plotted only for the time t_2, these observations were verified for all times during the thermal transient. First, it can be observed that all

Figure 8   Comparison of the Technique of Kobayashi et al (1974) with the Semi-Analytical Procedure of Merkle (1974). a = 10 mm.

values of $K_{I,tot}$ around the crack front are always bounded by the values at the apex (θ = 90°) and at the surface (θ = 0°). Since these two positions are at the extremes of the fracture toughness variation (the warmer material at the apex has a higher fracture toughness than the cooler material at the vessel surface), a choice of the maximum value at each location will ensure conservativeness of the evaluation. The apex and surface values are shown in graphical form in Figure 9 for the time t_2. The value of $K_{I,tot}$ at the vessel surface (θ = 0°) has its maximum value when a/b = 0.8 and then decreases as the crack becomes more elliptic. Conversely, the apex value of $K_{I,tot}$ (θ = 90°) is maximum when the aspect ratio is smallest. For a crack of depth a = 10 mm which exists in the largest thickness weld region possible, 2b would equal 60 mm and a/b would become 0.33.

If the dimension 2b of the crack becomes large, then the aspect ratio, a/b → 0 (see Figure 1). It can be noted from Figure 9 that the functional form of $K_{I,tot}$ (θ = 90°) vs a/b is linear. In order to obtain a $K_{I,tot}$ value for the infinite axial crack (a/b → 0), a linear extrapolation was made to a/b = 0. In all cases, the extrapolated value equaled the value one would obtain from the solution for an edge crack in a semi-infinite body (see Irwin (1958)). If the average stress on the crack surfaces is used

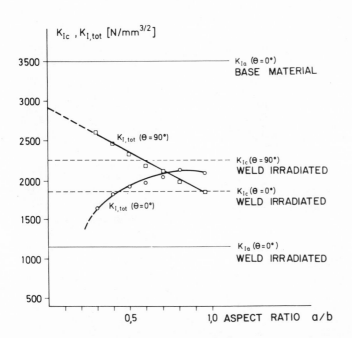

Figure 9 $K_{I,tot}$ Surface and Apex Values as a Function of Aspect Ratio with Corresponding K_{Ic} and K_{Ia} Values at the Surface and Apex for time t_2.

$$K_{I,tot} \ (a/b \rightarrow 0) = 1.12\sqrt{\pi a} \ \frac{\sigma_u + \sigma_{tot}}{2} \ .$$

Other aspects of Figure 9 will be discussed later in this paper.

In order to make a prediction of crack extension, a comparison must be made between the $K_{I,tot}$ value and the K_{Ic} value at the corresponding point on the crack front. In Figure 10 the maximum $K_{I,tot}$ values at the apex (a/b = 0.33), the surface (a/b = 0.8) and for a long axial crack (a/b → 0) calculated using the Kobayashi-Technique are given as a function of temperature for all times during the thermal event. The crack depth (a = 5 mm) was assumed to be the maximum which could occur in the non-detectable region. According to ASME Boiler and Pressure Vessel Code, Section XI, a 10 mm deep crack was then used in the calculations. In order to assess crack extension, the K_{Ic} values for the less-tough weld region are also given as a function of temperature. As can be seen in the figure, the $K_{I,tot}$ values for all three aspect ratio cracks start at low values since the effect of a sudden inner wall cooling has not yet caused a steep thermal stress gradient. However, later in the transient, the $K_{I,tot}$ values increase in magnitude as the thermal stresses become larger. The ratio, $K_{I,tot}/K_{Ic}$, controls crack extension. When $K_{I,tot}/K_{Ic} \geq 1$ the crack will begin to extend and this can be seen to occur for all three aspect ratio cracks at their respective locations at a time t_1 ($t_1 < t_2$, see Fig. 10). This crack extension is based on a comparison with the K_{Ic} values for an irradiated weld material and in actuality would occur only at the apex. A crack of aspect ratio of a/b = 0.33 would have a surface length of 2b = 60 and therefore would encounter irradiated base material which has a much higher fracture toughness at the same temperature (see Figure 2).

This can be better explained by again considering Figure 9. The $K_{I,tot}$ values at the surface location ($\theta = 0°$) and apex location ($\theta = 90°$) have been plotted as a function of aspect ratio a/b for the time t_2. Although plots of this type were made for all times, the main emphasis can be put on the time t_2 because at this time the $K_{I,tot}/K_{Ic}$ is maximum for the different crack geometries of a = 10 mm (see Figure 10, t = t_2). It can be noted from this figure that any elliptical shaped crack (including semi-circular) of depth a = 10 mm will extend at some location if situated in the weld region. However, it should be realized that the crack encounters tough base material when it reaches a surface dimension of 2b = 60 mm. In this study, it was assumed that this transition in toughness was abrupt whereas in actuality this is probably not true. Nevertheless, the crack will always extend at the apex since $K_{I,tot}(\theta = 90°)/K_{Ic}$ ($\theta = 90°$) is greater than one or will become greater than one as the crack extends at the surface. For this reason, a study of the crack behavior at the apex is required and is shown in Figure 11. If the crack arrests at the surface (2b = 60), then the $K_{I,tot}$ value at the apex can be computed using the Kobayashi-Technique up to a depth of approximately a = 30 mm (a/b = 1 ~ 0.95). It can be observed that as the crack is allowed to extend in the depth direction the $K_{I,tot}$ value begins to drop. If an extrapolation of the noted trend is made, then it can be seen that the crack will arrest at a ~ 35 mm. During this period of extension, the $K_{I,tot}$ surface value remained below the upper limit required for reinitiation. The crack "hinges" when it encounters the base material at the surface and is not allowed to become a

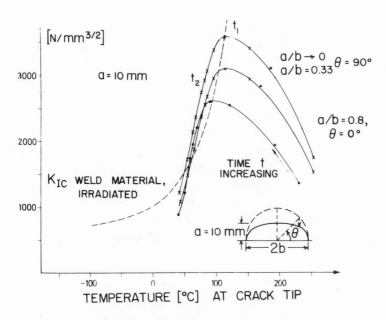

Figure 10 $K_{I,tot}$ and K_{Ic} Values as a Function of Temperature.
            $K_{I,tot}$ Calculated Using the Kobayashi-Technique.

long axial crack. This allows calculation of $K_{I,tot}$ values at the
apex of surface cracks which are smaller than those required if
the crack becomes very long (see Figure 10 for a/b → 0). Similar
calculations were peformed for earlier and later times in the
event but the apex arrest position of the crack for these times
was less than that for time t_2.

CONCLUSION

     The technique of Kobayashi et al (1974) for estimating K_I dis-
tributions for varying aspect ratio elliptical surface cracks subjected to a linear varying load has been applied to a thermal
stress problem which results in such a loading. The assumptions
necessary for applying the technique have been discussed and shown
to be justifiable for the crack sizes studied. The analysis gives
the procedure for predicting crack extension or arrest and shows
that it may be advantageous to consider such an analysis when material variations exist along the crack front in three dimensional
crack body problems.

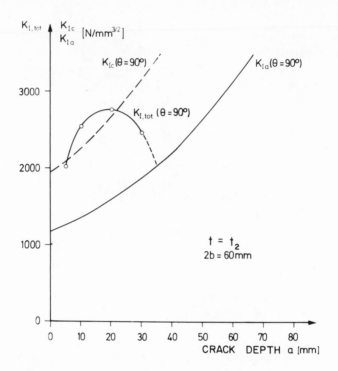

Figure 11 Comparison of $K_{I,tot}$ with K_{Ic} and K_{Ia} at the Apex of a Growing Surface Crack with 2b = 60 mm.

An analysis of this type should be of a preliminary nature only. For a detailed safety analysis, refined numerical and experimental studies which include the effects of plasticity and material property changes in the crack front region should be conducted.

ACKNOWLEDGEMENTS

The first author would like to thank his colleagues at the Fraunhofer-Institut für Werkstoffmechanik in Freiburg, West Germany, where this study was conducted. Special thanks are due Dr. Erwin Sommer, Institute Director, and Dr. Jörg Kalthoff for their valuable discussions during this study. The support of Prof. D. F. Frederick, Head of the Engineering Science and Mechanics Department at Virginia Polytechnic Institute and State University, is also appreciated.

REFERENCES

Ayres, D. S., 1975, Three-Dimensional Elastic Analysis of Semi-
    Elliptical Surface Cracks Subject to Thermal Shock, in: "Com-
    putational Fracture Mechanics", E. F. Rybicki and S. E. Benz-
    ley, eds., ASME, New York.
Cheverton, R. D., Iskander, S. K., Bolt, S. E., 1978, Applicability
    of LEFM to the Analysis of PWR Vessels Under LOCA-ECC Thermal
    Shock Conditions, Oak Ridge National Lab. NUREG/CR-0107,
    ORNL/NUREG-40.
Irwin, G. R., 1958, The Crack Extension Force for a Crack at a
    Free Surface Boundary, Report No. 5120, Naval Research Labora-
    tory.
Kantorovich, L. V. and Krylov, V. L., 1964, "Approximate Methods of
    Higher Analysis", Interscience, New York.
Kobayashi, A. S., Enetanya, A. N., Shah, R. C., 1974, Stress In-
    tensity Factors of Elliptical Cracks, in: "Proceedings of
    Conference on the Prospects of Advanced Fracture Mechanics",
    G. C. Sih, H. C. Van Elst, and D. Broek, eds., Noordhoff In-
    ternational Publishing, Leyden.
Kobayashi, A. S., Polvanich, N., Emery, A. F. and Love, W. J.,
    1977, Surface Flaw in a Pressurized and Thermally Shocked
    Hollow Cylinder, Int. J. Pres. Ves. & Pipings, 5(2): 103.
Kobayashi, A. S., Emery, A. F., Love, W. J. and Jain, A., 1979,
    Further Studies on Stress Intensity Factor of Semi-Elliptical
    Crack in Pressurized Cylinders, in: "Transactions of the 5th
    International Conference on Structural Mechanics in Reactor
    Technology - Vol. G", T. A. Jaeger, B. A. Boley, eds., North-
    Holland Publishing Company, Amsterdam.
Merkle, J. G., 1974, Quarterly Progress Report on Reactor Safety
    Programs Sponsored by the Division of Reactor Safety Research
    for July, ORNL/TM-4729, Vol. II, pp. 3-22.
Peters, W. H. and Blauel, J. G., 1979, Bruchmechanische Analyse
    des Vehalten von Axialrissen in einem RDB bei
    Kühlmittelverlust-Störfällen, Institut für Festkörpermechanik-
    Bericht V 7/79.
Schmitt, W. and Keim, E., 1979, Linear Elastic Analysis of Semi-
    Elliptical Axial Surface Cracks in a Hollow Cylinder, Int. J.
    Pressure Vessels and Piping, 7: 105.

ANALYSES OF MECHANICAL AND THERMAL STRESSES

FOR LOFT DENSITOMETER MOUNTING LUG ASSEMBLY

G. Krishnamoorthy and F. H. Chou

Professors of Engineering Mechanics
College of Engineering
San Diego State University
San Diego, California 92182

ABSTRACT

An analysis of mechanical and thermal stresses for LOFT Gamma Densitometer mounting lugs, bolts and pipe to which the lugs are attached is presented in this paper. The analysis does not include the densitometer of the supporting structure. The mechanical loads under consideration include internal pressure, seismic loads and dynamic loads due to blow down. Thermal transients associated with normal and upset operating conditions have been considered. Appropriate finite element models were constructed for obtaining stresses and deflection analyses due to mechanical loads and thermal transients. Existing computer code of SAPIV modified for Large core Memory (LCM) for CDC 7600 and SAASIII were used for the computation of mechanical and thermal stresses. Procedures to analyze the data obtained from computer analyses for ASME code evaluation are outlined. A stress summary is presented.

DESCRIPTION OF THE PROBLEM

The LOFT densitometer to be analyzed is mounted on a horizontal portion of primary coolant pipes of inner diameter 5.6 in. and outer diameter 7 in. made out of 316-S stainless steel. The installation includes four detectors within a tungsten alloy and lead shield/ collimator assembly and a shielded gamma source and actuator system as shown in Figure 1. Both the detector shield and the source shield are mounted on lugs (316 stainless steel) which are welded to the pipe. The purpose of the entire assembly is to monitor the flow in the coolant pipe. The total weight of inner shield and detector assembly is just under 450 pounds. Inner shield assemblies are

287

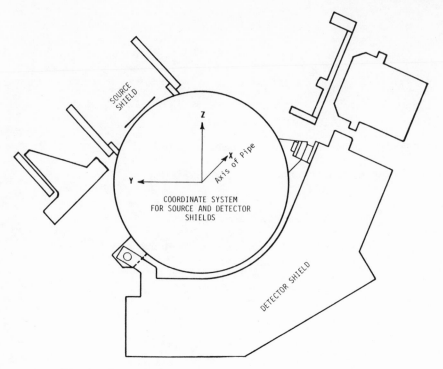

Figure 1. Schematic View of Loft Densitometer

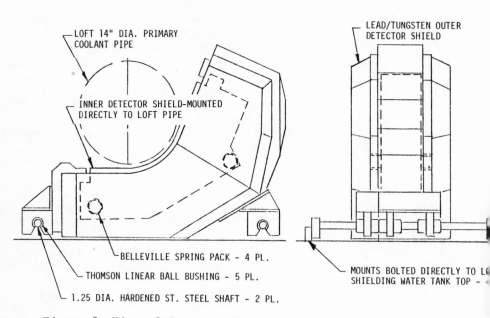

Figure 2. View of Support of Outer Shield and Inner Shield

attached to their mounts through water cooling plates and insulators
to reduce heat transfer from 650° LOFT pipe. The lower detector
shield incorporates a sliding feature which will eliminate thermal
expansion - induced stresses due to heat-up. Detector assemblies
are mounted to their inner shield with no metal to metal contact
and are thermally insulated.

An exterior shield, made of tungsten alloy and lead parts surrounds the interior detector shield but is mounted separately to
the "floor" as shown in Figure 2. A gap between inner and outer
shields allows the specified relative motion between the inner and
outer shields to accommodate the thermal expansion and Loss of
Coolant Event (LOCE). As the expected thermal expansion along the
axis of the pipe is very large the outer shields are mounted on
rails running parallel to the axis of the pipe via Thompson linear
ball bushings. The inner and outer shields are axially located
with respect to each other by opposed Belleville spring packs
mounted in the outer shield as shown in Figure 2. Under thermal
expansion, these springs will push the outer shields along with the
inner shields. However, during the high frequency vibration of the
pipe, the springs will deform and the inner shields mounted on the
pipe can follow the motion of the pipe freely without moving the
outer shield. The source is shielded into a tungsten alloy and lead
shield with a total weight of approximately 280 pounds. A water
cooled mounting plate incorporating the same features as the detector
mounts, attaches the source shield and the actuator to the main pipe.
For the purpose of computing stresses in the lugs and in the pipe,
three substructures are constructed. They are: (1) Inner shield
with lugs, (2) Source shield with lugs, and (3) LOFT primary coolant
pipe. Types of mechanical and thermal loads used in the analyses
are discussed in the following sections.

ANALYSIS OF STRESSES DUE TO MECHANICAL LOADS

The structural model including pipes and lugs is extremely
large even to accommodate the largest and fastest computer available
to date. Therefore the structural analyses were done in three parts
to reduce the size of the problem. (1) To compute the stresses in
the lugs and the bolts supporting the detector shield, (2) To compute the stresses in the lugs and the bolts supporting the source
shield, (3) To compute the stresses in the pipe. In both parts (1)
and (2), the lug-pipe interfaces were treated as the supports of the
structures under consideration. Resulting stresses at these interfaces were converted to loads for part (3) analysis. Overall structural models used in parts (1) and (2) analyses are shown in Figures
3 and 4. The structural models to compute pipes stresses in part
(3) are shown in Figures 5 and 6. It must be noted that the length
of the pipe was chosen on either sides of the lugs so that the
boundary effects at the ends of the pipe will have negligible effect
in the area of interest. Details of these three-dimensional models,

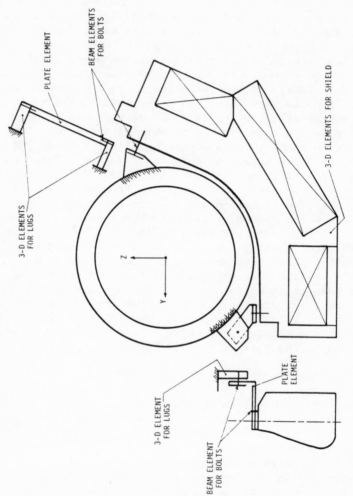

Figure 3. Structual Model for Detector Shield and Lugs

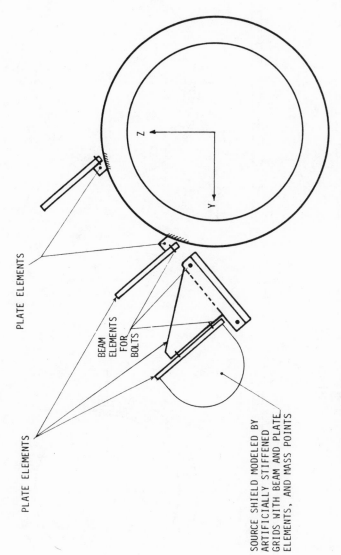

Figure 4. Structural Model for Source Shield and Lugs

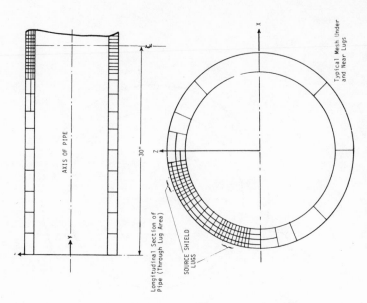

Figure 6. 3-D Model of Primary Coolant Pipe For Source Shield Lugs

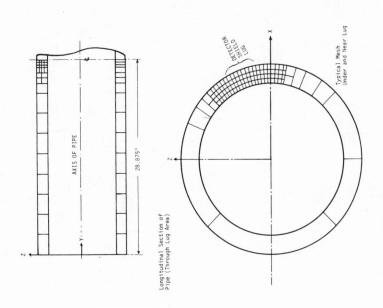

Figure 5. 3-D Model of Primary Coolant Pipe For Detector Shield Lugs

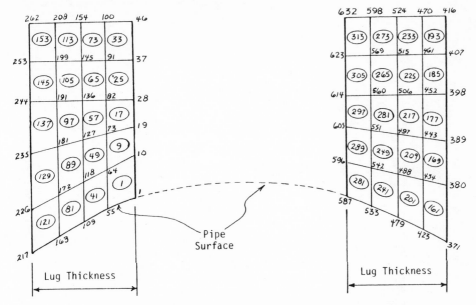

Figure 7. 3-D Model of Source Shield Lugs
(Typical Slice)

such as numbering of nodes and elements etc. are too lengthy to be
reproduced here. (See Reference 1). A sample slice of lug elements
and node members are shown in Figure 7. Loads used in this analysis
include deadweight, internal pressure, operating basis earthquake
(OBE), safe shut down earthquake (SSE) and loss of coolant event
(LOCE) mechanical. Spectra for OBE, SSE and LOCE mechanical were
properly chosen and "smoothed" out following NRC guideline (See
Appendix A). The SAPIV (Reference 2) program with extended core
version was used for spectrum analyses and computing stresses due
to various loads. All computations were performed by CDC 7600 at
Lawrence Radiation Berkeley Laboratory by remote terminal entry
available in San Diego.

ANALYSIS OF STRESSES DUE TO THERMAL TRANSIENT

     Finite element technique was also employed to analyze the
stresses in the lugs and the pipe. However, due to the availability
of computer code and the consideration of computing time, two dimensional analyses were performed. In order to perform the analyses
to duplicate the three dimensional model proper modifications were
performed in the properties of material and boundary condition of
the structural model. In analyzing the stresses in the pipe and
the lugs associated with the detector shield, an axisymmetric model
was assumed (See Figure 8) as the thickness of the lugs i.e. 1.5
inches is very small compared to the length of the pipe in the axial
direction. However, for source shield lugs and its associated

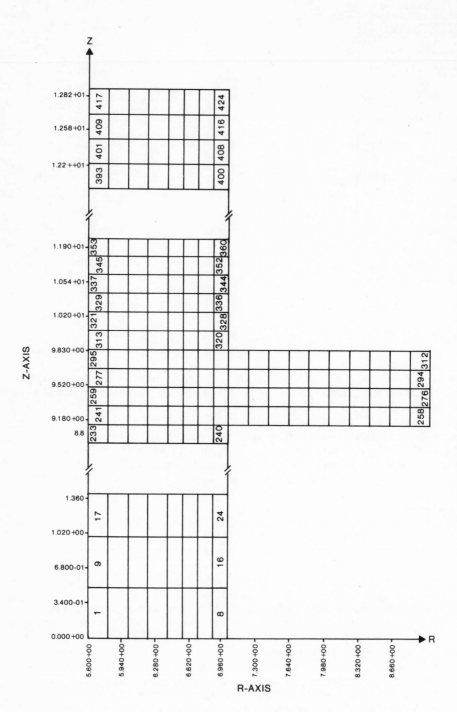

Figure 8. Axisymmetric Model for Thermal Analysis of Detector
Shield Lugs and Pipe

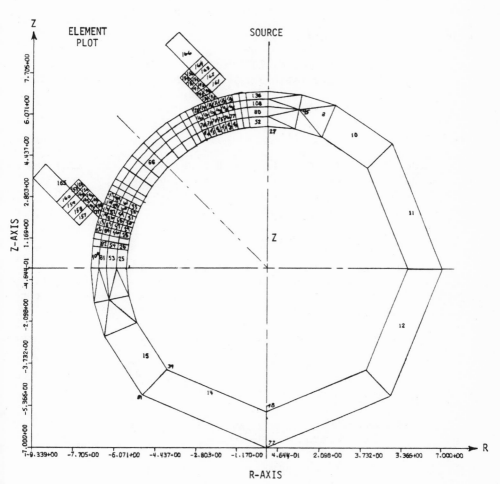

Figure 9. Plane Strain Model For Thermal Analysis of Source
Shield Lugs and Pipe

portion of the pipe, the lugs are 4.0 inches long in the axial direc
tion of the pipe and 0.75 inches thick and therefore, plane strain
model was used in this case (See Figure 9). It has been concluded
from previous data analyses (Reference 3) that these two dimensional
models were conservative and thus considered to be practical and
acceptable approximation. The thermal transients to be considered
include heat-up cycle, cool-down cycle and LOCE-thermal cycle. For
each cycle, temperatures at various locations and various time
slices are specified, stresses are computed based on the well-
established theory of thermal elasticity. These temperature-
location time data were established by EG & G Inc. Idaho based on
field measurements. SAASIII computer code (Reference 4) was used
in computing the thermal stresses for both models.

FATIGUE

     Stress levels obtained from mechanical and thermal analyses are
also used for the fatigue analysis required by the ASME code
(Reference 5). Load cycles for fatigue analysis are described as
follows:

Type 1 Load Cycle (1500 cycles)

     System goes from Ambient through heatup (at 100° F/Hr) to a
temperature of $T \simeq 540^\circ F$. Then 100 % load change occurs (reactor goes
from zero power to 100% power in 10 minutes) and steady-state conditions are established (p = 2250 psi, $T = 610^\circ F$). Then cooldown
occurs and system goes back to ambient. A total of 1500 cycles was
considered. 100% load change occurs several times for each heatup-
cooldown cycle. The additional 9500 load change cycles (11,000-
1500) were considered under Type 2 load cycles presented below.

Type 2 Load Cycle (9500 cycles)

     System goes from $T = 540^\circ F$ to $T = 610^\circ F$ in 20 min. due to 100%
load change, then returns to $T = 540^\circ F$.

Type 3 Load Cycle (120 cycles)

     This type is the same as Type 1 except that single control rod
drop transient is included which causes the system temperature to
drop from $T = 610^\circ F$ to $T = 470^\circ F$. Temperature then goes back up to
$T = 610^\circ F$ before cooldown occurs.

Type 4 Load Cycle (2840 cycles)

     This type is also the same as Type 1 except that SCRAM tran-
sient is included which causes system temperature to drop from
$T = 610^\circ F$ to $T = 530^\circ F$. Temperature then goes back to $T = 610^\circ F$
before cooldown.

Type 5 Load Cycle (280 cycles)

System goes from ambient tó steady-state conditions as for Type
1 Load Cycle then blowdown occurs (LOCE/ECC), then system returns to
ambient.

The number of permitted cycles for computed stress levels (5a) in
the lugs and the pipe can be found in Reference 3.

ANALYSES OF DATA

The results obtained from all the computer runs of both SAPIV
and SAASIII programs are components of stresses in the global directions of the particular structure under consideration. These stress
components are defined only at the centroids of each element.
Extensive data-manipulation and modifications were carried out in
order to check and satisfy the ASME B & PV code Section III allowables for class I components. The procedure to analyze these data
are described in the following steps: (1) From large volume of computer outputs, critical sections showing high stresses for various
individual loading conditions have to be identified. For both
mechanical and thermal loads, critical sections exist near or under
the lug-pipe interface as anticipated. As thermal stresses are one
magnitude higher than the other stresses examination of stress combinations was restricted to these critical areas. (2) wherever
global coordinates are different from local coordinates, components
of stresses obtained from computer analyses were transformed to local
coordinate systems. Local coordinates for pipe elements are three
axes in the axial, radial and tangential directions of the pipe.
Local coordinates for lug elements are three axes along the length,
width and thickness, respectively, of the lug. Detailed works of
extrapolization of stresses are presented in Appendix B. (3) Components of stresses in local coordinates due to various loading conditions, both mechanical and thermal, are superimposed following
ASME code specified (Reference 5) loading combinations. (4) Based
on the values of the superimposed stress components which are only
defined at the centroids of various elements, stresses at inner and
outer surfaces of the pipe, and at surfaces of the lugs are obtained
by a quadratic extrapolation. Stresses in particular those due to
thermal loads, are not linear across the thickness of the pipe or
lugs. Maximum stresses do exist generally at surfaces of various
cross-sections. (5) Principal stresses and stress intensities are
computed from critical stress components obtained in Step (4). It
should be repeated here that work of both Step (4) and Step (5)
are done for each load combination and for all critical sections.
From the results of Step (5), the most critical stresses and stress
intensities of each loading combination are chosen for evaluations
based on ASME code requirements (Reference 5). To handle the work
outlined by Step (1) through Step (5) which involves enormous data
handling, several computer programs in FORTRAN Language were prepared

and interconnected. Stress components across the thickness of the
pipe are also plotted. All detailed works can be found in Reference
1. Numerical procedures for transformation and extrapolation of
stresses are presented in Appendix B.

STRESS SUMMARY

The most critical stresses in the lugs and the pipe are summarized in the following table for evaluation based on ASME code
requirements. All notations used in the table are the standard
notations of Reference 5.

STRESSES IN LUGS AND PIPE

Item	Mat'l	Temp.	S_m	$P_m < S_m$	$P_L + P_B < 1.5 S_m$	$P_L + P_B + Q$
Lug	316	$500^{\circ}F$	17,900	5,937	13,103	$42,613 < 3 S_m$
Pipe	St Stl	$650^{\circ}F$	16,600	10,959	16,740	$74,633 > 3 S_m^*$

*Usage factor is computed to be 0.701

CONCLUDING REMARKS

A complete analysis of stresses in the lugs and pipe of a
Gamma Densitometer Assembly is presented. Although the finite element models for thermal stress analyses are two dimensional ones,
the combination of mechanical and thermal stress of critical sections
are judisciously chosen so that the resulting stresses will be conservative. It should be pointed out that a section by section plot
of all nodes and elements are made so that possible mistakes of input
data such as noding or connectivity are eliminated. The use of large
computer code such as SAPIV and SAASIII and the programs mentioned
earlier for data analyses provided an efficient combination necessary for stress analysis of problems of this type.

REFERENCES

1. G. Krishnamoorthy and F. H. Chou, "Stress Report for LOFT Den-
    sitometer Mounting Lug Assembly," Technical Report, Science
    Application Inc., La Jolla, California. Report No. SAI-018-79-
    602-LJ. (1979).

2. Klaus-Jurgen Bathe, SAPIV: A Structural Analysis Program for
    Static and Dynamic Response of Linear Systems, University of
    California, Berkeley, Earthquake Engineering Research Center

Report No. PB-221 967.   (June 1973).

3. R. A. Lindley, "LOFT Primary Coolant Loop Thermal Transient
    Analyses," Aerojet Nuclear Company, Report No. LTR 141-30
    (TR-588).   (Jan. 1975).

4. J. G. Crose and R. M. Jones, SAASIII, Finite Element Stress
    Analysis of Axisymmetric and Plane Solids with Different Ortho-
    tropic, Temperature Dependent Material Properties in Tension
    and Compression, Aerospace Corporation, Report No. TR-0059
    (56816-53)-1.   (June 1971).

5.   (i) ASME B & PV Code Section III, Division 1;
      (ii) ASME B & PV Code Nuclear Power Plant Components, Division
    1 Appendices.

APPENDIX A

Coordinate   systems and smoothed spectrum envelopes are shown
in the following figures.

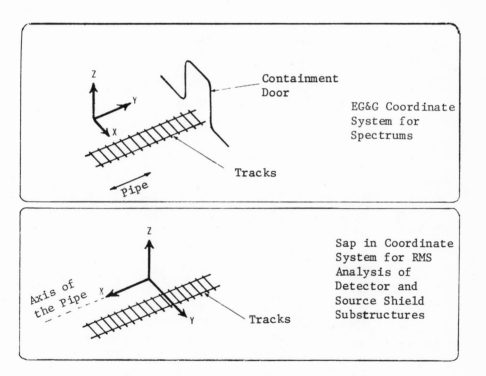

Figure A1 Coordinate System for Spectrums of EG&G and the
Substructures for Computer Analyses

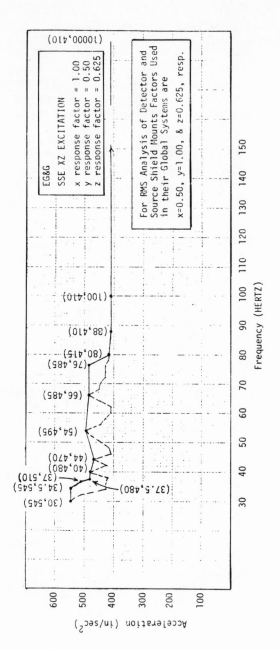

Figure A2.　Smoothed SSE-XZ Envelope

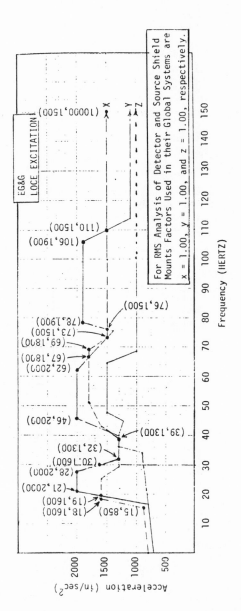

Figure A3. Smooth LOCE Envelope

APPENDIX B

EXTRAPOLATION AND LINEARIZATION OF STRESSES

The surface stresses in the pipe for a given cross section shown in Figure A4 were obtained using the following procedure.

The computer analysis of pipe for mechanical loads gave the stress components $(\sigma_{ij})_i$ in the global direction at the center (R_i) of the three dimensional finite element (where i = 1,2,3 or 4). These stress components are rotated to the local coordinate system as shown in Figure A4 to obtain the stress components $(\bar{\sigma}_{ij})_i$ in the local system.

$(\bar{\sigma}_{ij})$ at Radius = 7.00 in.

A second degree parabolic fit was obtained for each component of $(\bar{\sigma}_{ij})_i$ at any r_i could be computed

$$(\bar{\sigma}_{ij})_i = A_{ij}r_i^2 + B_{ij}r_i + C_{ij} \tag{A1}$$

where

$$A_{ij} = \frac{\begin{vmatrix} (\bar{\sigma}_{ij})_2 & R_2 & 1 \\ (\bar{\sigma}_{ij})_3 & R_3 & 1 \\ (\bar{\sigma}_{ij})_4 & R_4 & 1 \end{vmatrix}}{\Delta} \tag{A2}, \qquad B_{ij} = \frac{\begin{vmatrix} R_2^2 & (\bar{\sigma}_{ij})_2 & 1 \\ R_3^2 & (\bar{\sigma}_{ij})_3 & 1 \\ R_4^2 & (\bar{\sigma}_{ij})_4 & 1 \end{vmatrix}}{\Delta} \tag{A3},$$

$$C_{ij} = \frac{\begin{vmatrix} R_2^2 & R_2 & (\bar{\sigma}_{ij})_2 \\ R_3^2 & R_3 & (\bar{\sigma}_{ij})_3 \\ R_4^2 & R_4 & (\bar{\sigma}_{ij})_4 \end{vmatrix}}{\Delta} \tag{A4}, \qquad \Delta = \begin{vmatrix} R_2^2 & R_2 & 1 \\ R_3^2 & R_3 & 1 \\ R_4^2 & R_4 & 1 \end{vmatrix} \tag{A5},$$

where

R_2 = 6.2 in., R_3 = 6.550 in., and R_4 = 6.850 in.

Using the Eq. (A1), $(\bar{\sigma}_{ij})$ at the surface where R = 7.000 in. were found. In order to combine the thermal analysis later $(\bar{\sigma}_{ij})$ at r = 6.300, 6.475, 6.625, 6.775 in. were also evaluated.

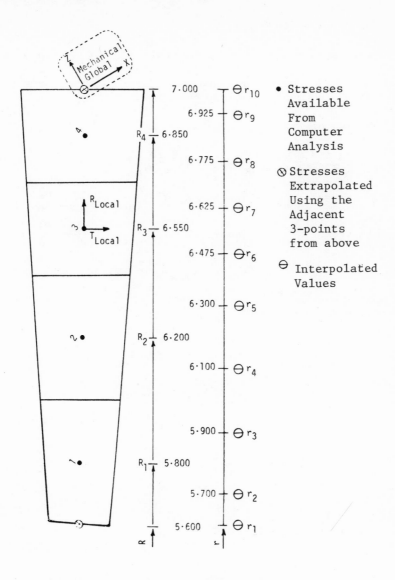

Figure A4   Local and Global Coordinates for
Extrapolation and Interpolation of σ_{ij}

$(\bar{\sigma}_{ij})$ at Radius = 5.600 in.

The inner surface stress components $(\bar{\sigma}_{ij})$ were evaluated using the values of the radius at points 1, 2, and 3 respectively. In order to combine the thermal analysis later $(\bar{\sigma}_{ij})$ at r = 5.700, 5.900, and 6.100 were evaluated.

Linearization of Stresses

Bending and membrane stresses were obtained using the following linearization technique on $(\bar{\sigma}_{ij})$ (See Fig. A4).

$$(\text{Bending stress})_{inner} = \text{Linearized } (\bar{\sigma}_{ij})_{inner} = \frac{c\alpha - A\beta}{BC - EA} \qquad (A6)$$

$$(\text{Bending stress})_{outer} = \text{Linearized } (\bar{\sigma}_{ij})_{outer} = \frac{B\beta - E\alpha}{BC - EA}$$

$$x \ (r_{outer} - r_{inner}) + (\bar{\sigma}_{ij})_{inner} \qquad (A7)$$

$$(\text{Membrane stress}) = (\bar{\sigma}_{ij})_{middle} = \frac{1}{2} \ x \ \left[\text{Linearized } (\bar{\sigma}_{ij})_{inner} \right.$$

$$\left. + \text{Linearized } (\bar{\sigma}_{ij})_{outer} \right] \qquad (A8)$$

$$\text{where} \quad \alpha = \sum_{i=1}^{i=9} \frac{1}{3} \ (r_{i+1}^3 - r_i^3)P_i + \frac{1}{2} \ (r_{i+1}^2 - r_i^2) \ (\bar{\sigma}_{ij} - r_i P_i) \qquad (A9)$$

$$\beta = \sum_{i=1}^{i=9} \frac{1}{4} \ (r_{i+1}^4 - r_i^4)P_i + \frac{1}{3} \ (r_{i+1}^3 - r_i^3) \ (\bar{\sigma}_{ij} - r_i P_i) \qquad (A10)$$

$$\text{and } P_i = (\bar{\sigma}_{i+1,j} - \bar{\sigma}_{i,j})/(r_{i+1} - r_i) \qquad (A11)$$

THERMAL STRESSES IN HEAT-ABSORBING BUILDING GLASS
SUBJECTED TO SOLAR RADIATION

Dieter Stahn

Fraunhofer-Institut für Werkstoffmechanik
Rosastrasse 9, D-7800 Freiburg
West Germany

INTRODUCTION

Correct design of building glass installations against failure
by fracture is an engineering problem of particular importance,
especially because of the increasing tendency of their all-out use
in modern building fronts. Existing guidelines for design ensure
satisfactory structural performance in most cases, as empirical
findings show, but they often lack a detailed theoretical foundation. This deficiency should be reduced by an intensified analysis
of the loading conditions which glass panes are subjected to in
the field. Such efforts will also help to avoid uneconomical, conservative solutions.

The complexity of the problem necessitates that the various
factors affecting the structural reliability of building glass
should be investigated separately, if possible. The present work
is intended to contribute to such an analysis. Before proceeding
to the special topic, it may be useful to outline at least some
of the aspects which must be considered in solving the over-all
problem. The assessment of the safety of building glass installations has to take into account the expected mechanically and thermally induced service stresses as well as the factors affecting
glass strength. Glass strength is mainly determined by the internal stress state, the edge and surface quality, the dimensions of
the panes, the environmental conditions, and the loading history.
The mechanical causes that may induce local stresses of a considerable magnitude include wind pressure, supersonic booms, building
settling, and the weigth of the panes in connection with setting
blocks within the frames.

Thermal stresses in building glass installations, the topic of this paper, originate because of nonuniform temperature distributions in the panes. Whether generated by sun exposure, radiant heaters on the roomside, heating registers directing air against the glass, or the laying of hot poured asphalt in a new building, thermal stresses are most serious in heat-absorbing glass[1-5]. Such glass is often used to provide a satisfactory indoor environment without large solar heat gains to the buildings.

FACTORS AFFECTING THE THERMAL STRESSES

The development of the thermal stresses in building glass installations is generally well understood. The central exposed areas of the panes increase in temperature due to the absorption of the incident heat radiation or convection, while the edges remain cool since they are shielded by the edge cover of the frames. Temperature differences similar to these may also result from additional partial shading of the panes by other screens. Thus, the areas of increased temperature try to expand against the restraint of the cooler edges and consequently edge tension stresses result.

These edge tension stresses depend on the temperature difference between the hottest and coldest areas of the glass, and on the detailed position of the temperature gradients across the panes. They are also a function of the size of the panes and of material properties, such as Young's modulus and the coefficient of thermal expansion. However, the latter dependencies will not be investigated here; the thermoelastic properties of building glass normally do not vary much anyway.

To investigate more closely on what factors the temperature difference between the hottest and coldest areas of the glass may depend, we consider for the moment a window consisting of a uniformly sunlit single pane. The intensity of the incident solar radiation whose absorbed energy heats the glass, is influenced by circumstances such as orientation of the window, time of the day, season, amount of cloud cover and atmospheric pollution, latitude and altitude of the building, reflections from the ground or adjacent structures, reflecting indoor blinds or curtains etc. The extent of the absorbed solar energy is determined by the type of glass, i.e. by its absorption factor governing the spectral absorption of the incident radiation, and by the thickness of the pane. As the absorbed solar energy is subsequently dissipated by radiation, convection, and conduction from the two glass surfaces, the rise of temperature in the exposed areas of the glass is also influenced by the temperatures and the velocity of the surrounding air and by adjacent insulating conditions. The importance of

thermal insulation from the building structure, especially in the
case of heavy masonry with large thermal inertia, is sometimes not
fully realized. It is also in this respect that selecting suitable
material and proper design of the framing system for building
glass installations deserve special attention[1,2,6,7].

As already mentioned, the position of the temperature gradients introduced by additional partial shading of the glass panes
may also affect the edge tension stresses. Such shadows may be
cast during sunshine hours, for instance, by protruding parts of
the structure, such as roofs, balconies, pillars, by fixtures,
such as awnings or louvers, by adjacent structures or near-by
trees, etc.

The present work represents an examination of the contributions of outdoor shading patterns to thermal stresses in heat-
absorbing building glass. Without a detailed consideration of the
shadow sources, the effect of a given temperature difference
between different areas of the panes is investigated varying the
shadow configurations systematically. These effects have scarcely
been studied in the past [1,4,8,9].

ASSUMPTIONS AND REMARKS RELATING TO THE NUMERICAL THERMAL STRESS
ANALYSIS

Since glass panes singly glazed in building fronts show no
significant thickness temperature variation, a nonuniform temperature distribution across such a pane only induces a two-dimensional
stress field. Consider a framed rectangular pane of annealed heat-
absorbing building glass to be 1000 x 500 mm^2 in size, unconstrained at the boundaries, and free from external forces. The
pane is supposed to be subjected to uniform solar radiation (or
heat radiation of any other source), but partially shaded by the
edge cover of the frame and additional screens of varying configurations. The width of the edge cover is taken as 20 mm. The temperature difference $\Delta \vartheta$ between the radiation exposed and the shaded
areas of the pane is assumed to be 20 degrees Kelvin. Temperature
differences of this magnitude frequently occur in building glass
installations in the field.

The assumption that $\Delta \vartheta$ at the shadow boundaries near the edges
of the pane (as in the case of the edge cover shading) is as high as
at the boundaries located within the central area, accounts for
the possibility of heat conduction towards a frame of large thermal inertia. This is a conservative consideration for most cases,
because the actual stresses may be overestimated.

In order to facilitate the provision of the temperature input
data for the thermal stress calculation, the temperature profiles

at the shadow boundaries are supposed to be in the form of steps,
instead of smoothly varying distributions. This may also lead to
an overestimate of the actual loading. In a later section the
effect of stepwise and measured temperature profiles will be compared.

The thermal stresses induced by different temperature distributions in the pane are calculated by the finite element method
using the program Solid SAP[10]. In a presentation of the results
it is sufficient to show only the edge stresses, instead of the
complete stress fields across the pane. This can be justified by
the following reasons. In the first place it is found, that the
relevant tensile stresses within the area of the pane are always
less than the maximum tension stresses at the edges. Secondly, the
edges of the pane represent a location of higher breakage potential because of the increased probability of weakening defects
there. Many minute flaws are introduced into the edges when glass
is cut to size, and further edge damage may occur during the
glazing procedure.

RESULTS AND DISCUSSION

Effect of the Edge Cover of the Frame

The first case considered here is illustrated in the small
figure inserted in Fig. 1. The pane is shaded by the 20 mm wide
frame only, and the temperatures within the central area are constant and 20 K higher than underneath the edge cover. The stress
components are calculated using a coordinate system also introduced in this small figure, and with Young's modulus $E = 7.43 \cdot 10^4$
MPa and the coefficient of linear thermal expansion $\alpha = 8.2 \cdot 10^{-6}$
K^{-1}.

Fig. 1 shows the distribution of the (normal) stress σ_y along
the longitudinal edges of the pane; σ_y acts parallel to the edges.
Because of symmetry, it is sufficient to show the stresses only
for the upper half of the pane ($0 \leq y \leq 500$ mm). One should note
that the tensile stresses are relatively constant at the maximum
value over a large part of the edges. Towards the corners σ_y
decreases to zero as expected, since the pane is supposed to be
free from external edge restraint.

Fig. 2 represents the distribution of the stress σ_x along
the upper and lower transverse edges of the pane ($y = \pm 500$ mm).
The magnitude of the maximum tension stress in the center of these
edges is almost the same as that at the longitudinal edges.

It was verified that the tension stresses σ_y and σ_x at the

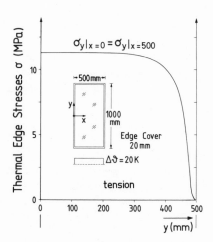

Fig. 1. Tension stress σ_y at the longitudinal edges of a partially
shaded pane. They temperatures within the central area are
assumed to be constant and 20 K higher than underneath
the 20 mm wide edge cover.

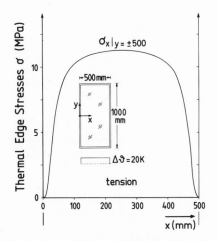

Fig. 2. Tension stress σ at the transverse edges of a pane
shaded by the edge cover of the frame only.

longitudinal and transverse edges of the pane are essentially
equal to the maximum principal tensile stress in each case, except
at the corners. Thus, since glass fracture is governed by the
maximum principal tensile stress, the σ_y- and σ_x-values can be
used directly in a safety assessment

In the simple case considered here the numerical results can
be checked by an analytic solution which is at least applicable
to certain regions of the thermally loaded pane [8]. Good agreement
was obtained.

Effect of an Additional Shading Parallel to the Longitudinal Edge

Extending the situation given in Fig. 1, an additional partial
shading with a shadow boundary parallel to a longitudinal edge of
the pane is now considered. Fig. 3 shows the distribution of the
edge stress σ_y along the left (x = 0) and right (x = 500 mm) longitudinal edge; the additional shadow on the right side of the pane
is 167 mm in width. The range of the maximum tension stress $\sigma_y|_{x=0}$
is somewhat narrower compared to the case of Fig. 1, but the magnitude is greater by about 27%. At the right longitudinal edge the
tensile stresses are relatively negligible and near the corners of
the pane compressive stresses occur.

As illustrated in Fig. 4, the maximum σ_x tension stress at the
transverse edges (y = $\pm$ 500 mm) is of almost the same magnitude as
the maximum σ_y at the left longitudinal edge (see Fig. 3). The
range of the maximum, however, is much more confined compared to
the case of Fig. 2.

With these results it is obvious that additional partial
shading of the glass pane can induce appreciably higher edge tension stresses than the shading effect of the frame only. By varying the width of the additional shadow parallel to a longitudinal
edge, it is found that the situation shown in Figs. 3 and 4 provides the highest stresses. Fig. 5 shows the tensile stress σ_y at
the center of the longitudinal edge lying opposite to the additional shadow as a function of the shadow width s. Compared to the
case of Fig. 1 (s = 20 mm) this stress is greater for all values
of s between 20 and 310 mm, with a maximum when s is one third of
the pane width.

Effect of an Additional Shading Parallel to the Transverse Edge

Fig. 6 summarizes the effects of an additional partial
shading with a shadow line parallel to a transverse edge of the
pane. The diagram shows the influence of the shadow height s' on
the maximum tension stress both at the longitudinal edges and at

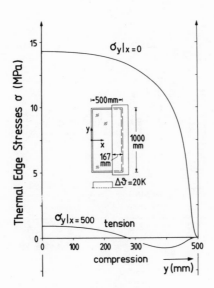

Fig. 3. Stresses σ_y at the longitudinal edges of a pane in the case of additional shading.

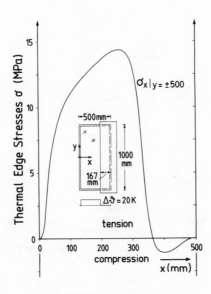

Fig. 4. Stress σ_x at the transverse edges of a pane in the case of additional shading.

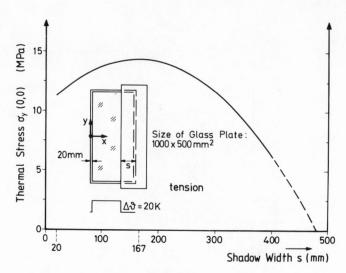

Fig. 5.   Tension stress σ_y at the center of the left longitudinal
          edge versus the shadow width s

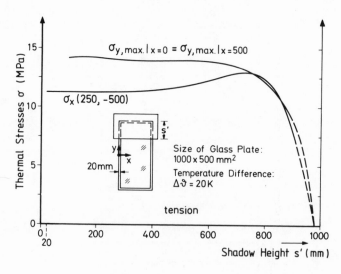

Fig. 6.   Maximum tension stresses at the longitudinal edges and
          the lower transverse edge versus the shadow height s'

the center of the transverse edge opposite to the shadow. Over a large range of s' the stress $\sigma_{y,max}$ is almost constant and greater by about 26% compared to the case of Fig. 1. This increase is about the same in magnitude as in the worst case of an additional shading parallel to a longitudinal edge. The distribution of σ_y along the longitudinal edges is not shown here, but is similar to that illustrated in the next example (see the stress distribution along the left longitudinal edge in Fig. 7). The maximum edge tension stress is located outside of the additionally shaded area close to the intersection point of the additional shadow line with the edge.

This additional partial shading at the upper transverse edge of the pane increases the maximum tension stress σ_x at the lower transverse edge only at relatively large values of the shadow height s'; the effect, however, is of minor significance.

Effect of a Combined Additional Shading Parallel to the Edges

A combination of the worst cases of the previous two sections, i.e. an additional shadow parallel to a longitudinal edge and one parallel to a transverse edge together, provides even higher maximum edge tension stresses than in the individual cases. Fig. 7 shows the distribution of the stress σ_y along the longitudinal

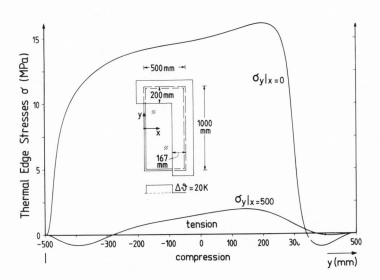

Fig. 7.   Stresses σ_y at the longitudinal edges of a pane in the case of additional shading

edges for such a combined shading pattern. The increase of the
maximum tension stress at the left edge is about 43% compared to
the case of Fig. 1.

Effect of an Additional Shading Pattern Bordered by Two Intersecting Diagonal Shade Lines

The configuration of the additional shading pattern considered
here first, is illustrated in the small figure inserted in Fig. 8.
The shadow has a wedge-shaped boundary with two diagonal shadow
lines intersecting at the middle of a transverse edge cover of the
frame. The resulting edge stresses in the pane were calculated for
different values of the included angle between the shadow lines α.
Fig. 8 shows the distributions of the stress σ_x along the upper
transverse edge of the pane for α = 45°, 70°, 120° and 145°. (The
case of α = 180° is the reference case with no additional shading,
i.e. with the shading effect of the frame only). The figure demonstrates that the maximum value of σ_x at the center of the edge
occurs when α = 120°. This maximum stress is larger than the
reference value by about 34%, but the range of the maximum is much
more confined.

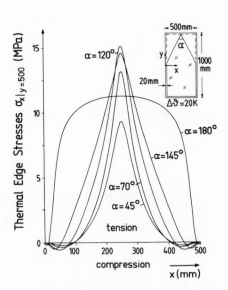

Fig. 8.   Stresses σ_x at the top edge of a pane for different
          angles α of the additional shading pattern

It should be noted that σ_x at the center of the upper transverse edge is not the (absolutely) highest edge tension stress
that can occur with the considered shadow configuration. In Fig. 9
the maximum edge tension stresses σ_x and σ_y at the top and the side
edges of the pane are compared for different values of the angle
α. It can be seen that the highest tension stress is induced at
the longitudinal edges for the case of $\alpha = 70°$. This means an
increase of about 38% compared to the reference case.

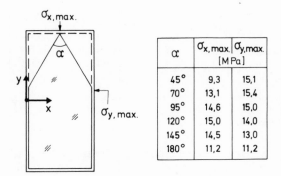

α	$\sigma_{x,max.}$	$\sigma_{y,max.}$
	[MPa]	
45°	9,3	15,1
70°	13,1	15,4
95°	14,6	15,0
120°	15,0	14,0
145°	14,5	13,0
180°	11,2	11,2

Fig. 9.   Maximum edge tension stresses versus the angle α of the
additional shading pattern

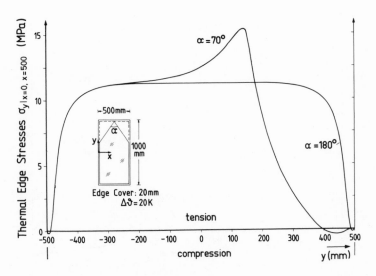

Fig. 10.   Stresses σ_y at the longitudinal edges of a pane for two
different shading patterns

The distribution of the σ_y - stress along the longitudinal edges in the worst case ($\alpha = 70°$) is illustrated in Fig. 10 in comparison to the reference case ($\alpha = 180°$).

Now we consider the case that the two diagonal shadow lines of the shading pattern intersect at the middle of a longitudinal edge cover of the frame. In Fig. 11 the maximum edge tension

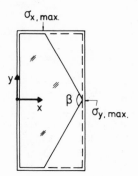

β	$\sigma_{x,max.}$	$\sigma_{y,max.}$
	[MPa]	
90°	1,3	15,9
120°	13,7	17,0
150°	14,2	15,7
180°	11,2	11,2

Fig. 11.   Maximum edge tension stresses versus the angle β of the additional shading pattern (size of the pane 1000x500 mm^2, width of the edge cover 20 mm, $\Delta\vartheta$ = 20 K)

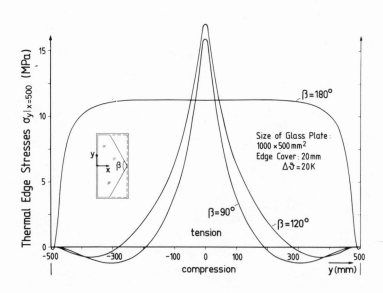

Fig. 12.   Stresses σ_y at the right longitudinal edge of a pane for different angles β of the additional shading pattern

stresses σ_x and σ_y at the transverse edges and the right longitudinal edge of the pane are compared for different values of the included angle between the shadow lines β.   The highest stress occurs
at the center of the right side edge when β = 120° and it is larger
than the reference value (when β = 180°) by about 52%.   The range
of the maximum is again very narrow in comparison to the reference
case (Fig. 12).   Of all the additional shading configurations considered here, this one produces the most severe condition in terms
of maximum tension stress.   On the other hand the effect is much
less significant than what was previously believed[1].

Effect of the Shape of Temperature Profiles

The effect of an idealized, stepwise and an actual, continuously varying temperature distribution on the edge tension stresses of
a glass pane is examined.   The results for two different shading
patterns are given below.   First, however, the apparatus applied in
the procedure of producing and determining actual temperature distributions will be described.

In Fig. 13 an infrared radiation wall (1) which exposes a heat-
absorbing glass pane (2) to a uniform heat radiation is shown.   The

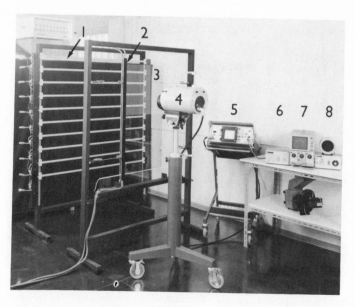

Fig. 13.   Apparatus used for producing and determining actual non-
           uniform temperature distributions on glass panes

pane is 1000 x 500 mm^2 in size. A fixture (3) is used to simulate
the edge cover of a frame. The resulting stationary temperature
field of the pane is scanned by an infrared–camera unit (4) and
made visible at the display unit (5) of an AGA Thermovision equipment Model 680. An additional device (6) provides the display of
single, selected temperature profiles on an oscilloscope (7). An
infrared radiation reference source (8) is used if the temperatures
are to be measured in absolute values.

A Shadow Boundary Remote from the Edges. A front view of the
simple example that is to be considered first is shown in Fig. 14.
The main part of the shadow boundary parallel to a longitudinal
edge is remote from the edges of the pane. No edge cover of a frame
is taken into account.

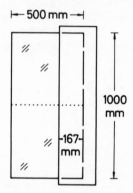

Fig. 14.   One of the examples of a partially shaded pane that was
           investigated experimentally

A top view of the arrangement is shown in Fig. 15 b). The
oscilloscope picture of Fig. 15 a) represents the measured stationary temperature distribution along the horizontal center line (dotted in Fig. 14) of the pane. The maximum temperature difference of
20K of the profile was obtained by an adjustment of the electric
power and, thus, of the intensity of the infrared radiation wall
of the apparatus (Fig. 13). It is obvious that the shadow boundary
located at x = 333 mm does not produce a temperature step, but a
continuously varying distribution extending to a length of about
200 mm. The temperature decrease at the left side edge of the pane
is caused by convection and radiation heat losses across the edge
surface, but will not be considered here. In Fig. 15 c), therefore, the temperature profile marked by a broken line represents a
modified distribution, in which only the measured temperature decrease at the shadow boundary is taken into account.

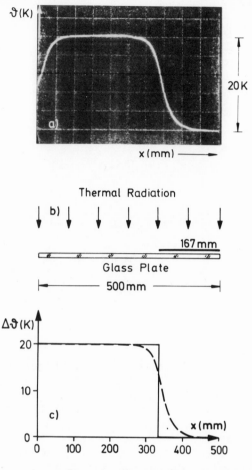

Fig. 15.   Stationary temperature profile in a glass pane due to the
thermal loading shown in the middle part (b))
a): Measured profile along the horizontal center line
(see Fig. 14)
c): Actual distribution (broken line) at the shadow
boundary and a stepwise temperature change (solid line)
used for stress calculations of Fig. 16

Stresses were calculated using the measured temperature profile and an idealized profile with a temperature step at x = 333
mm (Fig. 15 c)) assuming that both profiles are constant over the
pane height, i.e. that they do not depend on y. As demonstrated
in Fig. 16, the maximum tension stress at the left longitudinal
edge is affected very little by the shape of the temperature profile. This finding is not valid in the following case.

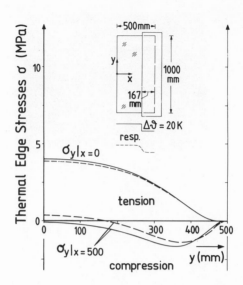

Fig. 16.   Stresses σ_y at the longitudinal edges of a partially
            shaded pane for different shapes of the temperature pro-
            file (see Fig. 15 c))

    A Shadow Boundary Close to an Edge.   Figure 17 b) illustrates
an extension of the situation given in Figs. 14 and 15 b) by additionally considering a 20 mm wide edge cover at the left pane edge.
The measured stationary temperature distribution along the horizontal center line is shown in Fig. 17 a).   From this, the following
findings can be inferred: The decrease at the shadow boundary at
x = 333 mm compares well with that of Fig. 15 a) as expected.   The
decrease at the shadow boundary at x = 20 mm is about nine tenths
in magnitude when compared to that to its right; this is attributed
to the combined effect of the shading and the heat losses across
the left edge surface.   The edge cover at the left side edge affects
the temperatures of the pane within a relatively large region of the
non-shaded central area.

    Stresses were calculated using the measured temperature pro-
file and an idealized profile with temperature steps at x = 20 mm
and x = 333 mm (Fig. 17 c)) assuming that both profiles are constant over the pane height.   As can be inferred from Fig. 18 in
connection with Fig. 16, a continuously varying temperature distribution at a small edge cover produces a significantly lower maximum
tension stress at the near-by edge of the pane than does a comparable stepwise distribution at the same location.   It should be noted
that at least under the assumptions of this paper the shading by the
edge cover of a frame provides a higher contribution to the maximum
edge tension stress than does any additional shading of the pane.

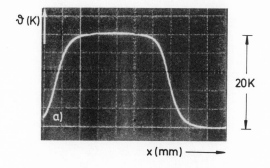

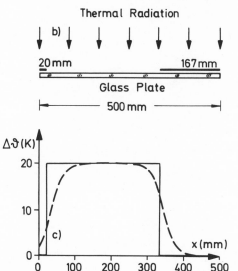

Fig. 17.   Stationary temperature profile in a glass pane due to the
thermal loading shown in the middle part (b))
a):   Measured profile along the horizontal center line
c):   Actual profile (broken line) and an idealized, stepwise temperature distribution (solid line) used for stress
calculations of Fig. 18

Thus, an optimization of the edge cover width and of the heat transfer conditions at the edge surfaces of a pane seems to be one of
the most promising goals in order to minimize dangerous thermal
stresses in heat-absorbing building glass.

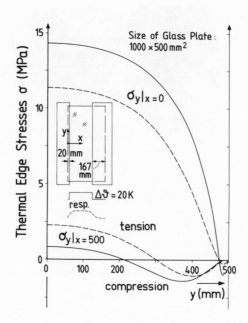

Fig. 18.   Stresses σ_y at the longitudinal edges of a partially
          shaded pane for different shapes of the temperature pro-
          file (see Fig. 17 c))

ACKNOWLEDGEMENTS

       The author would like to thank Dipl.-Phys. U. Seidelmann for
his help with the numerical computations and Mrs. I. Saalmann and
Mrs. B. Johnson for typing the manuscript.

       The financial support of this work by the Arbeitsgemeinschaft
Industrieller Forschungsvereinigungen e.V. (AIF), Köln, and by the
Hüttentechnische Vereinigung der Deutschen Glasindustrie e.V.,
Frankfurt (Main), is gratefully appreciated.

REFERENCES

1. Pittsburgh Plate Glass Industries, Installation Recommendations
    - Tinted Glass, Technical Service Report No. 104 C (1974), 27 p.

2. Pilkington Brothers Limited, The Application of Solar Control
    Glasses, Glass and Windows Bulletin 10 (1972), 29 p.

3. Pilkington Brothers Limited, The Fracture of Glass by Solar
    Radiation, Glass and Windows Bulletin 4 (1964), 3 p.

4.   Groupement Européen des Producteurs de Verre Plat (GEPVP),
     Comportement des vitrages absorbants aux effects du rayonnement
     solaire, Groupe de Travail "Constraintes Thermiques", Internal
     Report (1977), 30 p.

5.   F. Kerkhof, Über die Entstehung von Glasbruch durch Einbringen
     von heißem Gußasphalt, Glas + Rahmen 24 (1979), 1163-1166

6.   J.R. Sasaki, The Potential for Thermal Breakage of Insulating
     Glass Units is a Matter to be Mulled, Glass Digest 51 (1972),
     54-55, 90-91

7.   Y.W. Mai and L.J.S. Jacob, Thermal Fracture of Building Glass
     Subjected to Solar Radiation, in "Mechanical Behaviour of Mate-
     rials", K.J. Miller and R.F. Smith, eds., Pergamon Press,
     Toronto (1979), Vol. 3, 57-65

8.   D. Stahn, Wärmespannungen in großflächigen Verglasungen, Glas-
     techn. Ber. 50 (1977), 149-158.

9.   D. Stahn, Gefährdung großflächiger Verglasungen durch thermisch
     induzierte Spannungen, Proc. XIth Intern. Congress on Glass,
     Prague (1977), Vol. V, 415-425

10. E.L. Wilson, Solid SAP - Static Analysis Program for 3-Dimen-
     sional Solid Structures, Structural Engineering Laboratory,
     Univ. of California, Berkeley (1972)

EFFECT OF SPATIAL VARIATION OF THERMAL CONDUCTIVITY ON MAGNITUDE OF TENSILE THERMAL STRESSES IN BRITTLE MATERIALS SUBJECTED TO CONVECTIVE HEATING

K. Satyamurthy, D. P. H. Hasselman, J. P. Singh and
M. P. Kamat

Departments of Materials Engineering and Engineering
Science and Mechanics
Virginia Polytechnic Institute and State University
Blacksburg, Virginia

ABSTRACT

The results are presented for the effect of spatially varying thermal conductivity on the tensile thermal stress developed in a solid and a hollow circular cylinder subjected to different heating conditions. It is shown that the maximum tensile thermal stress in brittle ceramics can be reduced significantly by redistributing the temperature profile using (a) a spatial variation in thermal conductivity, (b) a spatial variation in pore content which in turn changes the density, thermal conductivity and modulus of elasticity and (c) by considering the effect of temperature on the thermal conductivity and specific heat. Possible methods for creating such variations in the material properties are discussed.

INTRODUCTION

The design of engineering structures or components operating under non-isothermal conditions must take into account the presence of thermal stresses which arise from, constrained non-uniform thermal expansion in addition to the normal operating stresses. For many structures or components operating at high-temperatures the transient thermal stresses during heat-up or cool-down can exceed the normal operating stresses by a considerable margin. Weakening of the structure (or complete failure) due to the thermal stresses should be avoided to assure long-term reliable operation. This is particularly critical for brittle ceramic materials in which because of the absence of non-linear stress-relieving mechanisms and low fracture toughness, thermal stress failure can be highly catastrophic with corresponding disastrous consequences for continued satisfactory performance.

   A number of options are available to the designer, working in
conjunction with the materials technologist for reducing the probability of failure due to thermal stresses. First, for many heat
transfer conditions, a redesign involving a change in the geometry
or a reduction in the dimensions of the structure or component will
reduce the magnitude of the thermal stresses. This approach, however, usually is incompatible with other performance criteria imposed on the structure or component. Secondly, the magnitude of
the thermal stresses can be reduced by reducing the rate of heat
transfer. Again, with few exceptions, this approach is impractical,
as the rates of heat transfer are beyond the control of the designer.
A third approach is to select the materials of construction of the
structure or component on the basis of figures-of-merit for thermal
stress resistance [1], which reflect the optimum combination of the
relevant material properties such as the coefficient of thermal expansion, Young's modulus, thermal conductivity, etc. It should be
noted, however, that thermal stress failure in even the optimum
materials selected in this manner, is not necessarily avoided, as
these figures-of-merit merely compare the relative thermal stress
resistance of potential materials of construction. Furthermore,
values of material properties which maximize thermal stress resistance may not be compatible with the property values required for
optimum performance of the structure or component in its intended
application.

   A fourth approach to improving thermal stress resistance is
based on the fact that the magnitude of the thermal stresses in a
structure or component not only is a function of the magnitude of
the temperature differences encountered within the structure or component, but also is a function of the temperature distribution. In
principle, therefore, it should be possible to decrease the magnitude
of thermal stress for a given heat transfer condition by modifications of the temperature distribution. Since the distribution of
the temperature within a structure or component is a function of
the thermal conductivity, modifications in those distributions can
be achieved by incorporating a spatially varying thermal conductivity. A number of studies carried out by the present writers and
co-workers [2,3,4,5] have shown that significant reductions in
the magnitude of the tensile thermal stresses in brittle ceramic
components can be achieved by decreasing the thermal conductivity
in the hotter parts of the structure. These results were obtained
for steady-state heat flow in a hollow-cylinder as well as in the
transient heating of solid circular cylinders. The reductions in
the tensile stresses were obtained not only by a direct spatial
variation of the thermal conductivity but also indirectly by a
spatial variation in density (which affects Young's modulus also),
and by the negative temperature dependence of the thermal conductivity of dielectric materials in which the conduction of heat

occurs primarily by phonon transport. It is the purpose of this
presentation to give an overview of these results.

CALCULATIONS, RESULTS AND DISCUSSION

A. Constant Heat Flow

    Spatially Varying Thermal Conductivity. The geometry selected
for the analysis of the effect of a spatially varying thermal conductivity under conditions of steady-state heat flow consisted of
an infinitely long hollow concentric cylinder with inner and outer
radius a and b, resp. The spatial variation of the thermal conductivity (K), for analytical convenience was selected as:

$$K = K_o \{1 + f_o (a/r)^n\}^{-1} \tag{1}$$

for radially outward heat flow and

$$K = K_o \{1 + f_o (r/b)^n\}^{-1} \tag{2}$$

for radially inward heat flow.

    In eqs. 1 and 2, K_O represents the spatially uniform thermal
conductivity and f_o and n are constants. The general form of eqs.
1 and 2 is such that the thermal conductivity is lowest in the regions of the cylinder which are at the highest temperature.

    From eqs. 1 and 2 and the general condition for steady-state
heat flow in a radial coordinate system [6], the temperature distribution within the cylinder was derived in terms of the temperature difference (ΔT) across the wall.

    For brittle materials the maximum tensile stress is of primary
interest. In a hollow cylinder undergoing steady-state heat flow,
the radial and hoop stresses are exceeded by the longitudinal stress,
which can be derived from [7]:

$$\sigma_z = \frac{\alpha E}{(1-\nu)} \{\frac{2}{b^2-a^2} \int_a^b rTdr - T(r)\} \tag{3}$$

where α is the coefficient of thermal expansion, E is Young's
modulus, ν is Poisson's ratio and T is the temperature. For the
purpose of this analysis, α, E and ν were assumed to be independent
of position or temperature.

    Figures 1 and 2 show the maximum value of the tensile longi-
tudinal thermal stress at r = b for radially outward heat flow and

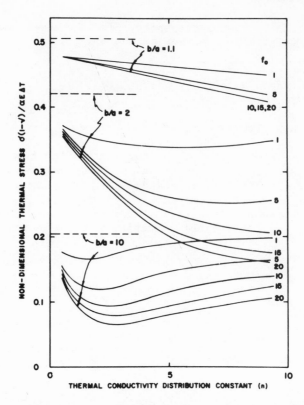

Fig. 1.   Maximum tensile thermal stress in a hollow cylinder under
          steady-state outward radial heat flow and spatially vary-
          ing thermal conductivity (equation 1). Dotted lines indi-
          cate values for uniform spatial distribution.

r = a for the radially inward flow for three values of the ratio of
outer to inner radius (b/a).   Included in the figure are the values
of stress for the spatially constant thermal conductivity.   Comparison of these stress values shows that a spatially varying thermal
conductivity can lead to a significant reduction in the magnitude
of thermal stress.   Since an order of magnitude variation in thermal
conductivity of materials can be easily found, these results should
be of practical significance.

     Spatially Varying Density.   The presence of a pore phase in
solid materials has a pronounced effect on many material properties
including the thermal conductivity.   In principle, therefore a
spatially varying thermal conductivity can be achieved by incorporating a spatially varying pore concentration in the material.   A
further advantage is achieved by the corresponding decrease in
Young's modulus.   Unfortunately, the tensile strength is also affected adversely by the pore phase which would lead to a decrease
in the thermal stress resistance.   However, this effect can be

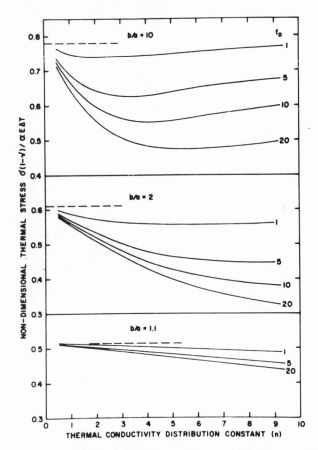

Fig. 2. Maximum tensile thermal stress in a hollow cylinder under steady-state inward radial heat flow and spatially varying thermal conductivity (equation 2). Dotted lines indicate values for uniform spatial distribution.

avoided by distributing the pore phase such that full density is maintained at the position of maximum tensile stress and having as high a porosity as practically possible in the region at higher temperature, which generally are in a state of compressive stress.

Accordingly, the porosity distribution for radially outward heat flow was arbitrarily taken as:

$$P = P_o \{(b - r)/(b - a)\}^n \tag{4}$$

and for radially inward heat flow as:

$$P = P_o \{(r - a)/(b - a)\}^n \tag{5}$$

where P_O is the porosity at $r = a$ and $r = b$ for radially outward and inward heat flow, resp. and n the porosity distribution coefficient Eqs. 4 and 5 indicate that $P = 0$ at the position of maximum tensile stress, i.e., $r = b$ and $r = a$ for the radially outward and inward heat flow, respectively.

The effect of the pore phase on the thermal conductivity, K and Young's modulus, E was chosen to be given by the well-known empirical equation [8,9]:

$$K, E = K_o, E_o e^{-bp} \tag{6}$$

where K_O and E_O represent the values of the thermal conductivity and Young's modulus for the non-porous material, p is the fractional porosity and b is a constant taken as 3 and 4 for K and E, respectively. Poisson's ratio was assumed to remain unaffected by the pore phase. In view of the analytical complexity of the variation of E and K as expressed by eqs. 4, 5 and 6, the thermal stresses were obtained by a finite element computer program compiled for this purpose.

For radially inward heat flow, Fig. 3 shows the maximum value of longitudinal tensile stress as a function of the constant n in eqs. 4 and 5 for three values of the ratio of the outer to inner radius with $P_O = 0.3$. Included in the figure is the value of stress corresponding to $P(r) = 0$. Similar results were obtained for radially outward heat flow. These results indicate that since the strength of the material remains unaffected in the critical regions, significant increase in thermal stress resistance is achieved. Additional calculations showed that the reduction in the magnitude of the maximum tensile stress to a good approximation was a linear function of the value of P_O.

Of interest to note in Fig. 3 is that the value of stress is relatively independent of the value of n. This occurs because the maximum reduction in stress due to the individual effects of the spatially varying porosity on Young's modulus and thermal conductivity occur at different values of n. Minimizing the stresses by reducing Young's modulus requires a value of $n \rightarrow 0$. This implies a uniform pore phase, which results in a uniform thermal conductivity with on reduction in the thermal stresses. As far as a spatial variation in thermal conductivity is concerned, optimum reduction in the thermal stresses is achieved by values of $n \gg 1$ such that a reduction in thermal conductivity is achieved in the immediate regions of the tube surface at the higher temperature.

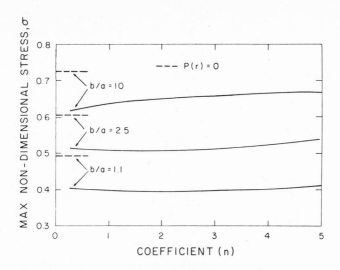

Fig. 3.   Maximum tensile thermal stress as a function of the porosity
distribution coefficient n in a hollow cylinder under steady-state radially inward heat flow with spatially varying
pore content (equation 5) with $P_o(r = b) = 0.3$.

B.   Transient Heating

An infinitely long solid circular cylinder of radius a was
selected as an appropriate geometry for the analysis of the effect
of a spatially varying thermal conductivity on the magnitude of
thermal stress under conditions of transient heating. The specific
heating conditions consisted of subjecting the cylinder initially
at thermal equilibrium at a temperature T_o, instantaneously to an
ambient temperature of T_∞. The transfer of heat was considered to
be Newtonian according to:

$$K(a)(\partial T/\partial r(a)) + h (T(a) - T_\infty) = 0 \qquad (7)$$

where h is the heat transfer coefficient and T(a) is the instantaneous surface temperature.

Calculations of the thermal stresses were carried out with a
finite element program which gave excellent agreement with stress
values for spatially uniform thermal conductivity obtained analytically by Jaeger [10].

The maximum tensile stresses were of primary interest which for the present heating conditions are the hoop stresses at r = 0.

For convenience, the numerical results are reported in terms of the non-dimensional thermal stress, $\sigma^* = \sigma(1-\nu)/\alpha E \Delta T$, the non-dimensional time, $t^* = \kappa t/a^2$, the non-dimensional radial coordinate, r/a and the Biot number, $\beta_O = ah/K_O$ in which $\Delta T = T_\infty - T_O$, κ is the thermal diffusivity and α is the coefficient of expansion. The Biot number is expressed in terms of the uniform thermal conductivity, prior to spatial modification.

Spatially Varying Thermal Conductivity. The radial dependence of the thermal conductivity was chosen to be of the form:

$$K(r) = K_0[1 - C(r/a)^n] \tag{8}$$

where n and C(0 < C < 1) are constants. Equation 8 indicates that at r = 0, K = K_O for any value of C and n. The lowest value of thermal conductivity occurs at r = a.

Fig. 4 shows typical results for the time dependence of the maximum tensile hoop stress at r = 0 for $\beta_O = 5$ and a range of values of C and n. The peak values of the stress for a spatially varying thermal conductivity are considerably less than the corresponding values for the constant thermal conductivity, included in Fig. 4 for comparison.

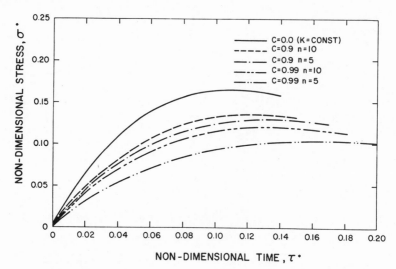

Fig. 4. Maximum tensile thermal stress as a function of time at the center of a solid circular cylinder subjected to sudden convective heating for various spatial distributions of the thermal conductivity (equation 8).

Figures 5a and 5b show the spatial variation of the temperatures and stresses, resp., at the time of maximum tensile stress for the same values of C and n as shown in Fig. 4.   These results indicate that the decrease in tensile stress at r = 0 is accompanied by a corresponding increase in the temperatures and compressive stresses in the surface (r = a).

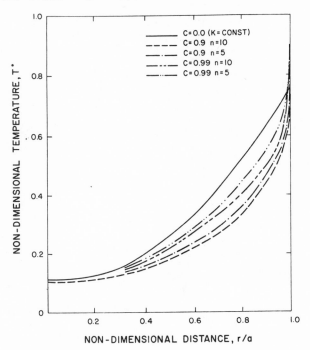

Fig. 5a.   Spatial distribution of the transient temperature at the time of maximum tensile stress, in a solid circular cylinder subjected to sudden convective heating for various spatial distributions of the thermal conductivity (equation 8).

Figure 6 shows the maximum values of tensile hoop stress at r = 0 as a function of n for a range of values of C.   Significant reductions in the magnitude of stress can be achieved, especially at the higher values of C.

Figure 7 compares the maximum values of tensile hoop stress for constant thermal conductivity and spatially varying thermal conductivity (C = 0.9, n = 3) as a function of Biot number.   The reduction in stress for varying thermal conductivity is significant, especially at the higher values of the Biot number, at which materials are most susceptible to thermal stress failure.

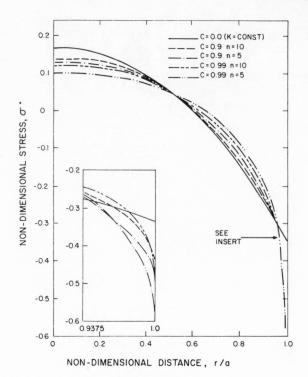

Fig. 5b. Spatial distribution of the transient thermal stress at
the tiem of maximum tensile stress, in solid circular cylinder subjected to sudden convective heating for various
spatial distributions of the thermal conductivity (equation 8).

Spatially Varying Porosity. The distribution of the pore
phase was chosen to be:

$$P = P_o(r/a)^n \tag{9}$$

The distribution for the thermal conductivity and Young's modulus
was assumed to be identical to the one chosen for the steady-state
heat flow given by eq. 6. Again, Poisson's ratio was assumed to
be independent of the pore phase. The thermal stresses were calculated by the finite element program referred to earlier.

Figures 8a and 8b show the spatial variation in the transient
temperatures and stresses for $\beta_o = 1$, 5 and 20, $n = 1$ and $P_o = 0.3$
at the instant of time the tensile hoop stresses at the center of
the cylinder reach their maximum value. A value of $P_o = 0.3$ was
chosen as it corresponds to a value realizable in practice. For
purpose of comparison, the temperatures and stresses for the fully

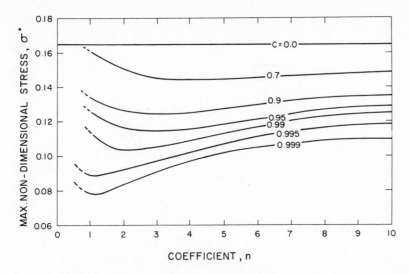

Fig. 6. Maximum tensile thermal stress as a function of coefficient, n in a solid circular cylinder subjected to convective heating for various spatial distributions of the thermal conductivity (equation 8).

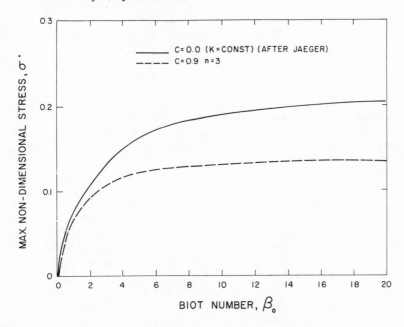

Fig. 7. Maximum tensile thermal stress as a function of Biot number in a solid circular cylinder with constant and spatially varying thermal conductivity (n=3, C=0.9) subjected to sudden convective heating.

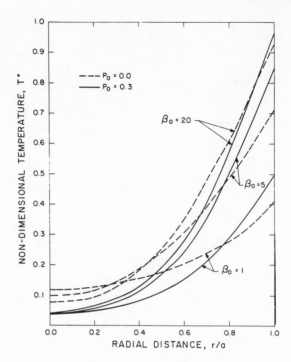

Fig. 8a. Spatial distribution of transient temperature at the time
of maximum tensile stress in a solid circular cylinder
subjected to sudden convective heating for spatially varying porosity described by equation 9 with n=1 and P_0=0.3.

dense material (P_0=0) are included in Figs. 8a and 8b. The numerica
results indicate that the spatially varying pore phase results in a
decrease in both the tensile and compressive thermal stresses in the
cylinder, accompanied by a simultaneous increase in the surface temperature.

Figure 9 gives the maximum tensile hoop stress as a function of
n for β_0 = 1, 5 and 20. The stress levels for the fully dense material are included for comparison. A major reduction in the magnitud
of the stresses as the result of the spatially varying pore phase is
evident. Again, this reduction is greatest at the higher values of
the Biot number, for which thermal stress fracture is most likely to
occur. The trend of the curves in Fig. 9 suggests that it may be
advantageous to have a porosity distribution corresponding to values
of n as low as possible. It should be noted, however, that porosity
also has a strong adverse effect on tensile strength. This implies
that for low values of n, a considerable decrease in strength occurs
very close to the center of the rod. In this case it is conceivable
that failure can occur at position in the rod away from the center,
so that the advantage of the spatial variation of the porosity can

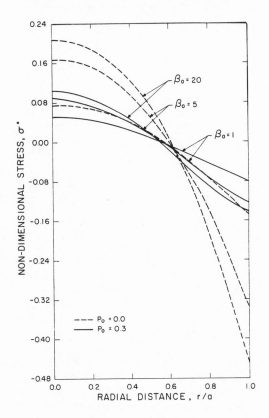

Fig. 8b. Spatial distribution of transient thermal stress at the
time of maximum tensile stress in a solid circular cylinder subjected to sudden convective heating for spatially
varying porosity described by equation 9 with n=1 and
$P_o = 0.3$.

be lost. Describing strength by eq. 6 with b = 6, values of n
should be less than unity, if failure at positions other than
r = 0 is to be avoided.

<u>Temperature Dependent Thermal Conductivity and Specific Heat</u>.
The discussion so far has concentrated on a spatial variation in
thermal conductivity, either introduced directly or indirectly by
incorporating a spatially varying pore phase.

A spatial variation of thermal conductivity can be caused by
another mechanism, namely the temperature dependence of the thermal
conductivity. It is well known, that for many dielectric materials
in which the conduction of heat occurs primarily by phonon transport,
the thermal conductivity exhibits a strongly negative temperature
dependence. For this reason, during transient heating, the sections

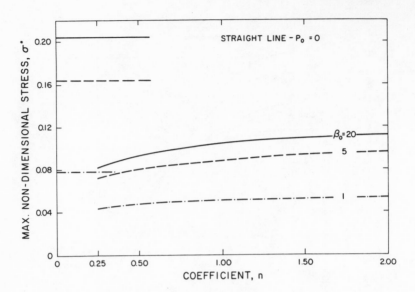

Fig. 9.   Maximum tensile thermal stress as a function of n in a
          solid circular cylinder subjected to sudden convective
          heating for spatially varying porosity described by equa-
          tion 9 with P_O = 0.3 for various values of Biot number.

in a structure or component at the higher temperature will have a
lower thermal conductivity than those regions still at a lower temperature. For a solid circular cylinder this should result in a
type of a transient spatially varying thermal conductivity, with
a distribution such that in analogy to the previous results, a
major reduction in the magnitude of the tensile thermal stresses
should be achieved.

     The validity of this hypothesis was verified by calculating
the maximum tensile thermal stresses in a solid circular cylinder
of radius a = 1 cm of high density aluminum oxide at an initial
uniform temperature of 300°K, subjected to convective thermal shock
by an instantaneous change in ambient temperature to 1000°C. The
transfer of heat was considered to be Newtonian. The temperature
dependence of the thermal conductivity of dense aluminum oxide to a
good approximation is described by [11]

$$K = 28/T - 0.01 \text{ cal.}°K^{-1}.cm.^{-1}sec^{-1}.$$

where T is the absolute temperature in °K. The temperature dependence of the specific heat was also taken into account [12]. All
other material properties were assumed to be independent of temperature.

     The thermal stresses were calculated with the finite element program referred to earlier, with the required modifications to include the necessary iterative procedure to take the temperature dependence of the thermal conductivity and specific heat into account.

Figures 10a and 10b compare the time dependence of the thermal stresses for temperature dependent and independent thermal conductivity and specific heat, resp., for a range of values of the Biot number. The temperature dependence of the thermal conductivity and specific heat causes a significant decrease in the magnitude of the maximum tensile thermal stresses.

Figure 11 shows the peak value of the transient tensile thermal stress as a function of the Biot number. The relative reduction in the magnitude of these stresses is significant, especially at the higher values of the Biot number.

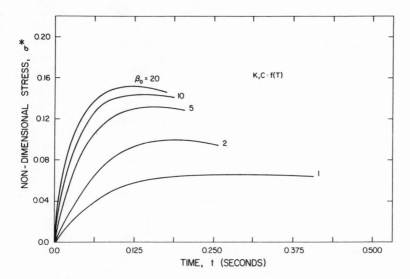

Fig. 10a.   Maximum tensile thermal stress as a function of time at the center of solid circular cylinder subjected to sudden convective heating for a range of Biot numbers with temperature dependent thermal conductivity and specific heat.

C.   Discussion

The results of the present study suggest that at least for the heat flow conditions considered, a spatially varying thermal conductivity can lead to significant reductions in the tensile thermal stresses. For all cases considered, this was achieved by decreasing the thermal conductivity in the sections of the geometries

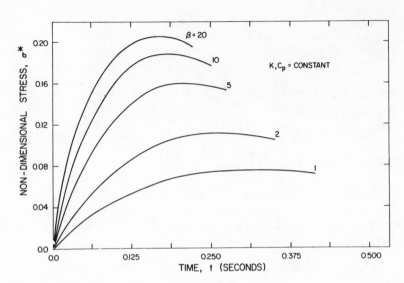

Fig. 10b.   Maximum tensile thermal stress as a function of time at
the center of solid circular cylinder subjected to sudden convective heating for a range of Biot numbers with
constant thermal conductivity and specific heat.

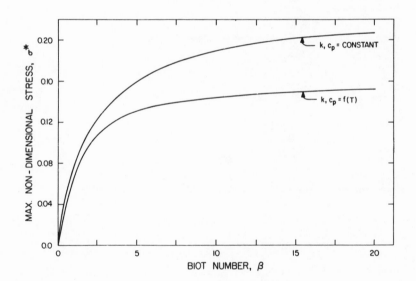

Fig. 11   Dependence of the maximum tensile thermal stress on Biot
number in a solid circular cylinder with temperature
dependent K, C_p and constant K, C_p subjected to sudden
convective heating.

studied which are at the higher temperature. At first sight, this latter fact contradicts the results of thermal stress theory, based on spatially uniform properties, that thermal stresses can be decreased by increasing the thermal conductivity. The results of the present study substantiates the significant effect of the nature of the temperature distribution on the magnitude of thermal stress. For the steady-state heat flow reported in the earlier part of this paper, based on a constant ΔT across the cylinder wall, the magnitude of the stresses is governed only by the temperature distribution which is controlled by the relative distribution of the thermal conductivity and not by its absolute magnitude.

For the transient heat transfer, a decrease in thermal stress is achieved by the spatial variation in thermal conductivity, inspite of the fact that decreasing the thermal conductivity in the surface, in effect, raises the Biot number, which should result in an increase in thermal stress. Again, the strong influence of the temperature distribution on the magnitude of thermal stress is indicated. In this respect, it should be noted that comparing the numerical data for the thermal stresses in terms of the Biot number, β_0, based on the original unmodified value of thermal conductivity, K_0, in fact, is conservative.

These general results of this study possibly can be of practical interest. Improvements in thermal stress resistance by a spatially varying pore phase already is common practice in the ceramic and refractory industry. Low thermal conductivity coatings can also be used, in which spatial variations in other properties should also be taken into account. Spatial variation in alloy content can be used very effectively to introduce a spatially varying thermal conductivity. Spatial variations in the degree of crystallinity for glass-ceramics can be used effectively for the same purpose, as well as other techniques appropriate for specific materials required for specific purposes.

ACKNOWLEDGEMENT

This study was conducted as part of a research program on the thermo-mechanical and thermal properties of brittle structural materials supported by the Office of Naval Research under contract N00014-78-C-0431.

REFERENCES

1. D. P. H. Hasselman, Figures-of-Merit For the Thermal Stress Resistance of High-Temperature Brittle Materials: A Review, Ceramurgia, 4:147 (1978).

2. D. P. H. Hasselman and G. E. Youngblood, Enhanced Thermal Stress Resistance of Structural Ceramics with Thermal Conductivity Gradient, J. Amer. Ceram. Soc., 61:49 (1978).

3. K. Satyamurthy, M. P. Kamat and D. P. H. Hasselman, Effect of Spatially Varying Thermal Conductivity on Magnitude of Thermal Stress in Brittle Ceramics Subjected to Convective Heating, J. Amer. Ceram. Soc., (in review)

4. K. Satyamurthy, J. P. Singh, M. P. Kamat and D. P. H. Hasselman, Effect of Spatially Varying Porosity on Magnitude of Thermal Stress During Steady-State Heat Flow, J. Amer. Ceram. Soc., 62:431 (1979).

5. K. Satyamurthy, J. P. Singh, M. P. Kamat and D. P. H. Hasselman, Thermal Stress Analysis of Brittle Ceramics With Density Gradients under Conditions of Transient Convective Heat Transfer, Trans. Brit. Ceram. Soc., (in review).

6. H. S. Carslaw and J. C. Jaeger, "Conduction of Heat in Solids," 2nd Ed., Oxford University Press, London (1959).

7. B. A. Boley and J. H. Weiner, "Theory of Thermal Stresses," J. Wiley, New York (1960).

8. R. M. Spriggs, Expression of Effect of Porosity on Elastic Modulus of Polycrystalline Refractory Materials, Particularly Aluminum Oxide, J. Amer. Ceram. Soc., 44:628 (1961).

9. F. P. Knudsen, Dependence of Mechanical Strength of Brittle Polycrystalline Specimens on Porosity and Grain Size, J. Amer. Ceram. Soc., 42:376 (1959).

10. J. C. Jaeger, On Thermal Stresses in Circular Cylinders, Phil. Mag. 36:418 (1945).

11. Y. S. Touloukian, R. W. Powell, C. Y. Ho and P. G. Klemens, Thermal Conductivity: Non-Metallic Solids (Thermophysical Properties of Matter, Vol. 2, p. 119, IFI) Plenum, New York (1970).

12. Y. S. Touloukian and E. H. Buyco, Specific Heat: Non-Metallic Solids (Thermophysical Properties of Matter, Vol. 5, pp. 27-29, IFI) Plenum, New York (1970).

THERMAL SHOCK OF REFRACTORIES

J. Homeny and R.C. Bradt

Ceramic Science and Engineering Section
Department of Materials Science and Engineering
The Pennsylvania State University
University Park, PA 16802

ABSTRACT

The quantitative description of the thermal shock of refractories is reviewed with emphasis focused on those physical properties pertinent to the calculation of damage resistance parameters. The excellent correspondence between theoretical damage resistance parameters and a variety of laboratory thermal shock tests of refractories is summarized. Microstructural features are discussed both from a crack propagation resistance viewpoint and their effects on other physical properties that affect thermal shock behavior.

INTRODUCTION

The problems of thermal shock damage have directly confronted refractory producers and consumers for many years[1]. However, only in the past several decades have their cooperative efforts produced truly significant gains in the performance of refractories subjected to severe thermal shock conditions. It has been even more recently that refractory design and application have transcended the gap from empirical concepts to quantitative principles. The progress has been truly remarkable as some vessel lining lives have been increased a hundredfold, often as a consequence of an increased understanding of the thermal shock damage and its interrelation with other forms of refractory degradation, including erosion and slag attack. The improvement has been a threefold process: (i) the application of fundamentals to the understanding of the important physical properties that are critical to thermal shock resistance, (ii) exhaustive laboratory thermal shock damage

343

testing by a number of different thermal shock techniques, and (iii) improved installation practices for refractory linings coupled with improved operating procedures.

This paper will primarily address the first two of these three features responsible for improved performance of refractories under severe thermal shock conditions. It will focus on brick rather than castables, plastics, etc. It will then address some of the micro-structural features of burned brick responsible for improved thermal shock damage resistance. This does not imply that refractory lining installation practices and operating procedures are not of comparable importance. To the contrary, in the final analysis of some cases, they are probably more important, for even the best physical proper-ties in a refractory that performs admirably in any, or all of the laboratory thermal shock tests will not insure satisfactory perfor-mance in an actual processing vessel unless that refractory is prop-erly installed and then appropriately maintained during its entire operational lifetime. For example, particularly great strides have recently been made in the installations of monolithics, particularly the placement of anchors and incorporation of fibers. Operating practice has made substantial progress in understanding the impor-tance of initial heat-up rates and continuing maintenance repairs to prolong lining lives. It may be that in certain future situa-tions, these factors will be even more crucial than the continued refractory property improvements.

PHYSICAL PROPERTIES OF REFRACTORIES

The applications and engineering designs of refractories in severe thermal shock environments are fundamentally based on the thermoelastic approach to the calculation of thermal stresses to avoid catastrophic fracture initiation and an energy balance concept to minimize thermal shock damage. The thermal stress equations are the classical ones reviewed by Kingery and modified to the specific geometries of the particular processing vessels under considera-tion[2]. The energy balance concepts are those initially advanced by Hasselman in the early 1960's which contrast the dissipation of the stored elastic strain energy of the thermal stress field with the work-of-fracture during crack propagation in the refrac-tory microstructure[3]. Because of the severe thermal shock con-ditions in most metallurgical applications, it has been generally accepted that it is practically impossible to avoid the initiation of thermal shock damage. A fair assessment is that many refrac-tories have been in the past, are currently, and in the future will continue to be designed on a damage resistance principle, accepting the inevitable initiation of fracture. Recently, however, it has become apparent that there exists a role for the limitation of thermal shock fracture initiation in refractory lining technology. That role occurs during the initial heat-up or burn-in cycle, and perhaps the next few cycles as well, times when refractory linings appear to be particularly susceptible to thermal shock cracking.

A common form of initial cracking is that of almost-planar continuous cracks parallel to the hot face, i.e., perpendicular to the thermal gradient. A single crack may occur, or a series of parallel cracks may eventually be generated. This can lead to a sheet-spalling of the hot face of refractory linings, often with a considerable portion of the lining falling-off during the initial cycle, or the first few cycles. In his review, White discusses this particular form of thermal shock damage[4]. It is generally, although not universally, accepted to occur as described by Kienow[5]. Ainsworth and Herron have experienced considerable success in applying these concepts to calculate safe heat-up rates for vessels[6]. Of course, the safest heating rates may be the slowest and are in direct opposition to achieving production goals, consequently a balance in heat-up rates must be the compromise, or some other techniques applied. One alternative approach is to apply a sacrificial spalling layer of a castable to the hot face of a burned brick lining. Shultz has reported that such installation practices can greatly reduce and perhaps even eliminate extensive planar crack formation parallel to the hot face of refractory brick linings[7].

If one examines the Kienow model, or others for thermal gradients necessary to initiate fracture, they can invariably be manipulated into a mathematical form resembling, or including the thermal shock resistance parameter R, where:

$$R = \frac{\sigma_f (1-\nu)}{\alpha E} \quad . \tag{1}$$

In Equation (1), σ_f is the fracture stress, ν is Poisson's ratio, α is the thermal expansion coefficient, and E is Young's elastic modulus. This simple parameter exemplifies the thermoelastic approach leading to a maximum stress criterion for fracture initiation. Numerous other thermal shock resistance parameters that are related to R have been subsequently developed[8]. These usually apply to special situations and frequently include other material's parameters, or properties such as the thermal conductivity, emissivity, viscosity, geometric parameters, etc. These other parameters will not be discussed here, for they often are not applicable to the operational conditions; however, the fact that they may be important should be kept in mind for reference.

The units of R are those of temperature and relate to a maximum rapid temperature change to reach the fracture stress, σ_f, at the surface of a very rapidly cooled object. It is of interest to examine the magnitude of R for a typical commercial refractory brick. For example, many high alumina refractories have σ_f's of about 20 MN/m^2, ν's of about 0.15, α's of about 7×10^{-6}/°C, and E's of about 30,000 MN/m^2. Incorporating these values into Equation (1) yields an R of about 70°C. Considering that molten metals, often above 1500°C, are directly introduced into considerably cooler refractory lined vessels, it is obvious why the design philosophy of fracture initiation acceptance and damage minimization is prevalent. Of course, thermal conductivity effects can reduce the apparent severity of the simple R estimate from Equation (1), but these do not negate the relevance of the result. Except for special situations, it does not appear that a design approach to completely avoid fracture initiation can be a very successful one for most refractory applications.

If the complete prevention of any fracture initiation in refractories were the goal of microstructrual design or processing of refractories, then it is important to examine the properties contained in R with regard to their potential for maximizing that parameter. If the premise is accepted that most refractories are chosen for their particular application based on principles of chemical reaction and phase equilibria, and then in a subsequent fashion the microstructure is optimized, the futility of any attempt to greatly increase R becomes immediately apparent. The thermal expansion, α, is fixed by the refractory's general chemistry, so it cannot be changed on any practical basis, nor can Poisson's ratio, ν, which is between 0.10 and 0.17 for virtually all refractories. This leaves only σ_f and E for potential microstructural modifications, and both are properties that are subject to a great deal of processing control. Practically, however, σ_f and E are usually directly coupled, so that improvements in one carry-over to the other. Furthermore, the nature of the manufacturing of refractories prevents any dramatic improvements in σ_f. Thus, the marginal increases possible in σ_f will be offset by increases in E, tending to keep the ratio low, and preventing any substantial increase of R.

It is significant that a reformulation of R in fracture mechanics terminology suggests that there may be some avenues of possible microstructural design to minimize extensive crack growth prior to catastrophic initiation. Incorporating the Griffith equation: $\sigma_f = K_{Ic} Y c^{-1/2}$, into R yields:

$$R = \frac{K_{Ic}(1-\nu)Y}{\alpha E\, c^{1/2}} , \qquad (2)$$

where K_{Ic} is the fracture toughness, Y is a geometry term, and c is a crack size parameter. Limited measurements of the K_{Ic}'s of refractories suggest they are low, and normally vary from about 0.15 to 1.5 MNm$^{-3/2}$, which then leaves only the flaw size parameter, c, as a parameter to control. There are two crucial aspects to the role of flaw size in thermal shock damage initiation: (i) the

initial, or intrinsic flaw size of the refractory microstructure and
(ii) the propensity for small flaws or cracks to grow into really
large cracks and thus greatly reduce R, or lead directly to spalling
themselves. There are not many avenues available to reduce the intrinsic flaw size of refractories, again because of the nature of
their processing. Large aggregates or grains tend to be flaw
sources; furthermore, the abuse during normal operational procedures
also introduces flaws. It must be concluded that the ways and means
for reducing or minimizing the effects of initial or intrinsic flaw
sizes in refractories are very limited indeed and perhaps even non-
existent.

Limited data on the effects of refractory structure, or microstructure on subcritical flaw growth do suggest that significant
improvements may be possible in this design area. During heating,
cooling, and thermal cycling, cracks in refractory linings are continually subjected to complex states of stress. Slow or subcritical
crack growth when these cracks are subjected to tensile stresses is
to be expected. Observations of cracks in metallurgical vessel
linings that grow or extend from one heat to another are common. It
is reasonable to assume that an initially small crack may be safe
for some R value in Equation (2) during one particular thermal cycle,
but eventually grows subcritically, reducing R, and leading to a
catastropic propagation event on a subsequent cycle. Slow or subcritical crack growth in brittle ceramics, including refractories
can be characterized by a slow crack growth parameter, N, where
small N-values are indicative of a high susceptibility to slow crack
growth and large N-values signify resistance to slow crack growth.

There are two common experimental methods for determining N-
values: (i) the stress rate effect on strength (σ_f vs $\dot{\sigma}$)[9] and
(ii) the double-torsion(K-V) technique[10]. They differ in princi-
ple in that the (σ_f vs $\dot{\sigma}$) method measures the subcritical growth of
small cracks or microcracks characteristic of strength specimens;
whereas, the (K-V) method measures large macrocracks, often several
inches long. There have been observations that the (K-V) method
yields higher N-values suggesting that macrocracks are more resistant to slow crack growth than are microcracks[11]. Adams, Landini,
Schumacher, and Bradt[12] have measured and compared N-values for
these two types of cracks for a number of alumina refractories,
Figure 1. They observed that the (K-V) double torsion macrocracks
do indeed have much larger N-values than the smaller (σ_f vs $\dot{\sigma}$) microcracks. On an N-value basis, the larger double torsion cracks are
3 to 8 times more resistant to slow crack growth. Another important
result of Figure 1 is that the microcrack N-values and macrocrack
N-values appear to be correlated. Furthermore, their order is not
predicted by alumina content, strongly suggesting a dominant effect
of microstructure. It appears that microstructure plays a very
important role in the inhibition of slow crack growth in

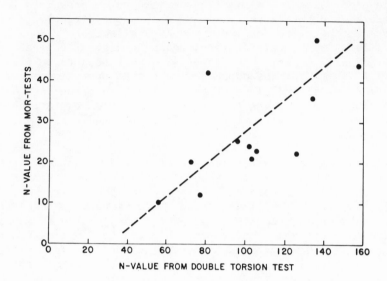

Figure 1.   A comparison of the N-values for the slow
            crack growth of aluminosilicate refractories
            from stressing rate-strength tests (MOR) and
            a fracture mechanics test specimen (double
            torsion)[12].

refractories. There appear to be considerable opportunities in
refractory design for improvements of this particular facet of
resistance to crack growth.

The resistance of refractories to thermal shock damage is
certainly the particular area where improvements have been readily
achieved in the recent past. From a physical properties viewpoint,
the energy balance concepts first advanced by Hasselman in the
early 1960's provide the fundamental basis for damage minimization
design;[3] unfortunately, even though Nakayama[13] also measured
the work-of-fracture in the early 1960's, the day-to-day application
and acceptance of the principles has rather lagged. This has occurred inspite of excellent reviews of the topic by Nakayama[14] and by
White[4], as well as a number of sequel papers by Hasselman[15-18].

Although the mathematics may have appeared rather formidable
at the time, Hasselman simply equated the elastic strain energy in
the thermal stress field at fracture initiation to the fracture
surface energy required by the crack propagation during damage, a
kind of Griffith energy balance applied to thermal shock. He thus
arrived at the thermal stress damage resistance parameter R'''',

where:

$$R'''' = \frac{E\gamma_{wof}}{\sigma_f^2 (1-\nu)} \quad , \tag{3}$$

The γ_{wof} term is the work-of-fracture measurement of the fracture surface energy. Hasselman originally used a $\gamma_{effective}$ for the fracture surface energy[3]; however, as a consequence of many measurements and numerous studies over the past two decades, these authors have taken the liberty of substituting the work-of-fracture for $\gamma_{effective}$. In one of the sequel papers[15], Hasselman defined the thermal shock damage resistance parameter R_{st}:

$$R_{st} = (\frac{\gamma_{wof}}{\alpha^2 E})^{1/2} \quad , \tag{4}$$

which he termed the thermal stress crack stability parameter. The damage resistance parameter R'''' was specifically derived for high strength, or short initial crack length situations that lead to kinetic crack growth. The R_{st} is for lower strength refractory bodies containing larger initial cracks, and thus subject to a quasi-static form of crack growth. The similarities of the two equations must be emphasized for both are natural consequences of Hasselman's unified theory energy balance. Thermal stresses are proportional to (αE), thus the term $(\alpha^2 E)^{1/2}$ in the denominator of Equation (4) is somewhat analogous to the (E/σ_f^2) of Equation (3). Both of the parameters R'''' and R_{st} are energy ratios, and both are dominated in their numerators by the work-of-fracture term, γ_{wof}. If simple rankings of refractories are made based on the usual groups, fireclays, high alumina, magnesite, etc., then it becomes evident that R'''' and R_{st} usually rank order the refractories the same within groups. In general, the similarities between R'''' and R_{st} far outweigh their differences, so much in fact, that they have, on occasion, been substituted interchangeably.

With regard to the physical properties relevant to thermal shock damage resistance, the only new parameter of concern with regard to damage is the work-of-fracture, γ_{wof}. However, before discussing γ_{wof}, it is important to contrast Equations (1) and (3) for they illustrate a crucial aspect of refractory design for thermal shock resistance. From Equations (1) and (3), it is apparent that the stress and elastic modulus terms are inverted in the two equations, thus attempts to maximize R will usually minimize R''''. Simultaneous maximization of both the resistance to initiation of fracture, (R), and the damage resistance, (R'''') is impossible. This is not altogether unfortunate for refractory microstructural design, since it is almost hopeless to prevent fracture initiation anyway, thus the damage resistance aspects can be maximized.

The work-of-fracture, γ_{wof}, of ceramics was initially measured by Nakayama in the early 1960's[19] and shortly thereafter by Tattersall and Tappin[20]. With this technique, it became immediately obvious that fracture processes in refractories consumed a considerable amount of energy. Rather than the 0.1 - 1 J/m^2 surface energies suggested by earlier ceramic single crystal studies as well as thermodynamic measurements, refractory's fracture energies usually exceeded these values by several orders of magnitude. The tremendous importance of γ_{wof} in Hasselman's energy balance was then obvious! Unfortunately, the significance of these measurements of the work-of-fracture and their direct application to refractories went practically unnoticed for a number of years, except for the classic study of Nakayama and Ishizuka on fireclays in 1966[21]. That study was probably coupled with the original measurements of γ_{wof} by Nakayama a number of years earlier. It was only in the latter 1960's and early 1970's that a widespread effort developed to extensively measure the work-of-fracture of refractories and apply the results to thermal shock situations and laboratory test thermal shock studies.

Coincident with the extensive measurement of the work-of-fracture of refractories, fracture mechanics techniques in general became more popular, as evidenced by the Sosman Lecture by Wachtman[22]. Measurement of the fracture toughness, K_{Ic}, naturally lead to the calculation of the surface energy for fracture initiation from:

$$K_{Ic}^2 = \frac{2E\gamma_{nbt}}{(1-\nu^2)} \quad , \tag{5}$$

where γ_{nbt} is the fracture surface energy for fracture initiation. The subscript (nbt) is a consequence of applying the notched beam test for the K_{Ic} measurement, applying the derivation of Brown and Srawley[23] to a single-edge-cracked type of specimen. Equation (5) can be combined with the Griffith equation: $\sigma_f = K_{Ic}Yc^{-1/2}$, and R'''', Equation (3) to yield:

$$R'''' \; \alpha \; (\frac{\gamma_{wof}}{\gamma_{nbt}}) \; c \quad , \tag{6}$$

illustrating that the thermal shock damage resistance parameter, R'''', is proportional to the ($\gamma_{wof}/\gamma_{nbt}$) ratio. This ratio has gained some acceptance as a general measure of the thermal shock damage resistance of refractories. Thermal shock damage prone materials have a ($\gamma_{wof}/\gamma_{nbt}$) very near unity; however thermal shock damage resistance refractories typically have ratios between 5 and 10, with some of the most thermal shock damage resistant refractories in excess of 20.

THERMAL SHOCK TESTS

A strictly academic approach suggests that the appropriate
physical property measurements such as those just discussed (σ_f,
E, α, γ_{wof}, K_{Ic}, etc.) would suffice on their own merits for refractory selection procedures. However, a more realistic view demands
an intermediate stage of thermal shock testing for evaluation of
refractory performance prior to actual large scale installations in
operating vessels. The economics of vessel linings strongly supports
such intermediate testing for verification of performance predictions which are based purely on theory and physical property measurements. The numbers of laboratory type thermal shock tests, as
well as the means of evaluating the degree, or amount of thermal
shock damage are substantial, for it appears that nearly everyone
has their own testing procedures. It is disappointing that only a
few of the tests have been applied in any fashion which permits a
direct correlation with the thermal shock resistance parameters
just discussed. Three specific types of tests which have been correlated will be reviewed including: (i) the panel spalling test,
(ii) various prism-like quench tests, and (iii) the ribbon test.
This does not imply that other types of thermal shock tests are not
valid, only that they have not been adequately correlated with physical properties and thermal shock damage resistance parameters in the
opinion of these authors. Of course, when any form of an intermediate scale laboratory thermal shock test is evaluated, it must
also be judged as to the success it has in simulating actual application conditions. Few, if any, scaled-down laboratory tests can precisely duplicate the conditions in actual metallurgical vessels
whose size may exceed that of the entire test laboratory; consequently, some reservation is in order when critically examining, or
when accepting any of these laboratory-scale tests results. However, in spite of any reservations, the continued insistance of
refractory consumers on some form of these tests strongly attests
to their merits.

The panel spalling test has not been looked upon very favorably
in recent times and is probably well on the way to obsolescence as
a routine evaluation method. In the past, it has served a vital
purpose in the area of fireclays for application in boilers and in-
cinerators; however, there currently exists justified widespread
dissatisfaction in its ability to discriminate between highly thermal
shock damage resistant refractories. For example, the development
of refractories over the past several decades has seen the assessment technique for refractories subjected to the panel spalling test
to progress from weight loss measurements, to visual assessment of
the degree of cracking, to practically no obvious damage at all.
Refractory producers have effectively developed thermal shock resistant refractories to the degree that they have surpassed the panel
spalling test's ability to seriously damage them.

Unfortunately, there has not been an exhaustive study where a single investigator studied a series of refractories, systematically measuring the physical properties, calculating the thermal shock damage resistance parameters, and then performing extensive panel spalling tests on the same refractories. Treusch[24], however, did measure the physical properties and calculate the damage resistance parameters. He then compared the results with sales product literature data for panel spalling test weight losses for the same four commercial fireclays. He observed excellent correlation between room temperature R'''' values and the percentage weight loss in the panel spalling test, with the higher R'''' values having the lower percentage weight losses. For R'''' values of approximately 7.5, 4.5, 4, and 2 (cm) the fireclay refractories had percentage weight losses of about 2, 3.5, 7, and 9. All of the fireclays had work-of-fractures between 50-60 J/m^2; however, their fracture stresses (MOR's) varied by over a factor of two, thus accounting for the differences in the thermal shock damage resistance parameters. One must conclude that the performance of fireclays, some of which experience up to 10% weight losses in the panel spalling test, is predictable from the thermal shock damage resistance parameter, R'''', as calculated from measured physical properties at room temperature.

For the purpose of discussion here, the various prism-type quench tests will be considered together inspite of the almost endless number of permutations of sample sizes, quenching or cooling procedures as well as thermal shock damage assessment techniques. For example: quenches have been performed from elevated temperatures into room temperature air, ice water, water sprays, liquid nitrogen, and molten metals. Specimens have also been rapidly heated by insertion into elevated temperature environments. Damage has similarly been assessed in a variety of ways, including weight loss measurements, elastic modulus changes, damping capacity increases, and strength losses. In spite of this endless number of combinations of test situations, every time that a parallel study of the measurement of physical properties and the calculation of thermal shock damage resistance parameters has been completed, there has been a strong correlation with either R'''' or R_{st}.

The initial of these prism-quench studies is the classic work of Nakayama and Ishizuka[21] on commercial firebricks. Their damage assessment procedure was measuring the weight loss from thermal shock, where they recorded the number of quench cycles to achieve a 5% spalling loss. As to be expected, they also measured the work-of-fracture and correlated the number of cycles with R''''. Their fireclays exhibited a wider range of fracture surface energies than those studied by Treusch[24], between 25 and 80 J/m^2; however, their R'''' values only ranged from about 1 to 4. Their calculated R'''' values did however exhibit a strong correlation with the number of cycles to a 5% weight loss. Fireclays near an R'''' of

1 cm required only 7 cycles, while those with 4 cm R'''' values
needed 14 cycles. When these results of a prism spall type of
thermal shock test are assessed along with Treusch's correlation
with the panel spalling test, one must conclude that these two types
of industrial tests for the thermal shock damage of refractories
produce the same ordering, or ranking of fireclay refractories. The
ranking is the same ranking predicted by the Hasselman damage resistance parameter, R''''. It is evident that calculation of R''''
based on physical property measurements and these two industrial
thermal shock damage tests yield very similar assessments for fireclay refractories.

    A logical progression from the fireclay refractories is to exam-
ine the higher alumina refractory compositions. A number of quenching prism-like thermal shock damage studies have been performed on
various alumina refractories including those by Ainsworth and
Herron[25], Larson, Coppola, Hasselman, and Bradt[26], Larson and
Hasselman[27], Stellan Persson[28], and Beals[29]. Every one of
these studies has revealed a strong correlation of thermal shock
damage resistance with either R_{st} or R''''. Each has utilized some
form of quench and then monitored the strength changes, either after
single or multiple quenches.

    Figure 2 illustrates a typical strength-loss plot for an alumina
refractory after a single thermal shock quench from the $\Delta T°C$ specified. The trend in Figure 2 is characteristic of a good thermal
shock damage resistant refractory, in that with increasing severity
of thermal shock, there is only a very gradual loss of strength.
Two approaches have been applied to analyze these curves, one is to
consider the strength losses for a number of refractories at a given
ΔT, and the other is to extend the results from the single thermal
shock quench to multiple thermal shock quenches for a given ΔT. The
results are quite different in that they appear to correlate with
different thermal shock damage resistance parameters. The R_{st},
Equation (4), appears to correlate best for single thermal shocks,
while R'''', Equation (3), appears to be preferable for multiple
thermal shocks.

    The Larson and the Stellan Persson studies[26-28] specifically
address the single thermal shock situation. Larson's studies were
quenching into room temperature water, but Stellan Persson's studies
were into molten lead at 700°C. Stellan Persson also included a
few magnesite brick[28]. The maximum single quench temperature
differentials were 1000°C for the Larson study and 700°C for that
of Stellan Persson, thus Stellan Persson quenched the alumina refractories from 1400°C into molten lead at 700°C. It is important to
note the elevated temperature which Stellan Persson quenched to,
for it is a unique study and has features which appear highly desirable from a practical vessel simulation viewpoint. Both studies

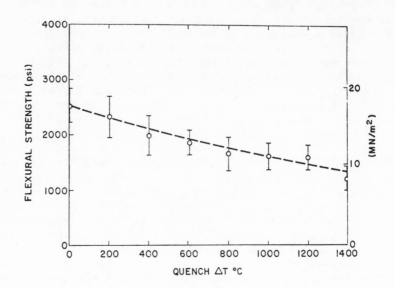

Figure 2.   The flexural strength, (MOR), of an alumina
refractory after being subjected to rapid
thermal shocks, quenches of ΔT into room
temperature water[29].

examined the percentage strength retained, Larson the flexural
strength (MOR) and Stellan Persson a ring-test type of tensile
strength. They correlated the percentage strength retained with R_{st}
in both cases, with correlation coefficients of 0.84 and 0.97
respectively, clearly indicative of the applicability of R_{st} to
predict single quench thermal shock damage resistance. The Larson
studies were considerably more extensive in the number of alumina
refractories studied, thus some additional observations are possible.
In the 38 different alumina refractories, there appeared to be some
general trends with alumina content; however, there is also considerable compositional overlap in the R_{st} values, clearly suggesting
that having established the alumina content of the refractory;
microstructural features probably assume a considerable role in determining the performance in severe thermal shock damage situations.

    Stellan Persson[30] introduced yet another concept into the
analysis of strength loss by thermal shock damage. It appears to
be an important contribution to the analysis of (strength) vs
(ΔT) curves, particularly for those examples where the crack growth
is completely quasi-static, as in Figure 2, and the slope is not a
function of the quench severity. When (dσ/dΔT) is constant, it is
possible to describe the linear decrease by the angle the straight

line makes with the horizontal, the abscissa. Stellan Persson defined the slope as a parameter ψ, so that a large negative value of
ψ means that the refractory has poor thermal shock resistance, while
a $\psi = 0$ indicates excellent thermal shock damage resistance.
Stellan Persson related the ψ parameter to R_{st} in the form:

$$\psi = - \frac{k}{(R_{st})^{2.2}} \quad , \qquad\qquad (7)$$

where k is some complex function of Biot's modulus and relates to
the severity of the quench media in the particular test. At the
present, there does not appear to be a strong interest in determining
ψ's; however, its direct relation to R_{st} points out a potentially
obvious advantage, for calculation of R_{st} demands measuring γ_{wof},
α, and E, while only determining ψ could be a much easier task under
some conditions.

The effect of only a single thermal shock quench on refractory
properties is an excellent laboratory study and a challenging basic
problem, but one that is rather far removed from the reality of
plant applications. Usually, refractories are repeatedly thermally
shocked and the damage is of a cumulative nature, thus, multiple
thermal shocks are a logical extension. A couple of multiple quench
thermal shock studies have been completed on alumina refractories,
the most notable being the pioneering effort of Ainsworth and
Herron[25], which also included some magnesite and some mag-chrome
refractories. Beals[29] later completed a related study focused
specifically on alumina refractories, where more severe quenches
were applied and a larger number of alumina refractories were tested.
The conclusion of these studies is that one may quite naturally
extend the single quench thermal shock theory developed by Hassleman
to situations involving multiple thermal shocks and that the R''''
damage resistance parameter is applicable.

The basis of these multiple thermal shock studies is that after
a large number of quenches, the thermal shock damage, as monitored
by the strength reduction or elastic modulus decreases, achieves a
constant level. Figure 3 illustrates this effect on the strengths
for quenching an alumina refractory into water for two different
quench severities, ΔT's of 800°C and 1400°C. The more severe
quench naturally causes a greater strength loss, but both eventually
exhibit the characteristic constant level of cumulative damage after
about a half dozen cycles. Subsequent quenches may produce a minor
amount of additional damage, but the existing results suggest that
it is minimal. Although Figure 3 shows this form of behavior for
an alumina quenched into water, Ainsworth and Herron[25] demonstrated
similar behavior for rapid air quenches on several magnesite and mag-
chrome bricks as well. The leveling-off of the cumulative thermal

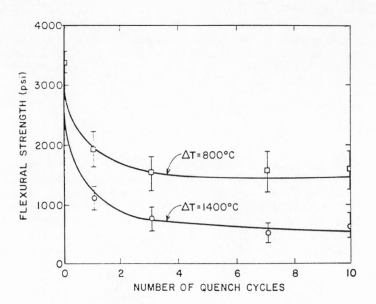

Figure 3.   The effect of multiple quenches on the
            strengths of an alumina refractory for
            two different severity quenches [29].

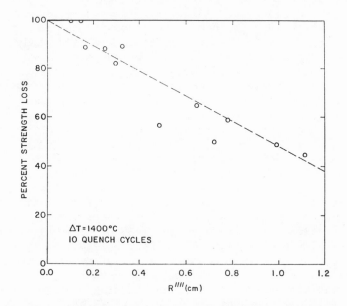

Figure 4.   A multiple quench (percentage-strength-
            loss) vs (R'''') for alumina refractories
            quenched from a ΔT of 1400°C [29].

shock damage is undoubtedly a universal trend, the level of which is dependent on the properties of the refractory brick.

Ainsworth and Herron[25] clearly demonstrated that the degree of cumulative damage, as measured in terms of percentage-strength-loss, could be related to the work-of-fracture through a modification of Hasselman's R'''' damage resistance parameter. They found a linear relation for (percentage-strength-loss) vs $(\gamma_{wof}E/\sigma_f^2V)$, where the V is the specimen volume. If this parameter is compared with R'''' in Equation (3), it is evident that this latter term is simply a volume compensated R'''', a quite logical extension since (σ_f^2/E) is related to the stored elastic strain energy per unit volume. Beals[29] in a later study directly related the percentage-strength-loss to R''''. Figure 4 illustrates the results of Beal's 1400°C quenches. Like the Ainsworth and Herron study, a linear relation is observed for alumina refractories. The magnesite refractories measured by Ainsworth and Herron fell close to the aluminas; however, their mag-chrome bricks exhibited considerably higher percentage-strength-losses for the same R'''' values. These multiple quench studies clearly demonstrate that the cumulative damage from a number of thermal shocks can also be quantitatively related to thermal shock damage resistance parameters, and specifically are related to the work-of-fracture. These studies substantiate the application of R'''' to the multiple thermal shocks of fireclays by Nakayama and Ishizuka[21], who only applied it with reservation to the multiple thermal shock situation since Hasselman basically derived it for only a single thermal shock.

A derivation for the form of the straight line behavior of Figure 4 was outlined by White[4] for a single quench and extended by Beals[29] to multiple quenches. White showed for a single quench that:

$$(\text{percentage-strength-loss}) = 1 - 4\ V(\frac{\gamma_{wof}E}{\sigma_f^2 V})(\frac{N}{\pi})^{1/2}, \qquad (8)$$

where (N) is the crack density of the refractory. Beals extended the derivation to multiple quenches and recast it into R'''' terminology for consistency with previous studies, arriving at:

$$(\text{percentage-strength-loss}) = 1 - f(q)(R''''), \qquad (9)$$

where f(q) is a function containing the number of quenches in a π-product fashion. The f(q) also contains the crack density parameter $(N)^{1/2}$, similar to Equation (8). It thus can be concluded that in addition to confirming the importance of R'''', studies such as those of Ainsworth and Herron, and Beals may be expected to reveal interesting features about the flaw density (N) of refractories as well as their thermal shock damage resistance. Beals' plot for the 800°C multiple quenches of the aluminas, Figure 5, indicates promise for this approach, as the refractories appear to split off into two distinct flaw density groupings. If Figures 4 and 5 are compared, it suggests either that flaw density, (N), may be an important parameter for moderate thermal shocks, but is relatively insignificant for very severe thermal shock quenches, or perhaps that Beals' heating to 1400°C before quenching altered the flaw distributions. In either case, it is evident from derivations and further encouraged by experimental results that these types of studies of multiple quenches may be able to provide flaw density information on refractories.

As a final point concerning the flaw density, (N), it should be emphasized that it is an extremely important parameter, but a rather poorly understood one in refractories. In his original unified theory thermal shock paper, Hasselman used (N) as an adjustable parameter to delineate the extent of individual crack growth under kinetic conditions. It invariably occurs in every theoretical assessment of thermal shock damage and is related to every damage measurement process. Unfortunately, whenever (N) values are actually calculated from experimental results on refractory bricks, unrealistically low numbers result. Since many refractories are microstructurally designed to contain cracks and flaws, these (N) difficulties will have to be eventually resolved, not only to further understand their fundamental role in the theory, but also to enable their quantitative incorporation into the basic aspects of microstructural design.

The other thermal shock testing technique which has demonstrated a strong correlation with the theoretical thermal shock damage parameter, R'''', is the ribbon test. Hawisher and Semler[31,32] have discussed the ribbon test in recent reports and presentations. Geometrically, any number of refractories of specified sizes are stacked side by side to present the appearance of a ribbon, below this ribbon of refractories is a line burner to heat one surface of the bricks. This test has the ability to examine refractories over a wide range of thermal gradients and thermal shock severity by adjusting the burner conditions, the on-off cycles, and the specimen sizes. It can be easily automated and readily achieves a variety of multiple thermal shocks through on/off cycles of the line burner. Thermal shock damage can be assessed by all of the usual methods, including measurements of weight losses, strength changes, elastic moduli, etc. Reproducibility has been excellent.

Hawisher, Semler, and Bradt[33] examined a number of the same alumina refractories which Beals[29] studied and some others. They varied the specimen size to achieve different thermal gradients and monitored the damage by measuring the elastic modulus of the test

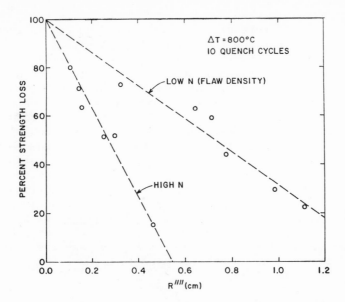

Figure 5. Multiple quench (percentage-strength-loss)
vs (R'''') plots for alumina refractories
quenched from a ΔT of 800°C into water[29].

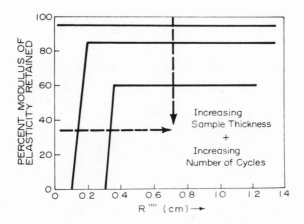

Figure 6. Composite results of the ribbon test for
alumina refractories, showing the effect
of specimen size and the number of thermal
shock cycles on the (percentage-modulus-
retained) vs (R'''')[33].

pieces after 1, 3, and 5 cycles. The composite trends are shown in
Figure 6. Both increased specimen thickness and an increased number
of cycles increase the thermal shock damage; however, another interesting feature is that there appears to be a critical R'''' value
for each set of test conditions. Refractories with R'''' values
below this critical R'''' value suffer catastropic failure during
the test, whereas, those above the critical R'''' value suffer only
minimal damage as measured by decrease in elastic modulus. Not only
does this laboratory test correlate with calculated R'''' values,
again confirming the value of the thermal shock damage prediction
from the Hasselman theoretical damage resistance parameter calculated
from measured physical properties, it also appears to introduce a new
concept of a critical R'''' value. This has additional importance,
for the existence of a critical R'''' for a specific thermal gradient
or test condition affords the test practitioner with the opportunity
to design the specimen and test variables to match his specific
application conditions. This promises to make the ribbon test extremely valuable as an accept-reject procedure for individual applications of refractories.

MICROSTRUCTURAL EFFECTS

     There have been several studies that specifically examined the
effects of microstructure on the thermal shock damage resistance
parameters of refractories. Schumacher[34] has studied commercial
alumina refractories, Uchno, Bradt, and Hasselman[35] have reported
on a number of magnesite refractories, Kuszyk and Bradt[36,37] have
examined mag-chrome brick and Landini has investigated dolomites[38].
Studies have also been made on castables, but they will not be
addressed here, other than to repeat that their thermal shock
spalling resistance has been greatly improved by the addition of
metal fibers[39].

     Only Uchno[35] for magnesites and Landini[38] for dolomites
have specifically addressed the microstructural facet of the work-
of-fracture and its effects on R_{st} and R'''' by consolidating specific microstructures based on selected grain sizings. They examined
refractories which ranged from a structure of a large fraction of
coarse sizing to a structure based exclusively on fines. Of course
most commercial refractories are intermediate in structure, but some
of both extremes are available. The trends for these two types of
basic refractories are that the coarser microstructures exhibit
higher γ_{wof}'s, and higher ratios of ($\gamma_{wof}/\gamma_{nbt}$), whereas the structures containing a significant amount of fines invariably exhibit
much lower values of each. The structures with many fines usually
consist of a considerable amount of continuous bonding after firing
and have relatively high strengths and high elastic moduli. Conversely the coarser grain structures are not extensively bonded,
but contain many pores and internal cracks, which, along with the

large grains interact very energetically with propagating cracks and
impart a high work-of-fracture to the microstructure. Undoubtedly,
for each type or class of refractories there exist optimum structures
for maximizing R_{st} or R''''; however, to date there have not been
sufficiently intense efforts to achieve that goal for any clay.

Refractories with good and poor thermal shock damage resistance
also exist in all of the commercial classes. Figure 7 illustrates
two alumina refractories and two dolomites, one of each with excellent and one with poor thermal shock damage resistance. The many
internal microcracks and pores are obvious in the damage resistant
structures, which have high work-of-fracture values, a major portion
of which can probably be related to the role the microcracks assume
in dissipating energy by promoting crack branching and multiple secondary crack propagation. By contrast, the damage prone brick are
neither crack nor pore free; however, they clearly possess a considerably larger extent of continuous bonding.

CONCLUSIONS

The basis for the quantitative understanding of the thermal
shock damage of refractories exists. Its formulation in terms of

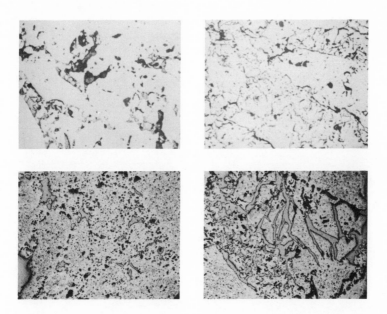

Figure 7. Microstructures of two aluminas (top) and
two dolomites (bottom), 40X, illustrating
typical highly thermal shock resistant
microstructures (right) and damage prone
microstructures (left).

thermal shock damage resistance parameters clearly illustrates the
dominant importance of the work-of-fracture of refractories. The
damage resistance parameters, R_{st} and R'''' exhibit excellent correlation with single thermal shock and cumulative damage multiple
thermal shock laboratory scale tests, including panel spalling,
prism, and ribbon tests. These tests have revealed some difficultie
in the analysis of the flaw density of refractories, but they also
indicated promising experimental routes to an understanding of the
effects of flaw density. Microstructure is one means of designing
refractories for high work-of-fracture values. Refractories with
high work-of-fractures usually contain substantial coarse grain
sizings and exhibit considerable porosity and an abundance of internal microcracks. Although these microstructural features are understood, they are not at the present fully quantified to enable microstructural design on an apriori basis.

ACKNOWLEDGEMENT

     The authors gratefully acknowledge the technical assistances
of T. Beals, J. Bodkin, G. Davenport, K. Ishler, D. Landini, and
C. Schumacher and also the current and past research support of
DOE (EF-77-S-01-2678) and AISI (46-339) on related refractory
fracture studies. Without those contributions, as well as others
to numerous to mention, this paper could not have been written.

REFERENCES

1.   J. R. Coxey, pp. 9-13, "Refractories", Pennsylvania State
       University (1950).
2.   W. D. Kingery, J. Amer. Cer. Soc., 38(1):3-15 (1955).
3.   D. P. H. Hasselman, J. Amer. Cer. Soc., 50(9):229-234 (1963).
4.   J. White, Refractories J., 52(11-12):10-19 (1976).
5.   V. S. Kienow, Ber. Dt. Keram. Ges., 47(7):426-430 (1970).
6.   J. H. Ainsworth and R. H. Herron, Proc. 9th St. Louis Refrac-
       tories Symp. (1973).
7.   R. L. Shultz, Private Communication.
8.   D. P. H. Hasselman, Bull. Amer. Cer. Soc., 49(12):1033-1037
       (1970).
9.   H. C. Chandan, R. C. Bradt, and G. E. Rindone, J. Amer. Cer.
       Soc., 61(5-6):207-210 (1978).
10. J. O. Outwater, M. C. Murphy, R. G. Kumble, and J. T. Berry,
       pp. 127-138, ASTM STP 559; ASTM (1974).
11. B. J. Pletka, E. R. Fuller, Jr., and B. G. Koepke, pp. 19-37,
       ASTM STP 678; ASTM (1979).
12. T. A. Adams, D. J. Landini, C. A. Schumacher, and R. C. Bradt
       (to be published).
13. J. Nakayama, J. Amer. Cer. Soc., 48(11):583-587 (1965).
14. J. Nakayama, pp. 759-778, Vol. 2, "Fracture Mechanics of
       Ceramics", Plenum Publ. Corp. (1974).

15. D. P. H. Hasselman, J. Amer. Cer. Soc., 52(11):600-604 (1969).
16. D. P. H. Hasselman, J. Amer. Cer. Soc., 53(9):490-495 (1970).
17. D. P. H. Hasselman, pp. 89-103, "Ceramics in Severe Environ-
     ment", Plenum Publ. Corp. (1971).
18. D. P. H. Hasselman, Ber. Dt. Keram. Ges., 54(6):195-201 (1977).
19. J. Nakayama, J. J. Appl. Phys., 3(71):422-423 (1964).
20. H. G. Tattersall and G. Tappin, J. Mat. Sc., 1:296-301 (1966).
21. J. Nakayama and M. Ishizuka, Bull. Amer. Cer. Soc., 45(7):666-
     669 (1966).
22. J. B. Wachtman, J. Amer. Cer. Soc., 57(12):509-519 (1974).
23. W. F. Brown and J. E. Srawley, pp. 3-15, ASTM STP 410; ASTM
     (1966).
24. A. L. Treusch, M. S. Thesis, The Pennsylvania State University
     (1973).
25. J. H. Ainsworth and R. H. Herron, Bull. Amer. Cer. Soc.,
     53(7):533-538 (1974).
26. D. R. Larson, J. A. Coppola, D. P. H. Hasselman, and R. C.
     Bradt, J. Amer. Cer. Soc., 57(10):417-421 (1974).
27. D. R. Larson and D. P. H. Hasselman, Trans. Brit. Cer. Soc.,
     74(2):59-64 (1975).
28. S. B. Stellan Persson, pp. 325-328, Proc. 3rd CIMTEC, Italy,
     May (1976).
29. T. A. Beals, B. S. Thesis, The Pennsylvania State University
     (1978).
30. S. B. Stellan Persson, pp. 37-55, 3rd Nordic High Temp. Symp.
     (1972).
31. T. H. Hawisher and C. E. Semler, Ref. Res. Center Report,
     April (1979).
32. C. E. Semler and T. H. Hawisher, Paper 27-R-79, Amer. Cer. Soc.
     81st Annual Meeting (1979).
33. T. H. Hawisher, C. E. Semler, and R. C. Bradt, Paper 28-R-79,
     Amer. Cer. Soc. 81st Annual Meeting (1979).
34. C. A. Schumacher, M. S. Thesis, The Pennsylvania State Univer-
     sity (1980).
35. J. J. Uchno, R. C. Bradt, and D. P. H. Hasselman, Bull. Amer.
     Cer. Soc., 55(7):665-668 (1976).
36. J. A. Kuszyk and R. C. Bradt, Ind. Heat., J. Heat. Tech.,
     52(3):61-64 (1975).
37. J. A. Kuszyk and R. C. Bradt, Ind. Heat., J. Heat. Tech.,
     52(4):24-26 (1975).
38. D. J. Landini, M.S. Thesis, The Pennsylvania State University
     (1980).
39. P. 42, Brick and Clay Record, October (1979).

EFFECT OF CRACK HEALING ON THERMAL STRESS FRACTURE

Tapan K. Gupta

Ceramics and Glasses Department
Westinghouse Research and Development Center
Pittsburgh, PA 15235

ABSTRACT

The general approach to the selection of materials with high
resistance to thermal stress fracture is to follow criteria which
would identify materials resistant to crack initiation or having
low crack propagation. A third approach is presented here having
the basis of retarding propagation wherein the propagating crack is
allowed to heal during service, thus arresting premature crack completion and specimen fracture. The discussion presented in this
paper suggests that in order to promote crack healing and delay
crack fracture completion, the material should be selected on the
basis of high surface diffusion coefficient, high surface energy,
small grain size and small crack width. Service temperatures where
grain growth and bulk diffusion are promoted should be avoided.

INTRODUCTION

The failure of a ceramic under the conditions of a thermal or
a mechanical stress can be visualized as a sequential combination
of four stages in the development of a crack: nucleation, initiation, propagation and fracture completion. Nucleation of a crack
is believed to occur spontaneously in ceramics, even during handling,
and exists prior to the application of any load. Initiation is the
growth of the existing nucleus as a result of the application of
load. Propagation is the intervening stage of the crack motion
between initiation and fracture completion. In an ideal brittle
material, such as glass, once a crack is initiated by the application of a load, propagation and fracture completion follow immediately. The approach to the selection of materials with high
resistance to thermal stress fracture has, therefore, been to

develop criteria which would identify materials resistant to crack
initiation, propagation and fracture completion.

According to crack initiation criteria,[1] materials are selected
on the basis of high values of strength and thermal conductivity,
combined with low values of thermal expansion, Young's modulus,
Poisson's ratio and emissivity. For application where crack initiation in ceramics cannot be avoided, such as refractory linings of
blast furnaces, the selection criteria has been to minimize the
extent of crack propagation.[2] According to these criteria, materials
are selected on the basis of high values of Young's modulus, Poisson's
ratio, surface fracture energy and low values of strength and small
specimen size. Another criterion for the selection of a material
is proposed herein, based on retarding crack propagation by healing
at a high temperature, thus arresting premature crack fracture
completion. This phenomenon, known as crack healing, has been
observed in a number of ceramics in recent years and can provide a
selection criterion for dealing with the last stage of crack behavior

GENERAL BACKGROUND ON CRACK HEALING IN CERAMICS

The presence of surface cracks in ceramics is known to cause
an overall decrease in room temperature strength. These surface
cracks are caused by grinding, cutting, machining, and even handling.
However, when such ceramics are subjected to heat treatment at high
temperatures, surface cracks disappear and strength values are
improved. This phenomena, known as crack healing, has been observed
in a number of ceramic materials. Thus, annealing[3] or flame
polishing[4] gave higher strength values in single crystal Al_2O_3, and
reheating[5] prior to testing increased the strength values of polycrystalline Al_2O_3. In all cases, it was assumed that the surface
cracks introduced during specimen preparation were healed by heat
treatment. During the past several years,[6-12] further evidence of
crack healing was obtained from controlled experiments in which
deep surface cracks were deliberately introduced into dense polycrystalline ceramics which were then annealed. The strength lost
by the deliberate introduction of cracks was recovered by the heat
treatment. The beneficial effect of heat treatment on strength was
observed in ZnO,[6] MgO,[8] Al_2O_3,[7,9] UO_2,[12-14] and $(U,Pu)C$[15] and was
attributed to crack healing. Recent studies[15-23] have indicated
that crack healing and strengthening are related to the void evolution caused by diffusive transport of atoms. The object of this
paper is to review this recent development in Al_2O_3 (and to some
extent in MgO) and develop a criteria for the selection of material
on the basis of existing knowledge.

STRENGTHENING AND TOUGHENING

Deep surface cracks were introduced by thermally shocking
MgO[8] and Al_2O_3[9] bars in water. The temperature from which the bars

were quenched was adjusted in such a manner that the strength after thermal shock was reduced to about 0.30 σ_i (σ_i = initial strength). The crack patterns were like crazed glaze when observed by dye penetrant,[6] and the crack depths were approximately 1/4 of the width of the specimen. The specimen size was approximately 38 x 3 x 3 mm. The density of the oxide bars was 99+% of the theoretical density.

Following thermal shock, the specimens were heat treated under controlled conditions of temperature and time. MgO bars were heated in air, whereas Al_2O_3 bars were heated in a vacuum. Strength data were obtained by four-point bend tests at room temperature. Flexural strengths were also obtained on as-cut (no thermal shock, no heat treatment) and thermally shocked (no heat treatment) specimens.

The nominal room temperature strengths of as-cut specimens and those after thermal shock are shown in Table 1. The normalized strength recovery data of thermally shocked, reheated oxides are illustrated in Fig. 1, where F_t represents the strength recovered at time, t (σ_t), over the initial strength (σ_i). It is seen that the recovery proceeds through several stages. During the stage of partial recovery of strength, the effect of temperature appears to be stronger than that of time. The strengthening rate actually diminishes with time as the strength approaches the stage of complete recovery. In the case of alumina, the stage of complete recovery is followed by a stage of "relapse" where the strength decreases with time. This decrease in strength with prolonged heating has been observed by other workers[7] and has been attributed to grain growth. As will be shown later, no conclusive explanation is available for this decrease at present.

The kinetics of strengthening has been treated theoretically by several authors.[12,13,20,22] Unfortunately, the equations thus derived do not fit the experimental data acceptably. On the contrary, the empirical equation derived by Roberts and Wrona,[12] when modified[10] by invoking the relation between initial strength, σ_i,

Table 1. Flexural Strength of As-Cut and
Thermally Shocked Specimens

Surface Condition	Strength (MN/m^2)	
	Al_2O_3	MgO
As-Cut Specimens	195	170
Thermally Shocked Specimens	64	54

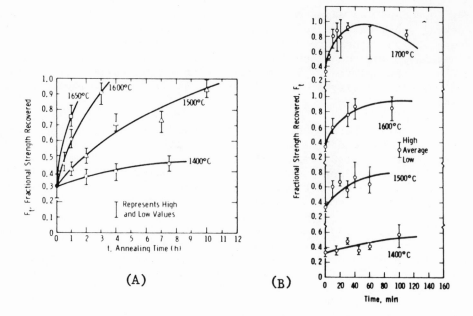

Fig. 1.   Fracture strength recovery of initial strength
of (A) MgO and (B) Al$_2$O$_3$ as a function of
annealing temperature and time.

and the strength after thermal shock, σ_o, can be fitted to experimental data for Al$_2$O$_3$ and MgO quite satisfactorily. A general
expression which describes the strength recovery with time can be
stated as follows:[10]

$$F_t^2 = F_o + At \qquad\qquad (1)$$

which is valid when $F_o < F_t < 1$, and where A = constant, F_o =
constant and $F_t = (\sigma_t/\sigma_i)$ = the fractional strength recovered at
time, t.

The isothermal strength recovery data for MgO and Al$_2$O$_3$ are
plotted in Fig. 2 according to Eqn. (1). In spite of the scatter
in the strength values, the data are well represented by Eqn. (1).
It is seen that the linear plot of F_t^2 vs t is followed to $F_t = 1$
(complete recovery case) for both MgO and Al$_2$O$_3$. Once the value
of $F_t = 1$ is reached (e.g., in Al$_2$O$_3$ at 1700°C), deviation from
linearity is observed, and strength decreases. The strength in
this region is believed to be governed by the most deleterious
inherent flaw in the material, and not by the critical crack length
from healing.

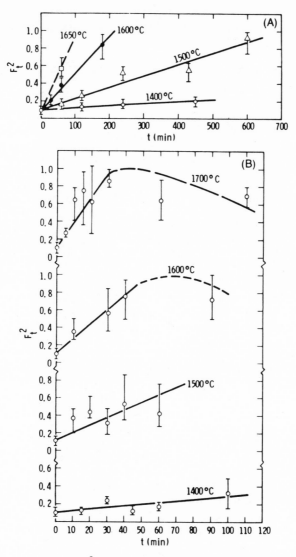

Fig. 2.   Plot of F_t^2 vs t for (A) MgO and (B) Al_2O_3

The activation energy for strengthening was obtained from Eqn. (1) by assuming that the constant A has the form of an Arrhenius equation: $A = A_o \exp(-\Delta H/RT)$, where ΔH is the activation energy and RT has its usual meaning. Eqn. (1) can then be expressed as:

$$F_t^2 = F_o + A_o \exp\left(-\frac{\Delta H}{RT}\right) t. \qquad (2)$$

Equation (2) was subjected to nonlinear regression analysis using the data for MgO and Al$_2$O$_3$. The best fit equations obtained from the analysis were as follows:

For MgO:

$$F_t^2 = 0.10 + 9.02 \times 10^6 \exp(-80.4 \text{ kcal/RT})t, \qquad (3)$$

and for Al$_2$O$_3$:

$$F_t^2 = 0.18 + 7.94 \times 10^4 \exp(-56.6 \text{ kcal/RT})t. \qquad (4)$$

The low activation energy for alumina suggests a surface diffusion mechanism as will be discussed later.

Finally, since the fracture stress, σ_f, and the stress intensity factor, K_c, are related by the following expression:

$$K_c = \sigma_f \, y\sqrt{c} \qquad (5)$$

where y is a geometric constant and c is a characteristic flaw dimension, it follows that an increase in strength by crack healing will also increase the fracture toughness in the partially and completely healed materials.

Evans and Charles[20] have measured the stress intensity factor of alumina as a function of the degree of crack healing (K_c') and compared the data to that of pristine material (K_c). They showed that K_c'/K_c approaches unity as the crack proceeds from the stage of partial to complete recovery (Fig. 3). The data suggests that the resistance to crack propagation increases as the healing proceeds. This would imply that crack fracture completion would be delayed as the propagating crack is healed.

GRAIN GROWTH AND MICROSTRUCTURE

Several workers[6,7,12] have observed grain growth during crack healing. From the preliminary data obtained on ZnO,[6] Al$_2$O$_3$,[7] and UO$_2$,[12] it was suggested that the strengthening of thermally shocked oxides might be related to grain growth. Gupta[9] has made a systematic study of the effect of grain growth on the strength recovery in Al$_2$O$_3$ (Table 2). It is seen that the grain size increases by merely ≈10% (from 27.1 μm to 29.8 μm) at the highest temperature (1700°C) and time (∼2 h) of crack healing and ∼5% or less (where data cannot be separated from error in measurement) at a lower temperature and shorter time with a corresponding strength degradation of 15 to 20%.

On the other hand, Lange and Radford[7] observed a decrease in strength of ≈18% after heating for 30 to 50 hr at 1700°C and they

Table 2. Strength and Grain Size of Thermally
Shocked and Annealed Al_2O_3 Specimens

Annealing Conditions		Avg. Strength	Grain Size[a]
Temp. (°C)	Time (min)	MN/m^2	(μm)
1700	110	162.7	29.8
	60	156.3	29.1
	30	181.5	29.4
	20	153.9	28.5
	10	158.6	28.1
	5	101.3	27.4
1600	90	165.9	28.6
	40	170.0	28.4
	30	147.7	28.0
	10	116.3	27.5
1500	60	124.9	28.0
	40	143.0	28.1
	30	109.6	28.0
	20	129.9	27.5
	10	117.6	27.0
1400	100	112.7	27.5
	60	80.6	27.6
	45	68.7	27.4
	30	96.3	27.2
	15	71.6	26.9
As-cut Specimens		195.0	27.1

[a]Planar measurement

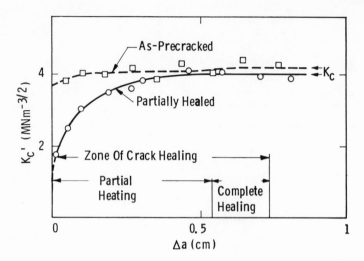

Fig. 3.   The effect of a change in crack length, a due to
crack healing on the critical stress intensity
factor, K_c', for a partially healed and an as-
precracked sample tested at a constant displacement rate.   (After Evans and Charles[20])

attribute it to the increase in grain size which was estimated to
be ⌃25%. In the present investigation, the 15 to 20% strength
degradation after heating for only 1 to 2 hr at 1700°C cannot be
explained by a mere ⌃10% increase in grain size. Thus, grain growth
does not appear to be the cause of strength decrease after complete
recovery. The most likely cause would be the presence of deleterious
defects in the material whose origin is not crack healing.

Similarly, an explanation of strength recovery based on grain
growth alone cannot be justified for several reasons: (1) the sharp
increase in strength (Fig. 1) occurs within a very short time before
any measurable grain growth can occur, (2) the crack healing temperature is well below the sintering temperature at which the specimens
were prepared (>1800°C), (3) the crack (void) space will inhibit
grain growth (the very small grain growth during crack healing may
be a reflection of this phenomenon), and (4) single crystals can
recover strength during annealing without grain growth (see next
section). Thus, grain growth appears to have, if any, a secondary
effect on the recovery of strength in alumina.

The microstructure changes involved in the removal of thermally shocked cracks are illustrated in Fig. 4 with a series of optical micrographs obtained on polished and etched surfaces of Al$_2$O$_3$. The thermally shocked cracks were formed along the grain boundaries (intergranular) as represented by heavy lines between the grains (Fig. 4(A)). As the specimens were heated at 1600°C, the crack began to heal as seen by the gradual disappearance of the heavy lines (Figs. 4(B) and (C)) with longer heating. After 90 minutes of heat treatment (Fig. 4(D)), most of the cracks were healed and the specimens regained an average of 90% of their original strength.

These observations suggest that healing occurs via the gradual disappearance of crack space (void) between adjacent grains, similar to the disappearance of pores during sintering of powder compacts. The driving force for void elimination appears to be the reduction of interfacial energy.

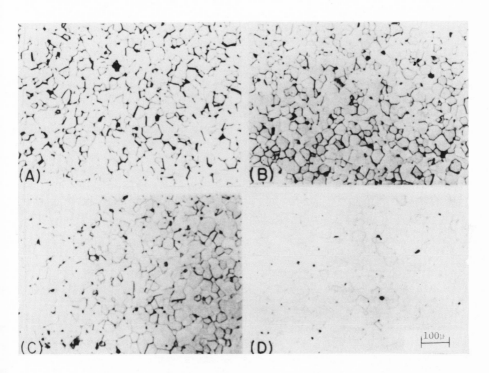

Fig. 4.   Crack-healing characteristics of Al$_2$O$_3$ as indicated by
           the disappearance of grain-boundary cracks (heavy lines).
           (A) Thermally shocked blank and samples annealed at
           1600°C for (B) 10 min, (C) 30 min, and (D) 90 min.

VOID EVOLUTION AND ITS KINETICS

The discussion presented thus far summarizes the effect of crack healing on strength, fracture toughness, grain growth and microstructure. The question remains, what happens to the crack through this entire period of crack healing? There is a general lack of details in this area, but a picture regarding the morphology changes in crack emerges from the study of alumina.[9,20]

The crack healing in alumina commences either with the regression of the primary crack as observed by Evans and Charles[20] or by crack surface pinch-off as observed by Gupta.[9] This difference can be explained by assuming that the material was stress-free in the former case, whereas a residual stress-field was retained in the latter. Thus, in a stress-free material, there will be a general regression of crack length on heating but a nonuniform release of stored energy will cause the initiation of crack pinch-off in thermally shocked materials with residual stress-field. Whatever the starting mechanism, the process is immediately followed by the formation of an array of cylindrical voids ("cylinderizing") in regions adjacent to crack tip. The cylinders, being unstable,[18] then quickly dissociate into spheres.

A series of photomicrographs describing the above morphological changes are shown in Fig. 5, where the healing commences with the typical "pinching" of long cracks (Fig. 5(A)), which then evolves into channels of cylindrical voids (Figs. 5(B) and (C)). Upon prolonged heating, the latter breaks up into a multitude of spherical voids.

Whereas the kinetics of crack "pinching" has not been investigated, Evans and Charles[20] studied the kinetics of primary crack recession in Al_2O_3. They found that the equation that best fit the experimental data is characterized by:

$$a = a_o - \lambda \left| \frac{e^{-Q/RT} t}{T} \right|^{\omega} \tag{6}$$

where a_o is the initial crack length, a the crack length at time t, and temperature T, λ and ω are constants equal to 54 and .6, respectively. The activation energy, Q, calculated from the crack recession data was found to be 65 kcal/mole, which they claim to be due to surface diffusion. This value is, however, close to the activation energy of strengthening as shown in Eqn. (4).

Extensive microstructural observations in Al_2O_3 by the present author revealed that the process of "cylinderizing" of a "flat" crack is extremely rapid and thus difficult to study. However, if any analogy with the metal system is permitted, the "cylinderizing" process in alumina can be visualized as similar to coarsening of

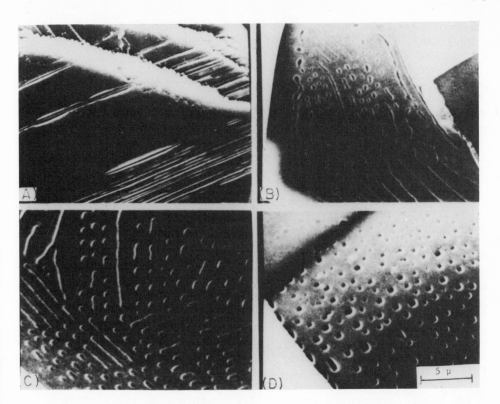

Fig. 5 - Void evolution sequence during crack healing. (A) Typical
pinching of thermally shocked cracks, (B) and (C) formation of
cylindrical voids and their breakup into spherical voids, and
(D) spherical voids.

manganese sulphide inclusions in rolled steel during homogeniza-
tion.[24] It was found that the kinetics of "cylinderizing" of
manganese sulphide inclusion can be described by the theory proposed by Nichols and Mullins[25] for the decay of circumferential
perturbations on an infinite rod and the rate of "cylinderizing"
is surface diffusion controlled. The rapidity with which the
"cylinderizing" takes place in alumina when subjected to crack
healing experiments also seems to support a surface diffusion
control mechanism.

The breakup of a cylindrical void in alumina during crack
healing has been studied in detail by Gupta.[18] He found that the
kinetics of void breakup can be described again by the theory of
Nichols and Mullins[25] for the decay of a longitudinal perturbation
on a semi-infinite cylinder. It was found that, depending on
aspect ratio, the cylinders can break up either by "ovulation" or

transform directly to sphere by "spheroidization." The general expression[18] for the kinetics of void breakup can be stated as:

$$\tau = C R_o^n \tag{7}$$

where τ is the time for ovulation, C a constant, R_o the cylinder radius and n is an exponent which determines the mass transport mechanism. The predominant void breakup on thermal anneal confirms[18] a model for surface-diffusion-control material transport in alumina. A least-square fit on the data gives the following expression for the surface diffusion coefficient:

$$D_s = 4.2 \times 10^6 \exp(-118 \pm 12.5 \text{ kcal/RT}). \tag{8}$$

The magnitude of diffusion coefficients obtained from the above equation agrees well with the data from the literature.[18] However, there is a range of literature values reported for activation energy, 56-133 kcal/mole,[23,26-32] for surface diffusion in alumina, and the crack healing data are in satisfactory agreement with this range.

VOID ELIMINATION AND ROLE OF GRAIN BOUNDARIES

In addition to monitoring the void evolution during crack healing, extensive microstructural investigation was conducted to follow the mechanism of void disappearance as the crack healing proceeded to completion. It was found that the final spherical voids are terminated in adjacent grain boundaries which act as sinks. This phenomena is illustrated by the scanning electron micrograph of Fig. 6, which depicts the spherical void configuration on an intergranular surface of a completely healed specimen. It is seen that the regions adjacent to grain boundaries are depleted of voids, suggesting migration and disappearance of voids in the neighboring grain boundaries, while some residual voids are still present in the middle of the grain. Moreover, a trail of voids (marked by an arrow) is also seen approaching the lower grain boundary. Indications are thus strong that there is a diffusional migration of voids towards grain boundaries which act as sinks for voids. This conclusion is consistent with the sintering theory.

The above observation suggests that the rate of crack healing and strengthening will be more rapid in polycrystalline alumina than in single crystal sapphire. Although a comparative experiment is difficult to conduct (because original strength and thermally shocked crack width are different in polycrystal and single crystal) the beneficial effect of grain boundaries on strengthening can be appreciated by comparing the strengthening data of single crystal sapphire as shown in Table 3 with that of polycrystalline alumina in Table 2. Note that heat treatment in sapphire was conducted at temperatures between 1500 and 1900°C and times between 2 and 5 hours, whereas in polycrystalline alumina the corresponding temperatures

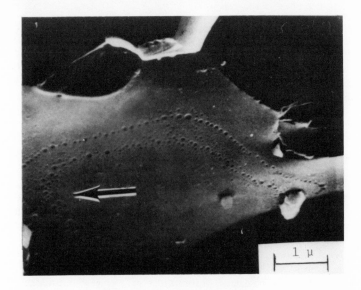

Fig. 6. Void elimination in grain boundaries.
Note that the area adjacent to grain
boundaries are depleted of voids.
(Annealing conditions: 1700°C, 110 min.)

Table 3. Strength of Thermally Shocked Sapphire Bars
Annealed at Temperature and Time Indicated

Annealing Conditions		Average Strength
Temp. (°C)	Time (h)	(MN/m^2)
1500	2	40.0
1500	5	73.0
1600	2	50.3
1600	5	82.0
1700	5	84.1
1800	5	90.3
1900	5	148.2
As-cut		323.4
As-cut and thermal shocked		35.9

and times were 1400–1700°C and <2 hours. In spite of this higher temperature and longer time for single crystal sapphire, the strength recovery was slow. At 1900°C for 5 hours, the strength recovered to <0.5 σ_i in sapphire, whereas in polycrystalline alumina, the strength recovered to >0.95 σ_i even at 1700°C for 5 hours. It can be concluded from these results that the presence of grain boundaries causes a rapid healing and strengthening in polycrystalline alumina.

CRITERIA FOR MATERIAL SELECTION

Surface diffusion appears to be a dominant mechanism in all stages of crack healing, whereas surface energy is the dominant driving force. Thus, crack healing will be promoted in materials with high values of surface diffusion coefficient and surface energy. Since ovulation is enhanced in thin cylinders (Eqn. 7), materials which generate narrow cracks upon thermal shouck should also promote crack healing. Since grain boundaries appear to act as sinks for voids, a greater abundance of grain boundaries would cause rapid annhilation of ovulated voids. This would suggest that polycrystalline materials with smaller grain size will be preferred to large grain size. An added advantage of small grain size would be to reduce the diffusion distance for the migrating voids. Needless to say, temperature and time will promote crack healing by enhancing the diffusion coefficient and allowing voids to diffuse further. Very high temperatures should be discouraged to avoid grain growth (enlarging diffusion distance) and bulk diffusion. The effect of environment is unclear at this stage.

Thus, to promote crack healing and delay fracture completion, the materials should be selected on the basis of high surface diffusion coefficient, high surface energy, small grain size, and small crack width.

REFERENCES

1. W. D. Kingery, "Factors Affecting Thermal Stress Resistance of Ceramic Materials," J. Amer. Cer. Soc. 38 (1) 3–15 (1955).
2. D. P. H. Hasselman, "Unified Theory of Thermal Shock Factor Initiation and Crack Propagation in Brittle Ceramics," J. Amer. Cer. Soc. 52 600–604 (1969).
3. A. H. Heuer and A. P. Roberts, "Influence of Annealing on the Strength of Corundum Crystals," Proc. Brit. Cer. Soc., 6, 17–27 (1966).
4. F. P. Mallinder and B. A. Proctor, "Preparation of High Strength Sapphire Crystals," Proc. Brit. Cer. Soc., 6, 9–16 (1966).
5. H. P. Kirchner, R. M. Gruver, D. R. Platts, P. A. Rishel and R. E. Walker, "Chemical Strengthening of Ceramic Materials," Summary Report, Contract N00019-68-C-0142, Naval Air Systems, (1969).

6. F. F. Lange and T. K. Gupta, "Crack Healing by Heat Treatment," J. Amer. Cer. Soc., 53 [1] 54-55 (1970).

7. F. F. Lange and K. C. Radford, "Healing of Surface Cracks in Polycrystalling Al₂O₃," J. Amer. Cer. Soc., 53 [7] 420-21 (1970).

8. T. K. Gupta, "Crack Healing in Thermally Shocked MgO," J. Amer. Cer. Soc., 58 [3-4] (1975).

9. T. K. Gupta, "Crack Healing and Strengthening of Thermally Shocked Alumina," J. Amer. Cer. Soc., 58 [5-6] 1976).

10. T. K. Gupta, "Kinetics of Strengthening of Thermally Shocked MgO and Al₂O₃," J. Amer. Cer. Soc., 59 [9-10] 448-449 (1976).

11. F. F. Lange, "Healing of Surface Cracks in SiC by Oxidation," J. Amer. Cer. Soc., 53 [5] (1970).

12. J. T. A. Roberts and B. J. Wrona, "Crack Healing in UO₂," J. Amer. Cer. Soc., 56 [6] 297-99 (1973).

13. G. Bandyopadhyay and J. T. A. Roberts, "Crack Healing and Strength Recovery in UO₂," J. Amer. Cer. Soc., 59 [9-10] 415-419 (1976).

14. G. Bandyopadhyay and C. R. Kennedy, "Isothermal Crack Healing and Strength Recovery in UO₂ Subjected to Varying Degrees of Thermal Shock," J. Amer. Cer. Soc., 60 [1-2] 48-50 (1977).

15. R. N. Singh and J. L. Routbort, "Fracture and Crack Healing in (U,Pu)C," J. Amer. Cer. Soc., 62 [3-4] 128-133 (1979).

16. T. K. Gupta, "Alteration of Cylindrical Voids During Crack Healing in Alumina," 6th Intl. Matl. Symp. on Ceramic Micro-structure, Univ. of California, Berkeley, CA, 354-365 (1967).

17. T. K. Gupta, "Diffusive Transport in Alumina During Crack Healing," presented at the Annual Amer. Cer. Soc. Mtg. in Detroit, Basic Sci. Div., Cer. Soc. Bull. 57 [3] 311 (1978).

18. T. K. Gupta, "Instability of Cylindrical Voids in Alumina," J. Amer. Cer. Soc., 61 [5-6] 191-195 (1978).

19. R. Raj, W. Pavinich and C. N. Ahlquist, "On the Sintering Rate of Cleavage Cracks," Acta Metall., 23 [3] 399-403 (1975).

20. A. G. Evans and E. A. Charles, "Strength Recovery by Diffusive Crack Healing," Acta Metall., 25 [8] 919-29 (1977).

21. S. M. Park and D. R. O'boyle, "Observation of Crack Healing in Sodium Chloride Single Crystals at Low Temperature," J. Mat. Sci., 12 840-841 (1977).

22. R. Dutton, "Comments on Crack Healing in UO₂," J. Am. Cer. Soc., 56 [12] 660 (1973).

23. C. F. Yen and R. L. Coble, "Spheroidization of Tubular Voids in Al₂O₃ Crystals in High Temperatures," J. Am. Cer. Soc., 55 [10] 507-509 (1972).

24. Y. V. Murty, J. E. Morral, T. Z. Kattamis and R. Mehrabian, "Initial Coarsening of Manganese Sulphide Inclusions in Rolled Steel During Hologenization," Met. Trans. A, 6A, 2031-2035 (1975).

25. F. A. Nichols and W. M. Mullins, "Morphological Changes of a Surface of Revolution Due to Capillarity-Induced Surface Diffusion," J. Appl. Phys., 36 [6] 1826-35 (1965).

26. W. M. Robertson and R. Chang, Role of Grain Boundaries and Surface in Ceramis, "Materials Science Research, Vol. 3," W. W. Kriegel and H. Palmour III, Plenum, New York (1966).

27. W. M. Robertson and F. E. Ekstrom, Kinetics and Reactions in Ionic Systems, "Materials Science Research, Vol. 4," T. J. Gray and V. D. Frechett, Plenum, New York (1969).

28. J. F. Shackelford and W. D. Scott, "Relative Energies of ($\overline{1}100$) Tilt Boundaries in Aluminum Oxide," J. Am. Cer. Soc., $\underline{51}$, [12] 688-92 (1968).

29. T. Maruyama and W. Komatsu, "Surface Diffusion of Single-Crystal Al_2O_3 by Sketch Smoothing Method," J. Am. Cer. Soc., $\underline{58}$, [7-8] 338-39 (1975).

30. Y. Moriyoshi and W. Komatsu, "Kinetics of Initial Combined Sintering," Yogyo Kyokai Shi, $\underline{81}$ [3] 102-107 (1973).

31. S. Prochazka and R. L. Coble, Surface Diffusion in the Initial Sintering of Alumina: I," Phys. Sintering, $\underline{2}$ [1] 1-18 (1970); "II," [2] 1-14, "III," 15-34.

32. W. R. Rao and I. B. Cutler, "Initial Sintering and Surface Diffusion in Al_2O_3," J. Am. Cer. Soc., $\underline{55}$ [3] 170-71 (1972).

IMPROVEMENT OF THERMAL SHOCK RESISTANCE OF BRITTLE STRUCTURAL

CERAMICS BY A DISPERSED PHASE OF ZIRCONIA

N. Claussen* and D. P. H. Hasselman**

*Max-Planck Institut für Metallforschung, Stuttgart
 West Germany
**Virginia Polytechnic Institute and State University
 VA, USA

ABSTRACT

This paper discusses the effect of unstabilized zirconia dispersions on the thermal shock resistance of brittle structural ceramics.

A general discussion is given of the preferred direction of modification of the pertinent material properties which affect thermal shock resistance. The nature of the crystallographic phase transformation in the zirconia and its effect on these properties is presented. The zirconia dispersed phase can lead to significant improvements in fracture toughness by transformation - or microcrack toughening. The volume change during the zirconia phase transformation also lowers the effective coefficient of thermal expansion. Surface compressive stresses which result from the phase transformation during surface grinding also are beneficial in improving thermal shock resistance. The zirconia phase also can change the unstable (catastrophic) mode of failure to the more preferred stable one with decrease in fracture stress. These effects are illustrated by experimental data for composites, consisting of zirconia dispersions in aluminum oxide, silicon nitride and zircon matrices.

INTRODUCTION

The advantageous properties of ceramic materials such as their excellent high-temperature strength, creep and corrosion resistance, make them ideal candidates for the replacement of such

materials as metal super-alloys for turbine engines and many other
engineering applications involving extreme temperatures. Unfortunately, ceramic materials display one unfavorable characteristic,
namely an extreme brittleness or notch-sensitivity. This high degree of brittleness renders such materials highly susceptible to
catastrophic failure under conditions of thermal shock. Because of
the absence of any kind of plasticity, the non-uniform thermal expansions which result from the non-uniform temperatures, cannot be
accommodated by non-linear deformation. The low value of fracture
toughness of brittle ceramics also contributes to extensive crack
propagation following thermal stress failure. This frequently
renders any structure or component made from brittle ceramics completely incapable of continued satisfactory performance following
the fracture due to thermal stresses.

In order to fully capitalize on the very favorable properties
of structural ceramic materials, it is imperative that their thermal
shock resistance be improved significantly. In this respect it
should be noted that ceramic composites exhibit greater thermal
shock resistance than single-phase ceramic materials (1,2,3). More
recent results have shown that the introduction of an unstabilized
ZrO_2 dispersed phase in a brittle matrix can lead to a significant
enhancement of many mechanical properties, such as strength and
fracture toughness, of importance to the resistance to thermal
stress fracture. This effect is attributed to the unique tetragonal-to-monoclinic phase transformation encountered in zirconia (4-
9).

The purpose of this paper is to summarize the role of this
phase transformation in enhancing the thermal shock resistance of
brittle ceramic materials.

FUNDAMENTALS

In order to establish the role of the zirconia dispersions,
a brief overview will be given of the pertinent material properties
which govern the thermal stress failure of brittle materials.

a.   Initiation of Thermal Stress Failure

For structures or components in which thermal stress failure
cannot be tolerated, the material properties for a given design
must be selected such that the maximum thermal stresses do not
exceed the failure stress of the materials. Generally, brittle
materials exhibit values of the tensile strength well below the
compressive strength. For this reason, thermal stress failure of
brittle materials usually occurs in tension.

The usual procedure followed in ceramic technology for establishing the role of the pertinent material properties which affect

thermal stress failure of a brittle material subjected to a given thermal environment, is to derive an expression for the maximum tensile thermal stresses developed. By equating this value of stress to the tensile strength, an expression is then obtained for the maximum thermal environment (i.e., temperature difference, heat flux, etc.) to which the material can be subjected without incurring thermal stress failure. As an example, the maximum temperature difference to which the ceramic structure can be subjected, can be expressed (10,11):

$$\Delta T_{max} = \frac{A\sigma_f(1-\nu)K}{\alpha E} \tag{1}$$

where A is a constant depending on the geometry of the structure, σ_f is the tensile fracture stress, ν is Poisson's ratio, K is the thermal conductivity, α is the coefficient of thermal expansion and E is Young's modulus.

In fracture-mechanical terms, the tensile strength can be expressed:

$$\sigma_f = YK_{Ic}/a_o^{1/2} \tag{2}$$

where Y is a constant related to the geometry of the failure-initiating crack, K_{Ic} is for critical stress intensity factor (fracture toughness) the mode I crack-opening and a_o is a measure of the crack size or depth.

Substitution of eq. 2 into eq. 1 yields:

$$\Delta T_{max} = \frac{AYK_{Ic}(1-\nu)K}{\alpha Ea_o^{\frac{1}{2}}} \tag{3}$$

If in the region of the failure initiating flaw, a compressive stress (σ_c) exists, ΔT_{max} can be expressed by:

$$\Delta T_{max} = \frac{AYK_{Ic}(1-\nu)K}{\alpha Ea_o^{\frac{1}{2}}} + \frac{A\sigma_c(1-\nu)K}{\alpha E} \tag{4}$$

b.   Crack Propagation in Thermal Stress Fields

In high-temperature technology, frequently the situation is encountered that even in materials with the highest thermal stress resistance, thermal stress failure cannot be avoided. In this case, thermal stress resistance is not measured in terms of the maximum thermal condition to which the materials can be subjected without failure. Instead, thermal shock resistance is measured by the ability of the material to render continued satisfactory

service in spite of having undergone thermal stress failure. One
such popular measure for this criterion is the strength retained
following thermal stress failure. The retained strength is a measure of the size of the arrested cracks which resulted from the
failure process.

A simple fracture-mechanical analysis of the nature of dynamic crack propagation in a thermal stress field has shown that the
size of the crack (a_f) after propagating and arrest is inversely
proportional to a ratio, which involves the energy (Y_f) required
to create unit area of new crack surface, Young's modulus (E),
the initial fracture stress (σ_f) and the number of cracks (N),
written as:

$$a_f = f(Y_f E/\sigma_f^2 N) \tag{5}$$

From eq. 2, the strength (σ_a) retained after thermal stress
failure is:

$$\sigma_a = YK_{Ic}/a_f^{\frac{1}{2}} \tag{6}$$

From eqs. 2, 5 and 6, it can be concluded that for a given
number of cracks, the strength retained after failure is an inverse
function of the initial strength or:

$$\sigma_a = f(\sigma_f^{-1}) \tag{7}$$

In terms of the objectives of this paper, eqs. 1, 3 and 7, indicate the problem faced by the materials technologist in selecting
or developing brittle materials subjected to high values of thermal
stress. As shown by eq. 3, for a given value of K_{Ic}, thermal
stress resistance as measured by ΔT_{max} can be increased by decreasing the flaw size. This, in effect, increases the initial
tensile strength. Unfortunately, however, as indicated by eq. 7,
if this increase in strength is not sufficient and failure still
occurs, the only effect would be to increase the size of the cracks
after fracture and to decrease the resulting load-bearing ability
of the material. In fact, for ultra-high-strength structural ceramic materials, thermal stress failure can be quite explosive, ren-
dering the structure or component completely useless for further
service. In fact, the preferred approach would be to increase the
original crack size. This has the further advantage, that if the
flaw size becomes large enough, crack propagation will occur in a
stable manner. At the same time high strength can be retained both
before and after failure, by increasing the fracture toughness as
much as possible. In addition, reductions in the coefficient of
thermal expansion and Young's modulus, if possible, results in further improved thermal stress resistance.

These conclusions are summarized in Fig. 1, which shows the
typical strength behavior following thermal shock of increasing

severity. The discontinuity of strength at ΔT_c is the result of the dynamic crack propagation. For stable crack propagation, no such instantaneous strength loss occurs with strength decreasing monotonically for values of $\Delta T > \Delta T_c$. Included in Fig. 1, are the ratios of the pertinent material variables which need to be modified to increase thermal stress resistance. Further improvements in thermal stress resistance as indicated by eq. 4 can also be achieved by creating a compressive stress, which opposes the tensile thermal stress, in the region of the material where the failure originates.

Experimental data have indicated that incorporating a zirconia dispersed phase in a brittle matrix material can lead to significant improvements in thermal stress resistance by increasing the fracture toughness, lowering the coefficient of thermal expansion, introducing compressive stresses and by changing the failure mode to the preferred stable mode of crack propagation as demonstrated in the following sections.

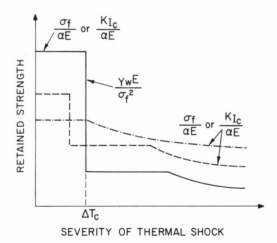

Fig. 1.   Variation of strength of typical brittle ceramic subjected to thermal quench of increasing severity and appropriate ratios of material properties relevant to thermal stress resistance. Solid curve: high-strength ceramic; dotted curve: low strength ceramic.

NATURE OF ZIRCONIA PHASE TRANSFORMATION

In order to understand the role of the zirconia dispersions in improving thermal stress resistance, the nature of the tetragonal-monoclinic phase transformation should be examined (12-15).

Figure 2 shows a diagram of a zirconia particle incorporated in a matrix before (t = tetragonal) and after (m = monoclinic) its transformation. The lower half of Fig. 2 shows the thermal expansion behavior of a brittle matrix containing zirconia inclusions. On cooling from high temperatures, the inclusions go through the transformation at a temperature T_R, which involves a 3-5% volume expansion. The tangential stresses in the matrix, caused by this expansion, can lead to microcrack formation when the particle diameter exceeds a critical value d_{Rc} (12). On reheating the particles can retransform to the tetragonal phase at a temperature T_H. Generally $T_H > T_R$ so that the expansion behavior shows a hysteresis. The height of the hysteresis curve is a function of the volume fraction of zirconia. The temperatures T_R and T_H are a function of

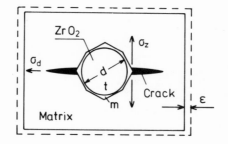

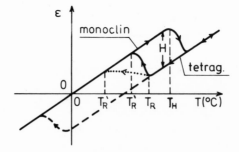

Fig. 2.   Zirconia particle contained within ceramic matrix, before (t) and after (m) the tetragonal-monoclinic phase transformation and associated thermal expansion behavior of the composite.

the zirconia particle size and the constraint provided by the matrix. T_R and T_H generally decrease with decreasing particle size.
Thus, by controlling the particle size, the nature of the microcracking in the matrix as well as the temperature range over which
the transformation occurs can be controlled.

EFFECT OF PHASE TRANSFORMATION IN ZIRCONIA ON MATERIAL PROPERTIES
AND STATE OF STRESS

The tetragonal-to-monoclinic phase transformation in the zirconia dispersions has a profound effect on a number of material properties which control thermal stress resistance as follows:

1. Fracture toughness (K_{Ic}) can be modified by two methods referred to as micro-crack toughening and transformation toughening.
At sufficiently large zirconia particle size, the hoop stresses
can cause extensive micro-cracking in the matrix. During failure,
these micro-cracks promote crack bifurcation and an enhanced dissipation of energy, thereby effectively increasing the fracture
toughness. Figure 3 indicates the increase in fracture toughness
which can be achieved in this manner for four different ceramic
matrix materials.

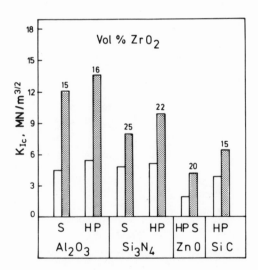

Fig. 3. Fracture toughness at room temperature of ceramic matrix
materials without (white bars) and with (grey bars) various
volume percentages of a dispersed zirconia phase indicated
by the number above the grey bars.

Transformation toughening can occur by zirconia dispersions of sufficiently small size, which as the result of the matrix constraint retain their tetragonal crystal structure at room temperature. However, during fracture, the high stresses in the immediate region of the crack tip cause a temporary removal of the constraint. This permits the phase transformation to take place. The particles now, however, have a larger volume, so that on passage of the stress field of the crack, a major modification in the matrix stress field is encountered, thereby effectively increasing the fracture toughness.

2. Compressive stresses. The zirconia phase transformation can cause the creation of compressive stresses which can oppose the tensile thermal stresses. This effect can be achieved by two mechanisms (16,17). Steady-state compressive stresses can be created by the stress-induced phase transformation. One very effective method by which this can be achieved is to cut or grind the surface. The volume expansion associated with the transformation leads to the formation of high compressive stresses in the immediate surface regions. Figure 4 shows the increase in bend strength and fracture toughness which can be achieved by this method. Figure 4

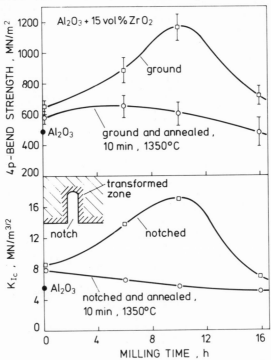

Fig. 4. Effect of anneal and milling time on flexural strength and fracture toughness of hot-pressed alumina with 15 vol.% ZrO_2, in which tetragonal-to-monoclinic phase transformation was induced by surface grinding and cutting.

also contains data for the effect of micro-crack toughening on
strength and fracture toughness, labelled "ground and annealed."
Annealing causes a relaxation of the surface stresses so that any
remaining improvements in strength and toughness must be attributed
to effects other than surface compression. In fact, at the very
short (milling) time for mixing the alumina and zirconia powders,
the observed increase in toughness is due to micro-crack toughening, because the zirconia particle size is still quite large. The
decrease in fracture toughness with increased milling time is due
to the accompanying decrease in particle size.

Thermal stresses arise from temperature non-uniformities. Because of this non-uniformity of temperature, the temperature dependence of the phase transformation permits the introduction of
transient compressive thermal stresses which oppose the tensile
thermal stresses. During cooling, for, instance, the maximum
thermal stresses occur in the surface, where also the transformation
will occur first. The volume expansion associated with this transformation created a compressive stress in the surface which opposes
the tensile thermal stress due to the greater degree of cooling.
It should be noted here that this mechanism for improving thermal
stress resistance, is effective only for cooling from temperature
immediately above or within the range of transformation temperature. A similar statement must be made for compressive stresses
created by surface grinding. Such stresses are useful only when
the tensile thermal stresses exist in the surface alone as in the
case for transient cooling.

3.   Coefficient of thermal expansion. As indicated in Fig. 2,
the volume expansion which accompanies the phase transformation on
cooling, in effect, reduces the effective coefficient of thermal
expansion. Since the range of transformation temperature is a
function of the particle size incorporating a bi-modal particle
size distribution results in a widening of this temperature range.
In fact, a coefficient of thermal expansion near zero can be
created between about 800°C and room temperature. This effect is
indicated schematically in Fig. 5.

4.   Stable crack propagation. When a high density of microcracks of sufficient size is introduced in a matrix such that its
elastic behavior is affected, crack propagation in a thermal
stress field will become stable. Strength loss with increasing
severity of thermal shock will be monotonic rather than discontinuous for unstable crack propagation as shown in Fig. 1. Crack
sizes resulting from stable crack propagation generally are smaller than those which result from unstable crack propagation. For
this reason, although perhaps initially rather weak, materials
which fail in a stable manner under conditions of thermal shock
will exhibit superior strength behavior following the thermal shock.
This fact is used to advantage in refractory structures in which
thermal stress failure generally cannot be avoided, but in which

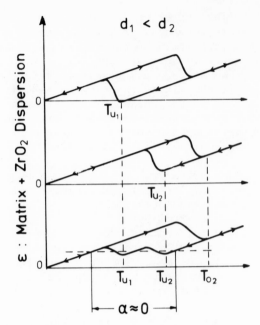

Fig. 5. Thermal expansion behavior of ceramic matrix with zirconia
         inclusions with bi-modal particle size distribition.

the structural materials nevertheless must render satisfactory
service in spite of being in a partially fractured condition.

The introduction of a zirconia dispersed phase can also be
used successfully to change the undesirable unstable failure mode
of a brittle matrix to the more preferred stable one. This, however, will require relatively large particle sizes and high volume
fractions to assure a significant decrease in Young's modulus, a
prerequisite for stable crack propagation. As an example, 20 vol.%
ZrO_2 particles with a diameter of approximately 10 μm in an alumina
matrix caused a decrease in Young's modulus and bend strength from
about 400 $GN.m^{-2}$ and 500 $MN.m^{-2}$ to about 150 $GN.m^{-2}$ and 170 $MN.m^{-2}$.
Of interest to note is, that the microcracking in the matrix had
little or no effect on the strain-at-fracture, frequently used as
a measure for thermal stress resistance. In fact, extensively
microcracked materials can exhibit strains-at-fracture an order
of magnitude above that for crack-free materials.

THERMAL SHOCK DATA

Experimental data were obtained to illustrate many of the a-
bove phenomena (18,19,20). For all materials studied, specimens

were subjected to thermal shock by quenching from higher temperature
into a water bath at room temperature followed by a strength determination in 4-point bending.

Figure 6 shows the strength behavior of an alumina matrix with
about 20 vol.% ZrO_2 dispersions of various sizes. The 100% alumina
with initial strength of approximately 500 MN.m^{-2} exhibited a
critical quenching temperature difference of approximately 200°C.
The introduction of the zirconia particles with a diameter of approximately 1 μm, as the result of microcrack toughening, raised the strength to about 650 MN.m^{-2} and ΔT_c to about 300°C, a
substantial improvement. Note that for the latter two materials
as indicated by the instantaneous strength loss at ΔT_c fracture
occurred by unstable crack propagation. This unstable mode of
failure can be changed to the stable one by incorporation of particles with sizes of 5 and 10 μm. Note that the incorporation of
these larger particles not only has improved ΔT_c, but also that
the strength for the composites for both particle sizes, at high
values of ΔT exceeds the corresponding values for the 100% alumina
as well as the alumina with the 1 μm dispersions.

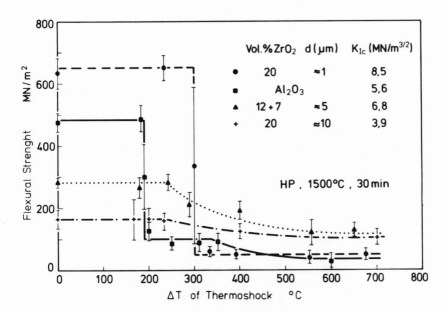

Fig. 6.   Effect of quenching temperature difference on strength of
          composites of alumina with zirconia dispersions of dif-
          ferent sizes, quenched into water at 20°C.

As an illustration of the decrease in the coefficient of thermal expansion or the existence of a transient compressive surface stress, Fig. 7 compares the strength behavior of silicon nitride (lightly dotted curve) and silicon nitride with 20 vol.% ZrO_2, subjected to a similar quench as the alumina of Fig. 6. The particle size of the zirconia was such that the temperature for the phase transformation ranged from about 600 to 850°C. Since the specimens were at room temperature prior to reheating below 850°C the zirconia is in the low temperature monoclinic modification. Therefore, in quenching from below this temperature, no advantage is obtained from the phase transformation. In Fig. 7 this is indicated by the data points in the form of the crosses. However, on heating to above 850°C, the high-temperature tetragonal phase is obtained, which upon quenching reverts back to the monoclinic form so that thermal stress failure is avoided as indicated in Fig. 7. Fracture by quenching from 600–850°C can be avoided only if the specimens are

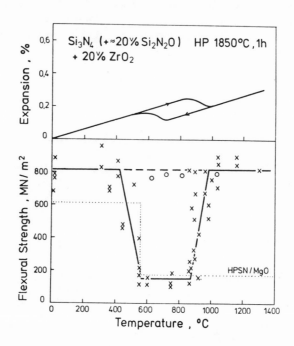

Fig. 7.   Thermal expansion and retained strength after thermal quench of silicon nitride with zirconia dispersions. For meaning of different data points the reader is referred to the text. Dotted lines indicate strength behavior of silicon nitride without zirconia dispersions.

first heated to higher temperature to achieve the tetragonal crystal structure, followed by slow cooling to the desired temperature from which the specimens are to be quenched. In Fig. 7 the feasibility of this procedure is shown by the data points indicated by circles.

Figure 8 shows the effect of the ZrO_2 dispersions on the thermal shock behavior of zirconium silicate. For the 100% zirconium silicate, the strength loss behavior is typical of unstable crack propagation shown in Fig. 1. The effect of the 10 wt.% zirconia dispersions is to modify the fracture mode to the stable one, as indicated by the monotonic decrease in strength for $\Delta T \geq \Delta T_c$. This effect must be atrributed to an increase in the size of the failure initiating flaws. Of interest to note is that this increase in the flaw size did not result in a decrease in the strength. This implies that the decrease in flaw size must have been accompanied by a simultaneous increase in fracture toughness, due to microcrack toughening. It is clear from the data of Fig. 8 that the zirconium silicate with zirconia dispersions is much preferred over the 100% zirconium silicate as a material for high–temperature structures susceptible to thermal stress failure.

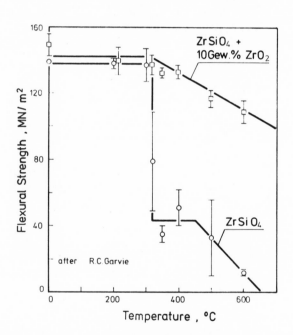

Fig. 8.   Effect of quenching temperature difference on strength of zirconium silicate with and without zirconia dispersions, quenched into water at 20°C.

CONCLUSION

The discussion and experimental data presented in this paper indicate that the incorporation of a zirconia dispersed phase, can lead to substantial improvements in the thermal stress resistance of brittle structural materials.

ACKNOWLEDGMENTS

The preparation of this manuscript was supported by the German Research Foundation (DFG) and the Office of Naval Research under contract No.: N00014-78-C-0413. The authors are indebted to J.P. Singh and G. Petzow for review of the manuscript.

REFERENCES

1. Alan Arias, "Thermal Shock Resistance of Zirconia with 15 Mole % Titanium," J. Amer. Ceram. Soc., 49 (1966) 334.
2. K. Chyung, "Fracture Energy and Thermal Shock Resistance of Mica Glass-Ceramics," Fracture Mechanics of Ceramics, edited by R. C. Bradt, D. P. H. Hasselman and F. F. Lange, Plenum Press, New York (1974).
3. R. C. Rossi, "Thermal Shock Resistance Ceramics with Second-Phase Dispersion," Amer. Ceram. Soc. Bull., 48 (1969) 736.
4. N. Claussen, "Fracture Toughness of Al_2O_3 with an Unstabilized Dispersed Phase," J. Amer. Ceram. Soc., 59 (1976) 49.
5. R. C. Garvie, R. H. Hannink, and R. T. Pascoe, "Ceramic Steel," Nature, 258 (1973) 703.
6. D. L. Porter, and A. H. Heuer, "Mechanism of Toughening Partially-Stabilized Zirconia (PSZ)," J. Amer. Ceram. Soc., 60 (1977) 183.
7. N. Claussen, J. Steeb, and R. F. Pabst, "Effect of Induced Microcracking on the Fracture Toughness of Ceramics," Amer. Cer. Soc., Bull., 56 (1977) 559.
8. D. L. Porter, A. G. Evans, and A. H. Heuer, "Transformation-Toughening in Partially-Stabilized Zirconis (PSZ)," Acta. Met., 27 (1979) 1649.
9. N. Claussen, "Stress-Induced Transformation of Metastable Tetragonal ZrO_2 Particles in Ceramic Matrices," J. Amer. Ceram. Soc., 61 (1978) 85.
10. D. P. H. Hasselman, "Figures-of-Merit for the Thermal Stress Resistance of High-Temperature Brittle Materials," Ceramurgia, 4 (1979) 147.
11. D. P. H. Hasselman, "Unified Theory of Thermal Shock Fracture Initiation and Crack Propagation of Brittle Ceramics," J. Amer. Ceram. Soc., 52 (1969) 600.
12. N. Claussen and J. Jahn, "Transformation of ZrO_2 particles in a Ceramic Matrix," Ber. Dt. Keram. Ges., 55 (1978) 487.
13. E. C. Subbarao, H. S. Maiti and K. K. Srivastava, "Martensitic Transformation in Zirconia," Phys. Stat. Sol., (a) 21 (1974) 9-40.
14. F. F. Lange in Fracture Mechanics of Ceramics, Vol. 2, R. C. Bradt, D. P. H. Hasselman and F. F. Lange, Plenum Publ. Corp., New York (1974).
15. R. C. Garvie, "The Occurrence of Metastable Tetragonal Zirconia as a Crystallite Size Effect," J. Phys. Chem., 69 (1965) 1238.

16. N. Claussen and R. Wagner, 1979, "Influence of Surface Condi-
     tion on Strength of Al_2O_3 Containing Tetragonal ZrO_2," Amer.
     Ceram. Soc. Bull., 58 (1978) 883.
17. R. T. Pascoe and R. C. Garvie, in R. M. Fulrath and J. A.
     Pask (Eds.), Ceramic Microstructures, '76, Westview Press,
     Boulder, Colorado (1977).
18. N. Claussen and J. Jahn, "Mechanical Properties of Sintered
     and Hot-Pressed Si_3N_4-ZrO_2 Composites," J. Am. Ceram. Soc.,
     61 (1978) 94.
19. R. C. Garvie, personal communication.
20. N. Claussen and G. Petzow, "Strengthening and Toughening Com-
     pacts in Ceramics," Proc. 4th CIMTEC, St. Vincent, Italy
     (1979).

DEPENDENCE OF THERMAL STRESS RESISTANCE ON MATERIAL

PARAMETERS: CERAMIC COMPOSITE SYSTEMS

P.F. Becher, D.Lewis III,
W.J. McDonough and R.W. Rice

Naval Research Laboratory
Washington, D.C. 20375

G.E. Youngblood and L. Bentsen

Montana Energy & MHD Research Institute
Butte, Montana

INTRODUCTION

The nature of the dependence of the thermal shock resistance of ceramics on the various thermo-mechanical properties of these materials can be used to design, in a systematic way, ceramics with improved thermal shock resistance. Great versatility in designing for thermal shock resistance and other desired properties can be achieved via ceramic composite approaches where properties of the two (or more) phases can be used to tailor the properties of the composite.

The desirable modifications to properties are suggested by the form of the two general thermal shock parameters which relate properties to the material's resistance to crack initiation under thermal stress[1] (Eqs. 1a and 1b) and to the material's ability to retain strength in thermal stress situations where crack initiation is unavoidable (Eq. 2). In the latter case, the two crucial factors are the limitation of crack growth and the minimization of strain energy at the initiation of crack propagation. In the former case, the critical factors are the magnitude of the thermally-generated stresses and the material's resistance to crack initiation. These equations are:

$$\Delta T_c \propto \frac{\sigma_i (1-\nu)}{E \alpha} \, f(k/ah) \qquad (1a)$$

or

$$\Delta T_c \propto \frac{\gamma_{IC}^{\frac{1}{2}}(1-\nu)}{(cE)^{\frac{1}{2}}\alpha} f(k/ah) \tag{1b}$$

and

$$\sigma_r/\sigma_i \propto \frac{E\gamma_{IC}}{\sigma_i^2} N^{\frac{1}{2}} \tag{2}$$

where E is Young's modulus, ν is Poisson's ratio, α is the coefficient of linear thermal expansion, σ_i is the initial strength, σ_r is the retained strength for ΔT ΔT_c, k is the thermal conductivity, h the surface heat transfer coefficient, a is a characteristic length related to sample size, N is the density of flaws of size c activated in the material, and γ_{IC} is the effective fracture surface energy.

Experimentally, the ΔT_c and σ_r/σ_i parameters can be determined by measuring the strength retained in test bars quenched from various selected high temperatures into a low temperature bath (e.g. water). The determination of σ_r/σ_i is complicated by the presence of other (not material) factors which may limit crack growth and hence control retained strength.[2] One such factor is specimen size, which affects both stress gradients and duration of the thermal stresses. Crack growth may be limited by negligible tensile stresses or compressive stresses within the interior of a specimen which arrest propagating cracks, or by the short duration of the thermal stresses, which may limit the extent of crack growth. Experimental procedures may be modified, when necessary,

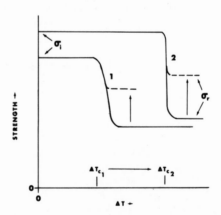

Figure 1.   Thermal Stress Resistance as Determined by Strength Loss from Quenching Sample (T_1) into Liquid (H_2O) Bath (T_2) Where $T_1 > T_2$. Increasing ΔT_c or σ_r achieved by modifying material properties (i.e. equations 1 and 2).

to minimize the effects/factors and produce an accurate estimate of the retained strength ratio, σ_r/σ_i. Using techniques such as the water quench test one can examine the effects of changes in mechanical and/or thermal properties of a composite material on its thermal stress resistance, either in the material's ability to avoid any damage from thermal shock (high or increased ΔT_c) or in its minimization of the damage from thermal shock [high or increased σ_r (or σ_r/σ_i)] as shown in Fig. 1.

Theoretical analyses can also be of benefit in tailoring materials for thermal shock resistance, both in providing estimates of the general dependence of thermal shock resistance on the various material parameters, as in Eqs. 1 and 2, and in providing guidance in the best experimental procedures. The theoretical analyses suggest, for example, that modification of certain material parameters, while the remainder are maintained (or improved), can have significant effects on thermal shock resistance in certain regimes. For example, increasing the conductivity of a ceramic will improve the thermal shock resistance, particularly for small values of the Biot modulus, $\beta = ah/k$, see Appendix I. Increasing the strength of an <u>alumina</u> ceramic will affect the quench test ΔT_c employing a 22°C water bath only slightly because of rapid changes in surface conductance, h, in the temperature range involved in degrading the strength of alumina in this specific case.[3a] Thus, while the <u>actual</u> thermal shock resistance may be improved, experimental measurements will indicate no apparent improvement. A brief analysis of the possible effects of variations in some of the material parameters is contained in Appendix I, together with the details of some of the theoretical calculations in the following sections.

As noted above, the principal advantage of ceramic composite systems is in the potential ability of the processor to alter one or more of the pertinent properties without undue degradation in other properties. In some cases, several of the significant material parameters are improved simultaneously in such composite systems, leading to greatly enhanced thermal shock resistance. For example, where simple mixture rules apply, e.g.

$$\frac{1}{k_c} = \frac{\nu_1}{k_1} + \frac{\nu_2}{k_2} \tag{3}$$

where k_c is the conductivity of the composite, k_1 and k_2 the conductivity of two phases of volume fractions, ν_1 and ν_2, the addition of a high conductivity second phase will lead to an increase in ΔT_c, as seen in the Al_2O_3-SiC system discussed here. On the other hand, addition of the second phase can also lead to an increase in the initial fracture energy, γ_{IC}, and strength

σ_i, as especially seen in the phase transformation associated toughening of materials with tetragonal ZrO_2^{4-9} second phases, and hence to increases in both ΔT_c (through increased K_{IC} and/or σ_i and σ_r/σ_i (through increased γ_{IC}). Note: $K_{IC}^2 = 2E\gamma_{IC}$. Additions of low modulus second phases (e.g. BN) can also decrease the Young's modulus of the composite leading to an increase in ΔT_c.[10] Finally, thermal expansion mismatches between the matrix and second phase can also be used to produce materials that are tougher and/or more thermal shock resistant by introduction of controlled microcracking which can lead to improved ΔT_c and/or σ_r/σ_i, also resulting in stable (i.e. σ decreases slowly with increasing ΔT) thermal shock crack propagation as opposed to catastrophic propagation (where σ decreases abruptly at ΔT_c). The results obtained with several of the approaches described above are presented below, with appropriate theoretical analyses.

I.   Composites With Controlled Thermal Properties, k and α :
     Al_2O_3-SiC*

As noted in Eq. 1, ΔT_c is a function of conductivity, k, thus where mixture laws such as the Maxwell-Eucken relationship for thermal conductivity[5] can be applied, ΔT_c can be increased by the addition of a high thermal conductivity second phase. The magnitude of the effect depends strongly on the heat transfer conditions, i.e. on the Biot modulus, $\beta = ah/k$. In the case of addition of SiC to Al_2O_3, where the conductivity follows the Maxwell-Eucken relationship, Fig. 2a, there is a concomitant increase in ΔT_c, which is partially attributable to the increase in conductivity, and partially due to the simultaneous decrease in thermal expansion coefficient, α.

Using Eq. (1) with f(k/ah) taken in the appropriate form for a rod (approximating a square cross-section bar with beveled edges)[3b], one calculates for 0 v/o SiC, with k ~ 0.02 w/cm-°C, that h = 0.62 w/cm² °C. For 80% SiC, with k ~ 0.05 w/cm-°C, h = 0.86 w/cm² °C. Thus, for this case, significant differences in h are not present and do not significantly affect the ΔT_c values, unlike the situation with Al_2O_3 and ΔT_c ~ 250° C. Considering the changes in k and α separately gives estimated increases in ΔT_c, going from 0-80% SiC, again using Eq. (1a) of 110°C for just the increase achieved in k, from 0.02 to 0.05 w/cm-°C (assuming h ~ 0.74 w/cm² °C and using literature values for other properties of SiC and Al_2O_3), and an additional increase of 140°C in ΔT_c for a calculated decrease in α from 8×10^{-6}/°C (alumina) to 5.2×10^{-6}/°C

---

*Ceradyne 7700 series microwave absorbing ceramics, Ceradyne,Inc., Santa Ana, California.

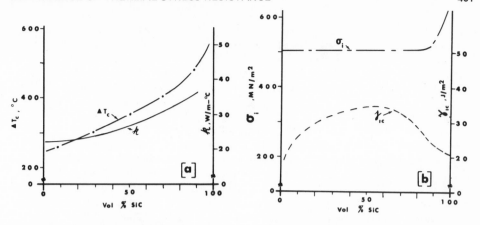

Figure 2.   Measured Thermal-Mechanical Properties of Al_2O_3-SiC
            Composites

(80% SiC), here assuming k $\simeq$ 0.35 w/cm-$^\circ$C. The sum of these two
(250°C) is quite close to the actual total increase in ΔT_c (Al_2O_3
ΔT_c ~ 225°C vs 450^{+o}C for SiC + 20% Al_2O_3). Thus, it would appear
that approximately 40-50% of the increase in ΔT_c in the Al_2O_3-SiC
composite results from the measured increase in conductivity, the
remainder from the expected simultaneous decrease in thermal expansion coefficient, α, (here assume that α follows the mixture
laws as does k).

      In this case, as with any composite system, changes in other
pertinent properties must be considered as well. In this instance,
there is a small increase in initial strength with increasing SiC
content, which will have a beneficial effect on ΔT_c, while there
appears to be a maximum in γ_{IC} for intermediate SiC levels (apparently with compensating changes in flaw sizes). On the other
hand, the values for E of these composites are nearly constant
(365 $\pm$ 35 GN/m^2) over the range of SiC content tested in the thermal shock tests. Apparently the modest changes in σ_i and γ_{IC} and
the small changes in E do not have nearly the significant effect on
ΔT_c, in <u>this</u> composite system, as do k and α. This is not the
case in the Al_2O_3-ZrO_2 system discussed later, where the influence
of γ_{IC} and σ_i are substantial.

II.   Composites with a Low Young's Modulus Second Phase

      As discussed by Hasselman et al[10], the Young's modulus, E,
of a matrix can be decreased by incorporation of a low E second
phase; as long as any reduction in initial strength, σ_i, is less,
relative to that achieved in E, the ratio σ_i/E (see Eq. 1a) will
be greater than that of the matrix phase, and ΔT_c should increase
(assuming negligible changes in other parameters). At the same

addition of the second phase may result, with appropriate choice
of the second phase material, in an increase in γ_{IC}, through mechanisms such as microcracking as fabricated or during crack propagation, crack branching and/or deflection and localized crack
arrest, in addition to the transformation toughening approach discussed later. Such crack propagation approaches to toughening
(increasing γ_{IC}) operate through interaction of the main crack
with: microcracks associated with 1) a phase transformation volume expansion/shear, 2) thermal expansion mismatch stresses, and
by interaction of the main crack with the second phase particles
through 1) thermal expansion mismatch stresses which deflects the
main crack, 2) second phase particles which cleave readily-acting
as pseudo porosity and blunting the crack, 3) tough second phase
particles which locally pin the crack front.

The incorporation of BN, having a low value of E, in an alumina matrix, with high E, is one example of the approach outlined
above. As seen in Table I, BN, which is highly anisotropic both
thermally and mechanically, has both a low E and high γ_{IC}. The
thermal expansion anisotropy of the BN, which may result in thermal stresses and/or microcracking, and the mica-like fracture of
the BN result in a composite with high fracture energy, which
together with the decreased Young's modulus produces a material
with a substantial increase in ΔT_c over that of the alumina matrix
(i.e. ΔT_c of 400 to 800°C for an Al_2O_3 v/o BN composite vs 220°C
for Al_2O_3 alone using a 22°C water quench).

The microstructures of typical Al_2O_3-BN composites are seen
in Figs. 3a and 3b which show fracture surfaces with the fracture
plane parallel to the HPA (vertical direction). Notable features
in the microstructure include the flakelike BN particles which
appear uniformly dispersed in a fine alumina matrix (grain size
5μm). There appears to be a significant degree of orientation of
the BN particles with C axes parallel to the HPA. Also evident
are regions of large Al_2O_3 grains, most probably resulting from
agglomerates present in the starting powder, which produce lense-
shaped clusters of large (ca 5μm) alumina grains flattened in
the direction parallel to the HPA. The more poorly mixed material
(3b) exhibits much poorer thermal shock response than the well-
mixed material Fig. (3a).

Calculation of the expected values for ΔT_c for the Al_2O_3-BN
system is complicated by the large thermal, mechanical and shape
anisotropy of the BN platelike particles, and by the large degree
of orientation of these particles in the hot-pressed composites.
The resulting composite structre then has signficant thermal and
mechanical anisotropy measured parallel and perpendicular to the
hot pressing axis (HPA). The measured properties are shown in
Table II.

Table 1. Properties of Al_2O_3- 30 v/o BN composites

Property	‖HPA	⊥ HPA
Bend Strength (MPa)*	160 (est.)	400
Fracture Energy (J/m²)*	25	40-100
Young's Modulus (GPa)*	136	190
Thermal Conductivity (w/cm ⁰C)	0.08	0.09

*Stress direction ⊥ or ‖ to HPA (hot pressing axis).

Table II. Comparison of Al_2O_3-BN Composites With Other Materials

Material	Bend Strength (MPa)	Young's Modulus, E (GPa)	γ_{IC} (J/m²)	ΔT_c[+] (⁰C)
Al_2O_3	350	⩾ 350	25	220
BN	85*	20*	70[++]	-
Corning Glass-Ceramic (Pyroceram 9606)	250	120	25	400
Al_2O_3-30 v/o BN	170	120	60	700-850
Al_2O_3-30 v/o BN	400	120	40-100	450

[+]Determined by water quench (22⁰C water) test using 3 mm sq. bars.

*Parallel to "a" direction, pyrolytic BN, Union Carbide Corp (Boralloy).

[++]Crack plane parallel to "C" axis in pyrolytic BN, Union Carbide Corp (Boralloy).

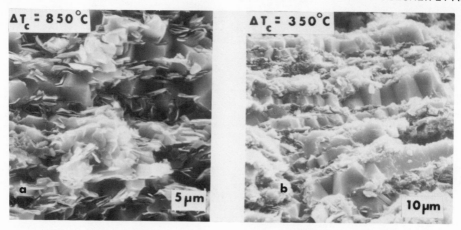

Fig. 3 Microstructure of Well-Mixed (a) and Poorly-Mixed (b)
Al_2O_3-30v/o BN composites.

Calculations of ΔT_c from Eq. (1a) may be done in a variety
of ways, with different assumptions about the orientation of the
BN platelets in the composite. If one assumes complete orientation (c-axis ‖ HPA) then one obtains (shown in Table II) grossly
different estimates for both thermomechanical properties and thermal shock resistance (ΔT_c) for stress parallel (‖) and perpendicular ($\perp$) to the HPA. Note that the experimental results shown in
Table I are for the perpendicular case. Also indicated are results of calculations for a randomly oriented BN phase (pseudo-
isotropic) and for the assumption that the BN particles act as
pores (pseudo-porous).

It is clear from the data in Tables II and III that the closer
fit to most of the measured properties comes from assuming the BN
flakes behave like pores.   This may be a plausible assumption
since the extremely large c-axis expansion ($\alpha \sim 40 \times 10^{-6}/°C$) of BN
would cause the particles to pull away from the Al_2O_3 matrix on
cooling from fabrication temperature. Moreover, no chemical reaction is present which would produce bonding between the two
phases. The large discrepancy in strength – 400 MPa measured for
30% BN vs approximately 40 MPa for 30% porous Al_2O_3 may arise from
the different scales of the porosity in these two cases. The
pseudo-porosity introduced by the BN particles may be fine enough
(ca. 1-10 µm) that it does not significantly affect the strength.
The most interesting effect is the dramatic increase in K_{IC} (from
3 $MN/m^{3/2}$ for Al_2O_3 to 5-10 for Al_2O_3-30 v/o BN) and γ (from
25 J/m^2 for Al_2O_3 to 40-100 for 30 v/o BN). This must be attributable to interaction of crack fronts with the highly heterogeneous,

Table III   Calculated Composite Properties

Property	∥ HPA	⊥ HPA	Pseudo-isotropic*	Pseudo-porous**
Young's Modulus (GPa)	67	200	77	190
Thermal Conductivity (w/cm °C)	0.05	0.25	0.07	0.08
Thermal Expansion (1/°C)	20×10^{-6}	6.8×10^{-6}	15×10^{-6}	9×10^{-6}
Thermal Stress Resistance T_c(°C)+	120 (470)	330 (650)	330 (1200)	45(360++) (105(840++))

* Using average properties for hot-pressed BN (no orientation dependence).

** Treating BN particles as porosity.

+ Results for h=10 w/cm^2 °C and h=1 w/cm^2 °C.

++ ΔT_c calculated assuming actual measured composite strength.

anisotropic microstructure, or with microcracks associated with the BN second phase. Introduction of microporosity alone would not produce this dramatic increase in fracture toughness.

III.   Phase Transformation Toughening: Al_2O_3-ZrO_2

In composites such as the Al_2O_3-ZrO_2 system, fracture toughness is improved by the incorporation of second phase particles in a metastable state (e.g. tetragonal ZrO_2). These particles are transformed to the stable state (monoclinic ZrO_2) when subjected to a tensile stress field associated with the fracture process or other phenomena. In the case cited, a substantial volume increase (and shear) occurs with the phase transformation, producing radial compressive and hoop tensile stresses in the surrounding matrix. These stresses can themselves deflect propagating cracks or can produce microcracking, either during fabrication or in advance of the crack-tip. The microcracking, which interacts with the crack front can cause significant increases in effective fracture surface energy as can the transformation process itself, which can absorb energy in the transformation of particles in the vicinity of the crack front. Toughening results from one or both of these effects: generation of compressive stresses, and energy absorption by the transformation process. In addition, transformation of the second phase particles can lead to microcracking through the local hoop tensile stresses developed around the particles. These microcracks may also interact with the main crack resulting in crack branching and an attendant increase in the energy required to continue crack growth. Regardless of the actual mechanism, the phase transformation toughening approach should also be applicable to the improvement of the thermal shock resistance of the matrix materials, as indicated in Eqs. 1b and 2, which suggest that improvements in fracture energy, γ_{IC} , should result in increases in both ΔT_c and σ_r/σ_i.

This has recently been demonstrated in the Al_2O_3-ZrO_2 system, where metastable, tetragonal ZrO_2 was added to an Al_2O_3 matrix to increase the fracture energy and strength of the matrix.[11] The observed effect of such toughening on the thermal shock behavior is summarized in Table V. The data in the table indicates that the greatest increases in ΔT_c and retained strength are observed in the composite exhibiting the greatest strength and toughness (fracture energy).

Table V.   Thermal Shock Behavior of Al_2O_3-ZrO_2 Composites

Vol. % ZrO_2	E (GN/m^2)	σ_i (MN/m^2)	γ_{IC} (J/m^2)	ΔT_c $(^\circ C)$	σ_r/σ_i
0.5	400	410	20	600	0.35
4	410	420	25	600	0.4
9	380	700	55	950	–
11.5	380	780	60	1150	0.6–0.75
14	375	680	35	800	0.65
19	350	350	35	800	0.65
30	165*	90	< 15	500	–
0+	400	325	20	275	0.40

*Sample highly microcracked in as-fired condition.

+Comparative data for alumina (General Electric Lucalox).

For Al_2O_3-ZrO_2 composites with $\leqslant 16$ v/o ZrO_2, where no spontaneous microcracking occurs, Greve et al[12] found a negligible effect of ZrO_2 content on conductivity, k. However, the presence of microcracks associated with the transformation of tetragonal ZrO_2 to monoclinic lowers k substantially (k decreased by approx. 50% in a microcracked Al_2O_3 - 16 v/o ZrO_2 composite). In the current study, spontaneous microcracking was confirmed in the 30 v/o ZrO_2 composite, where E values and x-ray analysis indicated the presence of the ZrO_2 as the monoclinic phase exclusively. Thus, a decrease in k is expected for this material, which when coupled with the low strength and fracture energy of the 30 v/o ZrO_2 composite, should result in relatively poor thermal shock performance as seen experimentally. On the other hand, pre-existent microcracks are absent in or have insignificant effects on the behavior of the composites with $\leqslant 14$ v/o ZrO_2, as the ZrO_2 is present principally as the tetragonal phase. Here the concepts of toughening through second phase additions offer substantial potential for improving the thermal shock resistance of these ceramics. One is particularly encouraged by the potential for improvement in both ΔT_c and σ_r/σ_i by the second phase toughening approach coupled with very high initial strengths.

SUMMARY AND CONCLUSIONS

It has been shown that several ceramic-ceramic composite approaches offer the potential for improving or modifying the response of ceramics in environments where thermal stresses are significant, while maintaining or improving other properties (electrical, dielectric, optical, mechanical) required for particular applications (radomes, irdomes, bearing and seal materials, high performance refractories, heat engine components). While the details (size, distribution and nature of second phase) and the composite properties (σ_i, σ_r, γ_{IC}, ΔT_c) are still being examined, and further optimization of properties is still probable, the results to date indicate that there are a wide variety of possible solutions to the problem of improved thermal stress resistance of ceramics. For example, the incorporation of low modulus, highly anisotropic BN particles in alumina results not only in a material which is much more resistant than alumina to a water quench thermal shock, but which also exhibits significantly greater fracture toughness, better machinability, and a much greater resistance to failure from the thermal stresses produced by high energy CW and pulse laser irradiation. While nearly all ceramics fail rapidly,

in a catastrophic manner (i.e. by complete fracture), from intense
laser irradiation, the Al_2O_3-BN composites survive for long times
and exhibit eventual failure by <u>controlled</u>, stable crack growth
rather than rapid (explosive-like) failure.

Other composite approaches, which seek to modify in appropriate directions, the pertinent thermal properties - α and k, decreasing α, increasing k then to increase the thermal stress resistance
of the composite system, are also successful, although possibly
more limited in application, as was the case with the Al_2O_3-SiC
system. This was used also in MgO-SiC composites where the ΔT_c
is approx. $375^\circ C$ vs $200^\circ C$ for MgO alone. In this last case, the
SiC addition also has the detrimental (for thermal shock resistance)
effect of raising the Young's modulus significantly, but the effect
of SiC additions is still beneficial overall.

An alternative approach, other than modification of thermal
properties, is the use of phase transformation toughening, where
a metastable second phase (e.g. tetragonal ZrO_2) undergoes a stress-
induced volume expansion. Both the strength and fracture toughness
(K_{IC}, γ_{IC}) can be significantly improved in this approach, resulting
as seen in the Al_2O_3-ZrO_2 system, in substantial increases in both
ΔT_c and retained strength in the composite, without significantly
affecting other physical properties. The combination of excellent
mechanical properties with excellent thermal shock resistance make
this last approach very attractive for improvement of ceramics
subject to both structural and thermal loads. However all of the
ceramic-ceramic composite approaches offer advantages in particular
situations and might provide viable techniques for producing better
thermal shock resistant ceramic materials.

APPENDIX I. DETAILS OF THEORETICAL CALCULATIONS

The transient stresses generated by the temperature gradients
in thermal shock vary with both time and position in a body. The
peak stresses, which usually must be calculated by numerical means,
are typically functions of specimen geometry and the heat transfer
conditions. Thus one finds results of the form:

$$\sigma_{max} = \Delta T \frac{E\alpha}{1-\nu} \bigg/ f(\beta) \qquad\qquad (A1)$$

where β = ah/k is the Biot modulus, and the functional dependence,
(f), is obtained from numerical calculations of maximum stresses.
If one equates the maximum stress to the minimum strength,
σ_{min}, of the material, as a criterion for determining

ΔT_c, the result is:

$$\Delta T_c = \frac{\sigma_{min}(1-\nu)}{E\ \alpha}\ f(\beta) \tag{A2}$$

Manson [13] has determined $f(\beta) \simeq 1.5 + 3.25/\beta - 0.5e^{-16/\beta}$ as an approximation to the numerical solution for an infinite plate. One of the authors has determined $f(\beta) = 1.5 + 4.67/\beta - 0.5e^{-51.1/\beta}$ as a reasonable approximation to the stress results for a circular rod. Most thermal shock test specimens have been either circular rods or square cross-section bars with beveled edges, which should closely approximate a circular rod. Thus, the more appropriate form for $f(\beta)$, and the one used in this study, is the latter one. A complication which arises in calculations of ΔT_c, as pointed out in a separate paper [3] is the variation of h with ΔT in the water quench thermal shock test. This variation makes simple calculation of ΔT_c considerably more difficult, as does the incorporation of the variation of material properties with temperature. A complete statement of Eq. A2 might be

$$\Delta T_c = T_c - 20^{\circ}C = \frac{\sigma_{min}(T_c)}{E(T_c)\alpha\ (T_c)}\frac{1-\nu(T_c)}{}\left[1.5 + 4.67\ k(T_c)/ah(Tc)\right.$$
$$\left. - 0.5\ \exp\ -51.1\ k(T_c)/ah(T_c)\right] \tag{A3}$$

Iterative solution of this equation, given h(T) and the variation of E, α, etc. with T permits accurate calculation of ΔT_c from material properties, or of h(T) from measured ΔT_c values and material properties. These procedures have been used extensively in the calculations for this study, and Eq. A3 above is used as the basis for the parametric study following.

For small changes in the values of the parameters, estimates of the resulting changes in ΔT_c can be made as indicated below:

$$\delta(\Delta T_c) \simeq \frac{\partial \Delta T_c}{\partial \sigma_f}\ \delta\sigma_f + \frac{\partial \Delta T_c}{\partial E}\ \delta E + \ .\ .\ . \tag{A4}$$

$$+ \frac{\partial \Delta T_c}{\partial k}\ \delta k + \frac{\partial \Delta T_c}{\partial h}\ \delta h \tag{A4}$$

This ignores second order effects and interrelationships between parameters, and more importantly, the effect of the variations of h with T. If, for example, one attempts to increase ΔT_c by increasing σ_f, but h increases rapidly with T in the region, then the actual increase in ΔT_c will be much smaller than that calculated. On the other hand, if the same thing is attempted in a temperature regime where h decreases rapidly with T, a much greater increase in ΔT_c will result than that predicted. However, neglecting these effects, Eq. A4 provides a basis for estimating the effects of changes in material properties. First, for the parameters: σ_{min}, ν, E, α, the result is relatively simple:

$$\delta(\Delta T_c) \cong \Delta T_c \left[\frac{\delta \sigma_{min}}{\sigma_{min}} - \frac{\delta \nu}{\nu} - \frac{\delta E}{E} - \frac{\delta \alpha}{\alpha} \right] \tag{A5}$$

which indicates that the possible magnitude of the improvements in ΔT_c is proportional to ΔT_c and the <u>relative</u> change in the material parameter. Thus, one has the most chance for improving T_c where the largest <u>relative</u> changes in property can be achieved, not with the largest <u>absolute</u> increases. The effects of changing the conductivity are somewhat more complicated, e.g.,

$$\delta(\Delta T_c) \cong \frac{\sigma_{min}(1-\nu)}{E \alpha} \left[\frac{4.67}{ah} - \frac{25.6}{ah} e^{-51.1k/ah} \right] \delta k \tag{A6}$$

which is not linear in $\delta k/k$, except in certain regimes. If, for example, $\beta = ah/k$ is small, $\beta < 10$ approx., then:

$$\delta(\Delta T_c) \cong 4.67 \frac{\sigma_{min}(1-\nu)}{E\alpha ah} \delta k$$

$$= 4.67 \frac{\sigma_{min}(1-\nu)}{E \alpha \beta} \frac{\delta k}{k} \tag{A7}$$

For large β, Eq. A6 simplifies to:

$$\delta(\Delta T_c) \cong 0 \tag{A8}$$

which would seem to indicate that conductivity changes in low conductivity materials subject to rapid heat transfer (high a.h) will have no effect on measured ΔT_c; conductivity change would be more significant in low Biot modulus regimes, e.g. for small β. However, as suggested by Eq. (A7), increases in conductivity have two opposite effects, decreasing β, which results in a larger $\delta(\Delta T_c)$, and decreasing the ratio $\delta k/k$ for the same absolute change in k, which produces a smaller $\delta(\Delta T_c)$. Thus, the net effect of conductivity changes, even in the low β regime, may be small.

REFERENCES

1. D. P. H. Hasselman, Figures-of-Merit for the Thermal Stress
    Resistance of High-Temperature Brittle Materials, Cera-
    murgia International, in press (1979).

2. A. F. Emery and A. S. Kobayashi, The Transient Stress Inten-
    sity Factors for Edge and Corner Cracks in Quench Test
    Specimens, J. Am. Ceram. Soc. , to be published (1979).

3a. P. F. Becher and D. Lewis, Thermal Shock of Ceramics: Size and
    Strength Effects, Bull. Am. Cer. Soc. 58:339 (1979).

3b. P. F. Becher, D. Lewis, R. Carman and A. Gonzalez, Thermal
    Shock Resistance of Ceramics: Size and Geometry Effects in
    Quench Tests, submitted to Am. Ceram. Soc. (1979).

4. R. T. Pascoe and R. C. Garvie, Surface Strengthening of Trans-
    formation-toughened Zirconia, in: "Ceramic Microstructures
    '76", F. M. Fulrath and J. A. Pask, eds., Westview Press,
    Boulder (1977).

5. T. K. Gupta, F. F. Lange and J. H. Bechtold, Effect of Stress-
    Induced Phase Transformation on the Properties of Poly-
    crystalline Zirconia Containing Metastable Tetragonal
    Phase, J. Mat. Sci.13:1464 (1978).

6. N. Claussen, Fracture Toughness of Al_2O_3 With an Unstabilized
    ZrO_2 Dispersed Phase, J. Am. Ceram. Soc.59:49 (1976).

7. N. Claussen and J. Jahn, Mechanical Properties of Sintered and
    Hot-Pressed Si_3N_4-ZrO_2 Composites, J. Am. Ceram. Soc. 61:94
    (1978).

8. N. Claussen, J. Steel and R. F. Pabst, Effect of Induced Micro-
    cracking on the Fracture Toughness of Ceramics, Bull. Am.
    Ceram. Soc. 56:559 (1977).

9. R. M. Cannon and T. D. Ketchum, Toughening in the ThO_2-ZrO_2
    System, Bull. Am. Ceram. Soc. 58:338 (1979).

10. D. P. H. Hasselmann, P. F. Becher and K. S. Mazdiyasni, An-
    analysis of the Resistance of High-E, Low-E Brittle Com-
    posites to Failure by Thermal Shock, to be published in
    J. Mat. Tech. (1979).

11. P. F. Becher, Thermal Shock Behavior of ZrO_2 Toughened Alum-
    ina, submitted to J. Am. Ceram. Soc. (1979).

12. D. Greve, N. E. Claussen, D. P. H. Hasselman and G. E. Young-
    blood, Thermal Diffusivity/Conductivity of Alumina With a
    Zirconia Dispersed Phase, Bull. Am. Ceram. Soc.56:514 (1977).

13. S. S. Manson, Thermal Stresses Part I: Appraisal of Brittle
    Materials, Machine Design, June:114 (1958).

THERMAL STRESS IN CYLINDRICAL GLASS SEALS IN MICROELECTRONIC

PACKAGES UNDER THERMAL SHOCK

Klod Kokini, Richard W. Perkins & Charles Libove

Department of Mechanical & Aerospace Engineering
Syracuse University
Syracuse, New York   13210

I. INTRODUCTION

In order to ensure the reliability of microelectronic packages, these packages are subjected to various tests which simulate the environments which the packages may experience in service. The present study concerns a detailed investigation of the Thermal Shock Test (Method 1011.2 in MIL-STD-883B [1]).

The purpose of the thermal shock test is to determine the resistance of the device to sudden exposure to extreme change in temperature. Physical damage may be experienced by the seals of a package during thermal shock as a result of thermal stresses which accompany the extreme temperature distribution which must exist during the transient heat transfer process associated with the sudden change in temperature.

Thomas [2] conducted an experimental study of packages submitted to thermal shock and found that leakage appeared to be directly correlated with the thermal shock. He suggested that thermal stresses in the vicinity of the lead-through seals may be large enough to permit leakage during the thermal shock which might go undetected during subsequent gross leak testing of the package. Other literature [3] indicates that stresses during thermal shock testing can lead to fracture of the glass seal.

In view of these findings, it appears desirable to have a procedure for predicting or estimating the magnitude of the thermal stresses which may be present in the lead-through glass seal of microelectronic packages which are subjected to thermal shock testing. The predicted stress conditions can be used by the indi-

413

vidual performing the thermal shock test to select a test level
consistent with the level of stress desired in the package. The
same analysis could also be used by the package designer to ensure
that the fabricated package will perform satisfactorily during the
anticipated environmental thermal shock conditions that it would be
subjected to.

The objective of the present study was to develop a mathematical
model for predicting the transient thermal stresses that occur in a
typical annular lead-through glass seal of a microelectronic package
during a thermal shock test.

II.   THERMAL STRESS ANALYSIS

Since microelectronic packages typically have many lead-through
seals and since the package geometry varies considerably from one
case to the next, it was necessary to develop a simplified model for
purposes of analysis. The approximate model which was adopted in
the present work was very similar to the one used by Borom and
Giddings [4] which focuses attention on the stresses in the vicinity
of a typical lead-through seal. The considerations underlying the
formulation of the model were as follows:

Consider a typical package such as the one shown in Fig. 1. In
this case the leads all pass through the base of the package. When
a package such as this is subjected to a thermal shock by immersing
it in a hot or cold liquid bath, the lead changes temperature most
rapidly due to its small mass and high thermal conductivity. The
base responds somewhat less rapidly because of its larger mass. Both
the lead and the base are generally made of Kovar (a Ni, Co, Si steel
alloy used in metal-to-glass sealing) and therefore have a high ther-

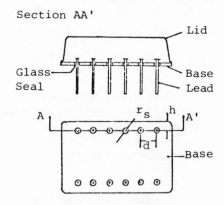

Fig. 1.   Typical microelectronic package showing lead-throughs and
          glass seals.

mal diffusivity. Because of this and because of the nature of the
heat transfer process from the liquid to the lead and the base, it
is quite reasonable to assume that the temperature of the lead and
the temperature of the base are uniform throughout. On the other
hand, the annular glass seal has a much lower thermal diffusivity
and has very little area exposed to the liquid bath environment.
The temperature distribution in the glass seal can be approximately
modelled as having a radial variation from the temperature of the
lead on the inner annular surface to that of the base on the outer
annular surface. The axial variation in temperature is neglected
in view of the small surface area in contact with the fluid bath in
comparison with the much larger areas of the glass seal in contact
with the lead and the base.

The state of thermal stress can be reasonably approximated by
using the simplified model of Borom and Giddings [4]. The model is
illustrated by Fig. 2 which shows a typical glass seal consisting
of a lead, the glass annulus, and an annular portion of the package
base. The outside radius r_0 is selected so that the radial stress
on the outer annular surface could be assumed to be zero. The model
therefore, consists of the lead, the annular glass seal and annular
segment of base material surrounding the seal as illustrated in
Fig. 3.

The thermal stress analysis applied to the three-cylinder model
follows the same basic assumptions as those made by Poritsky [5]
in his analysis of two concentric cylinders subjected to uniform
temperature changes (hence stresses arising from the mismatch of
thermal expansion coefficients of the glass seal and metal lead and
base materials). It should be noted that Borom and Giddings [4]
extended Poritsky's analysis to include three concentric cylinders.
The general problem of the thermal stresses in a system of many

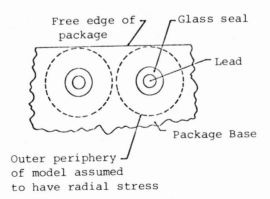

Fig. 2.   Typical glass seal consisting of lead, annular glass seal
          and part of package base in the form of an annulus.

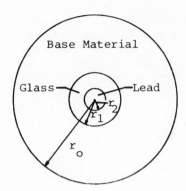

Fig. 3.   Model of the lead-through consisting of lead, glass seal
          and a part of the base as three concentric cylinders.

cylinders with nonuniform temperature distribution was studied by
Gatewood [6]. The thermal stress equations obtained by writing
the general equations of uncoupled thermo-elasticity for a typical
element of material in each of the lead, glass seal and base
cylinders are essentially the same as presented by Gatewood.

Assuming the materials to be isotropic, homogeneous, linearly
elastic and treating the problem as an axisymmetric one, the following relations are obtained for each of the three concentric
cylinders

$$
u_j = \frac{(1+\nu_j)\,\alpha_j}{(1-\nu_j)}\,\frac{1}{r}\int_{r_i}^{r}\theta r\,dr + C_{1j}r + \frac{C_{2j}}{r} \tag{1}
$$

$$
\sigma_{rj} = -\frac{E_j\alpha_j}{(1-\nu_j)}\,\frac{1}{r^2}\int_{r_i}^{r}\theta r\,dr + \frac{E_j}{(1+\nu_j)}\left[\frac{C_{1j}}{(1-2\nu_j)} - \frac{C_{2j}}{r^2}\right]
$$

$$
+ \frac{\nu_j E_j \varepsilon_z}{(1+\nu_j)(1-2\nu_j)} \tag{2}
$$

$$
\sigma_{zj} = E_j\varepsilon_z\left[\frac{(1-\nu_j)}{(1+\nu_j)(1-2\nu_j)}\right] + \frac{2\nu_j E_j C_{1j}}{(1+\nu_j)(1-2\nu_j)} - E_j\alpha_j\theta_j\left[\frac{1}{(1-\nu_j)}\right] \tag{3}
$$

where E = Young's modulus.
      ν = Poisson's ratio.
      α = Coefficient of thermal expansion.

       u = Displacement in the radial direction.
      σ_r = Stress in the radial direction
      σ_z = Stress in the axial direction.
      ε_z = Axial strain.
       θ = Temperature change.
  C_1, C_2 = Constants of integration.
      r_1 = Inner radius of the particular component.
       j = 0,1,2 designating respectively the base, the glass seal
           and the lead.

    The following boundary conditions apply:

    1) The radial displacement at the center of the cylinders has
to be finite.

    2) The respective radial displacement u_1 and u_2 at the lead
glass interface, as well as u_1 and u_0 at the glass-base interface
have to be equal to each other.

    3) The respective radial stresses at the same interfaces as
above should be equal to each other.

    4) The outermost boundary should be free of stress.

    5) The total force in the axial direction should be zero, i.e.,

$$\sum_{i=0}^{2} \int_A \sigma_{zi} dA = 0.$$

    These conditions yield 7 equations and 7 unknowns for which
numerical results are obtained by placing in each equation the
respective values of the known coefficients and solving for the
unknowns. Hence the radial and tangential thermal stresses can be
obtained at each increment of time assuming the temperature distribution through the three cylinders is known.

III. TEMPERATURE ANALYSIS

    The evaluation of the thermal stresses requires the knowledge
of the temperature distribution in the three cylinders at every
instant of time. For this purpose a temperature distribution model
is developed which takes into consideration the heat transfer and
geometric characteristics of the lead, the glass and the base
assembly.

    The physical characteristics of the temperature distribution
model consist of a single lead surrounded by its annular glass seal
which is in turn surrounded by a "base" which has 1/Nth the mass
and 1/Nth the outside surface area of the total base. N is equal

to the number of leads in the package. The lead and the base are considered to be lumped masses and as such conduct heat instantaneously. It is assumed that heat is transferred between fluid and lead and fluid and base but not between fluid and glass. It is also assumed that the temperature distribution in the glass is axisymmetric at every instant of time and varies in the radial direction only. The temperature is assumed to be continuous across the glass-metal interfaces.

The assumption of a lumped mass for the lead and the base can be justified by the facts that the metal from which the lead and the base are constructed (usually Kovar) has a high thermal diffusivity and the masses of the two components are small. The reason for assuming that no heat is transferred between the glass and the fluid is that the outside surface area of the glass is very small in relation to the outside surface areas of the lead and the base. In addition, the thermal diffusivity of glass is about one tenth that of the metal. Therefore the heat transferred between the fluid and the glass is small compared to the heat transferred between the fluid and the metal, and then, by induction, between the metal and the glass. The reason for temperature variation only in the radial direction is that no heat is transferred from the fluid to the glass. Also, since the lead and the base each are assumed to have a uniform temperature distribution, the only temperature gradient exists in the radial direction.

The temperature model described above is used to determine the transient temperature distribution in the glass. The transfer of heat is governed by the partial differential equation.

$$\frac{\partial T}{\partial \tau} = \frac{k_g}{\rho_g c_g} \left(\frac{\partial^2 T}{\partial r^2} + \frac{1}{r} \frac{\partial T}{\partial r} \right) \tag{4}$$

where k_g, ρ_g, c_g are respectively the thermal conductivity, the density and the specific heat of glass. τ and r indicate respectively the time and radial coordinate. T is the temperature in the glass which is a function of the radius and time.

The initial condition is

$$T(r,o) = T_o \tag{5}$$

where T_o is an initial uniform temperature for the lead-base-glass assembly corresponding to the beginning of the thermal shock.

The boundary conditions are

$$-k_g \left.\frac{\partial T}{\partial r}\right|_{r=r_2} = \frac{\bar{h}A_O}{A_2} \left.(T_\infty - T)\right|_{r=r_2}) - \rho_k c_k \frac{V_\ell}{A_2} \left.\frac{\partial T}{\partial \tau}\right|_{r=r_2} \tag{6}$$

$$k_g \left.\frac{\partial T}{\partial r}\right|_{r=r_1} = \frac{\bar{h}A_b}{A_1 N} \left.(T_\infty - T)\right|_{r=r_1}) - \rho_k c_k \frac{V_b}{A_1 N} \left.\frac{\partial T}{\partial \tau}\right|_{r=r_1} \tag{7}$$

where T_∞ is the temperature of the fluid and $\bar{h}$ is an effective heat transfer coefficient governing the rate of heat transfer between the lead or the base and the fluid. Also

ρ_k = the density of the metal (usually Kovar).
c_k = the specific heat of the metal.
A_O = the outside surface area of the lead in contact with the fluid.
A_b = the outside surface area of the base and lid.
A_1 = the surface area of the glass-base interface
A_2 = the surface area of the lead-glass interface
V_b = the total volume of the base and lid.[1]
N = the total number of leads

The differential equation (4) and conditions (5,6,7) were solved using a finite element analysis described in [7]. The solution of equation (4) yields the transient temperature distribution in the glass as well as the temperature-time histories of the lead and the base.

The differential equation (4) can be solved if the value of $\bar{h}$ in equations (6) and (7) is known. To determine the effective heat transfer coefficient, an experimental package was subjected to the thermal shock test condition A (for which the fluid is water and the temperature extremes are $32°F$ and $212°F$) [1] and the temperature-time histories of the lead and the base were recorded. The same temperature-time histories were evaluated theoretically and the heat transfer coefficient minimizing the error between the initial few seconds of the experimental and theoretical curves was determined. This procedure yielded an experimental $\bar{h}$ which was used in the determination of the thermal stresses during the thermal shock test.

---

[1] The microelectronic package used in this investigation consists of a base and a lid. Other packages have the base surrounded with vertical walls with a lid in the form of a plate. For these types of packages the walls should be included in the analysis. More generally, A_b, and V_b refer to all parts of the package except the glass seals and the leads.

IV.   RESULTS AND DISCUSSION

The calculation of the thermal stresses using the analysis described above was performed by means of an APL program presented in [7].   The results are presented in terms of the following non-dimensional parameters.

$$r^* = \frac{\bar{h} A_o}{k_g A_2} r \tag{8}$$

$$\tau^* = \frac{\bar{h} A_o}{c_k \rho_k V_\ell} \tau \tag{9}$$

$$E^* = \frac{E_k}{E_g} \tag{10}$$

$$\alpha^* = \frac{\alpha_k}{\alpha_g} \tag{11}$$

$$\sigma^* = \frac{\sigma}{E_g \alpha_g \theta_o} \tag{12}$$

where

   r^* = non-dimensional radial parameter.
   τ^* = non-dimensional time parameter.
   E^* = non-dimensional Young's modulus ratio.
   α^* = non-dimensional thermal expansion coefficient ratio.
   σ^* = non-dimensional stress.
   θ_o = difference between the final and initial temperature.
   k = subscript representing the metal (usually Kovar).
   g = subscript representing glass.

When the experimental package was taken from 32 F to 212 F the resulting transient non-dimensional radial and tangential stresses were computed and plotted in Figs. 4a, 4b, 5a, and 5b.   The absolute values of the stresses can be obtained from these graphs by using Eq. (12) which can be written as

$$\sigma = \sigma^* E_g \alpha_g \theta_o .$$

The dashed lines in each of these figures represents the stress distribution in the 3-cylinder model.   The three regions are identified on the figures as the lead, the glass and the base.   Each line shows the stress distribution at a different time.

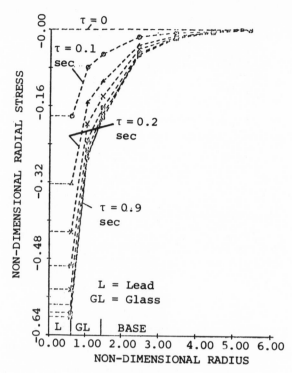

Fig. 4a. Non-dimensional radial stress at each radius as a function
          of time.

It is assumed in these figures that at $\tau = 0$, the entire package
is stress-free. Fig. 4a shows that as time increases, the radial
stresses become compressive for all values of radius and the magnitude of the compressive stress at each radius increases until
$\tau = 0.9$ sec. (At this time the maximum non-dimensional compressive
radial stress at the lead-glass interface is -0.617 which corresponds
to -2331 psi). Fig. 4b represents the continuation of the heating
process. The graph shows that the magnitude of the compressive
radial stress at each radial position decreases as time increases.
When finally thermal equilibrium is almost reached at $\tau \cong 12$ sec.,
the radial stress achieves a relatively small compressive value at
the lead-glass interface whereas it has a tensile value at the
glass-base interface. Inspection of Figs. 4a and 4b shows that the
largest radial stress magnitude occurs at the lead-glass interface
soon after immersion in the hot fluid bath. Investigation of
several examples having different values of r_o, r_1, r_2 showed that
this general behavior regarding the location of the highest magnitude of stress at the lead-glass interface was invariably obtained.
In subsequent analysis it has been assumed that the most severe
radial stress always occurs at the lead-glass interface. Of course,

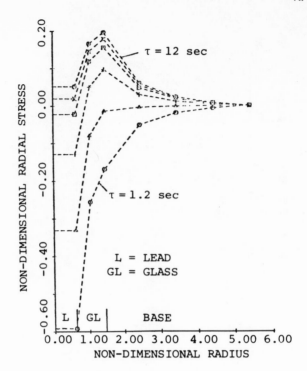

Fig. 4b.   Non-dimensional radial stress at each radius as a function
           of time.

the most severe stressing from the failure point of view would be
expected during immersion in the ice bath which would result in
large tensile radial stress at the lead-glass interface.

     The same process is shown in Fig. 5a and 5b where the tangen-
tial stress distribution in the heating process is plotted for
different times. Fig. 5a shows that a small portion of the glass
near the glass-base interface experien-es a compressive stress.
However, for most radial positions, the tangential stress is
tensile and the magnitude of the stresses at each radial position
becomes larger as time increases. Fig. 5b shows that the tangential stress at each radius decreases at the right hand side of the
glass-base interface and increases at the left hand side of the
glass-base interface until $\tau \cong 12$ sec., when the temperature of the
entire package is almost equal to the temperature of the fluid and
a steady-state value of the stress is reached. This process can be
seen better in Figs. 6 and 7 where the non-dimensional radial and
tangential stresses in the glass, at the lead-glass and glass-base
interface are plotted as a function of non-dimensional time.

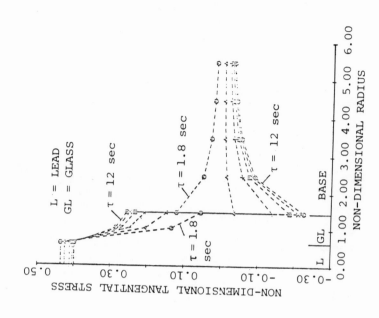

Fig. 5b. Non-dimensional tangential stress at each radius as a function of time.

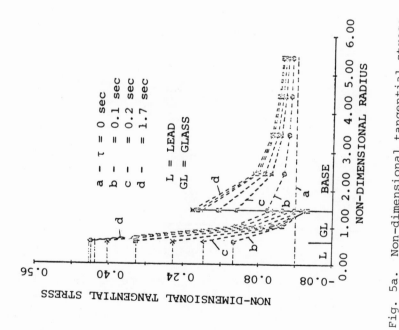

Fig. 5a. Non-dimensional tangential stress at each radius as a function of time.

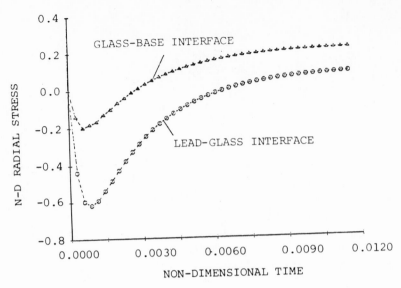

Fig. 6.   Non-dimensional radial stress at the glass-base and lead-
glass interface as a function of non-dimensional time.

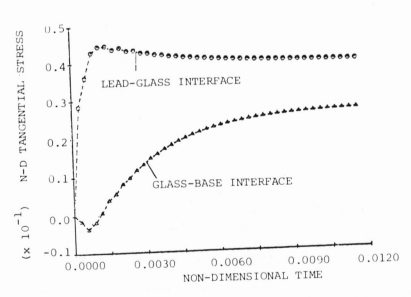

Fig. 7.   Non-dimensional tangential stress at the glass-base and
lead-glass interface as a function of non-dimensional
time.

The maximum and minimum stresses occur when the temperature difference across the 3 cylinders becomes the largest. As the heating process continues, this difference decreases causing the stresses to diminish. When the steady-state temperature is reached the stresses come to a steady-state value. This residual stress is caused by the mismatch in thermal expansion coefficients of the metal and the glass for this temperature range. Recall that it was assumed that the stresses are zero at the beginning of the heating period.

These results illustrate the stress behavior of the seal when it is submitted to the heating cycle of the thermal shock test. In the case of a cooling cycle, the numerical results for all the stresses would be the same as those illustrated for a heating cycle, however, the stresses would be of opposite sign.

In order to assess the importance of knowing the true value of the overall effective heat transfer coefficient $\bar{h}$ for the package insofar as the calculated maximum stress is concerned, calculations were made for a range of values of the $\bar{h}$ consistent with the probable range of uncertainty as determined by experiment. The average value of $\bar{h}$ was found to be 210 $Btu/ft^2-hr-°F$. However, the experimental values ranged from a low of 70 $Btu/ft^2-hr-°F$ to a high of 375 $Btu/ft^2-hr-°F$. The influence of changing the value of $\bar{h}$ in the calculation of the stresses is illustrated in Table 1, which records the maximum value of the non-dimensional stresses (at the lead-glass interface) for varying values of $\bar{h}$ ranging from 130 to 290 $Btu/ft^2-hr-°F$. As shown in the table, the maximum difference from the stress calculated for 210 $Btu/ft^2-hr-°F$ was about 10 percent. This is not judged to be a large difference, and it is felt that some uncertainty in the true value of $\bar{h}$ does not result in a very large uncertainty in the maximum value of the calculated stresses at the lead-glass interface. On the other hand, calculations have shown that if $\bar{h}$ is assumed to be ten times larger than its average value (210 $Btu/ft^2-hr-°F$), the maximum radial stress becomes about 37 percent larger and if $\bar{h}$ is ten times smaller than its average value, the maximum radial stress is about 48 percent smaller than the maximum radial stress corresponding to the average value of $\bar{h}$. This represents an important variation. However an effective heat transfer coefficient which is that much different than the average value would correspond to a different heat transfer environment. In fact, the heating and cooling procedure used in thermal cycling tests (as opposed to thermal shock tests) is characterized by a very slow heat transfer process for which $\bar{h}$ would be expected to be very small compared to the $\bar{h}$ obtained in the thermal shock experiments. Therefore it would be reasonable to conclude that the small stresses evaluated using an effective heat transfer coefficient 10 times smaller than the average experimentally obtained $\bar{h}$ could be indicative of the stresses occurring during a temperature cycling test.

TABLE 1

Maximum non-dimensional transient stresses at the lead-glass interface for different values of $\bar{h}$ compared to the maximum value of the stresses obtained when $\bar{h} = 210$ Btu/ft²-hr-°F

σ^*_{max} $\bar{h}$(Btu/ft²-hr-°F)	130	150	170	190	210	230	250	270	290
Radial	-0.554	-0.574	-0.587	-0.605	-0.617	-0.625	-0.638	-0.653	-0.665
%Δ ($\Delta = \| \sigma^*_{max} - \sigma^* \|_{\bar{h}=210} \|$)	10.2	7.6	4.9	1.9	0	1.3	3.4	5.8	7.8
Tangential	0.406	0.418	0.428	0.437	0.447	0.453	0.458	0.464	0.473
%Δ	8.8	6.1	3.8	1.8	0	1.8	2.9	4.3	6.3

The results which have been presented to this point were obtained by applying the mathematical model developed for the experimental package. However, other microelectronic packages which are commercially available exhibit different characteristics. The present study applies to all packages which are characterized by certain material property ratios (Young's modulus, thermal expansion coefficients, Poisson's ratio, heat transfer properties) similar to the ratios of the experimental package which was studied[1]. When this similar condition is fulfilled, the only other factor that affects the stress level is the geometrical characteristic of the package. Since the lead, the glass, and the base were modeled as three concentric cylinders, the ratios of the cylinder radius (r_o/r_1 and r_1/r_2) influence the stresses. The non-dimensional radial and tangential stresses at the lead-glass interface were computed for different radius ratios and plotted in Figs. 8 and 9. It is believed that these graphs provide the means for evaluating the effect of package geometry on the maximum stresses for most of the microelectronic packages which are commercially available.

Figs. 8 and 9 illustrate the influence of the choice made for the radius of the third cylinder which represents the base of the package. The experimental package which was used in this study was characterized by a ratio $r_1/r_2 = 2.417$. From Fig. 8 it can be determined that the percent change in the value of the maximum radial stress for $r_1/r_2 = 2.5$ when r_o/r_1 is varied from 1.5 to 4 is 12.6 percent. Fig. 9 shows that the percent change in the tangential stress corresponding to the same value of r_1/r_2 for the same variation in r_o/r_1 is 4.4 percent. Thus, it appears that for the specific package used for experimental purposes, the choice of the outside radius of the third cylinder is relatively unimportant insofar as the determination of the maximum stress is concerned. It should be noted, however, that for other values of the corresponding ratio r_1/r_2 it is possible that the choice of r_o/r_1 could significantly influence the calculated maximum stress.

The thermal shock test condition A($32°$F to $212°$F) of the MIL-STD-883B was applied to the experimental package. The maximum radial and tangential stresses occurring during this thermal shock test were presented in non-dimensional form. The thermal shock test procedure includes test levels which are more severe because they require larger temperature extremes. The maximum stress values depend linearly on the temperature difference between the cold and hot baths. Therefore, the results presented for test A can be easily calculated for any other test level provided that the

---

[1]In figures 8 and 9 the subscripts m and se indicate any metal and any sealing material, respectively. L is the total length of the lead, L_o, the length of the lead outside the base which is in contact with the fluid.

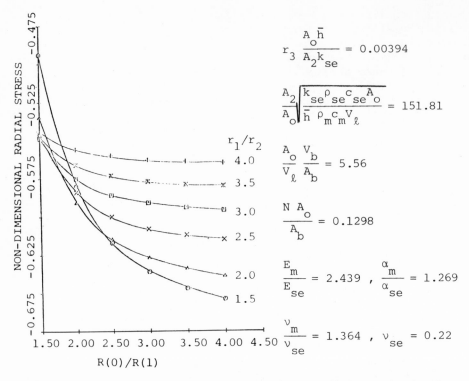

$$r_3 \frac{A_o \bar{h}}{A_2 k_{se}} = 0.00394$$

$$\frac{A_2}{A_o} \sqrt{\frac{k_{se} \rho_{se} c_{se} A_o}{\bar{h} \rho_m c_m V_\ell}} = 151.81$$

$$\frac{A_o}{V_\ell} \frac{V_b}{A_b} = 5.56$$

$$\frac{N A_o}{A_b} = 0.1298$$

$$\frac{E_m}{E_{se}} = 2.439 \quad , \quad \frac{\alpha_m}{\alpha_{se}} = 1.269$$

$$\frac{\nu_m}{\nu_{se}} = 1.364 \quad , \quad \nu_{se} = 0.22$$

Fig. 8.   Maximum non-dimensional transient radial stress in the glass at the lead-glass interface for the radius and material property ratios shown on the figure.

effective heat transfer coefficient is the same as determined for test level A. The maximum stress values corresponding to the various stress levels for the experimental package, based on this proviso are presented in Table 2. It should be noted that 5000 psi is a reasonable estimate of the radial tensile strength of the lead-glass interface. Thus, Table 2 indicates that the higher test levels could be damaging to this interface.

Acknowledgement

This work was sponsored by the Rome Air Development Center under Contract No. F30602-78-C-0083, Task No. PRN-8-5165.

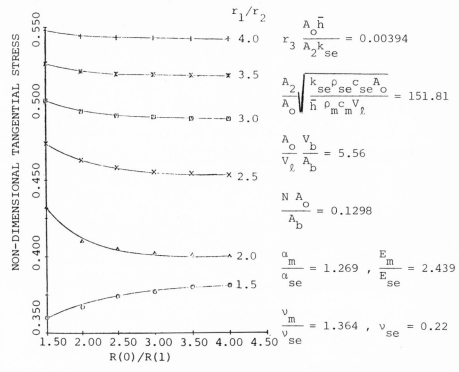

$$r_1/r_2$$

$$r_3 \frac{A_o \bar{h}}{A_2 k_{se}} = 0.00394$$

$$\frac{A_2}{A_o} \sqrt{\frac{k_{se} \rho_{se} c_{se} A_o}{\bar{h} \rho_m c_m V_\ell}} = 151.81$$

$$\frac{A_o}{V_\ell} \frac{V_b}{A_b} = 5.56$$

$$\frac{N A_o}{A_b} = 0.1298$$

$$\frac{\alpha_m}{\alpha_{se}} = 1.269 \ , \ \frac{E_m}{E_{se}} = 2.439$$

$$\frac{\nu_m}{\nu_{se}} = 1.364 \ , \ \nu_{se} = 0.22$$

Fig. 9.   Maximum non-dimensional transient tangential stress in the glass at the lead-glass interface for the radius and material property ratios shown on the figure.

TABLE 2

Maximum transient stresses at lead-glass interface for the different levels of thermal shock test conditions as described in MIL-STD-883B.

	A	B	C	D	E	F
T_i (°F)	32	−67	−85	−85	−319	−319
T_f (°F)	212	257	302	392	302	392
σ_{max} (Radial) (psi)	−2333	−4199	−5015	−6181	−8047	−9215
σ_{max} (Tang.) (psi)	1685	3033	3623	4465	5813	6656

REFERENCES

1.   MIL-STD-883B, Method 1011, "Test Methods and Procedures for
     Microelectronics," August 31, 1977.
2.   R.W. Thomas, "IC Packages and Hermetically Sealed-in Contami-
     nants." NBS Special Publication 400-9, pp. 4-19, (Dec. 1974).
3.   C.M. Johnson and L.K. Conaway, "Reliability Evaluation of
     Hermetic Dual in Line and Flat Microcircuit Packages,"
     McDonnell-Douglas Unclassified Report, Contract NAS8-31446,
     Oct. 1976 - Dec. 1977.
4.   M.P. Borom and R.A. Giddings, "Considerations in Designing
     Glass/Metal Compression Seals to Withstand Thermal Excursions,"
     Amer. Ceram. Soc. Bull., 55[12], (Dec. 1976).
5.   H. Poritsky, "Analysis of Thermal Stresses in Sealed Cylinders
     and the Effect of Viscous Flow During Anneal," Physics 5,
     pp. 406-411, (1934).
6.   B.E. Gatewood, "Thermal Stresses in Long Cylindrical Bodies,"
     Phil. Mag. Ser. 7, 32, pp. 282-301, (1941).
7.   K. Kokini, R.W. Perkins, C. Libove, "Thermal Stress Analysis
     of Glass Seals in Microelectronic Packages Under Thermal Shock
     Conditions." Final Technical Report, No. RADC-TR-79-201,
     Rome Air Development Center, New York, (July 1979).

RELIABILITY ANALYSIS OF THERMALLY STRESSED VISCOELASTIC

STRUCTURES BY MONTE CARLO SIMULATION

Thomas L. Cost*

The University of Alabama
Tuscaloosa
University, Alabama 35486

ABSTRACT

Probabilistic models for predicting the service life of missile structures subjected to random thermal loads have been developed. The models are applicable to both elastic and visco-elastic materials and utilize the finite element method. Statistical descriptions of the stresses, strains and damage parameters are generated using Monte Carlo computer simulation techniques and mathematical models of the thermal environments in arctic temperate, and desert geographical locations. The statistical descriptions of the induced stresses, strains, and damage are compared with similar descriptions of allowable parameters using interferrence techniques to compute a probability of failure. Realibility concepts are then used to generate service life estimates.

INTRODUCTION

During the service life of many tactical missile systems, the missiles, including the rocket motors used for propulsion, are stored in thermal environments which change with time. The environmental changes may result from changes in location, season, time-of-day, cloud cover, or weather front movement. These time-varying thermal conditions can produce thermal stresses and strains in solid rocket motors which result in motor failure.

*Professor, Department of Aerospace Engineering

The varying thermal environments can produce thermal stresses and strains in solid rocket motors which vary in a manner similar to the variation in the temperature. For example, in a typical case-bonded solid rocket motor the thermal expansion coefficient of the propellant grain is normally about 10 times greater than that for the motor case. Consequently, the grain tends to expand and contract but is restrained by the motor case. Thermal stresses result from the interaction of the grain and case and from non-uniform temperature distributions throughout the grain. Strains result from the thermal expansion and contraction but are also influenced by the grain-motor case interaction. The normal stress between the case and grain is usually of interest since an adhesive bond exists at the interface. The circumferential strain at the inner bore of the grain is also of interest since its magnitude has been related to failure. A typical history of the interface bond stress for a typical solid rocket motor is illustrated in Fig. 1.

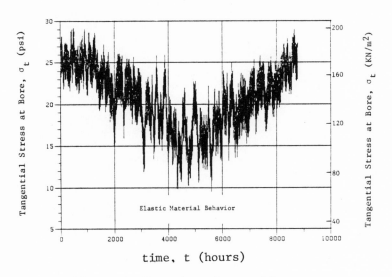

Figure 1. Transient Stress History

Since the environmental changes and thermomechanical response functions are approximately random, a probabilistic design and analysis approach appears justified. Such an approach permits statistical descriptions of the loads, response variables, and material allowables to be incorporated into a service life prediction method. The method involves application of the concepts of reliability or probability of failure under repeated cycles of load. The load cycles are produced by the varying thermal conditions over the life of the rocket motor. The period of the cyclic loading depends upon the specific character of the thermal

environment and is of fundamental importance in predicting motor
service life.

Prediction of service life is difficult since both the loads
and material properties vary with time. In a previous paper [1],
a method for treating variations in the thermal environment was
described based upon the assumption that the environmental
temperature was a random variable. The method employs a Monte Carlo
simulation procedure and a time-dependent model of the thermal
environment to generate statistical descriptions of the state
variables for the period of one year. The induced thermal stresses
and strains are compared with similar descriptions of experimentally
determined allowable values through a stress-strength interferrence
procedure and the probability of failure computed. The concept of
reliability under repeated load cycles [2] is then used to predict
service life. Since the Monte Carlo simulation procedure is based
upon a statistical description of the state variables, the influence
of statistical variations in the structure mechanical and thermal
properties on service life can easily be evaluated [3,4]. Heller
[5-7] has proposed a probabilistic approach based upon use of
frequency response functions and the power spectral density. This
differs from the methods described here and in Reference 1 in which
all calculations are made as a function of time.

In previous works [1,3,4] on this topic, details of the basic
method have been described. In the present paper, computer simulation results related to thermoviscoelastic material behavior are
described. In addition, the mechanical properties are assumed to
change with time due to aging effects. The form of the aging
model is based upon work by Layton [8].

BASIC ANALYSIS CONCEPTS

The time-varying thermal environments produce time-varying
thermal stresses of the type shown in Fig. 1. The stress shown
occurs at a critical failure position within the structure.
Assuming the stress magnitudes throughout a period of one day
follow a Gaussian distribution, both a mean value $\bar{s}$ and a standard
deviation σ_s may be calculated. These values are then compared
with experimentally measured allowable strength values $\bar{S}$ and σ_S.
By use of an induced stress-allowable strength procedure [1], the
probability of failure under a daily cycle of load P_f may be calculated using the expression

$$P_f = 1-P_s = 1 - \frac{1}{\sqrt{2\pi}} \int_{-\frac{\bar{\zeta}}{\sigma_\zeta}}^{\infty} \exp[-z^2/2] \, dz \qquad (1)$$

where

$$\bar{\zeta} = \bar{S} - \bar{s} \tag{2}$$

and

$$\sigma_\zeta = \sqrt{\sigma_S^2 + \sigma_s^2} \tag{3}$$

and where p_s is the reliability for one day of loading. This calculation may be repeated for each of the 365 days of a year. The probabilities of failure during the winter days are usually higher than in the summer. Due to continued application of load cycles, the probability of failure increases throughout the year. The annual probability of failure p_f may be expressed in terms of the daily probabilities of failure $p_f(i)$ as

$$p_f = 1 - \prod_{i=1}^{365} [1-p_f(i)] \tag{4}$$

If the material does not age, the probability of failure in M years would become

$$Q(M) = M \, P_f. \tag{5}$$

Conversely, if the maximum acceptable probability of failure is specified to be Q_{max}, then the service life N in years may be computed using the expression

$$N = Q_{max}/P_f \tag{6}$$

If the material ages, the probability from one year to the next will change and the service life should be calculated by determining the number of cycles N which allows the following equation to be satisfied:

$$Q_{max} = 1 - \prod_{j=1}^{N} [1-P_f(j)] \tag{7}$$

COMPUTATIONAL PROCEDURES

Details of both the elastic stress and strain and the transient thermal computational procedures used in the life simulation procedure have been presented in Reference 1. The stress and strain computations in elastic cylinders are obtained by evaluating well known mathematical equations applicable to hollow circular cylinders. All integrals appearing in the equations are evaluated numerically. The transient thermal calculations are performed using a finite element heat conduction code.

The viscoelastic stress and strain calculations were performed using a one-dimensional finite element computer cope applicable to cylindrical geometries. A code developed by Herrmann and Peterson [9] was modified for use in the simulation procedure. All visco-elastic calculations were made using this code; elastic calcula-tions can also be made using the same code by appropriately specify-ing the relaxation modulus.

MATERIAL DESCRIPTION

Service life predictions are presented in a later section for a realistic rocket motor. The calculations are based upon material properties for a solid propellant designated by the letters GBP. Experimental measurements were made to determine the tensile relax-ation modulus E(t). Assuming incompressibility, this data may be converted to the shear relaxation modulus G(t) by dividing the tensile data by 3. The tensile relaxation modulus data is illus-trated in Fig. 2 along with values calculated using an equation of the form

$$E(t) = E_e + \Sigma E_i e^{-t/\tau_i} \tag{8}$$

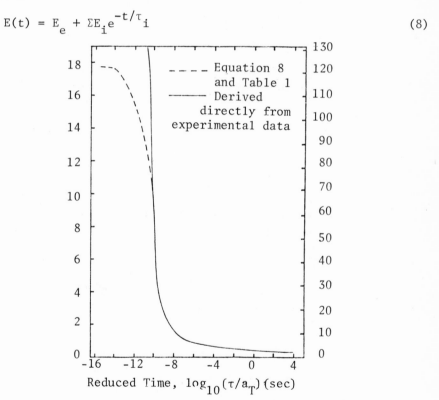

Reduced Time, $\log_{10}(\tau/a_T)$ (sec)

Fig. 2. Tensile Relaxation Modulus for GBP Propellant

Fifteen terms were used in the series expression. Values of the
constants in the equation are illustrated in Table 1. The equilibrium tensile relaxation modulus is E_∞ = 300 psi. Although the
glassy portion of the relaxation modulus curve is not well satisfied
by the equation, this result was found to have no effect on the
results due to the relatively slow loading times associated with
the thermal variations.

Table 1. Tensile Relaxation Modulus Constants

i	E_i, psi (KN/m^2)	τ_i (sec)
1	3,162.3718 (21,804.55)	3.16228 X 10^{-13}
2	5,145.4592 (35,477.94)	3.16228 X 10^{-11}
3	6,518.8814 (44,947.68)	3.16228 X 10^{-10}
4	1,130.7983 (7,796.85)	3.16228 X 10^{-9}
5	618.5605 (4,264.97)	3.16228 X 10^{-8}
6	285.6918 (1,969.85)	3.16228 X 10^{-7}
7	249.6106 (1,721.07)	3.16228 X 10^{-6}
8	132.6689 (914.75)	3.16228 X 10^{-4}
9	122.8246 (846.88)	3.16228 X 10^{-2}
10	16.5026 (113.79)	3.16228 X 10^{-0}
11	54.1111 (373.10)	3.16228 X 10^{+1}
12	73.2562 (505.10)	3.16228 X 10^{+3}
13	42.8024 (295.12)	3.16228 X 10^{+5}
14	20.7054 (142.76)	3.16228 X 10^{+6}
15	33.0961 (228.20)	3.16228 X 10^{+7}

The shift factor for the GBP propellant is illustrated in Fig. 3.

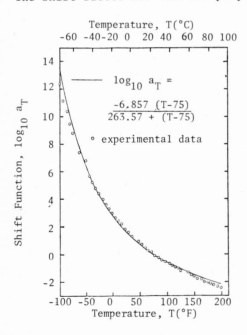

Figure 3. Shift Function For
GBP Propellant

To aid in analysis, a mathematical expression was determined to describe the shift factor dependence on temperature. The circled points in Fig. 3 represent experimental data while the curve was obtained by plotting values from the expression

$$\log a_T = \frac{-6.857(T-75)}{263.57 + (T-75)} \tag{9}$$

where T is the temperature in degrees Fahrenheit.

The maximum stress $\bar{S}$ of the GBP propellant is used as an allowable value in the analysis. The dependence of the maximum stress on the reduced time is illustrated graphically by the circled points in Fig. 4.

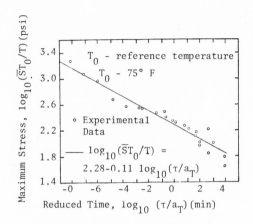

Fig. 4. Maximum Allowable Stress
for GBP Propellant

The expression

$$\log(\bar{S} T_0/T) = 2.28 - 0.11 \log (\tau/a_T) \tag{10}$$

is used to mathematically represent the data and is illustrated by the curve in Fig. 4.

Failure at the bore of the propellant grain is often predicted using a maximum normal strain failure criterion. This mode of failure is considered in the service life simulation model described in this report. The dependence of the maximum strain $\bar{\varepsilon}$ upon reduced time is illustrated graphically in Fig. 5. The circled points are experimental data and the curve is a graphical representation of the expression

$$\bar{\varepsilon} = 1.17 - 0.02595 [\log(\tau/a_T) + 1.4]^2 \tag{11}$$

Failure is assumed to occur in the case bonded rocket motors under consideration whenever the induced tangential stress or strain at the bore of the propellant grain exceed their allowable values. As indicated by Figs. 4 and 5, the allowable values depend upon the loading rate and temperature. For the cyclic thermal loadings under consideration here, a response time τ of 60 minutes is used in the allowable stress and strain calculations for all cycles.

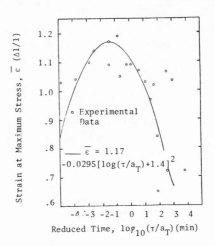

Fig. 5. Maximum Allowable Strain
for GBP Propellant

The allowable stress $\bar{S}$ and strain $\bar{\varepsilon}$ values are then calculated using Eqs. 10 and 11. The local temperatures at the grain-case interface and at the inner bore are used to compute the shift factor α_T and consequently the mean values of the allowable stress $\bar{S}$ and strain $\bar{\varepsilon}$.

AGING MODEL

On the basis of extensive test results, Layton [8] has discovered that certain propellants age linearly with the logarithm of time. If η (T,t) is used to designate the ratio of a propellant property which changes with age to the value of the property at zero time, then the propellant aging behavior may be expressed as

$$\eta(T,t) = 1 - \beta(t) \log t_{age} \tag{12}$$

All properties are assumed to age according to this same function $\eta(T,t)$. The temperature dependence of the parameter $\beta(T)$ is assumed to be of the form

$$\beta(T) = A \exp (-B/T) \tag{13}$$

where A and B are constants. Values of A and B determined for the GBP propellant are $A = 3.7124 \times 10^{13}$ and $B = 10.6902 \times 10^{3}$.

If the temperature changes continuously with time, the aging behavior can be predicted by assuming the temperature changes by constant amounts at times $t_1 = \Delta t$, $t_2 = 2\Delta t$, ..., $t_i = (i)\Delta t$. The aging factor $\zeta(t)$ may then be expressed as

$$\eta(t_n) = 1 - \sum_{i=1}^{n} \beta(T_i)[\log\tau_i - \log \tau_{i-1}^e \tag{14}$$

where

$$\tau_i = \tau_{i-1} + \Delta t \; ; \; \tau_{i-1} = (\tau_{i-1})^{\beta_{i-1}/\beta_i} \; ; \; t_n = n\Delta t \tag{15}$$

In Eqs. 14 and 15, τ plays the role of a reduced time.

For purposes of illustration, the aging factor $\eta(t)$ has been evaluated for the GBP propellant stored in the Huntsville, Yuma, and Barrow environments assuming the propellant reaches the same temperature as the ambient air. The aging behavior under varying thermal conditions is illustrated graphically in Fig. 6. Both seasonal and locational influences are apparent in the aging behavior; aging occurring most rapidly during the summer months and at the hotter geographical locations. Since many simulations are carried out in performing a complete service life analysis, Prony series expressions were fitted to the actual aging behavior to minimize computational effort. The series expressions are illustrated by the dashed lines in Fig. 6.

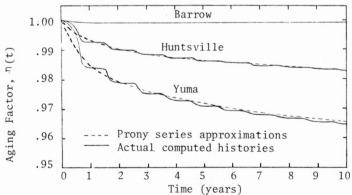

Fig. 6. Time History of GBP Propellant Aging
Factor in Three Different Climatic
Conditions

RESULTS

The viscoelastic and aging behavior described previously have
been integrated into a computer simulation procedure used to predict service life. All of the calculations presented in this
section employ the maximum principal stress and strain failure
criteria and assume the maximum acceptable probability of failure
Q_{max} is 0.0001.

Sample Problem Definition

The thermostructural response of the plane-strain, hollow,
circular, rocket motor model illustrated in Fig. 7 was simulated
to predict service life. Some GBP propellant properties have been
described previously; other pertinent properties are contained in
Table 2.

$$a = 0.925 \text{ in. } (2.35 \text{ cm})$$
$$b = 2.440 \text{ in. } (6.198 \text{ cm})$$
$$c = 2.500 \text{ in. } (6.35 \text{ cm})$$
$$h_c = 0.060 \text{ in. } (0.152 \text{ cm})$$

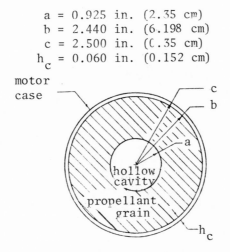

Fig. 7.   Rocket Motor Configuration

Viscoelastic Effects

The transient tangential stresses at the inner bores of the
elastic and viscoelastic motors are illustrated in Fig. 8, the
daily probabilities of failure for the two cases in Fig. 9, and
the cumulative probability of failures during the year in Fig. 10.
The annual probabilities of failure and service lives for the
elastic and viscoelastic cases are given in Table 3.

Table 2. Properties of GBP Propellant

Property	Value	Cov (%)
Thermal Conductivity	$K = 0.155$ BTU/Hr-Ft-$^{\circ}$F $= 6.41 \times 10^{-4}$ CAL/CM-S-$^{\circ}$C	6
Specific Heat	$C_P = 0.368$ BTU/lb-$^{\circ}$F $= 0.368$ CAL/g-$^{\circ}$C	3
Density	$\rho = 0.0624$ lb/in^3 $= 1727$ Kg/M^3	5
Coefficient of Thermal Expansion	$\alpha = 8.5 \times 10^{-5}$ in/in-$^{\circ}$F $= 15.3 \times 10^{-5}$ cm/cm-$^{\circ}$C	5
Temperature Shift Factor	$a_T = $ (see Fig. 6)	Unknown
Stress Relaxation Modulus	$E(t) = $ (see Fig. 5)	10
Maximum Normal Stress	$\bar{s} = $ (see Fig. 7)	10
Strain at Max Stress	$\bar{\varepsilon} = $ (see Fig. 8)	10
Glass Temperature	$T_G = -75^{\circ}$F$(-59.4^{\circ}$C$)$	Unknown
Strain Free Temperature	$T_f = +130^{\circ}$F$(\pm54.4^{\circ}$C$)$	Unknown
Cumulative Damage Parameters	$\sigma_{t_0} = 133$ psi $(917$ KN/m$^2)$ $t_0 = 0.02$ hr $\sigma_{cr} = 0$ $B = 8.90$	

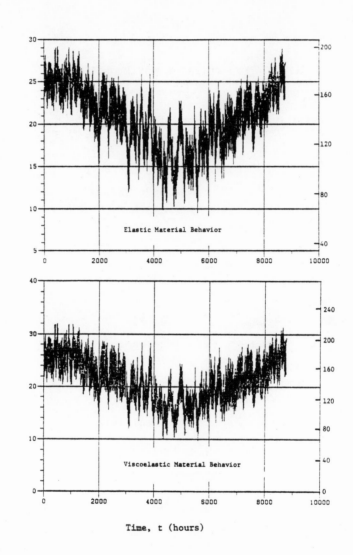

Fig. 8.   Comparison of Elastic and Viscoelastic
Transient Stress Responses

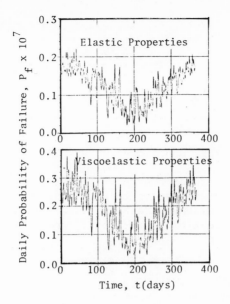

Fig. 9.   Daily Probability of Failure Variation
          Throughout Year for Elastic and Visco-
          elastic Materials

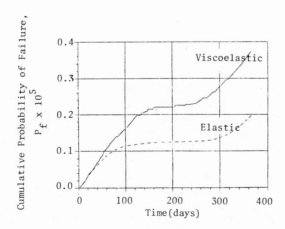

Fig. 10.   Cumulative Probabilities of Failure and
           Viscoelastic Materials

Table 3. Effect of Viscoelasticity on Service Life

	Viscoelastic	Elastic
P_f (annual)	0.3725×10^{-5}	0.1937×10^{-5}
N (life in years)	26.8	51.6

Influence of Climatic Conditions

The service life of the rocket motor described in Fig. 7 was
evaluated assuming storage in three locations· Yuma, Huntsville,
and Barrow as discussed in Section 1. The results shown in Table 4
illustrate the effect of thermal environment on strength and
stiffness.

Table 4. Service Life Dependence on Geographical Location

	Yuma	Huntsville	Barrow
P_f (annual)	0.2924×10^{-5}	0.3725×10^{-5}	1.28×10^{-5}
N (life in years)	34.2	26.8	7.8

DISCUSSION AND CONCLUSIONS

Monte Carlo simulation of the thermostructural response of
viscoelastic structures subjected to random thermal loads has been
shown to be a useful method for generating service life estimates.
The method permits such complications as material property variability, viscoelastic material behavior complex geometrical features,
and material degradation with age to be handled in a straightforward manner. Since the simulation is conducted in real-time and
service life is expressed in units of time, the analysis results
are expressed in easily understood concepts.

Based upon the results obtained, it appears that both viscoelastic strength and stiffness properties significantly influence
service life predictions. From the results presented, viscoelastic
solutions should be used when estimating service life; elastic
stiffness solutions produce less conservative estimates of service
life than viscoelastic ones. Also, temperature dependent strength
properties should be included in service life predictions to obtain
the best accuracy possible.

The elastic solutions presented in Figs. 8 to 10 were obtained using the equilibrium modulus of the propellant (300 psi). The corresponding viscoelastic solutions predicted significantly higher stresses for times as long as one year. Use of the equilibrium modulus in elastic solutions produces nonconservative results in this case. Since this is a common practice in the solid rocket industry, care should be exercised in the use of such results.

ACKNOWLEDGEMENT

Support of the U. S. Army Missile Command under Contract DAAH01-78-C-0971 is gratefully acknowledged.

REFERENCES

1. Cost, T. L., "Computer Simulation of Solid Rocket Motor Service Life for Thermal Loads", Athena Engineering Company, Northport, Alabama, Report No. T-CR-78-9, March 1978.

2. Freudenthal, A. M., Garrelts, J. M., and Shinozuka, M., "The Analysis of Structural Safety", Journal of the Structural Division of ASCE, v. 92, pp. 267-325, February 1966.

3. Cost, T. L., "Influence of Statistical Property Variations on Service Life Predictions", Proceedings of the 1978 JANNAF Service Life Committee Meeting, Naval Post Graduate School, Monterey, California, April 18-20, 1978.

4. Cost, T. L., and Dagen, J. D., "Probabilistic Service Life Prediction of Solid Rocket Motors Subjected to Thermal Loads Using Computer Simulation", Chemical Propulsion Information Agency, Pub. No. 283, 1977, JANNAF SMBWG Meeting, Vol. II, pp. 77-90, April 1977.

5. Heller, R. A., "Thermal Stress as a Narrow Bond Random Load", ASCE Engineering Mechanics Journal, EM5, No. 12450, pp. 787-805, October 1976.

6. Heller, R. A., "Temperature Response of an Infinitely Thick Slab to Random Surface Temperature", Mechanics Research Communications, v. 3, pp. 379-385, 1976.

7. Heller, R. A. and Singh, M. P., "Probability of Motor Failures Due to Environmental Effects", U. S. Army Missiles Research and Development Command, Report No. T-CR-78-11, March 1978.

8. Layton, L. H., "Chemical Aging Studies on ANB-3066 and TP-H1011 Propellants", Final Technical Report, AFRPL-TR-74-16, 1974.

9. Herrmann, L. R. and Peterson, F. E., "A Numerical Procedure for Viscoelastic Stress Analysis", ICRPG Mechanical Behavior Working Group, 7th Meeting, November 1968.

PROPAGATION OF PROPELLANT-LINER SEPARATIONS

IN ROCKET MOTOR GRAINS UNDER TRANSIENT THERMAL LOADING

Donald L. Martin, Jr.

Propulsion Directorate, US Army Missile Laboratory
US Army Missile Command
Redstone Arsenal, Alabama 35809

I. INTRODUCTION

Several areas in a given grain design often require special consideration other than the usual strength analysis. Linear elastic stress analysis often predicts infinite stresses at material discontinuities and notches associated with bond end terminations, roots of release flaps, and defects such as cracks and localized unbond regions in a rocket motor grain. Many of the limitations of strength analyses may be overcome by the application of the energy balance concept of fracture mechanics. The energy balance concept is based on an exchange of strain energy stored in a propellant grain under a specified loading condition and the energy expended due to the formation of new surface area by fracture and/or crack growth.

From thermodynamic principles, the propagation of a crack must result in a decrease in the free energy of the system if no additional work is input at the boundary.[1] In materials that do not exhibit plastic flow or viscoelastic dissipation of energy, a crack will advance instanteously when the incremental release in strain energy (ΔU) in the body is greater than the incremental increase in surface energy (γ) as new surface area is created by the fracture. This is stated mathematically as

$$\frac{\partial U}{\partial A} > \gamma \quad , \tag{1}$$

where

U = strain energy

447

A = surface area

γ = characteristic surface energy.

For crack stability conditions,

$$\frac{\partial U}{\partial A} \leq \gamma \quad . \tag{2}$$

Since the original application of the energy balance concept by Griffith[2] to brittle materials, Orwan[3] and Irwin[4] have applied it to relatively ductile materials, and Rivlin and Thomas[5] have applied it to fracture in rubber. Williams[6] extended the concept to include viscoelastic materials by replacing the elastic moduli by their corresponding viscoelastic components and separating the geometric parameters from the time-dependent viscoelastic parameters. Recently, the energy balance concept has been considered as a useful technique in determining the criticality of flaws in rocket motor grains. Much of the theory still remains to be developed before the full significance of this concept can be realized. Many cases have been reported of motors containing flaws that were fired with no adverse ballistic effects. One of the purposes of fracture mechanics is to determine whether particular motors with flaws can be successfully and safely fired.

The approach used in this investigation is based on the concept of strain energy release rates ($\partial U / \partial A$), which is the amount of strain energy stored in a propellant grain that is released in the formation of 1 inch^2 of new surface area by flaw propagation.

Fracture mechanics type analyses are applied to known defects such as scratches, gouges, cracks, and localized unbond regions in a rocket motor grain. The fracture analysis is used to determine how critical (depth and area) a fracture would be once it has been initiated. The first step in a fracture analysis is to determine the fracture location and geometry. This may be accomplished using nondestructive testing methods for existing flaws, or by determining the most probable locations and geometries from structural analyses of unfractured grains.

The strain energy release rate method has been used by several investigators to determine the stability of different types of flaws in propellant grains when the motor was subjected to uniform loading conditions. However, very little work could be found in the literature that considers the effects of transient loading conditions. It is the objective of this investigation to determine the stability propellant/liner bond separations under transient thermal loading and to determine an empirical relationship that describes the rate of propagation with time.

II. MATERIAL CHARACTERIZATION

The usual material properties required to conduct a structural analysis are also required for a fracture stability analysis. In addition to a time-dependent relaxation modulus, $E_r(t)$, Poisson's ratio, ν, and a linear thermal expansion coefficient, α, a time-dependent fracture energy, γ, is also required.

The experimental approach used to define the mechanical behavior of the propellant consisted of a series of constant strain-rate tests at crosshead displacement rates of 50.0, 5.0, 0.5, and 0.05 cm/minute (20.0, 2.0, 0.2, and 0.02 inch/minute) at test temperatures of $-62°$, $-51°$, $-40°$, $-18°$, $25°$, $71°$, $82°$, and $93°C$ ($-80°$, $-60°$, $-40°$, $0°$, $77°$, $160°$, $180°$, and $200°F$). The propellant mechanical property data indicate a strong dependence on strain rate and temperature. The function, a_T, used to describe the equivalence of loading rate and temperature in producing corresponding effects on the propellant properties was determined empirically from the experimental data. The a_T function was fitted to a WLF-type equation of the following form:

$$\text{Log } a_T = -\frac{23.0(T - T_s)}{148.85 + T - T_s} \quad , \tag{3}$$

where T = test temperature and T_s = reference temperature. The glass transition was found to be $-84°C$ ($-119°F$); therefore, a reference temperature of $-34°C$ ($-30°F$) or $239°K$ was used. The constants 23.0 and 148.85 were determined from the experimental data. The a_T function determined is presented in Figure 1 and was used throughout this investigation.

Once the a_T function was established, a master stress-strain curve was developed from all uniaxial tensile tests. A stress relaxation modulus curve was then constructed from the slope of the master stress-strain curve according to the theory for a linearly viscoelastic material, i.e.,

$$E_R(t_i) = \left. \frac{\sigma}{\varepsilon(t_i)} \right|_{\varepsilon=\text{const}}$$

Figure 2 presents the relaxation modulus curve for the propellant at the reference temperature of $-34°C$ ($-30°F$). The relaxation modulus thus obtained was fitted to the following modified power law:

$$E_r(t) = \frac{E_g - E_e}{\left[1 + \dfrac{t}{\tau_o}\right]^\beta} + E_e \quad , \tag{4}$$

where

E_r = relaxation modulus

E_g = glassy modulus = 1.32×10^8 PA (19,200 psi)

E_e = equilibrium modulus = 2.69×10^5 PA (39 psi)

τ_o = 2.89×10^{-6} minute

β = 0.164

t = reduced time of interest, minute.

Also presented in Figure 2 is a plot of the factor $[2t^{-2}$
$E_r^{(2)}(t)]$ versus reduced time log (t/a_T). These data are
required for the evaluation of the fracture energy as determined
from laboratory experiments which will be discussed in detail
later. $E^{(2)}(t)$ is used to indicate the second integral of the
time-varying relaxation modulus, and the other symbols are as
defined previously.

Fracture energy may be determined experimentally from several
different laboratory tests such as a biaxial sheet with a central
through crack or edge crack. The biaxial strip sample with a central through crack has been used by several investigators to
determine the cohesive fracture energy.[7] The stress analyses
were readily available for this sample, and the results reported in
the literature indicated this sample configuration yielded fracture
energy values consistent with existing theory.[8] Also, the
Griffith analysis, which applies to a small crack in an infinite
medium, can be corrected for the finite sample dimensions by the
factors $F_1(C_o/w)$ and $F_2(C_o/b)$. Therefore, the cohesive
fracture energy may be determined by the relation[9]

$$\gamma_c = \frac{C_o \pi \varepsilon_{cr}}{2(1 - \nu^2)^2} \left\{ 2t^{-2} E_r^{(2)}(t_o) \right\} F_1\left(\frac{C_o}{w}\right) F_2\left(\frac{C_o}{b}\right) , (5)$$

where
  γ_c = cohesive fracture energy

  C_o = original crack length divided by two

  γ = Poisson's ration = 0.5

  ε_{cr} = average strain at the start of crack propagation in the
       direction of the applied load

  F_1, F_2 = correction factors

  w = specimen width divided by two

  b = specimen height divided by two

t_o = time from the initial load application to the point at which crack propagation is observed to start

E_r = the relaxation modulus

$E_r^{(2)}(t)$ = the second integral of the relaxation modulus.

The Westgard factor[10] $F_1(C_o/w)$ was used to correct for finite sample width and the Knauss factor,[11] $F_2(C_o/w)$, as modified by Rice,[12] was used to correct the results for finite sample height. These correction factors, which are dependent on the aspect ratio of the specimen, are presented in Figure 3.*

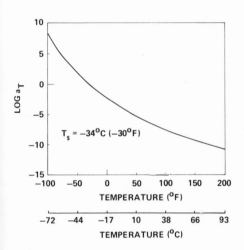

Figure 1.   Log a_T versus temperature for the propellant.

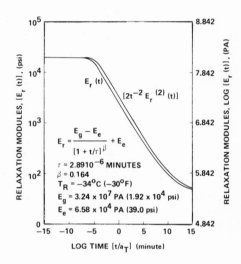

Figure 2.   Relaxation modulus versus reduced time for the propellant.

---

*Figure 3 is an excerpt from the JANNAF Solid Propellant Structural Integrity Handbook, September 1972, Edited by W. L. Hufferd and J. E. Fitzgerald, pp. 5-187.

The biaxial strip specimen used to determine the cohesive
fracture energy values for the propellant was approximately 10 cm
(4 inches) wide, 2.5 cm (1 inch) in height, and 0.938 cm (0.375
inch) thick. The specimen was bonded to two metal strips by a butt
end joint. A line crack approximately 1.5 cm (0.5 inch) in length
was cut in the center of the specimen. Each metal strip had a
centrally-located hole. The test fixture transmits the load to the
specimen through pin joints to minimize any effect of specimen misalignment. The ends of the specimen were displaced at a constant
rate in an Instron testing machine. The time at which the crack
was observed to propagate was indicated by a pip on the Instron
load-time chart. The critical strain, ε_{cr}, and time, t_o, was
determined from the Instron chart. The corresponding cohesive
fracture energy was then determined using Equation (5).

Only a limited number of cohesive fracture energy data points
was determined, as discussed previously, due to the limited amount
of propellant available at this time. However, the tear resistance
of the propellant was characterized previously according to the
standard method of test for tear resistance of vulcanized rubber
discussed in ASTM D 624-54.[13] Fracture energy is dependent on
sample size as well as strain rate, temperature, and stress state.
Therefore, the tear resistance as determined has not been theoretically related to the fracture energy as determined from biaxial strip samples. However, the two sets of data could be expected
to indicate the same general trends similar to the comparison of
uniaxial and biaxial tensile data. The tear resistance data
obtained from the original characterization work were replotted
versus reduced time and are compared with the biaxial fracture
energy in Figure 4. The tear resistance data plotted in Figure 4
were arbitrarily multiplied by the factor 8.0 to correspond with
the biaxial fracture energy data. As can be seen from Figure 4,
the biaxial fracture energy appears to be within the scatter of the
tear data. Therefore, the assumption that the biaxial fracture
energy follows the same trend and has the same time and temperature
dependence as the tear data appears to be justified. The solid
line in Figure 4 will be used in conjunction with the a_T function
in Figure 1 to determine the appropriate fracture energy for the
loading conditions investigated.

III. TRANSIENT THERMAL ANALYSIS

A transient thermal analysis was conducted on the propellant
grain shown in Figure 5 with an initial equilibrium temperature of
-37°C (-35°F) when subjected to a sudden change of environmental
temperature from -37° to 57°C (-35° to 135°F). The temperature
distributions at specified time intervals were determined using
AMGO 65, which is an axisymmetric finite element heat conduction

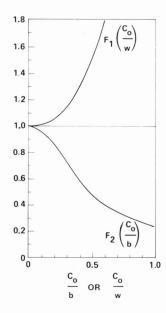

Figure 3. Aspect ratio correction factors.

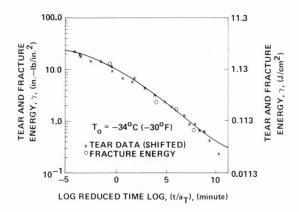

Figure 4. Tear and fracture energy versus reduced
          time for the propellant.

program.[14] The thermal property data used in the analysis were
as follows:

(a) Thermal expansion coefficient

α = 3.33 x 10^{-6} cm/cm, °C (5.5 x 10^{-5} in./in. °F)

(b) Thermal conductivity

K = 15.9 J, cm/°C, cm^2 (0.0209 Btu, in./°F, $in.^2$)

(c) Specific heat

Cp = 6.58 x 10^{-2} J/g (0.282 Btu/lb)

(d) Density

ρ = 7.76 kg/m^3 (0.064 lb/$in.^3$).

The liner and insulation were assumed to have the same thermal
properties as the propellant.

Some thermocouple data were available on the thermal response
of a similar grain configuration to changes in the environmental
temperature from −37° to 57°C (−35° to 135°F). Therefore, the outside film coefficient was determined using an iteration scheme to
match the predicted propellant−liner interface temperature with the
thermocouple data. Figure 6 shows a comparison of the predicted
and experimental propellant−liner interface temperature. The outside air heat transfer film coefficient was determined to be 0.382
J/cm^2, hour, °C (0.0125 Btu/$in.^2$, hour, °F). This film coefficient was used in the thermal analysis of the motor.

Figure 7 shows plots of the predicted temperature distributions for elapsed times of 1, 3, 5, 7, 10, and 18 hours, respectively. The indicated times were measured from the instant the
environmental temperature was suddenly changed from −37° to 57°C
(−35° to 135°F). The effect of these temperature distributions on
the propellant's response was investigated using the quasi-elastic
approach. The finite element method of analysis permits the use of
material properties dependent on position within the grain. Thus,
by determining the elapsed time and temperature at all points in
the propellant grain, the appropriate value of the modulus can be
used to reflect the viscoelastic response to instantaneous conditions at a given location within the grain. In this way, the
effect of viscoelastic material behavior in a transient thermal
field is included. Figure 8 shows the modulus versus radial
position at elapsed times of 1, 3, 5, 7, 10, and 18 hours.
respectively. These data represent the instantaneous material

behavior for the propellant grain when the motor is shock cycled
from −37° to 57°C (−35° to 135°F).

Figure 5.   Finite element model for the propellant grain
            in motor number 1.

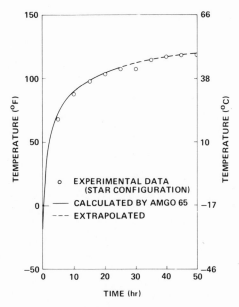

Figure 6. Case-grain interface temperature versus time at 57°C
          (]35°F) for the motors with an initial uniform
          temperature of −37°C (−35°F).

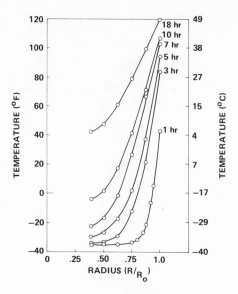

Figure 7.   Temperature distribution for motor number 1
            shock cycled from -37° to 57°C (-35° to 135°F).

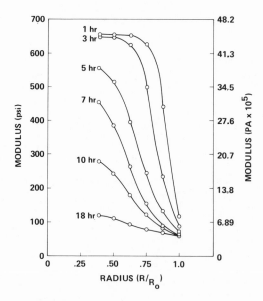

Figure 8.   Modulus variation for motor number 1 shock
            cycled from -37° to 57°C (-35° to 135°F).

The strain energy and the strain energy release rate have been shown to vary in direct proportion to the modulus for a given thermal load. A reasonable extension of this hypothesis is that the strain energy release rate varies directly proportional to the total strain energy for a given transient thermal condition. Based on this hypothesis, the thermal gradients most likely to cause a separation to become unstable and propagate may be determined by investigating the strain energy versus time for the unflawed grain geometry during a thermal shock cycle. Figure 9 presents the total strain energy plotted versus time for the propellant grain subjected to a thermal shock cycle from −37° to 57°C (−35° to 135°F). These data indicate that the thermal gradients in the propellant grain at an elapsed time of 1 hour [after environmental temperature is changed from −37° to 57°C (−35° to 135°)] should impose the most severe loading condition on the propellant-liner interface. The majority of the propellant grain is still at −37°C (−35°F), while the propellant-liner interface is approximately 6°C (43°F). Therefore, the majority of the strain energy should be contained within the propellant-liner interface region where the separations are located.

The thermal gradients determined by AMGO 65 were saved on permanent file and used as input to AMGO 74.[15] The crack stability was investigated using the quasi-elastic approach in conjunction with the thermal gradients obtained at one hour elapsed time. Because this represents the most severe loading condition for the thermal shock cycle, the results should indicate if the separations may be expected to propagate during repeated thermal shock cycles. Figure 10 presents the resulting strain energy release rate versus the crack length curve. These data indicate that at one hour elapsed time and an initial crack length of 0.34 cm (0.25 inch), the strain energy release rate is approximately 5.4 x 10^{-2} J/cm^2 (3.0 inches-pound/inches^2). The strain energy release rate decreases with increasing crack length to approximately 2.1 x 10^{-2} J/cm^2 (1.15 inches-pound/inches^2) at a crack length of 40 cm (16 inches). It was estimated that the motor would require approximately 5 days to 1 week to obtain an equilibrated temperature of −37°C (−35°F) from ambient conditions. One hour after the environmental temperature was changed from −37° to 57°C (−35° to 135°F), the propellant-liner interface temperature was 6°C (43°F). Therefore, the average fracture energy value was 1.17 x 10^{-2} J/cm^2 (0.65 inch-pound/inches^2) with $+3$ σ limits of 1.46 x 10^{-2} J/cm^2 to 8.82 x 10^{-3} J/cm^2 (0.81 inch-pound/inches^2 to 0.49 inch-pound/inches^2). Because the strain energy release rate from 0 to 40 cm (0 to 16 inches) crack length greatly exceeds this fracture energy value, there will be a strong tendency for the initiation and propagation of cracks near the propellant-liner interface at the base of the headend radial slot when the motor is thermally shocked from −37° to 57°C (−35°

to 135°F). Therefore, cracks would be expected to propagate under repeated thermal cycles of the type indicated.

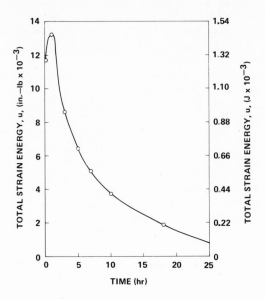

Figure 9.   Total strain energy versus time for motor number 1 shock cycled from -37° to 57°C (-35° to 135°F).

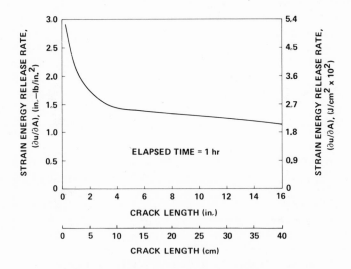

Figure 10.   Strain energy release rate versus crack length for motor number 1 thermally shocked from -32° to 57°C (-35° to 135°F).

To verify the hypothesis that for a given loading condition the strain energy release rate is proportional to the total strain energy in the body, a strain energy release rate curve was also established analytically for an elapsed time of 10 hours. Figure 11 shows the strain energy release rate versus crack length for the motor at an elapsed time of 10 hours when thermally shocked between −37° to 57°C (−35° to 135°F). The values of the strain energy release rate were also estimated from the data given in Figure 10 according to the previously stated hypothesis. The hypothesis is stated mathematically as follows:

$$\frac{\partial U}{\partial A} = \left(\frac{U}{U_o}\right) \frac{\partial U}{\partial A}_o \quad , \tag{6}$$

where U = total strain energy and o = subscript refers to reference time conditions. The remaining symbols are as indicated previously. The circled point values in Figure 11 were estimated according to the previous equation, and the solid line represents the values established analytically using a finite element analysis. The close agreement between the estimated values and the finite element analysis indicates the validity of the hypothesis expressed by Equation (6) for transient thermal conditions.

The material's fracture energy corresponding to an elapsed time of 10 hours is 5.04×10^{-3} J/cm^2 (0.28 inch−pound/inch^2) with a +3 σ variation of 6.48×10^{-3} J/cm^2 to 3.96×10^{-3} J/cm^2 (0.36 inch−pound/inch^2 to 0.22 inch−pound/inch^2). A comparison of these values with the strain energy release rate given in Figure 11 also indicates that there is still a tendency for cracks to propagate 10 hours after the environmental temperature is changed from −37° to 57°C (−35° to 135°F). The tendency for crack propagation decreases with elapsed time. From the data presented, it is estimated that for the first 15 to 25 hours in the cycle from −37° to 57°C (−35° to 135°F) there will be a tendency for cracks to propagate.

The crack stability was also investigated for a second motor using the qausi−elastic approach in conjunction with the thermal gradients obtained at one hour elapsed time. Figure 12 presents the finite element model for the second motor. The total strain energy versus time is presented in Figure 13. Figure 14 shows the resulting strain energy release rate versus crack length for the second motor which was thermally shocked from −37° to 57°C (−35° to 135°F). These data indicate that at one hour elapsed time and an initial crack length of 0.34 cm (0.25 inch) the strain energy release rate is approximately 7.2×10^{-2} J/cm (4.0 inch−pound/inch^2). The strain energy release rate decreases with increasing crack length to approximately 2.2×10^{-2} J/cm^2 (1.4 inch−pound/inch^2) at a crack length of 40 cm (16 inches).

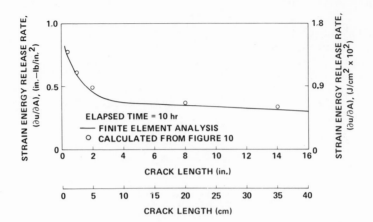

Figure 11.   Strain energy release rate versus crack length
of motor number 1 thermally shocked from −37°
to 57°C (−35° to 135°F).

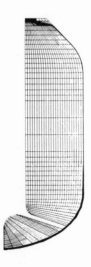

Figure 12.   Finite element grid for the second motor.

The propellant-liner interface temperature was 6°C (43°F) one hour after the environmental temperature was changed from 37° to 57°C (-35° to 135°F). Therefore, a corresponding average fracture energy value of 1.17×10^{-2} J/cm^2 (0.65 inch-pound/inch^2) was estimated for this condition with $+3$ σ limits of 1.46×10^{-2} J/cm^2 to 0.88×10^{-2} J/cm^2 (0.81 inch-pound/inch^2 to 0.49 inch-pound/inch^2). Because the strain energy release rate from 0 to 40 cm (0 to 16 inch) crack length greatly exceeds the lower 3-limit fracture energy value, there will be a strong tendency for the initiation and propagation of cracks near the propellant-liner interface at the base of the head-end radial slot when the second motor is thermally shocked from -37° to 57°C (-35° to 135°F). Therefore, cracks would be expected to propagate under repeated thermal cycles of the type indicated.

It was verified that the strain energy release rate was proportional to the total strain energy in the unflawed grain at different times within a given transient thermal cycle. To verify the hypothesis that for different transient thermal cycles (with different temperature limits) the strain energy release rate is still proportional to the total strain energy in the unflawed propellant grain, a strain energy release rate curve was also established for the second motor when shock cycled from -37° to 29°C (-35° to 85°F). Figure 15 presents the resulting strain energy release rate versus crack length for the second motor when shock cycled from -37° to 29°C (-35° to 85°F) at an elapsed time of 10 hours. The strain energy release rate values were also estimated from the data given in Figure 14 in accordance with the previously stated hypothesis.

The hypothesis is stated mathematically as follows:

$$\frac{\partial U}{\partial A} = \frac{U_1}{U_o} \left(\frac{\partial U}{\partial A} \right)_o , \qquad (7)$$

where

   U = total strain energy

   o = subscript indicating reference thermal cycle

   1 = subscript indicating new thermal cycle.

The remaining symbols are as defined previously.

The circled point values in Figure 15 were estimated according to Equation (7), and the solid line represents the values established analytically using a finite element analysis. The close agreement between the estimated values and the finite element

analysis indicates the validity of the hypothesis, expressed by
Equation (7), for transient thermal conditions with different
temperature limits.

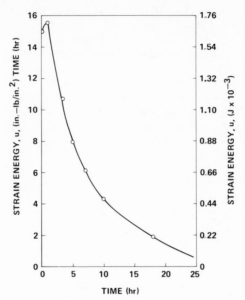

Figure 13.   Strain energy versus time for the second motor shock
cycled from −37° to 57°C (−35° to 135°F).

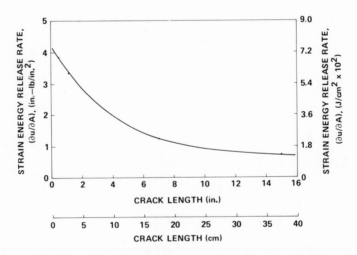

Figure 14.   Strain energy release rate versus crack length for the
second motor thermally shocked from −37° to 57°C (−35°
to 135°F) at an elapsed time of one hour.

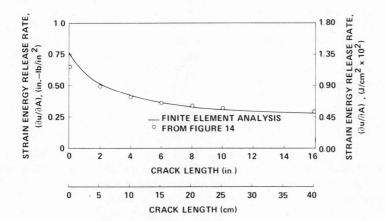

Figure 15.   Strain energy release rate versus crack length for the second motor shock cycled from −37° to 29°C (−35° to 85°F) at an elapsed time of 10 hours.

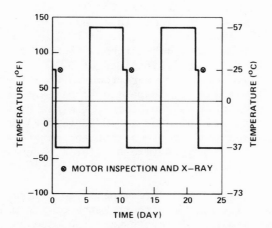

Figure 16.   Environmental temperature history temperature versus time for −37° to 57°C (−35° to 135°F) shock cycle.

The material's fracture energy corresponding to an elapsed time of one hour is 1.53×10^{-2} J/cm^2 (0.85 inch-pound/inch^2) with a $+3 \sigma$ variation of 1.26×10^{-2} J/cm^2 to 1.8×10^{-2} J/cm^2 ($\overline{0.70}$ inch-pound/inch^2 to 1.0 inch-pound/inch^2). A comparison of these values with the strain energy release rate given in Figure 15 also indicates that there is still a tendency for cracks to propagate one hour after the environmental temperature is changed from $-37°$ to $29°C$ ($-35°$ to $85°F$). The tendency for crack propagation decreases with elapsed time. From the data presented, it is estimated that for the first 10 to 12 hours in the cycle from $-37°$ to $29°C$ ($-35°$ to $85°F$) there will be a tendency for cracks to propagate.

IV. EXPERIMENTAL MOTOR DATA

The data presented previously indicated that a propellant-liner separation at the base of the head-end slot will be unstable and will tend to propagate with either the first or second motor when they are subjected to shock cycles between $-37°$ and $57°C$ ($-35°$ and $135°F$). Due to the various assumptions used in the analysis, it is desirable to verify the validity of these assumptions and the predictive technique by an experimental program using typical motor configurations. The experimental procedure used consisted of selecting one motor of each configuration with initial propellant-liner separations and subject those motors to thermal shock cycles between $-37°$ and $57°C$ ($-35°$ and $135°F$). The motors were to be subjected to an x-ray inspection once during each cycle.

Figure 16 shows the environmental temperature history of the motors versus time for the thermal shock cycle used. The motors, initially at ambient temperature [approximately $25°C$ ($75°F$)], were visually and x-ray inspected to determine the initial separation depth. The motors were then placed in a conditioning chamber at $-37°C$ ($-35°F$) for 5 days. After 5 days at $-37°C$ ($-35°F$), the motors were placed in another conditioning chamber in which the environmental temperature was maintained at $57°C$ ($135°F$). After 5 days at $57°C$ ($135°F$), the motors were removed from the conditioning chamber and visually and x-ray inspected at ambient conditions. After the x-ray inspection was completed, the motors were again placed in an environmental chamber at $-37°C$ ($-35°F$) for 5 days. The cycle was repeated several times with the motors subjected to x-ray inspection at ambient conditions after each cycle as indicated in Figure 16.

The separation length was determined from x-rays taken at 30-degree intervals around the circumference of the motor. An integrated average separation length was then used to determine the extent of propagation of the separation during each cycle. Figure

17 shows plots of the average crack length versus the number of cycles for the first and second motors when subjected to the thermal shock cycle from −37° to 57°C (−35° to 135°F). The first motor had an initial separation length of 2.68 cm (1.07 inches). The separation increased during each cycle and, after ten cycles, the average separation length had increased to 18.15 cm (7.26 inches). The experiment on the first motor was discontinued after ten cycles. The second motor contained an average initial separation length of 8.45 cm (3.38 inches). The separation increased during each subsequent cycle and, after five cycles, the separation in the second motor had increased to 18.7 cm (7.47 inches) in length. The experiment with the second motor was discontinued after five cycles.

It is unfortunate that the economic and time requirements dictated the necessity to terminate the experiments prior to the determination of the depth at which the separations would remain stable and not propagate with additional cycles. However, this limited amount of data does indicate the following important conclusions. The separations do propagate when the motors are subjected to thermal shock cycles between −37° and 57°C (−35° and 135°F). This is in accordance with the predictions discussed previously and tends to verify the assumptions and the predictive technique used. The rate of propagation decreases as the separation length increases, which indicates the rate of propagation is proportional to the quantity $(\partial U/\partial A - \gamma)$, where $\partial U/\partial A$ is the strain energy release rate or the driving force for crack propagation and γ is the material fracture energy or the ability of the material to resist crack propagation. These data suggest the following relationship could be used to predict separation length:

$$L = L_o + \int_o^t K \left(\frac{\partial U}{\partial A} - \gamma \right) dt \quad , \qquad (8)$$

where

   L = separation length

   K = proportionally constant

   t = time for cycles.

The remaining symbols are as defined previously.

The strain energy release rate was shown to be proportional to the total strain energy in the unflawed body for different times within a given thermal cycle and for similar cycles with different temperature limits. The $\partial U/\partial A$ for various times within the first

cycle were estimated from the data given in Figure 10 and 13 using
Equation (6). The fracture energy, γ, was determined from Figure 4
at a reduced time corresponding to the temperature of the propellant liner interface and the elapsed time from the start of the
cycling experiment. Figure 18 presents the $(\partial U/\partial A - \gamma)$ versus time
curve for the first motor with an initial crack length of 1.1
inches. The time integral of the quantity $(\partial U/\partial A - \gamma)$ was found to
be 1.014×10^3 J, hr/cm^2 (5.633 inch-pound hour/inch^2) for
the first cycle. The time integral was then estimated for
subsequent cycles by the following relationship

$$Q_2 = \frac{\left(\frac{\partial U}{\partial A}\right)_2}{\left(\frac{\partial U}{\partial A}\right)_1} \quad \frac{E_2}{E_1} \quad Q_1 \quad , \tag{9}$$

where

$$Q_i = \int_{t_i}^{t_f} \left\{ \left(\frac{\partial U}{\partial A}\right) - \gamma \right\} dt \tag{10}$$

$(\partial U/\partial A)_2$ = $\partial U/\partial A$ at the initial crack length for cycle
number 2

$(\partial U/\partial A)_1$ = $\partial U/\partial A$ at the initial crack length for cycle
number 1

E_1 and E_2 = modulus values at the beginning of the heat-up
portion of cycle numbers 1 and 2, respectively.

The total driving force for a complete cycling experiment is given
by the relationship:

$$Q = \sum_{i=1}^{N} Q_i \quad . \tag{11}$$

The quantity Q is correlated with the crack length at the end of
the corresponding cycles. The results are presented in Figure 19.
These data indicate Equation (8) with K = 68.13 cm^3/J, hour
(0.1515 inch^3/inch-pound, hour) may be used to describe the crack
growth in the first motor. Figure 20 presents a comparison of the
experimentally determined values of crack length and those calculated by Equation (8) with K = 68.13 cm^3/J, hour. The data
presented in Figures 19 and 20 indicate an initial crack length of
approximately 5.0 cm (2.0 inches) would be expected instead of the
2.75 cm (1.1 inches) value reported. It is possible that, during
storage of the motors, prior to the experiment the crack partially

closes or reheals, and the full crack length is not detectable by
x-ray until after the crack has opened again. Because Equation (8)
[with an initial crack length of 5.0 cm (2.0 inches)] correlates
the data on all subsequent cycles, an error in the initial crack
length measurement is indicated.

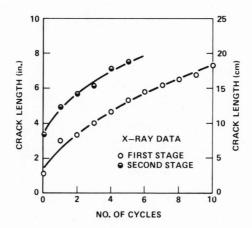

Figure 17.   Crack length versus number of cycles from $-37°$ to $57°C$
($-35°$ to $135°F$) for first and second motors.

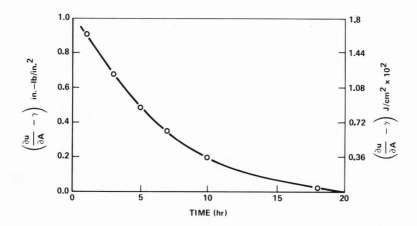

Figure 18. $(\partial U/\partial A - \gamma)$ versus time for the first motor thermally
shocked from $-37°$ to $57°C$ ($-35°$ to $135°F$), cycle number 1.

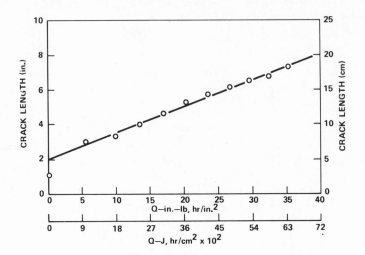

Figure 19. Crack length versus Q for the first motor thermally
            shocked from −37° to 57°C (−35° to 135°F).

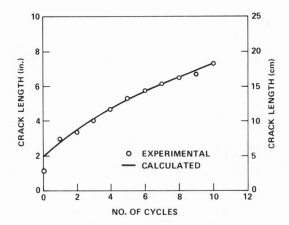

Figure 20. Comparison of experimental and calculated crack length
            versus time for the first motor thermally shocked from
            −37° to 57°C (−35° to 135°F).

Figure 21 presents the $(\partial U/\partial A - \gamma)$ versus crack length for the second motor. The time integral of the quantity $(\partial U/\partial A - \gamma)$ was found to be 1.22×10^3 J, hour/cm^2 (6.7 inch-pound, hour/inch^2) for the first cycle. Figure 22 presents the crack length versus Q for the second motor when thermally shocked between $-37°$ and $57°C$ ($-35°$ and $135°F$). These data indicate Equation (8), with K = 79.19 cm^3/J, hour (0.1760 inch^3/inch-pound, hour), may be used to describe the crack growth for the second motor. A comparison between the experimental values of crack length and those calculated by Equation (8) with K = 79.19 cm^3/J, hour (0.1760 inch ^{3}/ inch-pound, hour is shown in Figure 23. Therefore, the average K for this propellant appears to be 73.71 cm^3/J, hour (0.1638 inch^3/inch-pound, hour). The proportionality factor in terms of crack area is expected to be a constant for different motor sizes. The crack area proportionality factor for the motors used in this investigation was found to be 4.54×10^4 cm^4/J, hour (40.32 inch^4/ inch-pound, hour). Additional investigations will be required to verify if the proportionality factor will be constant for the propellant regardless of motor size, crack location, or crack trajectory. The area factor may be used in conjunction with Equation (8) with A substituted for L in Equation (8).

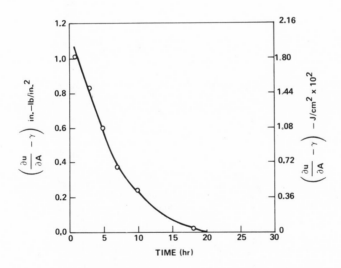

Figure 21. $(\partial U/\partial A - \gamma)$ versus time for the second motor shock
            cycled between $-37°$ to $57°C$ ($-35°$ to $135°F$)
            first cycle.

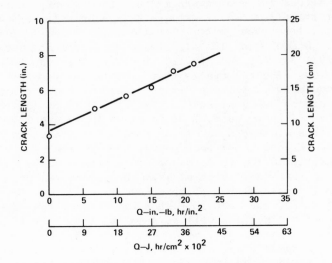

Figure 22.   Crack length versus Q for the second motor thermally
shocked from −35° to 57°C (−35° to 135°F).

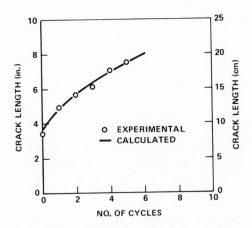

Figure 23.   Comparison of experimental and calculated crack length
versus number of cycles for the second motor thermally
shocked from −37° to 57°C (−35° to 135°F).

V. CONCLUSIONS

The stability of separations that may occur near the propellant-liner interface of grain termination points is a matter of
increased concern in several rocket systems using case-bonded
propellant grains. The strain energy release rate method of fracture mechanics was used to investigate the propagation of bond
separations in a rocket motor when subjected to transient thermal
cycling. Fracture stability analyses were conducted for motors
subjected to transient thermal shock cycles between the temperature
limits of −37° to 57°C and −37° to 29°C (−35° to 135°F and −35° to
85°F). The time dependence of the thermal gradients occurring in
the propellant grain during the cycle is accounted for in the technique used.

Fracture stability analyses were previously conducted for
motor storage at various constant temperatures. Although only the
transient thermal condition is discussed in this paper, conclusions
from the previous work form the basis for this investigation.
These conclusions were as follows: when conditions result in
similar states of stress, the strain energy release rate for
propellant-liner separations was found to vary in direct proportion to the strain energy in the unflawed propellant grain. Therefore, once a strain energy release rate curve is established for
one storage temperature, firing temperature, or transient thermal
cycling condition, the strain energy release rate may be estimated
for other storage temperatures, firing temperatures, or transient
thermal cycling, respectively. The strain energy release rate for
propellant-liner separations varies directly proportional to the
stiffness ratio of the propellant to liner. With the propellant
modulus remaining constant, the strain energy release rate varies
inversely proportional to the liner modulus.

Two motors were used in the experimental procedure. The
motors were initially visually and x-ray inspected to determine the
initial separation length. The first motor contained an initial
separation length of 2.68 cm (1.07 inch)which propagated to 18.15
cm (7.26 inches) after ten thermal shock cycles from −37° to 57°C
(−35° to 135°F). The second motor contained an initial separation
length of 8.45 cm (3.38 inches) which propagated to 18.7 cm (7.47
inches) after five thermal shock cycles from −37° to 57°C (−35° to
135°F). The separation length was determined after each cycle by
x-ray inspection at ambient conditions. The correlation of experimental results indicated that the separations propagate in a
predictable manner. The rate of propagation decreases with
increasing crack length and is proportional to the time integral of
the quantity $(\partial U/\partial A - \gamma)$ during the cycling experiment. The propagation rate constant for the conditions investigated was found to
be 73.71 cm^3/J, hour (0.1638 $inch^3$/inch−pound, hour) based on
crack length, which corresponds to 4.54 x 10^4 cm^4/J, hour

$3x$ .32 inch^4/inch-pound, hour) based on crack area. The use of either of these quantities in Equation (8) closely describes the crack propagation during the thermal cycling experiments.

Additional investigations will be required to verify if the proportionality constant is a propellant property and if it is independent of motor size, crack location, and crack trajectory. If the proportionality factor is a propellant property, analog motors or specialized test specimens may be designed for determining this property.

REFERENCES

1. M. L. Williams, "Initiation and Growth of Viscoelastic Fracture," International Conference on Fracture, Sendai, Japan, 1965.
2. A. A. Griffith, "The Theory of Rupture," Proceedings of the First International Congress of Applied Mechanics, DELFT, 1924, pp. 55-63.
3. E. Orwan, "Fundamentals of Brittle Behavior in Metals, "Fatigue and Fracture of Metals, W. M. Murray, – Editor, John Wiley and Sons, 1952, pp. 139-167.
4. G. R. Irwin, "Fracturing of Metals," Fracture Dynamics American Society for Metals, Cleveland, Ohio, 1968, pp., 147-166.
5. R . S. Rivlin, and A. G. Thomas, "Rupture of Rubber, Part I, Characteristic Energy for Tearing," Journal of Polymer Science, Vol. 10, No. 3, 1953, p. 291.
6. M. L. Williams, "Fatigue-Fracture Growth in Linearly Viscoelastic Materials," Journal of Applied Physics, Vol. 38, No. 11, October 1967, pp. 4476-4480.
7. S. J. Bennett, The Use of Energy Balance in Rocket Motor Grain Integrity Studies, Thiokol Chemical Corporation, Wasatch Division, Brigham City, Utah, CPIA Publication No. 193, Vol. 1, March 1970.
8. S. J. Bennett, and L. H. Layton, A Fracture Mechanics Approach to Surveillance, Thiokol Chemical Corporation, Wasatch Division, Brigham City, Utah, CPIA Publication No. 193, Vol. 1, March 1970.
9. ANON., Final Report Grain Stress Analysis, SLIM Program – Task 2, Thiokol Chemical Corporation, September 1969, TWR-3532.
10. H. M. Westgard, "Bearing Pressures and Cracks," Transactions American Society of Mechanical Engineers, Journal of Applied Mechanics, 1939.
11. W. G. Knauss, "Stresses in an Infinite Strip Containing a Semi-Infinite Crack," Journal of Applied Mechanics, American Society of Mechanical Engineers, Vol. 33, No. 2, June 1966.
12. J. C. Rice, "Stress in an Infinite Strip Containing a Semi-Infinite Crack," Transaction of American Society of Mechanical Engineers, March 1967, pp. 248-249.

13. "Standard Test Method for Tear Resistance of Vulcanized Rubber," ASTM D 624-54.

14. J. J. Brisbane, Heat Conduction and Stress Analysis of Solid Propellant Rocket Motor Nozzles, Rohm and Haas Company, Redstone Research Laboratories, Huntsville, Alabama, February 1969, Report No. S-198.

15. J. J. Brisbane, Advances in Stress Analysis of Solid Propellant Grains, Rohm and Haas Company, Redstone Research Laboratories, Huntsville, Alabama, September 1970, Report No. S-268.

THERMOVISCOELASTIC INTERACTION EFFECTS IN FILLED POLYMERS

*W. L. Hufferd **E. C. Francis

*University of Utah, Salt Lake City, Utah
**Chemical Systems Division, United Technologies
Corporation, Sunnyvale, California

ABSTRACT

Experimental data indicate that stress response predictions
based on the normal assumption of thermorheologically simple
material behavior can underpredict observed stress response by a
factor of two or more when applied to combined thermal and
mechanical load histories. In highly filled polymeric systems,
such as solid propellants, combined thermal and mechanical syner-
gistic effects induce high local stress gradients in the polymer
binder between filler particles that are not accounted for in
conventional analysis methods. An experimental and analytical
methodology is demonstrated which accounts for these interaction
effects and is applied to slow thermal-mechanical loading of a
solid propellant rocket motor.

INTRODUCTION

Many filled polymeric systems exhibit nonlinear viscoelastic
mechanical response behavior due to complex interactions between
the relatively hard filler particles and the soft binder matrix;
even under isothermal test conditions and at low strain levels.
In many instances, however; particularly for monotonic mechanical
loading histories, acceptable stress predictions are often obtained
assuming linear viscoelastic material behavior. The observed
nonlinear behavior is more pronounced under combined thermal and
mechanical loading histories, such as, for example, simultaneous
cooling and straining of a uniaxial tensile bar. This type of
loading history is typically considered within linear thermoviscoelasticity theory assuming thermoheologically simple
material (TSM) behavior. From a continuum mechanics point of

view this assumption implies an equivalence between time and temperature and the existence of a monotonic "time-temperature shift-
function", A_T defining this equivalence (see, e.g.,[1]). It has
been widely demonstrated that the assumption of TSM can under
predict stress response during simultaneous cooling and mechanical
straining of uniaxial bar by a factor of two or more [2-5] and
several analysis procedures have been proposed in an attempt to
account for the observed "thermorheologically complex material
(TCM) behavior" [2,3,5-7]. The most rigorous theoretical approach
is that developed by Schapery [6,7]; although it has only recently
been applied to any real material behavior [8]. This paper
summarizes the development of the approach used in [8] and compares
stress predictions with measured stresses in a structural analog
device.

The typical magnitude of the discrepancy between linear
thermoviscoelastic stress response predictions and measured
stresses in a uniaxial bar simultaneously cooled and strained
mechanically is shown in Figure 1 to illustrate the significance
of the problem.

THEORETICAL DEVELOPMENT

A thermoviscoelastic analysis procedure is described for a
TCM in which the temperature is transient, but spatially uniform.
For our present purposes it is sufficient to consider one-
dimensional loading.

The stress response of a uniaxial tensile bar simultaneously
subjected to a mechanical strain history, ε, and an applied temperature change, $\Delta T = T - T_o$, is given by

$$\sigma(t) = E_e \ (\varepsilon - \alpha \Delta T) + A_F \int_o^t \Delta E(\xi - \xi') \frac{d}{d\tau} \ (\varepsilon - \alpha \Delta T) \ d\tau \qquad (1)$$

where
$\quad$ E_e = Equilibrium value of relaxation modulus
$\quad$ $\Delta E(\xi) = E(\xi) - E_e$ = transient component of relaxation modulus
$\quad$ ξ, ξ' = "temperature reduced times", defined by

$$\xi - \xi' = \int_\tau^t \frac{dt'}{A_T \ [T(t)]} \qquad (2)$$

A_T = time-temperature shift factor, and

A_F = temperature and mechanical loading history coupling
$\qquad$ coefficient (A_F = 1 for thermorheologically simple material
$\qquad$ behavior)

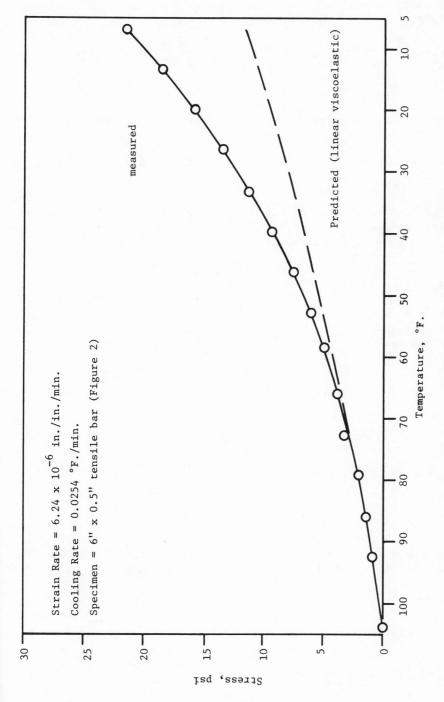

Strain Rate = 6.24 x 10^{-6} in./in./min.

Cooling Rate = 0.0254 °F./min.

Specimen = 6" x 0.5" tensile bar (Figure 2)

measured

Predicted (linear viscoelastic)

Temperature, °F.

Stress, psi

Figure 1. 64-Hour Straining-Cooling of TP-H1011 Solid Propellant

Assuming a constant (uniform) cooling rate and a constant mechanical straining rate, the effective modulus, E_{eff} is given by

$$E_{eff}(t) = \frac{\sigma(t)}{\varepsilon(t) - \alpha \Delta T} = E_e + \frac{A_F}{t} \int_0^t \Delta E(\xi - \xi') d\tau \tag{3}$$

The coefficient A_F is defined to be unity under isothermal conditions. Also, $\xi - \xi' = t/A_T - \tau/A_T$ for isothermal conditions. Thus, at constant temperature, (3) may be written, with a change of variables,

$$E_{eff}(t) = \frac{\sigma(t)}{\varepsilon(t)} = E_e + \frac{1}{\xi} \int_0^\xi \Delta E(u) \, du \tag{4}$$

which is the secant modulus, E_S at the given time and constant temperature. Under transient thermal conditions with $A_T = A_T[T(t)]$ and $A_F = 1$, thermorheologically simple behavior results, and

$$E_{eff}(t) = E_e + \frac{1}{t} \int_0^\xi \Delta E(u) \, A_T(\tau) \, du \tag{5}$$

If the equilibrium modulus is temperature dependent, that is, $E_e = E_e(T)$, then (3) may be rewritten

$$E_{eif} = A_F \left[E_e + \frac{1}{t} \right] \int_0^t \Delta E(\xi - \xi') \, d\tau = A_F E_s(t) \tag{6}$$

where

$$E_s(t) = \frac{\sigma(t)}{\varepsilon(t) - \alpha \Delta T(t)} \tag{7}$$

In general A_F may be a function of temperature, T, rate of change of temperature, $\dot{T}$, strain rate, $\dot{\varepsilon}$, and possibly strain level, ε:

$$A_F = A_F(T, \dot{T}, \dot{\varepsilon}, \varepsilon) \tag{8}$$

The effective modulus is determined from (6) most readily by assuming a power law representation for the transient component of the relaxation modulus,

$$\Delta E(t) = E_o \, t^{-n} \tag{9}$$

and a modified power law representation for the temperature shift factor [7-8]

$$A_T = \left(\frac{T_R - T_a}{T - T_a} \right)^m \tag{10}$$

Using (9) and (10) and making a change of variables from time to temperature such that

$$\Delta T_n = \frac{T - T_o}{T_o - T_a} \tag{11}$$

equations (2) and (6) can be integrated to give

$$E_{eff} = A_F \left\{ E_e + I_T (E_s - E_e) \right\} \tag{12}$$

where

$$I_{T_c} = \frac{1-n}{(1+m)^{1-n}} \left(-1 - \frac{1}{\Delta T_n}\right)^{1-n} \int_{(1+\Delta T_n)^{1+m}}^{1} (1-x)^{-n} (x)^{\frac{m}{1+m} + n-2} \, dx \tag{13}$$

for cooling below the stress free temperature T_o, (i.e., $-1 < \Delta T_n < 0$), and

$$I_{T_H} = \frac{1-n}{(1+m)^{1-n}} \left(1 + \frac{1}{\Delta T_n}\right)^{1-n} \int_{1}^{(1+\Delta T_n)^{1+m}} (x-1)^{-n} (x)^{\frac{m}{1+m} + n-2} \, dx \tag{14}$$

for heating above the stress free temperature T_o, i.e., $0 < \Delta T_n < 1$), and

$$E_s - E_e = \frac{\Delta E(t/A_T)}{1 - n} = \frac{E_o(t/A_T)^{-n}}{1 - n} \tag{15}$$

If the equilibrium modulus is not temperature dependent then E_{eff} is given by

$$E_{eff} = E_e + A_F \, I_T (E_s - E_e) \tag{16}$$

with (13), (14) and (15) unchanged.

The stress response for temperature loading is given by

$$\sigma(t) = E_{eff}(t) \left[\varepsilon(t) - \alpha \, \Delta T(t) \right] \tag{17}$$

The influence of the temperature history is incorporated in E_{eff} through the incomplete elliptic integrals (13) or (14).

LABORATORY CHARACTERIZATION PROCEDURES

The procedure for calculating the effective modulus, E_{eff}, involves determining if the equilibrium modulus is temperature dependent and conducting simultaneous cooling and straining tests from which the coefficient A_F is determined by comparing stress response predictions with the experimentally observed stress response.

The temperature dependence of E_e is determined in an iterative fashion. Stress relaxation tests are conducted at various temperatures spanning the anticipated temperature range following standard procedures. Temperature time shift factors, A_T, are next determined empirically and a plot of master relaxation modulus versus temperature-reduced time, $\xi = t/A_T$, constructed. A reasonably straight line of the master modulus on a log-log plot indicates that E_e is not strong function of temperature, whereas significant curvature indicates that $E_e = E_e (T)$.

In the former case, that is, $E_e \neq E_e(T)$, the master relaxation modulus is curve-fit to the power law representation

$$E(t) = E_e + E_o \ \xi^{-n} \qquad\qquad (18)$$

Since (18) contains three parameters the curve-fit is normally obtained by trial and error. One procedure is to assume several values for E_e and perform at least squares curve fit to

$$\Delta E(\xi) = E(\xi) - E_e = E_o \ \xi^{-n} \qquad\qquad (19)$$

and selecting the value for E_e that gives the best correlation coefficient. Alternatively, E_e may be estimated graphically by plotting $E(t)$ versus ξ^{-n} and extrapolating to $\xi^{-n} = 0$, or some small number.

If $E_e = E_e (T)$ the above curve-fit procedure is applied at each test temperature before constructing the master relaxation modulus curve. A plot of E_e versus temperature may then be constructed which may be curve-fit to a polynomial or exponential function if desired. The temperature shift factors are determined experimentally by shifting $\Delta E(t) = E(t) - E_e (T)$ at each temperature to form the master modulus curve at some reference temperature T_R. Note that the resulting shift factor curve is not the same as that obtained by shifting the modulus data before subtracting $E_e(T)$ at each temperature. The master modulus is then curve-fit to the power law

$$\Delta E(\xi) = E_o \ \xi^{-n} = E_o (t/A_T)^{-n} \qquad\qquad (20)$$

Simultaneous cooling and straining tests are required to evaluate the shift function A_F. These tests should be conducted at several cooling rates and strain rates which encompass the anticipated temperature range, cooling rate, and low temperature strain levels.

Linear thermoviscoelastic predictions of the stress response for simultaneous cooling and straining are made using (17) with E_{eff} given by (12) or (16) with $A_F = 1$, and E_e the appropriate value at each temperature. The shift function A_F is then determined by comparing the predicted response with the observed response. For the case of a temperature independent equilibrium modulus,

$$A_F = \frac{\sigma(t) - E_e \ (\varepsilon - \alpha\Delta T)}{I_T(E_s - E_e) \ (\varepsilon - \alpha\Delta T)} \tag{21}$$

If $E_e = E_e \ (T)$, then

$$A_F = \frac{\sigma(t)}{E_e + I_T(E_s - E_e) \ (\varepsilon - \alpha\Delta T)} \tag{22}$$

The shift function, A_F, may either be tabulated or curve-fit for use in motor stress predictions. Acceptable results have been obtained using a constant value if the variation in A_F is of the order of 10 to 15% over the temperature range of interest [8].

EXAMPLE CALCULATIONS

In order to illustrate the previous analysis and characterization procedures, tests were conducted on TP-H1011 solid propellant and stress response predictions were made for slow cooling of the structural analog device shown in figure 2 and compared with stress measurements obtained from embedded stress transducers.

Stress relaxation tests were conducted over a temperature range of -45 to $152^{\circ}F$ on TP-H1011 propellant that had been stored at $100^{\circ}F$ for two years. The master stress relaxation modulus curve versus log t/A_T is shown in figure 3. The strong curvature, which negates fitting it to a single power law, suggests a temperature dependent equilibrium modulus. Therefore several values of E_e were assumed at each test temperature and a least squares curve fit to

$$E \ (\xi) = E \ (\xi) - E_e = E_o \ (t/A_T) \tag{23}$$

was performed at each temperature. The values of E_e selected were those that gave the best squares correlation coefficient. Figure 4 shows E_e as a function of temperature. To facilitate calculations on a programmable calculator, the equilibrium modulus was curve fit to the following polynomial

$$E_e = 365 - 7.08T + 0.115T^2 - 1.16 \times 10^{-3}T^3 + 4.65 \times 10^{-6}T^4 \tag{24}$$

A new master stress relaxation modulus curve was constructed which was curve fit to (9) and the corresponding shift factor curve was curve fit to (10):

$$\Delta E = 465(t/A_T)^{-0.24} \tag{25}$$

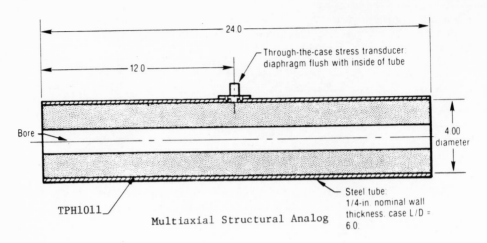

Multiaxial Structural Analog

Uniaxial Tensile Bar

Figure 2.   Simultaneous Straining-Cooling Test Specimens

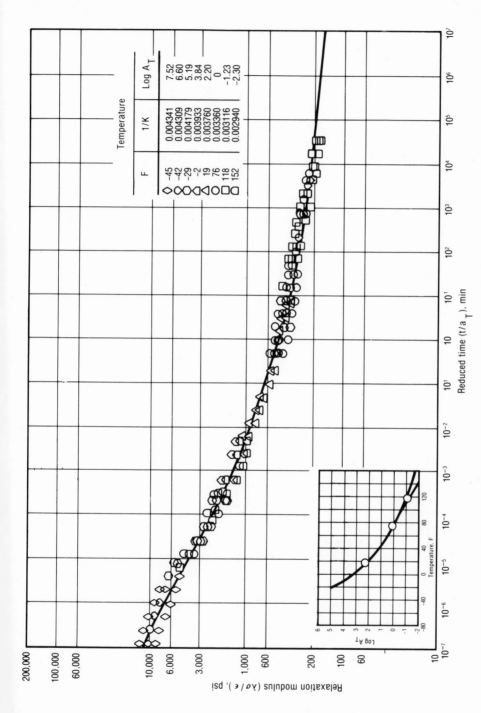

Figure 3. Master Modulus Data for TP H1011 at 3% Strain

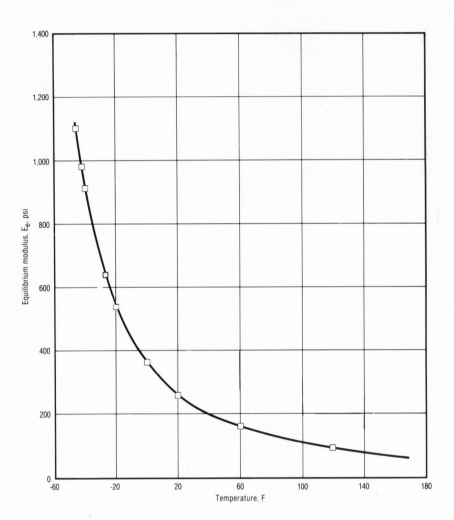

Figure 4.   Equilibrium Modulus versus Temperature,
TP-H1011

$$A_T = \left(\frac{146}{T + 70}\right)^6 \tag{26}$$

Simultaneous cooling and straining tests were conducted on uniaxial tensile bars with the results shown in figures 5, 6 and 7 for 6, 40 and 64-hour tests respectively.

The thermal-mechanical coupling coefficient, A_F, was calculated from the data presented in figures 3 through 7. The A_F's calculated varied from 1.48 to 1.76 with a mean value of 1.62, which was used in subsequent structural analog device stress predictions.

Once the coupling coefficient, A_f, is determined from laboratory tests, the effective modulus can be calculated for any temperature history. It has been demonstrated [8] that acceptable results may be obtained for slow cooling of the structural analog shown in Figure 2 using the quasi-elastic method of analysis. In this case, assuming incompressible material behavior, the stress response for any cooling history can be determined from a single elastic finite element analysis using

$$\sigma_{ij}(t) = (\sigma_{ij})_{FEM} \frac{E_{eff}(t)}{E_{FEM}} \frac{\alpha_p}{\alpha_{FEM}} \frac{T(t)}{\Delta T_{FEM}} \tag{27}$$

where the subscript FEM indicates the values used in the original elastic finite element analysis.

The radial bond stress at the mid-plane of the circular port structural analog was measured for slow cooling from $82^\circ F$ to $0^\circ F$ using through-the-case stress transducers [9,10,11]. The predicted stress response was calculated using (27) and the results are compared in Figure 8. Linear thermoviscoelastic stresses ($A_F = 1$) are also shown in Figure 8.

CONCLUSIONS

It may be seen from Figure 8 that the transient and peak stresses are predicted quite well using the procedure described herein, where as the stresses in both cases are significantly underpredicted using linear thermoviscoelasticity theory.

The relaxation behavior following the cooldown is over-predicted by the procedure described herein. This may be in part due to inherent nonlinearities in the material and in part due to the gradual decay of A_F from 1.62 to 1, once the constant temperature of $0^\circ F$ is achieved. This consideration was neglected in making the calculations shown in Figure 8. The agreement is still, however, within 15% of the observed stress relaxation behavior.

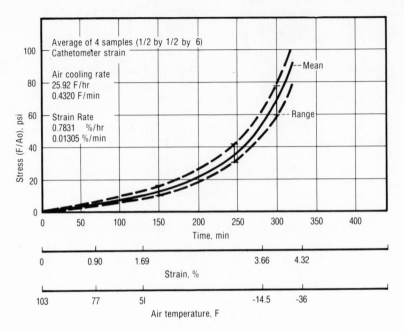

Figure 5.   Straining–Cooling Test on TP
            H1011 (6 hours)

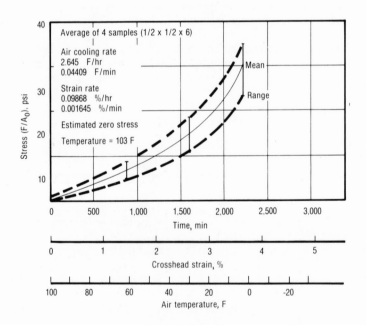

Figure 6.   Straining–Cooling Test
            on TP H1011 (40 hours)

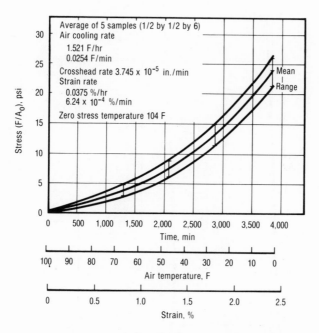

Figure 7. Uniaxial Straining–Cooling
Test on TP H1011 (64 hours)

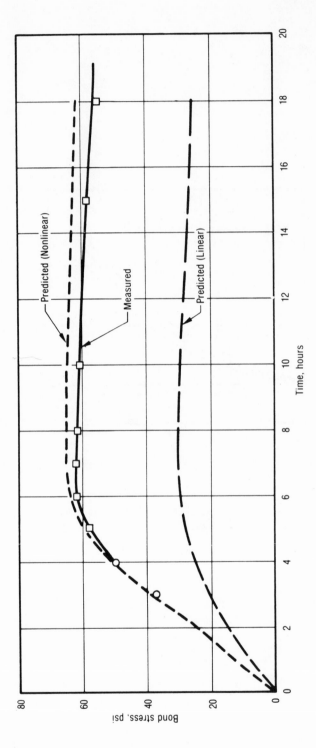

Figure 8.   Comparison of Measured and Predicted Bond Stresses for
Cooling from 82°F to 0°F of 0.875-in. Circular Port,
Case-Bonded Analog Motor

REFERENCES

1. Morland, L.W. and Lee, E.H., "Stress Analysis for Linear Viscoelastic Materials with Temperature Variation", _Trans. Soc. Rheology_, Vol. 4, pp. 233-263, 1960.

2. Bornstein, G.A., "Transient Thermoviscoelastic Analysis of A Uniaxial Bar", _Bulletin of the 7th JANNAF Mechanical Behavior Working Group Meeting_, CPIA Publ. No. 177, pp. 23, 1968.

3. Martin, D.L., "An Approximate Method of Analysis of NonLinear Transient Thermoviscoelastic Behavior", _Bulletin of the 8th JANNAF Mechanical Behavior Working Group Meeting_, CPIA Publ. No. 193, Vol. 1, pp. 45-52, 1969

4. Leeming, H., et. al., "Solid Propellant Structural Test Vehicle and Systems Analysis", AFRPL TR-70-10, March 1970.

5. Lee, T.Y., "Coupled Thermomechanical Effects in High Solids Propellant", _AIAA Journal_, Vol. 17, pp. 1015-1017, 1979

6. Schapery, R.A., "A Theory of Nonlinear Thermoviscoelasticity Based on Irreversible Thermodynamics", _Proc. 5th U.S. Nat'l. Congress Appl. Mech._, pp. 511-530, 1966.

7. Hufferd, W.L. and Fitzgerald, J.E., (Editors), "_JANNAF Solid Propellant Structural Integrity Handbook_, CPIA Publication No. 230, September 1972.

8. Francis, E.C., et. al., "Predictive Techniques for Failure Mechanisms in Solid Rocket Motors", AFRPL TR-79-87, January 1980.

9. Francis, E.C. and Thompson " Solid Propellant Stress Transducer Evaluation", 24th Industrial Instrumentation Symposium, Instrument Society of America, Albuquerque, New Mexico, 1978.

10. Francis, E.C. and Chelner, H. "Improved Stress Transducer Design and Stability Evaluation" 10th Transducer Workshop, Sponsored by Transducer Committee of the Telemetry Group Reporting to Range Commander Council, U.S. Air Force Academy, Colorado Springs, Colorady, June 1979.

11. Francis, E.C., Thompson, R.E. and Briggs, W.E., "The Development of Improved Normal Stress Transducer for Propellant Grains", AFRPL-TR-79-34, June 1979

ALLOWABLE STRENGTH OF VISCOELASTIC MATERIALS UNDER VARIABLE THERMAL LOADS

Robert A. Heller and Mahendra P. Singh

Department of Engineering Science and Mechanics
Virginia Polytechnic Insttitute & State University
Blacksburg, Virginia   24061

INTRODUCTION

Solid propellant rocket motors stored outdoors without thermal protection are subjected to variable amplitude thermal stresses. It has been shown[1,2] that the storage life of such motors can be calculated from the probability of failure that in turn is determined from a stress-strength interference analysis. Such a techique compares the statistical distribution of induced thermal stresses with the distribution of material strength. For visco-elastic materials both the induced stress and the allowable strength are functions of temperature and rate of loading as well as aging and cumulative damage.

To determine the allowable strength of the material, experiments are usually conducted at several temperatures under a number of constant strain-rate levels.

Because temperature variations are quite rapid, as has been shown by Cost[1], the resulting thermal deformations are induced under conditions of variable strain-rate.

The purpose of this paper is to propose a rational method for the evaluation of the allowable strength under such variable strain-rates.

Linear damage accumulation under variable strain-rate

The linear accumulation of fatigue damage has been suggested by Palmgren[3] and Miner[4] for aircraft structural materials under cyclic loading. A similar damage accumulation rule has been proposed for rocket propellants under piece-wise constant loads by Bills[5]. These rules state that the damage produced by a constant amplitude stress in a single load cycle, in the case of fatigue, and by a constant load applied for unit time, for a propellant, is inversely proportional to the number of stress cycles to failure and the total time to failure respectively,

$$\Delta_f = \frac{1}{N_f} \quad \text{(fatigue)} \tag{1}$$

and

$$\Delta_p = \frac{1}{t_f} \quad \text{(propellant)} \tag{2}$$

It is further postulated that damage is accumulated linearly and failure under variable loads will occur when total damage becomes 100%. Hence

$$1 = \sum_{i=1}^{m} \frac{p_i n_v}{N_{fi}} \tag{3}$$

and

$$1 = \sum_{i=1}^{m} \frac{p_i t_v}{t_{fi}} \tag{4}$$

where p_i is the fraction of total cycles n_v, or total time t_v to failure under a sequence of m load levels.

In a similar manner a failure stress will be calculated for variable strain-rates. Tests conducted at constant strain rates, $\dot{\varepsilon}$, to failure at various temperature levels result in data that can be plotted in the form of a master curve with the aid of the viscoelastic time-temperature shift function, a_T. Typically such a curve when plotted on double logarithimic scales forms a straight line as shown schematically in Fig. 1. Its equation may be written as

$$\log \dot{\varepsilon} a_T = \rho \log \left[\frac{T_o}{T} (S_f - S_o) \right] + C \tag{5}$$

or $\quad S_f = AT(\dot{\varepsilon} a_T)^\nu + S_o \tag{6}$

where S_f is the failure strength at the constant strain rate $\dot{\varepsilon}$, S_o

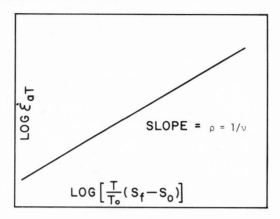

Fig. 1   Allowable Strength Under Constant Strain Rate.

is the failure stress at a very low strain rate, T is the absolute
temperature, T_o is the absolute reference temperature at which the
master curve is constructed, ρ is the slope of the line and C is
the intercept at log $\dot{\varepsilon}a_T = 0$. In the alternate form of the relation the constant A incorporates the antilog of C and T_o and $\nu=1/\rho$.

If a test is conducted at a strain rate $\dot{\varepsilon}_1$ at which the failure stress is S_{f1} but a stress $S_1 < S_{f1}$ is applied, the material
will be partially damaged as shown in Fig. 2. The partial damage,
Δ_1 is assumed to be proportional to the applied stress S_1 and inversely proportional to the failure stress (100% damage), S_{f1}.
From proportional triangles

$$\Delta_1 = \frac{S_1}{S_{f1}} \tag{7}$$

a change of strain rate to $\dot{\varepsilon}_2$ will change the slope of the damage
line in Fig. 2. Additional stress, ΔS, may be applied until 100%
damage is reached as indicated in Fig. 2. The additional damage
$1-\Delta_1$ is proportional to total damage $\Delta = 1$ as the additional
stress, ΔS, is to the failure stress S_{f2}. Hence

$$1 - \Delta_1 = \frac{\Delta S}{S_{f2}} \tag{8}$$

Using Eqs. 7 and 8 the failure strength in the two strain-rate

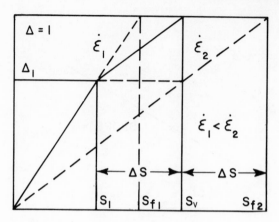

Fig. 2   Cumulative Damage Under Constant Strain Rate.

test is consequently

$$S_v = S_1 + \Delta S = S_1 + S_{f2}(1 - \frac{S_1}{S_{f1}}) \tag{9}$$

The same result is obtained when partial damages are added

$$\frac{S_1}{S_{f1}} + \frac{(S_v - S_1)}{S_{f2}} = 1 \tag{10}$$

Hence for variable strain-rates with monotonically increasing stress, at failure ($\Delta = 1$)

$$\sum_{i=1}^{m} \frac{p_i S_v}{S_{fi}} = 1 \tag{11}$$

results, with p_i the fraction of the cumulative failure stress S_v reached under constant strain-rate $\dot{\varepsilon}_i$.

For continuously varying stain-rates the fraction p_i is replaced by the probability that the strain rate is between two adjacent levels $\dot{\varepsilon}$ and $\dot{\varepsilon} + d\dot{\varepsilon}$ and the summation is replaced by integration

$$S_v \int \frac{f(\dot{\varepsilon})d\dot{\varepsilon}}{S_f} = 1 \tag{12}$$

Strain and strain rate in the storage environment.

Ambient temperature variations consist of three major components; the annual mean, T_m, and two superimposed cyclic temperatures, the seasonal and diurnal cycles. At any point within the motor temperatures will be alternated and delayed but will have a similar form

$$T = T_m + T_y \cos\omega_y t + T_d \cos\omega_d t \tag{13}$$

with $\omega_y = 2\pi/8760$ and $\omega_d = 2\pi/24$. The resulting thermal strains will also consist of three terms as shown in Fig. 3.

$$\varepsilon = \varepsilon_m + \varepsilon_y \cos\omega_y t + \varepsilon_d \omega_d \cos\omega_d t \tag{14}$$

The strain rate, obtained by differenciation may be expressed as

$$\dot{\varepsilon} = -\varepsilon_y \omega_y \sin\omega_y t - \varepsilon_d \omega_d \sin\omega_d t \tag{15}$$

During a 24 hour period the strain rate attributed to the seasonal cycle changes only slightly and can be considered to be constant, that is, for a one day period

$$\dot{\varepsilon} = \dot{\varepsilon}_c + \dot{\varepsilon}_d \sin\omega_d t \tag{16}$$

as indicated in Fig. 4.

The probability of finding strain-rate between $\dot{\varepsilon}$ and $\dot{\varepsilon} + d\dot{\varepsilon}$ is equivalent to the fraction of time spent in the time interval dt (Fig. 4)[6]. Differentiating Eq. 16

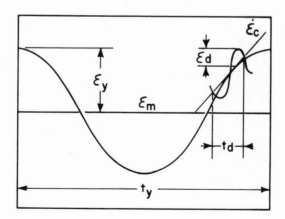

Fig. 3 Seasonal and Diurnal Strain Cycles.

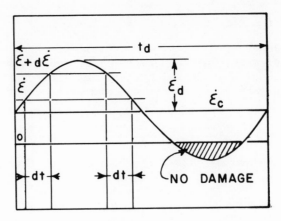

Fig. 4 Diurnal Strain Rate Variation.

$$d\dot{\varepsilon} = (\dot{\varepsilon}_d \omega_d \cos\omega_d t)dt \tag{17}$$

from which

$$dt = \frac{d\dot{\varepsilon}}{\dot{\varepsilon}_d \omega_d \cos\omega_d t} \tag{18}$$

strain rate will pass through the $d\dot{\varepsilon}$ region twice during each cycle. Hence the fraction of time per period, t_d becomes, with $t_d = 2\pi/\omega_d$.

$$\frac{2dt}{t_d} = \frac{2d\dot{\varepsilon}}{2\pi\dot{\varepsilon}_d \cos\omega_d t} \tag{19}$$

which is the probability $f(\dot{\varepsilon})d\dot{\varepsilon}$. Using the trigonometric identity $\sin^2\alpha + \cos^2\alpha = 1$ and recognizing from Eq. 16 that

$$\sin\omega_d t = \frac{\dot{\varepsilon}-\dot{\varepsilon}_c}{\dot{\varepsilon}_d} \tag{20}$$

$$f(\dot{\varepsilon})d\dot{\varepsilon} = \frac{d\dot{\varepsilon}}{\pi\dot{\varepsilon}_d \sqrt{1-(\frac{\dot{\varepsilon}-\dot{\varepsilon}_c}{\dot{\varepsilon}d})^2}} \tag{21}$$

The probability density function, $f(\dot{\varepsilon})$, is shown in Fig. 5.

For monotonically increasing stresses failure under negative strain-
rates is undefined; the portion of the density function for $\varepsilon < 0$
has to be truncated. Since the area under the truncated distribution has to be unity, the normalizing constant, C, is calculated
as

$$C \int_0^{\dot\varepsilon_c + \dot\varepsilon_d} \frac{d\dot\varepsilon}{\dot\varepsilon_d \sqrt{1-(\frac{\dot\varepsilon-\dot\varepsilon_c}{\dot\varepsilon_d})^2}} = 1 \tag{22}$$

With a change of variables

$$\frac{\dot\varepsilon-\dot\varepsilon_c}{\dot\varepsilon_d} = \cos\theta, \quad d\dot\varepsilon = -\dot\varepsilon_d \sin\theta d\theta \quad \text{and} \quad \dot\varepsilon = \dot\varepsilon_d\cos\theta + \dot\varepsilon_c \tag{23}$$

Eq. 22 becomes

$$\frac{C}{\pi} \int_0^{\cos^{-1}(-\frac{\dot\varepsilon_c}{\dot\varepsilon_d})} d\theta = 1 \tag{24}$$

from which

$$C = \pi/\cos^{-1}(-\frac{\dot\varepsilon_c}{\dot\varepsilon_d}) \tag{25}$$

as a result the truncated density function of positive strain rates
may be written as

$$f'(\dot\varepsilon) = \frac{1}{\cos^{-1}(-\frac{\dot\varepsilon_c}{\dot\varepsilon_d})\dot\varepsilon_d\sqrt{1-(\frac{\dot\varepsilon-\dot\varepsilon_c}{\dot\varepsilon_d})^2}} \quad 0 < \dot\varepsilon < (\dot\varepsilon_c + \dot\varepsilon_d) \tag{26}$$

The average positive strain rate $\dot\varepsilon_{AV}$ may now be calculated as the
expected value of the truncated density function, $\dot\varepsilon_{AV} = E[\dot\varepsilon]$ or

$$\dot\varepsilon_{AV} = \frac{1}{\cos^{-1}(-\frac{\dot\varepsilon_c}{\dot\varepsilon_d})} \int_0^{\dot\varepsilon_c + \dot\varepsilon_d} \frac{\dot\varepsilon d\dot\varepsilon}{\dot\varepsilon_d\sqrt{1-(\frac{\dot\varepsilon-\dot\varepsilon_c}{\dot\varepsilon_d})^2}} \tag{27}$$

Utilizing the changed variables of Eq. 23

$$\dot{\varepsilon}_{AV} = \frac{1}{\cos^{-1}(-\frac{\dot{\varepsilon}_c}{\dot{\varepsilon}_d})} \int_0^{\cos^{-1}(-\frac{\dot{\varepsilon}_c}{\dot{\varepsilon}_d})} (\dot{\varepsilon}_d \cos\theta + \dot{\varepsilon}_c) d\theta \qquad (28)$$

from which

$$\dot{\varepsilon}_{AV} = \dot{\varepsilon}_c + \dot{\varepsilon}_d \frac{\sin[\cos^{-1}(-\frac{\dot{\varepsilon}_c}{\dot{\varepsilon}_d})]}{\cos^{-1}[-\frac{\dot{\varepsilon}_c}{\dot{\varepsilon}_d}]} \qquad (29)$$

Failure strength under sinusoidal straining.

Substituting the probability density function, Eq. 26 and the failure strength relation, Eq. 6 into the cumulative damage expression, Eq. 12

$$\frac{S_v}{\cos^{-1}(-\frac{\dot{\varepsilon}_c}{\dot{\varepsilon}_d})\dot{\varepsilon}d} \int_0^{\dot{\varepsilon}_d + \dot{\varepsilon}_c} \frac{d\dot{\varepsilon}}{\sqrt{1-(\frac{\dot{\varepsilon}-\dot{\varepsilon}_c}{\dot{\varepsilon}_d})^2}\ [AT(\dot{\varepsilon}a_T)^\nu + S_0]} = 1 \qquad (30)$$

results.

The lower limit of integration is zero, because negative strain rates are not considered to contribute to tensile damage. The upper limit is the largest strain rate as shown in Fig. 5. Introducing the ˹hange of variables of Eq. 23., Eq. 30 can be rewritten as

$$\frac{S_v}{\cos^{-1}(-\frac{\dot{\varepsilon}_c}{\dot{\varepsilon}_d})\dot{\varepsilon}_d} \int_{\cos^{-1}(-\frac{\dot{\varepsilon}_c}{\dot{\varepsilon}_d})}^0 \frac{-\dot{\varepsilon}_d \sin\theta d\theta}{\sin\theta\{ATa_T^\nu[\dot{\varepsilon}_d\cos\theta + \dot{\varepsilon}_c]^\nu + S_0\}} = 1 \qquad (31)$$

Interchanging the limits and simplifying

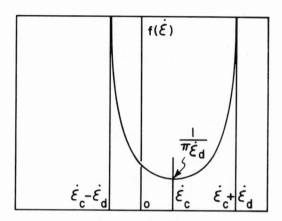

Fig. 5 Probability Density Function of Strain Rates per Diurnal Cycle. Cycle.

$$
\frac{S_v}{\cos^{-1}(-\frac{\dot{\varepsilon}_c}{\dot{\varepsilon}_d})AT(\dot{\varepsilon}_d a_T)^\nu} \int_0^{\cos^{-1}(-\frac{\dot{\varepsilon}_c}{\dot{\varepsilon}_d})} \frac{d\theta}{(\cos\theta + \frac{\dot{\varepsilon}_c}{\dot{\varepsilon}_d})^\nu + \frac{S_0}{AT(a_T\dot{\varepsilon}_d)^\nu}} = 1
$$

(32)

After numerical integration for given values of the parameters, S_v can be calculated.

It is recognized that a_T and T as well as $\dot{\varepsilon}$ are functions of time; however, in Eq. 32 the variations of the shift function and the temperature during the day are not considered. An average daily temperature and corresponding a_T value is suggested. It is possible to include temperature variations during the day into the integration of Eq. 31 with an increased amount of computational effort.

Numerical Examples

The failure strength for the hottest and coldest days of the year will be calculated. For these two days $\dot{\varepsilon}_c = 0$ and Eq. 32 reduces to

$$\frac{2S_v}{\pi A T (\dot{\varepsilon}_d a_T)^\nu} \int_0^{\pi/2} \frac{d\theta}{\cos\theta^\nu + \dfrac{S_0}{A T (a_T \dot{\varepsilon}_d)^\nu}} = 1 \tag{33}$$

For a typical solid propellant the shift function has the WLF form

$$\log a_T = \frac{-6.8(T-T_0)}{1.46+(T-T_0)} \tag{34}$$

for a reference temperature $T^\circ = 298^\circ K$. The parameters of Eq. 6 are taken as $A = 600$, $\nu = 1/8$, $S_0 = .35$ MN/M^2 also for a reference temperature of $298^\circ K$. Assuming further that the average temperatures on the hottest and coldest days of the year are $40^\circ C$ (desert) and $-40^\circ C$ (arctic) respectively and that the daily amplitude of temperature is $10^\circ C$ an approximate strain amplitude, ε_d, can be calculated from the simple relation

$$\varepsilon_d = \alpha \Delta T \tag{35}$$

where $\alpha = 1.5 \times 10^{-4}$ $M/M/^\circ C$ is the thermal coefficient of expansion. Hence

$$\varepsilon_d = 1.5 \times 10^{-3} \text{ M/M and } \dot{\varepsilon}_d = \varepsilon_d \omega_d = \varepsilon_d \frac{2\pi}{24}$$

and

$$\dot{\varepsilon}_d = 3.93 \times 10^{-4} \text{ (M/M)/HR}$$

Eq. 29 yields an average positive strain rate of $\dot{\varepsilon}_{AV} = 2/\pi \dot{\varepsilon}_d = 2.5 \times 10^{-4}$ M/M/HR. An identical average positive strain rate can be obtained by evaluating the slope of the chord of the cosine function (Fig. 4) at one quarter of the period, $t = 6$ HRS.

$$\dot{\varepsilon}_{AV} = \frac{\varepsilon_d}{6} = \frac{1.5 \times 10^{-3}}{6} = 2.5 \times 10^{-4} \text{ M/M/HR}$$

Calculations have been carried out in order to compare allowable strength values for three strain rates: $\dot{\varepsilon}_{AV}$, $\dot{\varepsilon}_{CUMUL}$ and $\dot{\varepsilon}_d$ and for S_{AV}. Results are presented in Table 1.

Table I. Comparison of Allowable Strength Values

T° K	a_T	$\dot{\varepsilon}_d$ M/M/HR	$\dot{\varepsilon}_{AV}$ M/M/HR	S_v MN/M^2 for $\dot{\varepsilon}_d$	$\dot{\varepsilon}_{AV}$	$\dot{\varepsilon}_{CUMUL}$	S_{AV}
233	2.86×10^5	390×10^{-6}	250×10^{-6}	.602	.588	.581	.583
313	2.33×10^{-1}	393×10^{-6}	250×10^{-6}	.409	.406	.404	.404

Considering $\dot{\varepsilon}$ as a random variable, the various values of S_V are mathematically defined as

$$S_V(\dot{\varepsilon}_d) = AT\ a_T^\nu(\dot{\varepsilon}_d)^\nu + S_0 \tag{36}$$

$$S_V(\dot{\varepsilon}_{AV}) = S_f(\dot{\varepsilon}_{AV}) = AT\ a_T^\nu(\dot{\varepsilon}_{AV})^\nu + S_0 \tag{37}$$

$$S_{VC} = S_V(\dot{\varepsilon}_{CUMUL}) = 1/E\{1/S_f(\dot{\varepsilon})\} \tag{38}$$

$$S_{AV} = E[S_f(\dot{\varepsilon})] = \int S_F(\dot{\varepsilon})f_{\dot{\varepsilon}}(\dot{\varepsilon})d\dot{\varepsilon}$$

$$= AT\ a_T^\nu \int_{\dot{\varepsilon}_c}^{(\dot{\varepsilon}_c+\dot{\varepsilon}_d)} (\dot{\varepsilon})^\nu f_{\dot{\varepsilon}}(\dot{\varepsilon})d\dot{\varepsilon} + S_0$$

$$= AT\ a_T^\nu\ m_\nu + S_0 \tag{39}$$

with m_ν the νth statistical moment of $\dot{\varepsilon}$. If the first order approximation is used to calculate the expected values in Eq. 37, 38 and 39, these three values will be numerically identical. The small differences in the values given in Table I are due to more exact evaluation of the expected values by numerical integration.

It is seen that S_{VC} based on the cumulative damage concept presented here is slightly lower than strength for a constant average strain rate and the average strength.

As a consequence it may be concluded that the use of these will result in little error but will produce unconservative strength values. It should be remembered that these allowables will vary daily throughout the year. An equivalent strain rate $\dot{\varepsilon}_{CUMUL}$ can be calculated from Eq. 6 (Table I).

Acknowledgement

The work reported here has been supported by the U.S. Army, USAMICOM under Contract No. DAAK-40-79-C-0231, with Mr. T. H. Duerr as project engineer. Their cooperation is gratefully acknowledged.

References

(1) Cost, T. L., "Reliability Analysis of Thermally Stressed Visco-Elastic Structures by Monte Carlo Simulation", Proc. Int. Conf. on Thermal Stresses, Editors: D. P. H. Hasselman and R. A. Heller, pp. 431-446, Plenum Press, N.Y., 1980.

(2) Heller, R. A., Kamat, M. P. and Singh, M. P., "Probability of Solid Propellant Motor Failure Due to Environmental Tempera-

tures", J. Spacecraft and Rockets, Vol. 16, No. 3, pp. 140–146, 1979.

(3) Palmgren, A., "Die Lebensdauer von Kugellagern", VDI, Zeit., Vol. 68, No. 14, p. 339, 1924.

(4) Miner, M. A., "Cumulative Damage in Fatigue", J. Appl. Mech., Vol. 12, p. A-159, 1945.

(5) Bills, K. W., Jr., and Wiegand, J. H., "The Application of an Integrated Structural Analysis to the Prediction of Reliability", Annals of Reliability and Maintainability, p. 514, 526, 1970.

(6) Newland, D. E., "Random Vibrations and Spectral Analysis", p. 3-4, Longman, N.Y., 1975.

FAILURE PROBABILITY EVALUATION OF AN ANISOTROPIC BRITTLE STRUCTURE

DERIVED FROM A THERMAL STRESS SOLUTION

J. Margetson

Research Division
Propellants, Explosives and Rocket Motor Establishment
Westcott, Aylesbury, Bucks, UK

ABSTRACT

A method is presented which will allow the failure probability
of a brittle structure to be evaluated from a thermal stress
solution. It is based on a statistical approach, a generalisation
of the simple Weibull distribution, and takes into account material
variability, component size and anistropic strength. The principal
stress values at the nodes of a finite element mesh are assumed to
be known for the structure. The stress distribution within each
element is then expressed in terms of these nodal values through
suitably derived shape functions. Certain stress volume integrals
are evaluated and the failure probability of each element, and
hence that of the whole structure, is calculated. The accuracy of
the method is assessed for various types of finite elements by
analysing a simple structure.

1   INTRODUCTION

The design of structures which operate in high temperature
environments has led to the increasing use of brittle materials.
Radically new design concepts are required and the method of designing with certainty, assuming a safety factor, no longer applies.
In a brittle material flaws are randomly distributed both in severity and in orientation so failure does not always take place at the
point of maximum stress. The combination of flaw severity and
stress orientation determines the point of failure. Furthermore,
nominally identical components fail under differing loads. These
observations illustrate the inadequacies of previous methods of

design and have led to the introduction of the idea of failure
probability.

Many failure theories have been devised during recent years.
The most successful are based on a statistical approach and make use
of the simple Weibull distribution. The theory[1] which forms the
basis of this work relies on a modified Weibull distribution function to predict the failure probability of the structure. Material
variability, component size and anisotropic strength are taken into
account and allowance for different compression and strength values
is included. The theory makes use of the 'weak link' hypothesis.
That is, failure occurs when the stress intensity at any flaw
reaches the critical value for crack propagation. A further assumption, the criterion of independent action, which states that the
weakening of a flaw with respect to stress in one direction is not
affected by stresses in the transverse direction, is also made.
With these assumptions the failure probability calculation is
reduced to the product of three terms: a factorial function of a
quantity called the Weibull modulus, a component size factor, and
a stress volume integral. Incorporated into the stress volume
integral is a quantity called the unit volume strength which is
related to the mean failure stress of the material in uniaxial
tension. The Weibull modulus is an inverse measure of the
variability of the material.

The failure characteristics of a material are defined by its
Weibull modulus and unit volume strength, both of which may be
determined from simple mechanical tests. Evaluation of the stress
volume integrals requires complete knowledge of the state of stress
within the structure. This type of detailed stressing information
may be readily obtained from a finite element analysis. Failure
probability calculations for each element could, in principle, be
carried out within the stressing routines as the finite element
analysis proceeds. There are distinct disadvantages in this method
of approach. The finite element computer programs are continually
being superseded as additional elements and improved methods of
solution become available. With each new program generated the
failure analysis coding must be repeated. Many finite element
programs have now become so complex that major efforts are required
to make minor modifications to the computer codes.

In any finite element analysis the nodal co-ordinates are
known and the nodal stresses are calculated. With this information
the element shape functions may be derived and the stress distribution within each element expressed in terms of the stress values at
the nodes. The stress volume integrals required in the failure
probability calculations may be evaluated. These procedures may be
carried out as an independent operation after the finite element
analysis has been completed.

The procedures described above are presented for three types of commonly used finite element. They are the three noded triangular element, the four noded quadrilateral, and the eight noded isoparametric element. The accuracy of the method is assessed by evaluating the failure probability of a brittle hollow cylinder subjected to internal pressure.

2   FAILURE PROBABILITY ANALYSIS

The failure probability, P_f , of a brittle specimen subjected to a uniaxial stress σ is given by the Weibull equation[2],

$$P_f = 1 - \exp \left\{ - \left[\frac{1}{m}! \right]^m \left[\frac{\sigma}{\bar{\sigma}_f} \right]^m \right\} \quad . \qquad (2.1)$$

The quantity m , called the Weibull modulus, is a reciprocal measure of the variability of the material and $\bar{\sigma}_f$ is the average failure stress taken over the whole batch. The term $\left[\frac{1}{m}! \right]$ is the factorial function of $\frac{1}{m}$[3]. If the uniaxial tensile specimen, volume V , is considered to be made up of N unit volumes v (Ref. 4)

$$\left(\frac{\bar{\sigma}_f}{\bar{\sigma}_{fv}} \right)^m = \frac{v}{V} \quad . \qquad (2.2)$$

Elimination of $\bar{\sigma}_f$ between equations (2.1) and (2.2) yields

$$P_f = 1 - \exp \left\{ - \left[\frac{1}{m}! \right]^m \frac{V}{v} \left[\frac{\sigma}{\bar{\sigma}_{fv}} \right]^m \right\} \quad . \qquad (2.3)$$

By treating the structure as an assembly of parts, and applying the 'weak link' hypothesis[4], it can be shown that the generalisation of equation (2.3) for a non-uniform stress field σ is given by[4]

$$P_f = 1 - \exp \left\{ - \left(\frac{1}{m}! \right)^m \left(\frac{1}{\bar{\sigma}_{fv}} \right)^m \frac{V}{v} \iiint\limits_V \sigma^m \frac{dV}{V} \right\} \quad . \qquad (2.4)$$

In general a structure is subjected to a multiaxial stress state. Equation (2.4), which has been derived for a single component of stress, must be modified to account for the multiaxial state. It is assumed that the three principal stresses acting on a small volume contribute independently to the failure probability of that volume (the criterion of independent action). Since the

unit volume strength $\bar{\sigma}_{fv}$ is dependent on direction for an anisotropic material equation (2.4) modifies to[1]

$$
\left.
\begin{aligned}
P_f &= 1 - \exp\left\{ -\left(\frac{1}{m!}\right)^m \frac{1}{v} S(V) \right\} \quad, \\[2mm]
S(V) &= S_1(V) + S_2(V) + S_3(V) \quad, \\[2mm]
S_1(V) &= \iiint\limits_V \left[\frac{\sigma_1}{\bar{\sigma}_{fv}^{(1)} H(\sigma_1)} \right]^m dV \quad, \\[2mm]
S_2(V) &= \iiint\limits_V \left[\frac{\sigma_2}{\bar{\sigma}_{fv}^{(2)} H(\sigma_2)} \right]^m dV \quad, \\[2mm]
S_3(V) &= \iiint\limits_V \left[\frac{\sigma_3}{\bar{\sigma}_{fv}^{(3)} H(\sigma_3)} \right]^m dV \quad.
\end{aligned}
\right\} \qquad (2.5)
$$

In the above equations σ_1, σ_2, σ_3 are the principal stresses
and $H(\sigma)$ is the step function

$$
\begin{aligned}
H(\sigma) &= 1 \quad, & \sigma &\geqslant 0 \quad, \\
H(\sigma) &= -\lambda \quad, & \sigma &\leqslant 0 \quad,
\end{aligned}
\qquad (2.6)
$$

relating the compressive strength and tensile strength of the
material.

The quantities $\bar{\sigma}_{fv}^{(1)}$, $\bar{\sigma}_{fv}^{(2)}$, $\bar{\sigma}_{fv}^{(3)}$ are the respective unit
volume strengths in the directions of the principal stresses σ_1,
σ_2, σ_3. The Weibull modulus has been shown experimentally
to be independent of direction[1].

The value of $\bar{\sigma}_{fv}$ in the direction of the principal stress
axes throughout the material volume is not generally known. A
transformation law which relates $\bar{\sigma}_{fv}$ at any orientation to the
three principal values of

$$
\sigma_{fv} \left[\bar{\sigma}_{fv}^{(m_1)}, \bar{\sigma}_{fv}^{(m_2)}, \bar{\sigma}_{fv}^{(m_3)} \right] \qquad (2.7)
$$

aligned with the principal axes of the material is required. Experimental results have shown that the unit volume strengths $\bar{\sigma}_{fv}$ lie on a triaxial ellipsoid with semi-major axes given by the principal values

$$\left[\bar{\sigma}_{fv}^{(m_1)} , \bar{\sigma}_{fv}^{(m_2)} , \bar{\sigma}_{fv}^{(m_3)} \right] .$$

In terms of the direction cosines l_1 , l_2 , l_3 the required transformation law becomes[1]

$$\bar{\sigma}_{fv} = \left[\left[\frac{l_1}{\bar{\sigma}_{fv}^{(m_1)}} \right]^2 + \left[\frac{l_2}{\bar{\sigma}_{fv}^{(m_2)}} \right]^2 + \left[\frac{l_3}{\bar{\sigma}_{fv}^{(m_3)}} \right]^2 \right]^{-1/2} . \quad (2.8)$$

The failure probability calculation has been reduced to equations (2.5) and (2.6). In this calculation the stress states of all points within the structure and the failure characteristics m , $\bar{\sigma}_{fv}^{(m_1)}$, $\bar{\sigma}_{fv}^{(m_2)}$, $\bar{\sigma}_{fv}^{(m_3)}$, are required. Stresses may be calculated from a finite element analysis. The failure characteristics are obtained by analysing the fracture data of a batch of rectangular beams which are each subjected to a three point bend test[1].

It is convenient to introduce the related survival probability, $P_s = 1 - P_f$, given by

$$P_s(V) = \exp \left\{ - \left[\frac{1}{m}! \right]^m \frac{1}{v} \left(S_1(V) + S_2(V) + S_3(V) \right) \right\} . \quad (2.9)$$

If the volume V is divided into n parts V_1 , V_2 , ... V_n such that

$$V = V_1 + V_2 + ... V_n \quad (2.10)$$

it follows that

$$S_1(V) = S_1(V_1) + S_1(V_2) + ... + S_1(V_n) ,$$

$$S_2(V) = S_2(V_1) + S_2(V_2) + ... + S_2(V_n) , \quad \left. \right\} \quad (2.11)$$

$$S_3(V) = S_3(V_1) + S_3(V_2) + ... + S_3(V_n) .$$

Substitution from equation (2.11) into equation (2.9) yields

$$P_s(V) = P_s(V_1) \quad . \quad P_s(V_2) \ldots P_s(V_n) \quad , \qquad (2.12)$$

where

$$P_s(V_i) = \exp \left\{ - \left(\frac{1}{m}! \right)^m \frac{1}{v} \left[S_1(V_i) + S_2(V_i) + S_3(V_i) \right] \right\} . \quad (2.13)$$

The failure probability of the whole structure is then given by

$$P_f = 1 - P_s(V_1) \quad . \quad P_s(V_2) \ldots P_s(V_n) \quad . \qquad (2.14)$$

If the n parts of the structure correspond to n finite elements of volumes V_1, $V_2 \ldots V_n$ the survival probability of the ith element is given by equation (2.13). Repetition of this procedure for all elements and substitution of the survival probabilities of each element into equation (2.14) gives an expression for the failure probability of the whole structure.

3   STRESS VOLUME INTEGRAL EVALUATION

3.1   Constant stress elements

The three noded triangular element and the four noded quadrilateral, which is usually an assembly of four triangular elements, are only capable of producing a constant stress state within each element. Stress discontinuities between elements often lead to crude approximations in the final analysis. These elements are popular for various reasons and may be used also to carry out a failure analysis on a structure provided that sufficient elements are taken. Since the associated stress distribution is constant the stress volume integrals appearing in equation (2.13) may be trivially evaluated. Because the stress values are defined to act at the element centroid the related stress volume integrals for both elements take on the form

$$S_i(V_p) = h\Delta_p \left(\frac{\sigma_{ip}}{H(\sigma_{ip}) \, \bar{\sigma}_{fvp}^{(i)}} \right)^m , \quad i=1,2,3 \quad . \qquad (3.1)$$

In equation (3.1) the subscript p denotes the pth element, h its thickness, and Δp its area. In the axisymmetric case

$$h = 2\pi \bar{r}_p \qquad (3.2)$$

where $\bar{r}_p$ is the radial distance of the element centroid from the axis of rotation of the structure.

3.2 Isoparametric elements

The isoparametric family of elements, which has been developed for both two and three dimensional situations, takes into account stress variations within an element. These elements are more accurate than the triangular and quadrilateral elements and their curved sides are particularly convenient in the description of structural boundaries and interfaces. Rapid developments in this field make it too tedious to list the numerous elements which have been developed. For the purpose of illustration, and without loss in generality, failure probability procedures are described for the eight noded isoparametric elements, which has four nodes at its vertices and four nodes along the curved sides joining these vertices. The four vertices lie in a plane and the so called mid-side nodes, also lying in that plane, may be placed arbitrarily on the curves joining the vertices.

When a finite element analysis is carried out on a structure it is usual to evaluate the principal stress values of the nodes of each element. If the failure probability of the structure is required the stress distribution throughout each element must be known. It is possible, by deriving suitable shape functions, to obtain an expression for the state of stress within an element in terms of its nodal values.

In practice the eight noded isoparametric element is a quadrilateral with slightly curved sides. To carry out the following analysis it is convenient to transform this element from the x_1, x_2 plane into the square of sides two units in the ξ, η plane (see Fig. 1). The functional relationships describing the transformation law are[5]

$$x_1 = C_1 + C_2 \xi + C_3 \eta + C_4 \xi\eta + C_5 \xi^2 + C_6 \eta^2 + C_7 \xi^2\eta + C_8 \xi\eta^2 ,$$

$$(3.3)$$

$$x_2 = D_1 + D_2 \xi + D_3 \eta + D_4 \xi\eta + D_5 \xi^2 + D_6 \eta^2 + D_7 \xi^2\eta + D_8 \xi\eta^2 ,$$

$$(3.4)$$

where C_i, D_i, $i=1,2 \ldots 8$ are arbitrary constants. The above transformation law may be conveniently expressed in matrix notation as

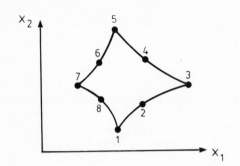 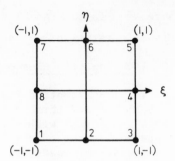

Figure 1. Eight Noded Isoparametric Element in the (x1, x2)
          plane and transformed (ξ, η) plane

$$x_1 = [\,P\,][\,C\,] \quad , \tag{3.5}$$

$$x_2 = [\,P\,][\,D\,] \quad , \tag{3.6}$$

where $[\,P\,]$ is the row matrix containing the polynomial terms in
$(\xi \, , \, \eta)$ and the vectors $[\,C\,]$ and $[\,D\,]$ contain the unknown
constants. Insertion of the x_1 co-ordinates $[\,x_1^e\,]$ of all the
nodes and their corresponding $(\xi \, , \, \eta)$ values into equation (3.5)
yields

$$[\,x_1^e\,] = [\,A\,][\,C\,] \tag{3.7}$$

Each row of $[\,A\,]$ corresponds to the insertion of a particular
nodel value $(\xi \, , \, \eta)$ into the matrix $[\,P\,]$. If the nodel
ordering shown in Fig. 1 is adopted the matrix $[\,A\,]$ becomes

$$[\,A\,] = \begin{bmatrix} 1 & -1 & -1 & 1 & 1 & 1 & -1 & -1 \\ 1 & 0 & -1 & 0 & 0 & 1 & 0 & 0 \\ 1 & 1 & -1 & -1 & 1 & 1 & -1 & 1 \\ 1 & 1 & 0 & 0 & 1 & 0 & 0 & 0 \\ 1 & 1 & 1 & 1 & 1 & 1 & 1 & 1 \\ 1 & 0 & 1 & 0 & 0 & 1 & 0 & 0 \\ 1 & -1 & 1 & -1 & 1 & 1 & 1 & -1 \\ 1 & -1 & 0 & 0 & 1 & 0 & 0 & 0 \end{bmatrix} \tag{3.8}$$

The constant vector $[\,C\,]$ follows when both sides of equation
(3.7) are multiplied by the inverse of $[\,A\,]$, i.e.

$$[c] = [A^{-1}][x_1^e] \quad . \tag{3.9}$$

Elimination of $[c]$ between equations (3.5) and (3.9) gives

$$x_1 = [P][A^{-1}][x_1^e] \quad . \tag{3.10}$$

A similar analysis performed on equation (3.6) leads to

$$x_2 = [P][A^{-1}][x_2^e] \quad , \tag{3.11}$$

where $[x_2^e]$ denotes the x_2 co-ordinates of the element nodes.

The stresses vary over the element in a manner corresponding exactly with the transformation assumption made for the co-ordinates. If the principal stress values at a point are denoted by σ_1, σ_2, σ_3 and their respective nodal values at the nodes by matrices $[\sigma_1^e]$, $[\sigma_2^e]$, $[\sigma_3^e]$, it follows that

$$\sigma_1 = [P][A^{-1}][\sigma_1^e] \quad ,$$

$$\sigma_2 = [P][A^{-1}][\sigma_2^e] \quad , \tag{3.12}$$

$$\sigma_3 = [P][A^{-1}][\sigma_3^e] \quad .$$

The stress volume integral associated with the ith finite element and the σ_1 principal stress is given by

$$S_1(V_i) = \iiint_{V_i} \left[\frac{\sigma_1}{\bar{\sigma}_{fv}^{(1)} H(\sigma_1)} \right]^m dV \quad , \tag{3.13}$$

where V_i is the element volume. In a two dimensional idealisation the above expression reduces to

$$S_1(V_i) = \iint_{A_i} \left[\frac{\sigma_1}{\bar{\sigma}_{fv}^{(1)} H(\sigma_1)} \right]^m h dx_1\, dx_2 \quad , \tag{3.14}$$

where A_i is the element area. For the plane stress or plane strain situation the quantity h denotes element thickness. In an axisymmetric case, with Ox_2 as the axis of rotation,

$$h = 2\pi x_1 \quad . \tag{3.15}$$

Evaluation of the surface integral appearing in equation (3.14) is complicated by the curvature of the element sides. This problem is solved by transforming the element into a square and carrying out the integration in the (ξ, η) plane. Changing the variables of integration yields[6]

$$S_1(V_i) = \int_{-1}^{1} \int_{-1}^{1} f(\xi, \eta)\, d\xi d\eta \quad , \tag{3.16}$$

where

$$F(\xi, \eta) = \left[\frac{\sigma_1(\xi, \eta)}{\bar{\sigma}_{fv}^{(1)} H(\sigma_1)} \right]^m h(\xi, \eta)\, ||\, J(\xi, \eta)\, || \quad . \tag{3.17}$$

In equation (3.17) $||\, J\, ||$ is the Jacobian determinant

$$||\, J\, || = \begin{bmatrix} \partial x_1/\partial \xi & \partial x_2/\partial \xi \\ & \\ \partial x_1/\partial \eta & \partial x_2/\partial \eta \end{bmatrix} \quad . \tag{3.18}$$

Changes of variables for the co-ordinates and stresses are given by equations (3.10) and (3.12). Spatial derivatives of the co-ordinates x_1 and x_2 with respect to ξ and η are found from the formulae

$$\left. \begin{aligned}
\frac{\partial x_1}{\partial \xi} &= [\, \partial P/\partial \xi\,][\, A^{-1}\,][\, x_1^e\,] \quad , \\[2mm]
\frac{\partial x_1}{\partial \eta} &= [\, \partial P/\partial \eta\,][\, A^{-1}\,][\, x_1^e\,] \quad , \\[2mm]
\frac{\partial x_2}{\partial \xi} &= [\, \partial P/\partial \xi\,][\, A^{-1}\,][\, x_2^e\,] \quad , \\[2mm]
\frac{\partial x_2}{\partial \eta} &= [\, \partial P/\partial \eta\,][\, A^{-1}\,][\, x_2^e\,] \quad .
\end{aligned} \right\} \tag{3.19}$$

Similar expressions may be derived for the stress volume integrals $S_2(V_i)$ and $S_3(V_i)$.

A method of integration, which gives high accuracy for a relatively low number of terms, is given by the Gaussion quadrature formula[7]

$$\int_{-1}^{1} \int_{-1}^{1} F(\xi , \eta) \, d\xi d\eta = \sum_{i=1}^{N} W_i \sum_{j=1}^{N} W_j \, F(\xi_i , \eta_j) \qquad . \qquad (3.20)$$

The weighting coefficients W_i , W_j and the argument values ξ_i , η_j are derived from certain orthogonality relationships of the Legendre polynomials. Values of these coefficients for various values of N are shown in Table 1. Experience indicates that $N=4$ gives adequate results for most applications.

4   NUMERICAL ILLUSTRATION

For the purpose of illustration the failure probability of a hollow cylinder, subjected to internal pressure at its bore, was calculated. The problem was solved using the three types of elements described in the previous section and their relative accuracy was assessed by comparing results with the analytical solution. Plane strain conditions applied and the ends of the cylinder were assumed to be free from normal stress. The cylinder geometry, material properties, failure characteristics, and applied pressure at the bore are given in Table 2.

TABLE 1

Argument values and weighting coefficients
for the Gaussian quadrature formula

N	ξ_i , η_i	W_i
2	± 0.57735	1.00000
4	± 0.86113 ± 0.33998	0.34785 0.65214
6	± 0.93246 ± 0.66120 ± 0.23861	0.17132 0.36076 0.46791

TABLE 2

Cylinder properties and loading data

Internal radius	R_i	= 2.0 in
External radius	R_o	= 3.0 in
Cylinder length	L	= 0.25 in
Young's modulus	E	= 10^6 lb/in^2
Poisson's ratio	ν	= 0.3
Unit volume strength	$\bar{\sigma}_{fv}$	= 262.8 lb/in^2
Weibull modulus	m	= 15
Compression tension ratio	η	= 8
Internal pressure at R_i	p_i	= 80-120 lb/in^2
External pressure at R_o	p_o	= 0.0 lb/in^2

When a cylinder is subjected to radial loading the principal stress axes coincide with the radial, axial and hoop directions. The corresponding stresses expressed as a function of radial position r are given by the formulae[8]

$$
\left.
\begin{array}{l}
\sigma_r = \dfrac{P_i R_i^2}{R_o^2 - R_i^2}\left(1 - \dfrac{R_o^2}{r^2}\right) \quad , \\[3em]
\sigma_\theta = \dfrac{P_i R_i^2}{R_o^2 - R_i^2}\left(1 + \dfrac{R_o^2}{r^2}\right) \quad , \\[3em]
\sigma_z = 0 \quad .
\end{array}
\right\}
\qquad (4.1)
$$

The quantities σ_r, σ_z, σ_θ denote the radial, axial and hoop stresses, respectively.

Since the axial stress is zero it makes no contribution to the failure probability. The radial stress is entirely compressive and its contribution is insignificant. The failure probability is dominated by the tensile stress in the hoop direction. Substitution of the stress values from equation (4.1) into equation (2.5), and evaluation of the stress volume integrals numerically by Simpson's

rule, yields the failure probability P_f . In the calculation the
integration step length was continuously reduced until a relative
error of less than 0.0001 was obtained between successive computed
values of P_f .

The failure probability was also calculated by taking the
finite element stress solution and applying the procedures described
in the previous sections. Calculations were made using triangular
and isoparametric elements. Relative accuracy for the different
types of elements used was assessed by varying the number of elements
through the thickness of the cylinder. Under conditions of plane
strain there was no variation of stress in the axial direction and
the length of the cylinder could be chosen arbitrarily. For the
following calculations the length of the elements and that of the
cylinder were taken to be equal. In the stress evaluation it was
necessary to ensure that the element aspect ratio (element breadth
divided by length) was not decreased to the extent that the stress
analysis became ill conditioned.

Failure probability calculations were made with the cylinder
subjected to a bore pressure of P_i = 100 lb/in^2 . The results
for different element geometries are given in Table 3. Computations
show that the isoparametric element is far superior to the other
two elements. With the meridional plane of the cylinder divided
into one isoparametric element the failure probability was calculated to within 2.1% of the exact value, compared with 79.8% for
one rectangular element and 64.9% for two triangular elements.
Twelve triangular elements and six rectangular elements are
required to attain the same degree of accuracy as one isoparametric
element.

The failure probability was also evaluated for the range of
bore pressures P_i = 80 - 120 lb/in^2 . In this calculation four
triangular, two rectangular and two isoparametric elements were
taken through the thickness of the cylinder. The results are given
in Table 4. Considerable inaccuracies occur when rectuangular and
triangular elements are used. This is particularly noticeable in
the low failure probability range where percentage errors reach
values of 49%. Results are improved significantly when the isoparametric elements are used and there is then excellent agreement with
the exact failure probability curve. Maximum errors for the isoparametric element are less than 2%.

5   SOLUTION CHART

The failure probability calculation may be made as a separate
and independent operation after the finite element analysis has
been completed. The solution chart describing the sequence of

TABLE 3

Failure probability corresponding to an applied pressure of $p = 100 \text{ lb/in}^2$ for various finite element mesh geometries

Number of elements through the cylinder thickness	Triangular elements		Rectangular elements		Isoparametric elements		Exact P_f value
	Computed P_f	% error	Computed P_f	% error	Computed P_f	% error	
1			0.1001	79.8	0.4857	2.1	0.4962
2	0.1742	65.0	0.2908	41.4	0.5004	0.8	0.4962
3			0.4018	19.0	0.4989	0.5	0.4962
4	0.3512	29.2	0.4280	13.7	0.4987	0.5	0.4962
6	0.4326	12.8	0.4813	3.0			0.4962
8	0.4509	9.1					0.4962
12	0.4870	1.8					0.4962

TABLE 4

Variation of the failure probability with pressure for various
finite element mesh geometries

Pressure	Analytical solution	Triangular elements four through the thickness		Rectangular elements two through the thickness		Isoparametric elements two through the thickness	
	Exact P_f	Computed P_f	% error	Computed P_f	% error	Computed P_f	% error
120	0.999	0.9987	− 0.12	0.9949	− 0.50	0.9999	0
115	0.9962	0.9704	− 2.58	0.9389	− 5.75	0.9964	+0.02
110	0.9429	0.8359	−11.34	0.7619	−1919	0.9449	+0.21
105	0.7596	0.5933	−21.89	0.5105	−32.79	0.7637	+0.53
100	0.4962	0.3512	−29.22	0.2908	−41.39	0.5008	+0.84
95	0.2721	0.1826	−32.89	0.1471	−45.9	0.2749	+1.02
90	0.1316	0.0852	−35.25	0.0683	−28.10	0.1331	+1.13
85	0.0581	0.0369	−36.4	0.0296	−49.0	0.0588	+1.20
80	0.0238	0.0151	−36.5	0.0120	−49.5	0.0251	+1.25

steps involved is shown below.

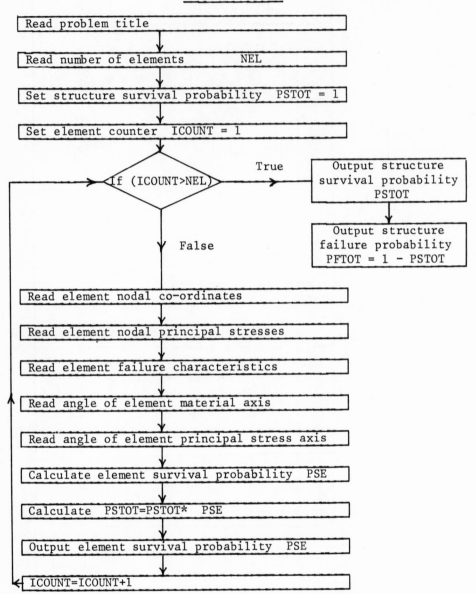

SOLUTION CHART

REFERENCES

1   J. Margetson, "A statistical theory of brittle failure for an
     anisotropic structure subjected to a multiaxial stress
     state", RPE Tech. Report No. 48 (1976).

2   W. J. Weibull, J. App. Mech., 1951, $\underline{18}$ (3).

3   E. J. Whittaker, "A course of modern analysis", Cambridge,
     C.U.P., 1963.

4   P. Stanley, H. Fessler and A. D. Sivell, "An engineers approach
     to the prediction of failure probability of brittle
     components", Proc. Brit. Ceram. Soc., 1973, $\underline{22}$.

5   O. C. Zienkiewicz, "The finite element method in engineering
     science", N.Y., McGraw-Hill, 1971.

6   E. G. Phillips, "A course of analysis", Cambridge, C.U.P.,
     1960.

7   F. Scheid, "Numerical analysis. Schaum's outline series", N.Y.,
     McGraw-Hill, 1968.

8   S. Timoshenko, "Theory of elasticity", N.Y., McGraw-Hill, 1934.

ESTIMATE OF STORAGE LIFE FOR M392/M728 PROJECTILES BASED ON FINITE ELEMENT THERMAL STRESS ANALYSIS

James O. Pilcher, Aaron Das Gupta, Thomas R. Trafton

Ballistic Research Laboratory
ARRADCOM
Aberdeen Proving Ground, MD   21005

INTRODUCTION

M392 projectiles have exhibited a high percentage of erratic exterior ballistic flight trajectories, especially in firing series which had been conditioned for elevated temperature testing. This erratic flight behavior has resulted in unacceptably large dispersion patterns on target. The sources of the erratic rounds tested were stockpile lots after extended storage and lots shipped to and tested in hot, dry climatic areas. The poor performance has been observed to occur with greater frequency in gun tubes with high secondary wear characteristics. Examination of the vulcanized fiber rotating bands indicated a shrinkage of the material from the dimensions specified in the technical data package. Failure of the bands in shear was also reported. Significantly different results have been obtained with rotating bands manufactured from vulcanized fiber produced by different suppliers. Preliminary tests of the M728 projectile indicated that there was a high probability of the same problem existing, particularly after extended storage.

In response to a request for investigative and consultative assistance from the Standard Tank Munition Task Force of LCWSL, ARRADCOM, the BRL investigated the possible causes of erratic flight behavior of the M392 and M728 projectiles. During this investigation, the relative merits of rotating band materials from different sources were examined particularly the effects of thermal cycling on rotating band dimensional stability. The BRL has extended the analysis of the thermal cycling behavior of the M392 and M728 projectile rotating bands to determine the relative effects of long term storage in various climate zones, cartridge case

521

throat size, secondary land wear and vulcanized fiber material
properties.

FINITE ELEMENT THERMO–STRESS ANALYSIS

    Finite element thermo–stress analyses were performed on the
M392/M728 projectile cross sections using the BRL version of the
SAAS II finite element computer program. Figure 1 shows the general
cross section of the assembled projectile and cartridge case at the
mid plane of the rotating band.

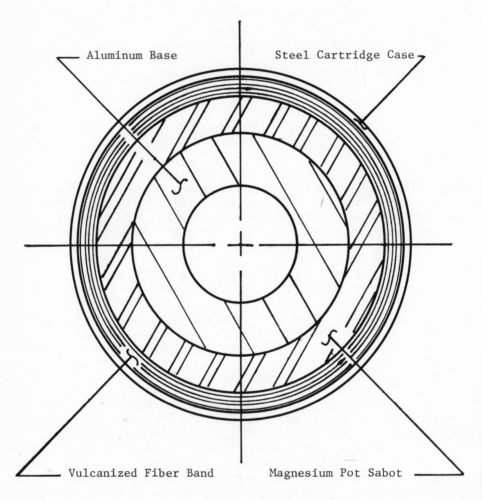

Figure 1. General Cross Section of the M392 and M728 Projectile/
           Cartridge Case Assembly at the Mid Plane of the Rotating
           Band.

The investigators examined the thermal strains and stresses for the following material combinations:

CASE	ALUMINUM	MAGNESIUM	VULCANIZED FIBER	STEEL
1	7075-T6	A261A	NVF	Low Carbon
2	7075-T6	A261A	Spaulding	Low Carbon
3	7075-T6	A261A	Spaulding(UK)	Low Carbon

The following material properties were used in the analyses.

PROPERTY		7075-T6	AZ61A	L.C. STL.	UK	NVF	SPAULDING
Thermal Expansion	θ	12.8	14.5	6.5	15	13.6	18.9
$(10^6/°F)$	R	12.8	14.5	6.5	32	13.6	48.6
Density (lb/in^3)		.101	.065	.284	.052	.055	.052
Elastic Mod.	θ	10.4	6.5	29.5	0.14	0.20	0.32
$(10^6$ psi$)$	R	10.4	6.5	29.5	0.10	0.14	0.21
Yield Strength $(10^3$ psi$)$		72	29	82			
Total Elongation (%)					60	44	29
Elastic Elongation (%)					9.41	6.95	4.64
Poisson's Ratio		.33	.35	.29	.33	.33	.33
Ultimate Strength	θ				12.41	13.23	12.63
$(10^3$ psi$)$	R				9.41	9.74	9.69

The above properties are from a compilation of data gleaned from laboratory tests performed by BRL and LCWSL and various handbooks and specifications of the ASM, ASTM and ASME.

The calculations show that the vulcanized fiber rotating band undergoes plastic deformation during thermal cycling. The extent of the plastic deformation is dependent on the material properties of the rotating band and maximum temperature of the assembly.

The plastic deformation in the rotating band causes a reduction in the diameter of the rotating band as the assembly cools to an ambient temperature of 75°F, which was used as the reference temperature for all calculations. Although the material will creep at low temperatures and cause the rotating band to approach its original dimensions, the rotating band will gradually shrink in thickness to some minimum value after repeated thermal cycling. The values of plastic deformation calculated in the finite element computer calculations represent the limit to which the rotating band should shrink after many cycles of thermal variation. The following table shows the maximum reduction ratio for each rotating band material at four different temperatures. The ratio R_u is the ratio of the shrunken band diameter to the original band diameter

MATERIAL	R_u, 95°F	R_u, 105°F	R_u, 125°F	R_u, 145°F
Spaulding	.9979	.9954	.9904	.9855
NVF	None	.9991	.9950	.9910
UK Spaulding	None	None	.9994	.9961

ESTIMATE OF STORAGE LIFE

In order to estimate the effective storage life for the M392 and M728 projectiles, an accumulative damage law was constructed and critical limits were established.

Damage Law

For the purpose of this problem a rational function type of damage law was used.

$$y = \frac{1+ax}{1+bx} \tag{1}$$

where   y = ratio of the shrunken band diameter to the original
              band diameter
          a = arbitrary constant
          b = arbitrary constant
          x = number of months of storage

The reason for choosing the above function as a model is that its behavior at the limits of x=o and x=∞ closely follows the physical behavior of the band material, that is

$$\lim_{x \to o} y = 1 \tag{2}$$

and

$$\lim_{x \to \infty} y = a/b \tag{3}$$

The rotating band that does not experience thermal cycling does not suffer a reduction in diameter as shown in Equation 2.

The rotating band that experiences an infinite number of thermal cycles suffers a reduction in diameter equal to the ratio of the arbitrary constants, a/b.

The ratio a/b then can be defined as the ultimate reduction ratio, R_u, as derived from the finite element thermo-stress analysis for a given temperature extreme.

$$a/b = R_u \tag{4}$$

Then Equation (1) can be recast as follows

$$y = \frac{1+ax}{1+\dfrac{ax}{R_u}} \tag{5}$$

thus requiring the determination of only one arbitrary constant, a.

From measurements on 162 of M392 projectiles provided by the Standard Tank Munitions Task Force of LCWSL, ARRADCOM, it was determined that after storage of 40 months in an environment with high temperature extremes of 105°F a mean reduction ratio of 0.9993 was achieved for Spaulding bands. Because no other data were available at the time of this analysis, the arbitrary constant, a, was calculated from these data and assumed to be the same for all three vulcanized fiber bands. Although this assumption is not necessarily correct, the differences in the arbitrary constant, a, from one material to another is not as significant as the differences in the ultimate reduction ratio R_u .

Critical Ratio

In addition to an accumulative damage law, one needs a critical
ratio beyond which performance degrades to unacceptable levels.   In
this case the critical ratio of reduction of band diameter is determined by the ratio of the throat diameter of the cartridge case (in
essence the original band diameter) to the diameter of lands of the
gun tube at the allowable secondary wear.

The allowable land diameter for a worn tube was originally
4.254 inches. Test firings showed that when the rotating band
diameter became less than the land diameter plus .005 inch the
projectile performance became unacceptable. All projectiles with
rotating band diameters in excess of this value had acceptable
performance. The Critical Ratio, R_c, can be defined as follows

$$R_c = \frac{D_L + W + .005}{D_{CT}} \tag{6}$$

where D_L = tube land diameter in inches

           W = allowable diametral secondary wear in inches

           D_{CT} = diameter of the cartridge case throat in inches.

Figures 2 through 4 show plots of the expected reduction ratio of
specific rotating band materials versus storage time in months;
superimposed on the plot is the critical ratio (dashed line) for
cartridge case throat diameter of 4.265 inches and secondary wear
of 0.120 inch on the land diameter. The intersection of the dashed
line with the curves gives the expected storage life. A summary of
results is shown below.

MATERIAL	PANAMA (95°F)	MILAN (105°F)	YUMA (125°F)	MOJAVE (145°F)
UK	No limit	No limit	No limit	28 Mo.
NVF	No limit	No limit	24 Mo.	16 Mo.
Spaulding	106 Mo.	25 Mo.	14 Mo.	8 Mo.

Figures 5 through 7 show the plots of the expected reduction ratio
versus storage time for the three rotating bands with a critical
reduction ratio based on a cartridge case throat diameter of 4.270
inches and a secondary wear of 0.100 inch on the land diameter. A
summary of the results is shown below

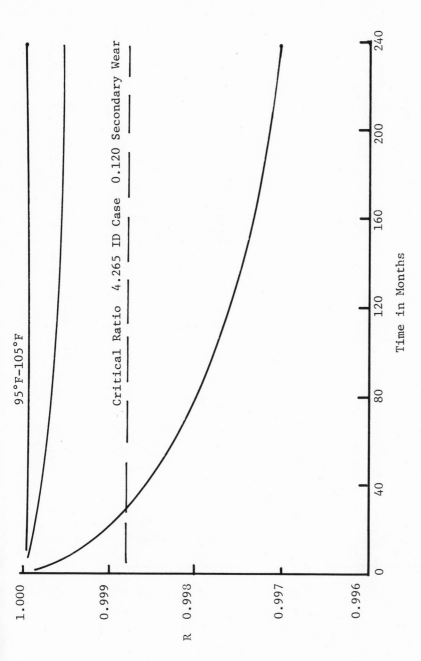

Figure 2. Ratio of Rotating Band Reduction vs Storage Time for UK Spaulding Vulcanized Fiber.

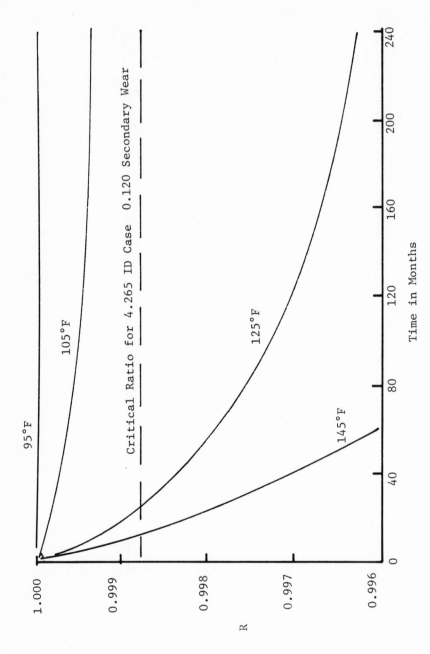

Figure 3. Ratio of Rotating Band Reduction vs Storage Time for NVF Vulcanized Fiber.

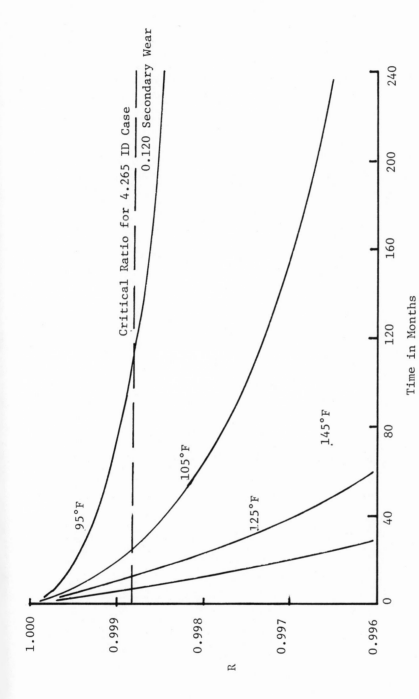

Figure 4. Ratio of Rotating Band Reduction vs Storage Time for Spaulding Vulcanized Fiber.

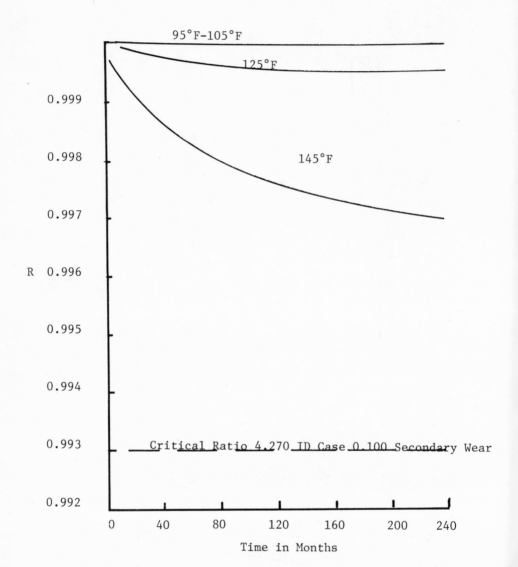

Figure 5.   Ratio of Rotating Band Reduction vs Storage Time for
UK Spaulding Vulcanized Fiber.

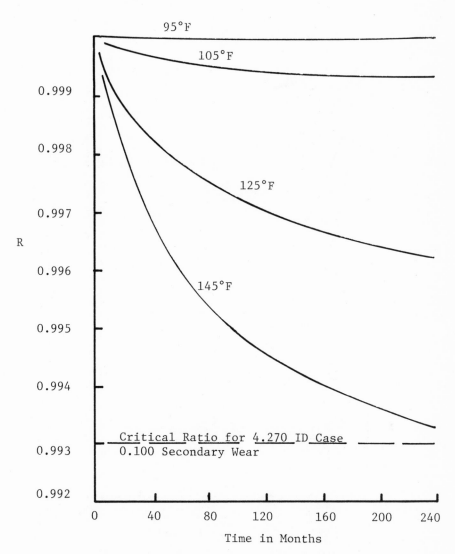

Figure 6. Ratio of Rotating Band Reduction vs Storage Time for NVF Vulcanized Fiber.

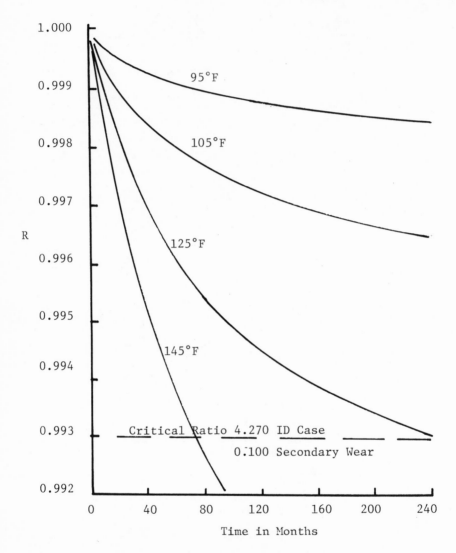

Figure 7. Ratio of Rotating Band Reduction Vs. Storage Time
for Spaulding Vulcanized Fiber.

MATERIAL	PANAMA (95°F)	MILAN (105°F)	YUMA (125°F)	MOJAVE (145°F)
UK	No limit	No limit	No limit	No limit
NVF	No limit	No limit	No limit	No limit
Spaulding	No limit	No limit	No limit	76 Mo.

The above analysis and calculations does not account for variations of moisture content with time.

MOISTURE CONTENT

The property most affected by moisture content is total elongation capacity of the vulcanized fiber. This property governs the ability of the projectile to withstand free run conditions that may occur at the origin of rifling and in the secondary wear region of the tube. The greater total elongation capacity of the rotating band will tend to lessen reengagement torsional shock at the end of free run and thus lessen the forces that tend to cause anomalies in projectile dynamics such as clutch slippage, and subprojectile misalignment. Unfortunately insufficient theory and data exist that would allow accurate calculation of these effects. However, it is felt that if the moisture content is maintained in the rotating band at a minimum of $5\frac{1}{2}\%$ this problem will be minimized greatly for the Spaulding material which has the worst problem in this respect. Likewise, the UK and NVF bands should be maintained at a $5\frac{1}{2}\%$ moisture content.

CONCLUSIONS

1. The band shrinkage problem can be alleviated by proper selection of band material, cartridge case throat internal diameter, and reduction of the wear criteria. All three corrections must be done in combination since one solution is not sufficient in itself to solve the problem.

2. Moisture content must be maintained in order to enhance the operability of the band during free run conditions.

RECOMMENDATIONS

1. Use either the UK Spaulding vulcanized fiber or the NVF vulcanized fiber for rotating band material on the M392, M728, L36 and like projectiles.

2. Resize all cartridge cases to a minimum throat diameter of 4.270 inches for a minimum depth of 2.750 inches.

3.   Limit the wear life of the M68 gun tube to a maximum of 0.100 inch secondary wear on the lands.

4.   Stabilize the moisture content of the rotating bands to a minimum of 5½% prior to final assembly to the cartridge case.

5.   Surveillance procedures should include measurements of the rotating band and cartridge case throat diameters.

REFERENCES

1. ASTM Designation: D 710-63, Standard Specifications for Vulcanized Fiber Sheets, Rods, and Tubes Used for Electrical Insulation.

2. George S. Brady, Materials Handbook, Ninth Edition, McGraw-Hill Book Company, 1963, p. 810.

3. Store Specification No: L10554, United Kingdom Ministry of Defence Specification for Manufacture and Quality Assurance of Vulcanized Fibre Tube Blanks, SX625 and SX816, 1972.

4. MIL-R-50792 (MU), Military Specification, Rotating Band Tubing, Vulcanized Fiber, 1973.

5. MIL-F-1148, Military Specification, Fibre Vulcanized, Electrical and Mechanical Grades, 1973.

6. R. M. Jones and J. G. Crose, "SAAS II Finite Element Stress Analysis of Axisymmetric Solids with Orthotropic, Temperature-Dependent Material Properties", Aerospace Report No. TR-0200 (S4980), 19 September, 1968.

7. S. Timoshenko and J. N. Goodier, "Theory of Elasticity", McGraw-Hill Book Company, 1951, pp. 399-427.

DETERMINATION OF THE THERMAL SHOCK FRACTURE TOUGHNESS OF REACTOR GRAPHITE SUBJECTED TO NEUTRON IRRADIATION AT HIGH TEMPERATURE

S. Sato*, H. Awaji**, Y. Imamura*, K. Kawamata*, and
T. Oku***

*The University of Ibaraki, Hitachi, 316, Japan
**Fukushima College, Iwaki, Fukushima, 970 Japan
***Japan A.E.R.I., Tokai, Ibaraki, 319-11, Japan

ABSTRACT

The arc discharge method was applied to determine the effect of neutron irradiation on the thermal shock fracture toughness and thermal shock resistance of reactor graphites. Experimental results show that the fracture toughness and resistance to thermal shock of four varieties of reactor graphite decrease remarkably when the materials are subjected to neutron irradiation of $(1.6 \sim 2.3) \times 10^{21}$ neutrons/cm^2 (>0.18 Mev) at 600 $\sim$ 850°C.

INTRODUCTION

Graphite has been used widely as a core material in high temperature gas-cooled reactors, because of its excellent resistance of high temperature, in addition to its appropriate nuclear physical properties. The reactor graphite is subjected to thermal stress due to internal radioactive heating and mechanical loading in severe thermal and neutron environment. A proper selection of high strength graphite is prerequisite to insure reactor safety. A so-called "high strength material" is, however, rather over-sensitive to defects in spite of its high strength. Here it is important to select material with great fracture resistance to thermal stress, especially under conditions of severe neutron irradiation.

Previously, we described a method for determining the thermal shock fracture toughness by subjecting the central area of a disk to thermal shock by arc discharge heating[1,3]. By this method, the parameter $K_{Ic}k/E\alpha$, in which K_{Ic} is the mode I fracture toughness, k is the thermal conductivity, E is Young's modulus and α is the thermal

expansion coefficient, could be determined en bloc by measuring the
threshold electric power of the arc discharge heating required for
the propagation of a crack in the edge of the disk. We have defined
this parameter as the thermal shock fracture toughness[1] which cor-
responds to the thermal shock resistance $\sigma_\theta k/E\alpha$ in which σ_θ is the
tensile strength, measured previously[2].

In this paper, we present results on how much change occurs in
the thermal shock fracture toughness of four varieties of reactor
graphite subjected to neutron irradiation of $(1.6 \sim 2.3)\times10^{21} n/cm^2$
(E > 0.18 Mev) at $600 \sim 850°C$. We demonstrate that the neutron irradiation significantly decreases the thermal shock fracture toughness significantly in reverse proportion to the mechanical strength.

METHOD

We have analyzed the stress intensity factor of a disk with an
edge crack caused by thermal shock at the central area[1,3], as shown
in Fig. 1, based on the analysis of Riney[4], and Tweed and Rooke[5].
The circumferential thermal stress in a thin disk of radius R heated by an internal heat source of q per unit volume and unit time at
a central area $0<r<a$, under conditions of zero initial temperature
and insulated outer boundary, in non-dimensional form is:

$$F(x) = \frac{\sigma_\theta}{E\alpha Q} = 2(\frac{a}{R})\sum_i [\{J_1(m_i\frac{r}{R})\frac{R}{m_i r} - J_0(m_i\frac{r}{R})\} \frac{J_1(m_i\frac{a}{R})}{m_i^3 [J_0(m_i)]^2}$$

$$\{1 - \exp(-\tau m_i^2)\}], \tag{1}$$

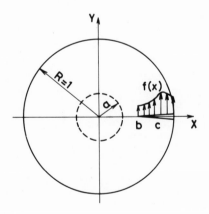

Fig. 1.   Thermal shock testing of disk by arc discharge heating.

where the summation extends throughout the positive roots of $J_1(m_i) = 0$, σ_θ is the circumferential stress, $Q = qR^2/k$, τ is a non-dimensional time expressed as $\kappa t/R^2 = (k/\eta\rho)t/R^2$, κ is thermal diffusivity, η is specific heat, ρ is density and t is time. Other stress components may not contribute to the thermal fracture of the disk.

The calculation results of Eq. (1) for several values of the time parameter τ are shown in Fig. 2 for a central heating radius $a/R = 0.3$. As shown, the circumferential thermal stresses are compressive at the central area, but they change into tensile stresses outside the heat source perimeter. For $\tau > 1/4$, the distribution of thermal stress saturates to the maximum. The maximum tensile stress depends on the value of a/R.

Tweed and Rooke[5] analyzed the problem of the stress intensity factor of the edge crack in a disk, is shown in Fig. 1. Based on this analysis, we calculated the stress intensity factor of a disk with an edge crack caused by thermal shock at the central area[1,3], where $f(x)$ in Fig. 1 was replaced by Eq. (1). These results were determined as a function of the radius a/R and c/R as is shown in Table 1, where c is a crack length. The stress intensity factor K_I is:

$$K_I = \sqrt{\pi c}\, F_I \tag{2}$$

is non-dimensional, since $f(x)$ has been expressed in the form of a non-dimensional thermal stress $\sigma_\theta/E\alpha Q$.

The heating quantity q is represented in terms of the electric power W, in kW, by

$$q = \frac{kQ}{R^2} = \frac{860W}{\pi a^2 h} \; [\text{kcal/m}^3\text{hr}] \tag{3}$$

where h and a are the thickness and the heating radius respectively, of the disk in m. Therefore, the fracture toughness becomes

$$K_{Ic} = \sqrt{\pi c}\, F_I E\alpha Q$$

$$= \sqrt{\pi c}\, F_I \frac{E\alpha}{k} \frac{860\beta W}{\pi h (a/R)^2} \tag{4}$$

where β is the factor of heat loss, for simplicity assumed to be unity. From the mechanical, thermal and fracture mechanics properties of the materials, the thermal shock fracture toughness parameter ∇ can be defined[1],

$$\nabla = K_{Ic} k/E\alpha = \sqrt{\pi c} \cdot F_I \frac{860\beta W}{\pi h (a/R)^2} \tag{5}$$

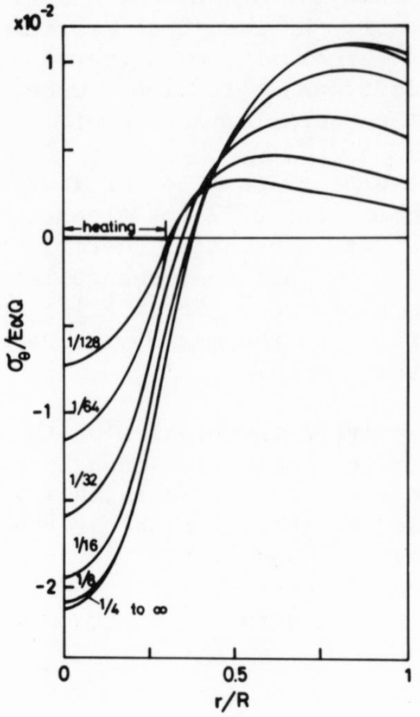

Fig. 2. Distribution of circumferential thermal stress for a heating radius a/R=0.3 for a range in values of $\tau=\kappa t/R^2$.

Table 1. Non-dimensional stress intensity factor F_1 for the thermal shock fracture testing

a/R \ c/R		Heating radius 0.1 (×10⁻²)	0.2 (×10⁻²)	0.3 (×10⁻¹)	0.4 (×10⁻¹)	0.5 (×10⁻¹)	0.6 (×10⁻¹)	0.7 (×10⁻¹)
Edge carck depth	0.05	0.148	0.574	0.124	0.201	0.280	0.344	0.372
	0.1	0.157	0.608	0.130	0.212	0.295	0.362	0.389
	0.15	0.166	0.641	0.136	0.223	0.309	0.376	0.402
	0.2	0.174	0.672	0.143	0.233	0.321	0.388	0.408
	0.25	0.181	0.701	0.149	0.241	0.331	0.395	0.407
	0.3	0.189	0.727	0.154	0.248	0.337	0.397	0.398
	0.35	0.195	0.751	0.158	0.253	0.341	0.393	–
	0.4	0.201	0.771	0.161	0.257	0.340	0.382	–
	0.45	0.206	0.788	0.164	0.257	0.334	–	–
	0.5	0.210	0.799	0.165	0.255	0.322	–	–

The value ∇ corresponds to the thermal shock resistance parameter[2] $\sigma_\theta k/E\alpha$. The thermal shock fracture toughness can be determined by measuring the critical electric power W required to initiate crack propagation of an edge crack in the disk at the heating time, $\tau > 1/4$.

The most important stress in the disk without an edge crack which decides the fracture strength is the ·maximum tensile stress σ_θ which appears near the disk edge. The maximum saturation stress at the time $\tau > 1/4$ was defined as the non-dimensional thermal stress S^*2,

$$S^* = \sigma_{\theta max}/E\alpha Q \qquad (6)$$

Table 2 shows the calculated results of S^* as a function of a/R. The values of S^* increase rapidly as a/R increases, and the maximum occurs at a/R $\simeq$ 0.7. The positions of S^* occurs on the radius slightly inside the edge.

Table 2.   Specific non-dimensional thermal stress as a function of a/R for the thermal shock resistance testing.

a/R	$S^* = \sigma_\theta/E\alpha Q$
0.05	3.390×10^{-4}
0.1	1.353×10^{-3}
0.2	5.210×10^{-3}
0.3	1.100×10^{-2}
0.4	1.791×10^{-2}
0.5	2.483×10^{-2}
0.6	3.024×10^{-2}
0.7	3.277×10^{-2}
0.8	2.882×10^{-2}
0.9	1.925×10^{-2}

If fracture of a disk occurs by electric heating power W, then the thermal shock·resistance of the disk at $\tau > 1/4$ is expressed by the following equation including en bloc the mechanical and thermal properties.

$$\Delta = \sigma_\theta k/E\alpha = S^* \frac{860\beta W}{\pi h (a/R)^2} \qquad (7)$$

EXPERIMENT

Graphite Specimens

Disk specimens of diameter 19 mm and thickness 2 mm consisted of four varieties of graphite considered for the high temperature gas-cooled reactor project, namely, isotropically molded IM2-24, 7477, and anisotropically extruded H327 and SMG. The properties of these graphites are listed in Table 3. Specimens 6 mm in diameter and 60 to 12 mm long were prepared for strength measurement in bending, and compression, respectively, were prepared as well. Graphite is generally anisotropic due to the molding or extruding pressure. In this paper, we have measured the strength and fracture toughness for the molded and extruded graphites in a direction transverse of the molding and extruding direction. The disk specimens were cut so as to be parallel to the direction of forming pressure.

Table 3.   Room Temperature Properties of Tested Graphites

Graphite	Direct.	IM2-24	7477	H327	SMG
Maker	-	AGL	Le Carbon	GLC	SDK
Coke	-	gilsonite	petroleum	petroleum	petroleum
Forming	-	M	IM	E	E
Density(g/cm^3)	-	1.78	1.75	1.78	1.75
Young's modulus E (kg/mm^2)	‖ ⊥	1.21 1.30	1.16 1.16	- 0.705	- 0.846
Tensile strength σ_t (kg/mm^2)	‖ ⊥	1.98 2.18	2.68 2.61	1.23 0.70	- 0.99
Thermal conduct. k (Kcal/mh°C)	‖ ⊥	 93.6	 78.7	 93.6	 108
Thermal expans. α ($\times 10^{-6}$/°C)	‖ ⊥	 4.24	 3.40	 2.08	 2.50

M=molded, E=extruded, IM=isostatically molded

Neutron Irradiation

Neutron irradiations were carried out using two instrumented capsules inserted in irradiation holes at the fuel zone of JMTR. Initially, the temperature was chosen to be about 1000°C, typical for a high temperature gas-cooled reactor. The real temperature attained, however, was about 850°C at the center and about 600°C at the ends of the capsule. Table 4 shows the details of irradiation conditions. The neutron fluxes and the energies were measured by the reaction of Fe^{54}(n,p)Mn^{54} of iron monitors. Strictly speaking, the conditions of irradiation of four varieties of graphite specimens inserted in the capsules were different, but we evaluated them under the same conditions in each capsule, since the specimens were placed at random and the direction of neutron flux along the axis of each capsule were comparatively flat.

Table 4. Conditions of Neutron Irradiations of
HTGR Graphites in the JMTR

Capsule	(1) 74M-58U	(2) 75M-12U
Position	H93 hole	H74 hole
Neutron fluence	$(1.6 \sim 1.7) \times 10^{21}$ n/cm^2(>0.18MeV)	$(2.1 \sim 2.3) \times 10^{21}$ n/cm^2(>0.18MeV)
Irr. temp.	$600 \sim 850°C$	$600 \sim 850°C$

Experiments

Post-irradiation tests were carried out at the hot laboratory
in the facilities of Tohoku University at Oarai, Ibaraki. Since the
radiation activities of graphite specimens were at a low level, the
experiments were carried out in a glove box. Thermal shock testing
was carried out with our recently[2] developed thermal shock testing
apparatus shown in Fig. 3.

Thermal Shock Resistance

Thermal shock resistance of a disk without edge crack was measured for heating radius a/R = 0.3. In this case, the specific non-
dimensional thermal stress $S^* = 1.100 \times 10^{-2}$. The heating time t
which reaches the distribution of the maximum saturation stress was
determined from $t = \tau \eta \rho R^2/k$, where $\tau = 1/4$, R = 9.5 mm. The values
of k, η and ρ took into account the irradiation effects on these
properties at the mean temperature of about 500°C. Using this estimation t was found to be 3 sec. Some uncertainty in t hardly affects the final results because the saturating stress distribution
was not sensitive to variation in τ as shown in Fig. 2.

Figure 4(a) and (b) show the typical recorder charts of electric power near the site of fracture in the graphite disk specimens
before and after irradiation, respectively. Some fluctuations of
electric power occurred during arc discharge. The values of electric power indicated in Fig. 4 represent critical power levels. The
threshold values of electric power required for the initiation of
fracture were determined by testing seven unirradiated specimens and
six irradiated specimens. Comparison of the threshold values of
electric power shown in Figures 4(a) and 4(b), readily shows that
irradiation significantly decreases the thermal shock resistance.

Figure 5 shows a crack in the graphite disks resulting from the
arc discharge heating. Crackings occur at and proceed from the

Fig. 3.   Thermal shock testing apparatus.

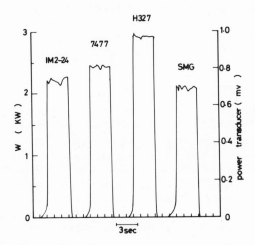

Fig. 4(a).   Typical recorder charts of arc discharge electric
             powers for HTGR graphites.   (before irradiation)

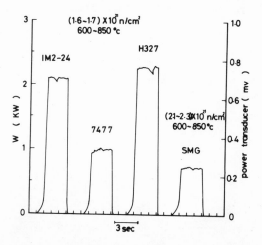

Fig. 4(b). Typical recorder charts of arc discharge electric power for HTGR graphites.(after irradiation)

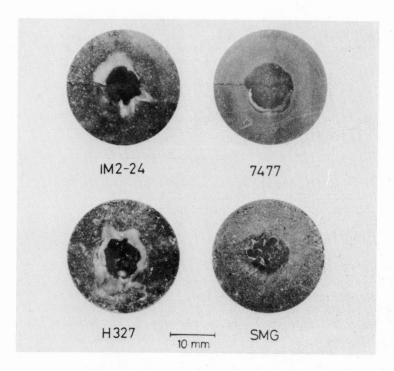

Fig. 5. Fracture appearances in the thermal shock resistances of HTGR graphites.

outer periphery to the center of the disk. These thermal stress
cracks are clearly seen at high temperature, but are difficult to
recognize at lower temperature due to crack closure.

Figure 6 shows the path of crack propagation during thermal
shock testing in greater detail. The cracks were observed to proceed through layered boundaries of filler cokes, linking void with
void. No noticeable effect of irradiation on fracture path was
recognizable.

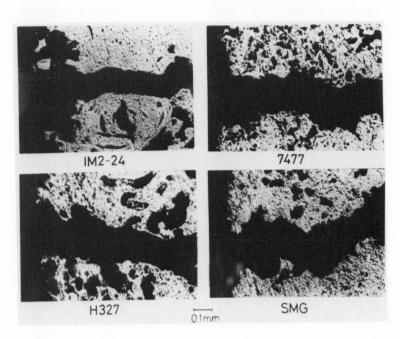

Fig. 6. Path of crack propagation during the thermal shock of
HTGR graphites.

Figure 7 shows the comparison of thermal shock resistance before and after irradiation of four varieties of graphite. The
hatched black parts show the ranges where they always fracture in
the upper limits, never under the lower limits. The average thermal
shock resistance of all graphite as the result of irradiation decreases about 32%. In graphite 7477 and SMG it decreases about 50%.
However, in IM2-24 made from gilsonite coke, the resistance decreases only slightly.

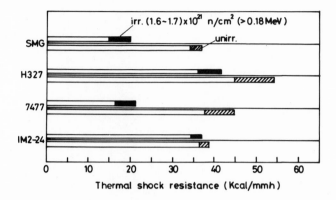

Fig. 7.   Thermal shock resistance of HTGR graphites before and
          after neutron irradiation.

Thermal Shock Fracture Toughness

       Figures 8(a) and (b) show charts of electric power consumed by
the arc discharge heating in the thermal shock toughness testing of
a disk with edge crack before and after irradiation, respectively.
These show the heating power near the tip of the edge crack.   The
neutron irradiation clearly reduces the electric power required to
initiate crack propagation.   As expected the disks with the edge

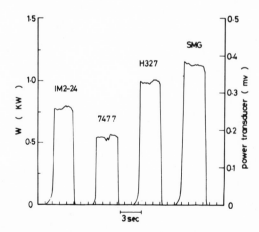

Fig. 8(a).   Typical recorder charts of arc discharge electric powers
             for HTGR graphites. (before irradiation)

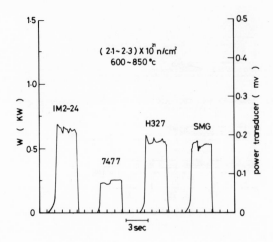

Fig. 8(b). Typical recorder charts of arc discharge electric powers
            for HTGR graphites. (after irradiation)

crack required much less power to initiate fracture than the disks
without edge crack.

Figure 9 shows a typical fracture path which shows that crack
propagation initiates at the tip of the edge crack after the pattern
of mode I cracking. Enlarged pictures in cracking areas adjacent to
the tip of the edge crack are shown in Fig. 10.

Figure 11 show the changes of the thermal shock fracture toughness due to irradiation. These figures show the ranges within which
the cracks do not propagate perfectly, along with the upper limits
within which cracks always propagate. Comparisons of the before and
after irradiations show that the thermal shock fracture toughness of
the irradiated specimens decreases 45% on the average for the four
varieties of graphite. In graphite IM2-24 made from gilsonite coke,
however, the decrease count is no greater than about 21%.

DISCUSSION

The thermal shock resistance Δ and the thermal shock fracture
toughness ∇ of the four graphites investigated in this study all
decreased due to the neutron irradiation. The tensile strength,
fracture toughness, and Young's modulus are known to increase due
to irradiation. Although thermal conductivity k was not measured
in this study, the report by Brinkele[7] on gilsonite graphite and
matrix graphite after irradiation at 500°C confirmed that it decreased by a factor of 0.42 and 0.47 times, respectively. The

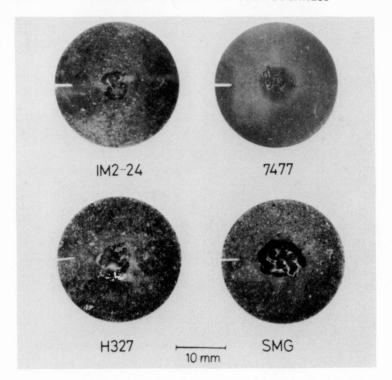

Fig. 9. Fracture appearances in the thermal shock fracture toughness of HTGR graphites.

coefficient of thermal expansion as shown by Helm[8] and Kelley[9], is
similarly affected because of a change in the thermal expansion
coefficient of the graphite grain crystal due to irradiation
and also due to change in density caused by dimensional changes.
Putting all the effects together, the changes in Δ and ∇ due to
irradiation can be deduced.

Graphite has many defects of various shape as is seen in the
microscopic micrographs of Fig. 6 and 10. These defects cause
cracks and affect the strength of the graphite materials. Now, if
we consider the thermal shock resistance Δ and the thermal shock
fracture toughness ∇ instead of the tensile strength σ_t and the
fracture toughness K_{Ic} of the material, respectively, we can deduce
the equivalent length of crack $2a_e$ existing in an infinitely wide
plate under tensile stress σ_t, represented by the following equation:

$$a_e = \frac{1}{\pi} \left(\frac{K_{Ic}}{\sigma_t}\right)^2 = \frac{1}{\pi} \left(\frac{\nabla}{\Delta}\right)^2 \tag{8}$$

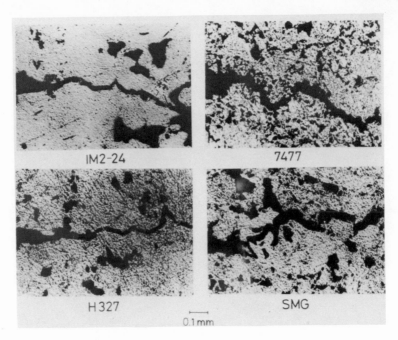

Fig. 10. Paths of crack propagations during thermal shock of precracked specimens of HTGR graphites.

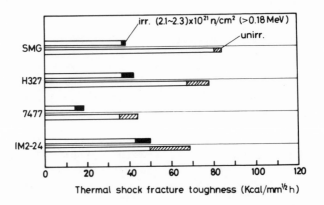

Fig. 11. Changes in the thermal shock fracture toughness of HTGR
graphites before and after neutron irradiation.

Table 5 shows the calculated results when differences in the irradiation conditions between two capsules are ignored. The lengths of
a_e of unirradiated graphites are indicated in each photograph of

TABLE 5. Experimental Results of HTGR Graphites

Graphite	(A) IM2-24			(B) 7477		
	unirr.	irr.(1)*	irr.(2)*	unirr.	irr.(1)	irr.(2)
Thermal shock resist. Δ(Kcal/mmh)	37.7	36.0 (0.95)		41.5	19.7 (0.47)	
Thermal shock fract. tough. ∇(Kcal/mm$^{\frac{1}{2}}$ h)	59.6		47.1 (0.79)	39.8		16.6 (0.42)
∇/Δ (mm$^{\frac{1}{2}}$)	1.58	1.31		0.96	0.846	
Equiv. crack length a_e (mm)	0.796	0.545 (0.68)		0.293	0.288 (0.78)	

Graphite	(C) H327			(D) SMG		
	unirr.	irr.(1)	irr.(2)	unirr.	irr.(1)	irr.(2)
Thermal shock resist. Δ(Kcal/mmh)	49.5	39.2 (0.79)		35.5		18.5 (0.52)
Thermal shock fract. tough. ∇(Kcal/mm$^{\frac{1}{2}}$ h)	72.7		39.5 (0.54)	82.6		37.6 (0.46)
∇/Δ (mm$^{\frac{1}{2}}$)	1.47	1.01		2.33	2.03	
Equiv. crack length a_e (mm)	0.687	0.323 (0.69)		1.72	1.31 (0.76)	

(*1) irr.(1):(1.6$\sim$1.7)x10^{21}n/cm^2(>0.18 Mev), irr.(2):(2.1$\sim$2.3)x10^{21} n/cm^2(>0.18 Mev)
(*2) values in bracket show the ratio to irradiated data

Fig. 12. The lenghs $2a_e$ are apparently larger than actual inherent lengths of defects. An explanation for this tensile strength
is governed by interaction of different defects. It is noteworthy
that a_e is decreased by about 27% on the average by the irradiation.
This decrease is much larger than the radiation-induced shrinkage of
graphite, which is less than 1%[10],[11]. A decrease of 27% in effective crack size corresponds roughly to the extent of increase in the
mechanical properties such as tensile strength, Young's modulus and
fracture toughness. Such decrease in effective crack length confirms the increases in the mechanical properties due to neutron
irradiation, deduced by the fracture mechanics approach.

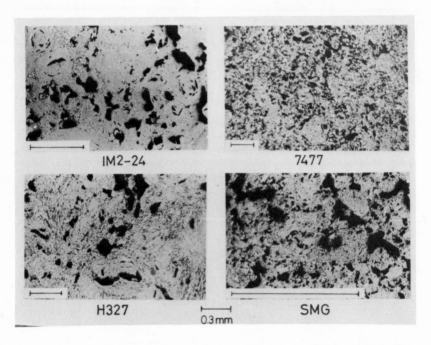

Fig. 12. Microscopic structures of graphite specimens. Each in-
        dicated length shows the equivalent length of crack a_e.

Although the mechanical strength of reactor grade graphite
generally increases due to neutron irradiation, neglecting of the
irradiation effect on thermal shock resistance and the thermal
shock fracture toughness of reactor grade graphite could endanger
the safe operation of a nuclear reactor. These effects should be
included in the design criteria. Fitzer and Vohler[11] as well as
Moore[12] emphasize the importance of the thermal stress resistance
of graphite for reactor design. Graphites which have large values
of thermal shock resistance are, as measured by the parameter Δ

(eq. 7), do not always, however, have large values of the thermal shock fracture toughness, as indicated by Fig. 7 and 11 or Table 5. In case of graphites with preexisting cracks, we prefer design on the basis of the thermal shock fracture toughness. Individual evaluations of mechanical strength and the thermal properties relating to the thermal stress are relatively complex. For this reason, the method described in this study is recommended for establishing performance criteria for graphites indeed for nuclear reactors.

REFERENCES

1. S. Sato, H. Awaji and H. Akuzawa, Carbon, Pergamon Press, 16:103 (1978).
2. S. Sato, K. Sato, Y. Imamura and J. Kon, Carbon, Pergamon Press, 13:309 (1973).
3. H. Awaji and S. Sato, J. Japanese Soc. Strength and Fracture of Materials, 28:631 (1978).
4. T. D. Riney, J. Appl. Mech., 38:631 (1961).
5. J. Tweed and D. P. Rooke, Int. J. Engng. Sci., 11:65 (1973).
6. T. Oku, M. Eto and K. Fujisaki, JAERI-M7647, (1978).
7. L. Binkele, High Temperature-High Pressure, 4:401 (1972).
8. J. W. Helm, BNWL-1056B, (1968).
9. B. T. Kelley, J. E. Brocklehurst and J. H. Gittus, "Graphite Structures of Nuclear Reactors", Inst. Mech. Engineers, 19 (1972).
10. G. B. Engle, W. P. Eatherly, High Temperature-High Pressure, 4:119 (1972)
11. E. Fitzer and D. Vohler, Radiation Damage in Reactor Materials IAEA, 593 (1963).
12. R. W. Moore, Extended Abstracts of SCI 4th London International Conf. Carbon and Graphite, 216 (1974).

THERMAL STRESS TESTING OF ADVANCED

OPTICAL CERAMICS BY A LASER TECHNIQUE

S.G. Schwille, R.A. Tanzilli and S. Musikant

General Electric Company
Re-entry Systems Division
Philadelphia, Pennsylvania 19101

INTRODUCTION

Thermal stress tests are conducted to determine the failure
strengths of a material in a thermal stress environment. This
can only be done if an accurate thermostructural analysis of the
test can be made. An accurate analysis of the test requires a
precise definition of both the heat flux and the time of failure.
Other thermal stress tests often lack one or both of these requirements. A test technique has been developed, using a flat top
laser to provide the heat flux, which offers both a well defined
thermal environment and an accurate indication of the time of
catastrophic failure. This paper describes the test procedure
and the corresponding analytical technique. Additionally the
results of some recent tests are presented demonstrating the
test applicability.

TEST DESCRIPTION

The technique used to induce thermal stresses in the
material is to irradiate a centrally located circular spot on
a circular disk. The disk is thin so that through the thickness
gradients are small and, since the temperature is axisymmetric,
the stress distributions are axisymmetric. The disk is free to
expand and rotate so that no boundary constraints are induced.
This test configuration induces primarily uniaxial tensile
stresses (hoop) in the cool portion of the disk. Figure 1 shows
the test specimen and the irradiated area.

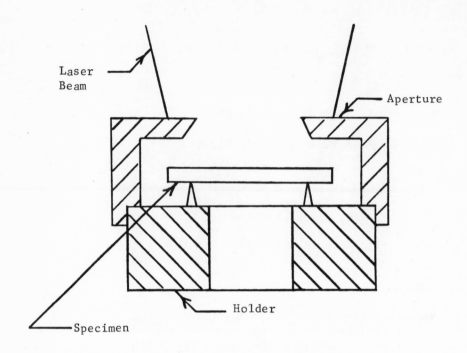

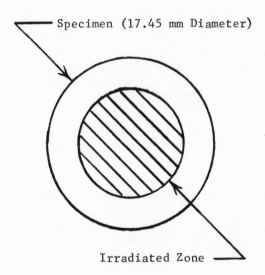

Figure 1. Specimen Configuration

The size of the heated spot is critical to a successful test. If the spot is too small then the compressive stresses will be much higher than the tensile stresses and an undesirable compressive failure may result. A large spot will produce tensile stresses high in relation to the compressive stresses, but if the spot is too large only the very outer rim will be in tension. This is not desirable since it represents only a small volume of material which is not as precisely machined as the faces of the disk.

A spot whose radius is two thirds the radius of the disk was selected. With this size spot the irradiated area is approximately equal to the unradiated area. Such a balance re-sults in the tensile stresses being approximately equal in magni-tude to the compressive stresses. The maximum tensil stresses occur slightly in from the outer rim of the specimen.

Figure 2 (A) shows the fixture used to support test disks (17.45 mm dia. x 1 mm thick). Test disks were aligned by two reference rods and supported by three pyramidal tips which were ground near the rim of a circular viewing hole placed in a larger diameter graphite support disk. This viewing port permitted direct viewing of the rear of each specimen using a suitably angled flat mirror. The irradiated area was controlled for each test by a copper aperture Figure 2 (B) which rested upon a second underlying graphite aperture. The cavity surrounding the speci-men was blanketed with helium gas to prevent oxidation of the graphite coating which was applied to both surfaces of each test disk. The coating provided uniform thermal bounday conditions insuring equality of net absorbed laser energy for each test disk, and also, equality of front and rear surface total hemispherical emittances.

Figure 3 (A) is a schematic of the optical path of the CO_2 laser beam which, for this test, provided a uniform flat-top beam profile of 300 W/cm^2. Figure 3 (B) is a schematic of the sample holder and associated test diagnostics including front surface pyrometry, high-speed movies and an ultrasensitive foil-gage pyrometer which was used to measure fracture time of test speci-mens. Upon fracture, a significant increase in thermal radiation from the rear of each specimen was directed to the foil gage (Gardon) resulting in immediate (millisecond) electrical response of the gage electronics. Figure 4 is a trace of a typical Gardon gage output showing a discontinuous rise in voltage output upon specimen fracture. Also shown in Figure 4 is a trace of the pyrometer output whose response drop-off was found to coincide with the time of fracture.

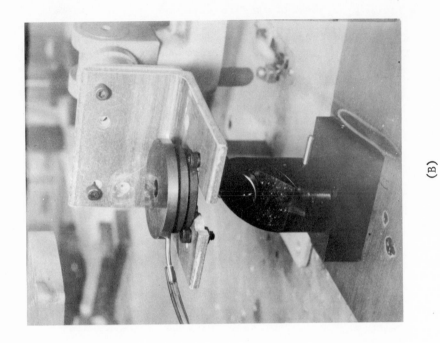

(A)

(B)

Figure 2.   Specimen support fixture for laser thermal shock test.

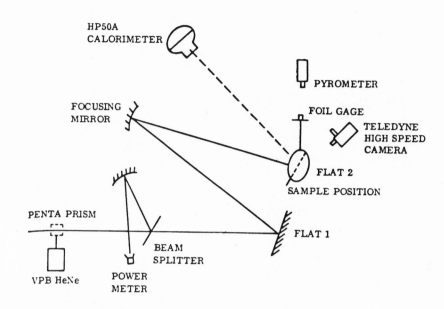

(A) OPTICAL CONFIGURATION FOR THE LASER AND DIAGNOSTICS

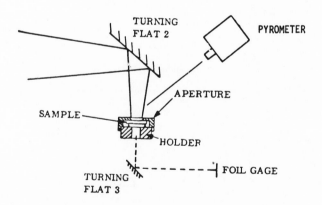

(B) DETAIL OF THE SAMPLE POSITION

Figure 3. Schematic of laser test set-up.

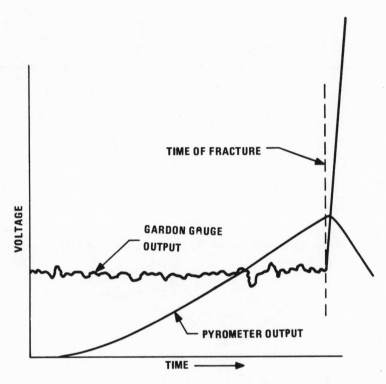

Figure 4.   Laser test diagnostics indicating time-of-fracture
measurement technique.

ANALYTICAL TECHNIQUE

The analytical proecure employed an uncoupled thermodynamic and structural analysis. Both analysis techniques, described below, have been used extensively and are well checked out.

The thermodynamic analysis was performed using a transient heat transfer computer program. This program employs an iterative solution of simultaneous equations for node temperatures derived from finite difference analysis. Convergence is recognized when the maximum change in any node temperature is equal to or less than the specification tolerance between two successive iterations. For this analysis a three dimensional pie slice model was created. The primary interest was the radial temperature gradient although through-the-thickness temperature histories were also generated. The thermal conductivity was allowed to vary with temperature however, the heat flux was assumed independent of temperature. The instantaneous heat flux of the laser could not be matched so the heat flux was ramped up over a 0.01 second time period. For most of the specimens tested, this time period is small when compared with the fracture time.

The structural analysis utilized the finite element method to determine the stress levels. A computer program was used which employed an isoparametric axisymmetric finite element. Although the model has several elements running through the thickness, only the radial temperature gradient was applied. Previous work indicated that the through the thickness gradient had little effect on the stresses. The capability exists for temperature dependent orthotropic material properties, however, the limited test data on candidate isotropic materials required the use of room temperature elastic properties at elevated temperatures. Thermal expansion data were available to about 1000°C and were linearly extrapolated beyond this temperature.

The result of the above described thermostructural analysis is a plot of the analytically predicted maximum hoop stress as a function of time. Since the time of failure is known from the test, the failure stress can be inferred from this plot.

RESULTS

As an example of the combined analytical and experimental procedures, detailed results for spinel are presented. Figure 5 shows the predicted temperature distribution at three selected times. The temperature distribution shows the effect of irradiating the inner zone and the conduction of the heat energy to the outer zone. Figures 6 through 8 are the stress distribution at six selected time points. The distribution all have similar characteristics; the hoop and radial stresses are equal and

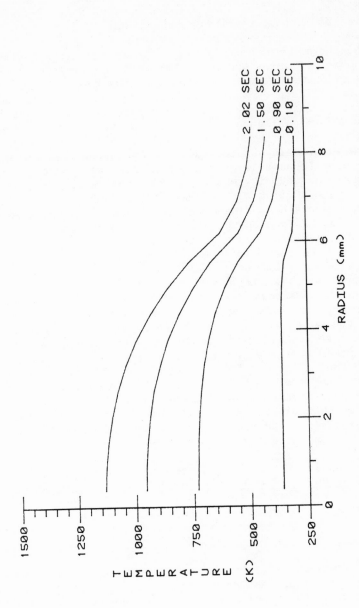

Figure 5.   Predicted temperature gradients as a function of time for single-crystal spinel.

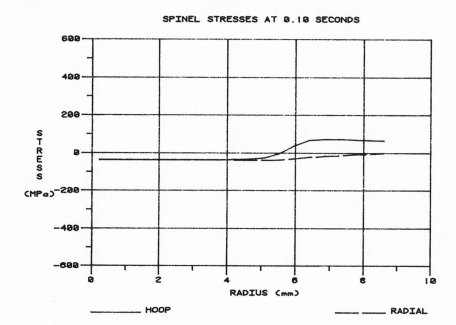

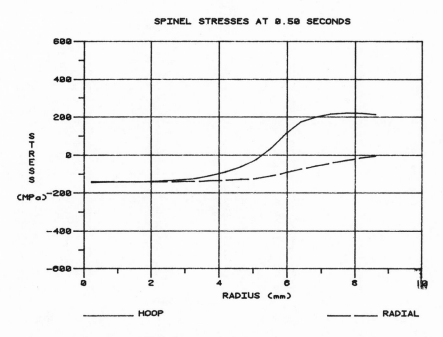

Figure 6. Predicted hoop and radial stress distributions for single crystal spinel at 0.10 and 0.50 seconds.

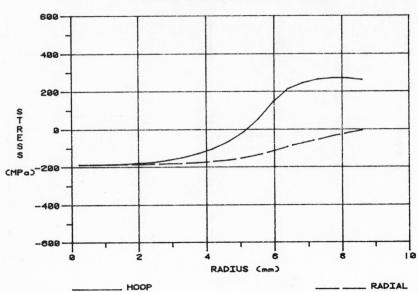

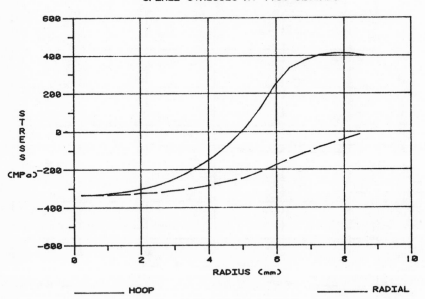

Figure 7.   Predicted hoop and radial stress distributions for
single crystal spinel at 0.70 and 1.50 seconds.

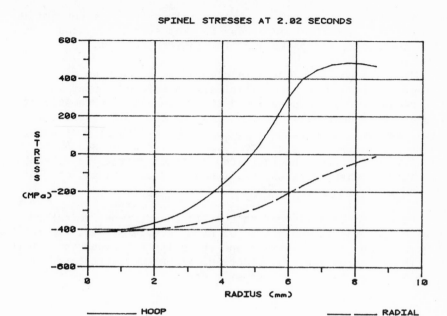

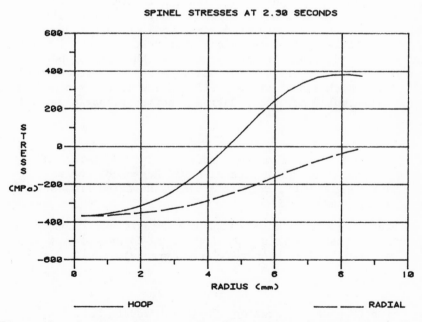

Figure 8. Predicted hoop and radial stress distributions for single crystal spinel at 2.02 and 2.30 seconds

compressive at the centerline, the radial stress goes to zero at the outer rim, and the hoop stress becomes tensile at around five milimeters radius and peaks just inward from the rim.

The maximum hoop stress is plotted as a function of time in Figure 9 along with the analytically predicted center and outer temperatures. Also plotted in Figure 9 is a schematic of the analytical heat pulse profile showing the 0.01 second ramps and the two second duration. Six specimen of spinel were tested with fracture times ranging from 0.53 to 0.77 seconds. Based on these fracture times, the indicated tensile strength is 230-285 MPa. This compares with the published flexure strength of 276[1] MPa.

Fractography performed on the broken specimen indicates failure occurring in the region of high tensile stress. Figure 10 illustrates one failure origin and shows its relationship with the rim. The failures are thus occurring in the area predicted in the analysis as having high tensile stress.

CONCLUSIONS AND FURTHER WORK

Based on the results to date, this test procedure is concluded to be a valid means of determining the tensile strength of a material in a thermostructural environment. Two areas of future work need to be pursued. First, more samples need to be tested of each material of interest so that a statistically ample data bank can be generated. Secondly, a technique is already under investigation to extend this test to biaxial failures induced by irradiating the rim rather than the central portion of the disk. In this configuration maximum tensile stresses will occur at the disk center thereby eliminating concerns about edge induced flaws which may occur during specimen preparation.

ACKNOWLEDGEMENT

This work is based on the "Advanced Optical Ceramics" program sponsored by the Office of Naval Research, contract number N00014-78-C-0466, Defense Advanced Research Projects Agency, order number 3387.

The authors also acknowledge the experimental support of C. Durken (thermal Analysis); J. Hanson, T. Horne and D. Roncin (laser testing); and J. D'Andrea (fractography).

REFERENCE
1.   Union Carbide Corporation, Crystal Products Bulletin, Czochralski Spinel Material Data Sheet.

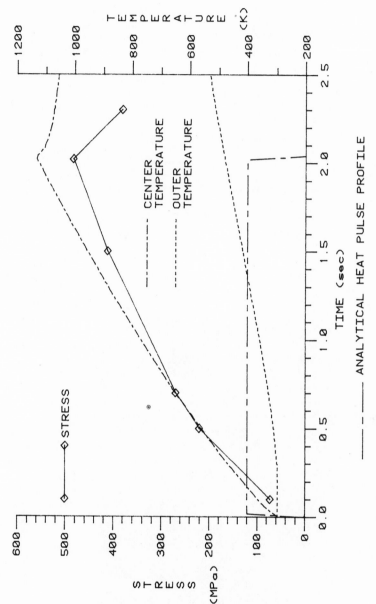

Figure 9. Predicted temperature profiles and maximum hoop stress as a function of time for single-crystal spinel.

LOCATION OF FRACTURE INITIATION POINT (1)
IS ADJACENT TO RIM (2) OF SAMPLE.
INITIATION POINT IS LOCATED BY TRACING
THE FRACTURE PATH DIRECTION (INDICATED
BY CLEAVAGE FEATURES) TO THE ORIGIN.

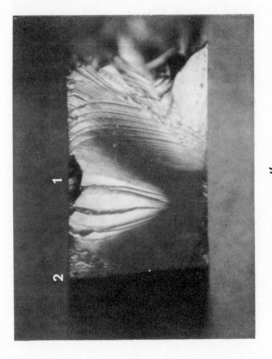

Figure 10. Fracture origin for spinel.

LASER INDUCED THERMAL STRESSES IN BRITTLE MATERIALS*

J. J. Mecholsky, P. F. Becher, R. W. Rice, J. R. Spann,
and S. W. Freiman

Sandia Laboratories,** Albuquerque, NM 87185; Naval
Research Laboratory, Washington, DC 20375; and National
Bureau of Standards, Washington, DC 20234.

ABSTRACT

This paper presented the theoretical and experimental background for understanding of opaque and transparent brittle materials
subjected to continuous wave laser irradiation. At irradiations
above ~ 1 kw/cm^2, the time to failure of opaque materials was proportional to the thickness squared and inversely proportional to the
thermal diffusivity. The time to fracture of transparent materials
was nearly independent of thickness but was highly dependent on the
absorption coefficient. The burnthrough time of both opaque and
transparent materials was proportional to the thickness and thermodynamic parameters and inversely proportional to the irradiance.

Laser irradiation can be used to rank the thermal stress resistance of opaque brittle materials. This ranking was the same as
that obtained from thermal quenching into water, as long as the
thickness, wavelength, and heating rates were equivalent. The reason
for the similarity was that the analytical expressions describing
both behaviors were similar.

INTRODUCTION

There is increasing interest in the application of high energy
lasers to technology. Metals have been laser scribed, cut, and

---

*This work performed at the Naval Research Laboratory.
**A U.S. DOE facility.

welded for several years.[1,2] The possibility of applying laser
technology to ceramic materials has also been shown to be quite
feasible;[3] several investigators have cut glass using this technique.[4-6] Also, applying high energy laser irradiation as a means
of simulating thermal shock in the laboratory is quite convenient.

Laser irradiation can be divided into two categories, continuous
wave (CW) and pulsed irradiation. Continuous wave irradiation of
ceramic materials results in the production of thermal stresses. The
exact nature of the thermal stress state is dependent upon whether
the material is transparent or opaque to the incident irradiation.
Pulsed irradiation, on the other hand, can produce mechanical and
thermal stresses. However, this paper will only consider thermal
stresses induced by either continuous wave irradiation or repetitive
pulse irradiation. For convenience, we will only refer to continuous
wave irradiation. There are two failure mechanisms as a result of
continuous wave irradiation: fracture and removal of material by
melting and ablation until burnthrough results.

This paper reviews the understanding of laser induced failure of
opaque and transparent brittle materials. Failure by laser irradiation will then be compared to conventional thermal shock fracture by
quenching. It will be shown that ranking of materials by laser induced thermal shock is analogous to that by quench down in water.

EXPERIMENTAL PROCEDURE

Ceramic and glass plates, typically 76.2 x 101.6 x 6mm, were
irradiated with 3.8 and 10.6 micron CW lasers of powers ranging from
300 watts to 60 kilowatts with beam sizes ranging from 1 to 50 sq.
cm. Samples that fractured from thermally induced stress were examined optically in order to identify the fracture surface features
(i.e. the fracture initiating flaw and fracture mirror boundaries).[7]
Several samples were instrumented with temperature compensated strain
gages. Front surface temperatures were monitored with an optical
pyrometer, and rear surface temperatures were monitored with chromel-
alumel thermocouples. Beam powers were measured with ball calorim-
etry techniques in most cases, and spacial beam shape was obtained
from the impressions burned into polymethalmethacrylate (PMMA)
blocks. Burn patterns were titrated with a soapy water solution to
determine the quantity of material lost and peak to average power
densities obtained from this technique. All tests were photograph-
ically recorded with at least one 4-frame per second camera and
usually two other 100-frame per second cameras. Thermal and mechanical properties of the materials were obtained from published literature.

ANALYSIS OF CW IRRADIATION OF CERAMIC MATERIALS

Temperature Distribution

Continuous wave laser irradiation of brittle materials results
in a temperature differential between the front and rear surfaces
which can lead to stresses that induce fracture. The determination
of whether the material will melt before fracture, melt, and not
fracture or will melt and fracture is dependent on the thermodynamic
and mechanical properties of the material, the intensity and wavelength of the laser irradiation, and the material thickness. Obviously, the thermal description of this complicated process involves
more than one mathematical expression. We will first approximate
the temperature profile resulting from irradiation of a finite,
nearly opaque plate to gain insight into laser induced failure and
then analyze the case for materials transparent to the incident
irradiation.

Assuming that the ceramic is irradiated so that the beam is
much larger than the thickness of the plate, the heat balance integral method of Goodman[8,9] can be used on a one-dimensional heat
conduction (through the thickness) problem that does not involve a
change in phase. The solutions derived depend on the relative
position of the "thermal layer," $\delta(t)$, i.e. the depth to which the
temperature diffuses in time, t. However, for the purpose of this
paper, we make the assumption that one temperature description will
suffice for the times being considered in this analysis. The temperature profile, $T(x,t)$, for an insulated rear surface is given by:

$$T - T_o = \frac{AQL}{2K} \left(1 - \frac{x}{L}\right)^2 + \frac{AQ}{K}\frac{D}{L}\left(t - \frac{L^2}{6D}\right) \qquad \begin{array}{c} 0 \leq x \leq L \\ t_\delta < t < t_m \end{array} \qquad (1)$$

where T_o is the ambient temperature on the rear surface, AQ is the
incident irradiance (with A being the absorbtance coefficient), K
the thermal conductivity, D is the thermal diffusivity, t is time,
x is the coordinate variable through the thickness, L, and $\delta(t) = \sqrt{6Dt}$. In reality for many materials at high fluxes, the front surface melts quite rapidly; however, it will be shown that the solution
with the assumption of no melting provides reasonable predictions of
time to fracture and stress profiles. It is further assumed that Eq.
(1) applies for all times before fracture.

Fracture of Opaque Materials

There have been a number of mathematical and numerical descriptions of the stress state in a brittle material which has been irradiated by continuous radiation. The actual mathematical description
is quite complicated; however, the problem is reduced in magnitude

if one notes that in most cases the temperature distribution through
the thickness only is the most important and that partial irradiation
of a front surface is likened unto the radiation of a circular plate
that has moment constraints about the edge. In reality, if the beam
does not flood the sample face, there is a radial distribution of
the temperature which leads to radial and circumferential stresses
on the front surface[10]--these can be important at low power densities. The problem is further complicated by the fact that, as the
material melts and vaporizes, the temperature distribution in depth
is nonuniform because the melt profile is not uniform. However,
ignoring these variations and noting that for a temperature distribution through the thickness, no matter what the boundary condition,
the maximum stress, σ_{max}, is expressible as:

$$\sigma_f = \sigma_{max} = \eta \frac{\alpha Y \Delta T_{max}}{(1 - \nu)} \qquad (2)$$

where σ_f is the fracture stress, Y is the elastic modulus, ν is
Poisson's ratio, η depends on the particular boundary condition and
temperature, and $T_{max} = T - T_0$. Assuming a maximum temperature just
below melting, we substitute Eq. (1) into Eq. (2) with $\eta = \frac{1}{2}$ and
x = 0.5L, and upon rearranging, we obtain:

$$t_f = \frac{L^2}{24D} + \frac{2\sigma_f(1 - \nu) \, KL}{\alpha \, Y \, DAQ} \qquad . \qquad (3)$$

This equation gives the time to fracture for brittle materials subjected to laser irradiation. We immediately notice that the first
term is independent of the incident power density. Thus, in the
limit, for very high power densities, i.e. $AQ > 2\sigma_f(1 - \nu)/\alpha Y$, the
time to fracture is proportional to the square of the thickness and
inversely proportional to the thermal diffusivity. At first, this
may seem contrary to what would be expected in conventional thermal
stress problems; however, it can be shown that the time to reach 95%
of the maximum stress is essentially the first term in Eq. (3).[12]
However, in conventional thermal stress problems, we are not usually
concerned with the time to reach maximum stress but the value of
the maximum stress that was reached. In laser irradiation problems,
the maximum stress, i.e. the fracture stress, is always reached in
the range of irradiation usually considered; it is just a question
of time.

For temperature distributions through the thickness, there is a
lower limit of irradiation at which stress levels no longer become
higher than that required to fracture the material. However, the
stresses on the front surface become much more important and fractur-
can occur due to hoop stresses.[10,11]

The stress distribution that results from the above treatment is presented for Pyroceram 9606 in Figure 1. There are two important points: (1) the maximum tensile stress moves from about the center toward the rear and (2) the rear surface stress is at first compressive, and then becomes tensile, eventually approaching the fracture stress of the material. Thus, fracture is expected to occur at or near the rear surface.

Fracture of Transparent Materials

For transparent ceramics, the incremental energy density, dE, in time, dt, at a depth, x, absorbed during laser irradiation can be expressed in terms of the absorbed power density, dQ_a:

$$dE = dt \, dQ_a \; . \tag{4a}$$

This in turn may be related to the change in temperature $dT(x)$

$$dE = \rho \, C_p \, dT(x) \, dx \tag{4b}$$

where ρ is the density, C_p is the heat capacity and dx is an incremental dimension through the thickness. For transmitting materials the transmitted intensity, Q', at a depth x into the material is attenuated exponentially through the thickness according to the Bouger-Lambert[13] law:

$$Q' = Q(1 - R) \, e^{-\beta x} \tag{5a}$$

where Q is the incident irradiance, R is the reflectivity at a surface, β is the absorption coefficient and x is the distance into the material. The change in the irradiance through the increment dx is given by differentiation of (5a):

$$dQ' = Q(1 - R) \, (-\beta) \, e^{-\beta x} \, dx \; . \tag{5b}$$

But this change must be equal and opposite to the increment absorbed

$$dQ' = -dQ_a \; . \tag{5c}$$

Thus, the description of the temperature profile through the thickness during irradiation may be obtained by combining (4a) and (4b) and substituting the differential dQ_a/dx obtained from Eqs. (5a, b, c).

$$dT(x) = \frac{dt}{\rho C_p} \frac{dQ_a}{dx} = \frac{dt}{\rho C_p} Q(1 - R) \, \beta e^{-\beta x} \tag{6a}$$

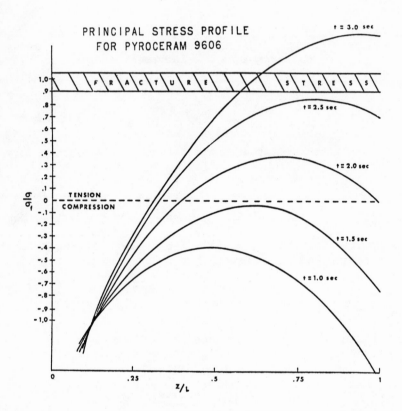

Figure 1.   Stress as a function of distance from the front surface,
z, for increasing times. Notice that the maximum tensile
stress moves toward the back surface with increasing
time.

or

$$T = (t/\rho C_p) \, Q(1 - R) \, \beta e^{-\beta x} \quad . \tag{6b}$$

The time to fracture, t_f, for transmitting materials, then is obtained by substituting Eq. (6b) into (2) and solving for t. The final result is:

$$t_f = \frac{\sigma_f(1 - \nu) \, \rho \, C_p}{\alpha Y(1 - R) \, A'Q} \tag{7}$$

where

$$A' = \beta e^{-\beta x} \tag{8}$$

with $T_o \approx 0$. The time to fracture is thus highly dependent on the absorption coefficient, as well as the thermal and mechanical properties of the material.

Penetration

The time to penetrate a material by laser irradiation can be easily derived. The total energy absorbed per unit area, qt_{bt}, must be balanced by the energy required to melt, vaporize, and remove the ceramic material, i.e.

$$qt_{bt} = \rho L \, C_p (T_m - T_o) + \rho L \, \Delta H_m + \rho L \, C_p (T_v - T_m) + \rho L \, \Delta H_v \tag{9}$$

where

$$q = AQ - h(T_{max} - T_o) - K_B \, \epsilon \, T_{max}^4$$

so that

$$t_{bt} = \frac{\rho L \, C_p (T_m - T_o) + \rho L \, \Delta H_m + \rho L \, C_p (T_v - T_m) + \rho L \, \Delta H_v}{q} \tag{10}$$

where A is the absorptivity, ρ is the density, C_p is the specific heat capacity at constant pressure, ΔH is the enthalpy, T_m, T_v, and T_{max} are the melt, vaporization, and maximum temperatures, respectively, h is the convective heat transfer coefficient, K_B is the Stephen-Boltman constant, and ϵ is the surface emissivity.

In general, burnthrough occurs at much higher power densities for transmitting materials, but Eq. (10) is still applicable. At

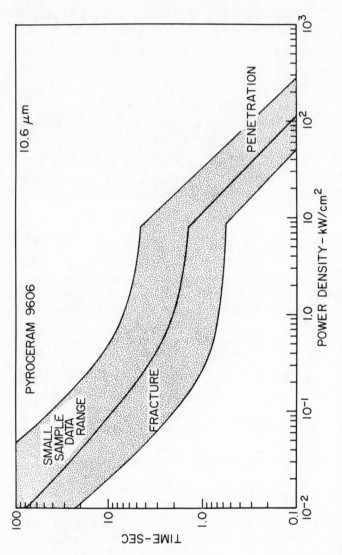

Figure 2.   Failure time from laser induced thermal stress as a function of irradiance (power density) for Pyroceram 9606. The shaded area represents the scatter for ~ 100 failures occurring during irradiation. The central solid line is Eqs. (3) and (10). Notice the crossover of the failure mechanism from fracture to burnthrough.

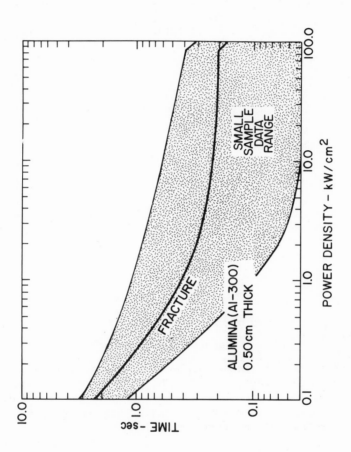

Figure 3. Laser induced thermal stress failure of Al-300 alumina as a function of irradiance. The shaded area represents ~ 50 failures occurring during irradiation. Notice that burn-through only occurs at extremely high power densities.

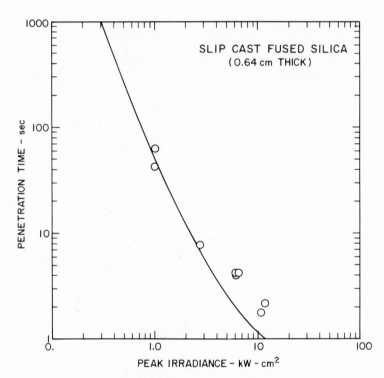

Figure 4. Burnthrough time as a function of laser irradiance for
           slip cast fused silica. The prediction of fracture
           (Eq. (7)) is well above the curve shown (Eq. (10))
           consistent with the fact that this material does not
           fracture under these conditions.

very high power densities, the materials behave as if they were opaque. The lowest value obtained between either Eqs. (3) or (7) and Eq. (10) determines failure time at a particular power density.

RESULTS AND DISCUSSION

Failure of Opaque Materials

The results of 10.6 micron laser irradiation on several opaque materials with a wide variety of properties are presented in Figs. 2 thru 4. The first three figures represent materials that: (1) melt, vaporize, and fracture (Fig. 2 - Pyroceram 9606); (2) have little or no melting and fracture (Fig. 3 - Sintered (A1300) Alumina); and (3) melt, but do not fracture (Fig. 4 - Slip Cast Fused Silica). Comparison of the data with the predictions of Eqs. (3) and (10) show that there is good agreement within the scatter of the data and the range of power density tested. The inherent scatter in the experimental data is shown by the crossed hatched sections which represent 50 to 100 plates. The fracture curve for slip cast fused silica, which is not shown in Fig. 4, is well above the penetration curve shown, as expected, because of its low thermal diffusivity and expansion coefficient. Many other glasses, e.g. soda-lime - silica glass, and other polycrystalline ceramics show the same behavior as described by Eq. (3), that is, at high power densities the time to fracture becomes almost independent of the power density and the fracture strength (or fracture toughness).

Effect of Mechanical Strength on Fracture

Eq. (3) shows that increase of fracture strength only significantly increases the fracture time at low power densities in agreement with limited observations. The exact stress distribution is quite complicated because of several sequential events which occur. Initially, a compressive stress developes on the front and rear surfaces (indicating an unconstrained plate condition). The compressive stress on the rear surface becomes tensile after a short time (approximately 0.1-1 second) and increases in agreement with observations from strain gages (Fig. 5). The maximum principal stress, σ_1, reaches a maximum just before failure. The reason for the shape and value of σ_2 is unknown, but typical. This tensile stress indicates that some constraint is occurring, probably due to bending of the plate. Meanwhile, the front surface in many cases has started to melt, and relief of the stress occurs. Strain gage measurements show decreases in back surface stresses before fracture, probably indicating this stress relief. In the case of Pyroceram, this relief continues throughout the thickness and there is competition between the relief and the loss of material due to melt. This competition is

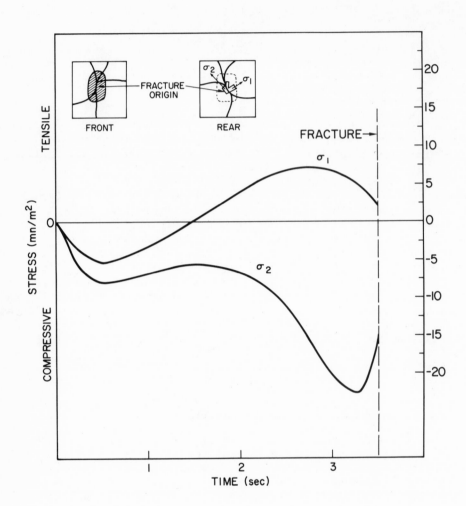

Figure 5.   Principal stresses, σ_1 and σ_2, as a function of time under
            laser irradiance derived from strain gage data.   The frac-
            ture originated within the bulk, close to the rear surface,
            under the irradiated area.   The crosshatched area on the
            front of the fracture specimen (upper left corner) is the
            irradiated area.   The solid lines on the specimen indicate
            the fracture pattern.

demonstrated by the fact that eventually, at high power densities, the material melts through before fracture.

Fracture surface analysis indicated that the failure in Pyroceram occurred within the bulk near the rear surface, in agreement with predictions. The stresses obtained from measurement of fracture mirrors indicated that fracture occurred at higher average values of stress, ~ 38,000 psi (262 MPa), than observed from the strain gage measurements on the rear surface, in agreement with the expected behavior. In most other materials, failure occurs on the rear surface because surface flaws are usually much more severe than bulk defects.[7]

Effect of Beam and Sample Geometry on Failure

Effect of Thickness. At high power densities, Eq. (3) indicates that the time to fracture is directly proportional to the thickness squared and inversely proportional to the thermal diffusivity. This relationship was tested for several brittle materials at power densities of 1 kw/cm^2 and greater by comparing the first term in Eq. (3) with data for several thicknesses as shown in Fig. 6. The agreement with the data indicates that Eq. (3) is a good qualitative representation of the behavior. At power densities lower than 1 kw/cm^2, the second term of Eq. (3) becomes more important for most materials and both terms must be used in order to obtain the true thickness dependence. This behavior is in contrast to other thermal stress models which show that the time to failure is inversely dependent on thickness or dependent on whether the front surface melts or not. Fig. 6 shows the same general behavior regardless of melting, sublimation, or no melting. The time to penetration is approximately linear with thickness as predicted so that the energy density of irradiation is approximately constant at high power densities. The low power density regime may be nonlinear due to radiation and convection losses, but this is only important in those materials that do not fracture.

Effect of Beam Size. Two extreme beam conditions should be considered as limits to the models presented (for opaque materials). When the beam diameter is small compared to the diffusion depth, $\sqrt{6Dt_f}$, then the one-dimensional thermal diffusion equation does not apply, and the proceeding analysis is no longer applicable. For small beam diameters (beam/sample area ≤ 0.15), the actual time to fracture can be much longer than predicted (Fig. 7). At the other extreme, when a sample is flood loaded, there is no longer a constraint to repress the thermal expansion, and stress relief from heating dominates the process and again longer times to fracture are observed (Fig. 7). In practice, there usually is a boundary either of material or mounting constraints, and it is very difficult to truly get flood loading.

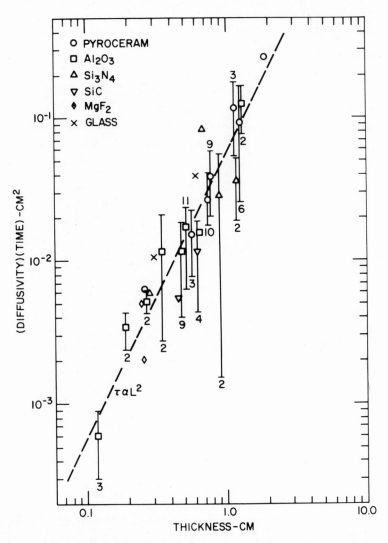

Figure 6.   Normalized fracture time, $\tau = Dt_f$, as a function of thick-- ness, L, for a variety of ceramic materials. $L^2/Dt_f = $ con- stant shows that a relationship like Eq. (3) represents the behavior of opaque brittle materials for high power density before burnthrough occurs (1 kw/cm^2 $\lesssim$ Q $\lesssim$ 10 kw/ cm^2).

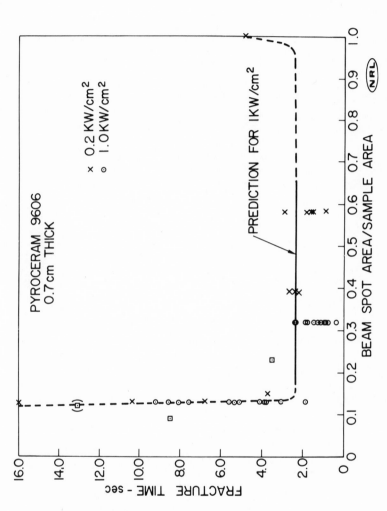

Figure 7. Fracture time as a function of beam/sample area for Pyroceram. In general, Eq. (3) predicts fracture for $0.15 \gtrsim$ beam area/sample area $\gtrsim 0.9$.

Failure of Transparent Materials

Irradiation of ceramics at wavelengths to which they are transparent can change the behavior due to the change in the absorption
coefficient and temperature distribution. Materials that are intrin-
sically transparent at lower wavelengths than 10.6 µm can still be
opaque due to scattering, e.g. alumina, Pyroceram 9606, Si_3N_4, and
slip cast fused silica all behave similarly at 10.6 µm and 3.8 µm.
Materials like MgF_2, ZnS, ZnSe, and infrared transmitting glasses
behave differently at different wavelengths depending on their
absorption coefficient.

In general, glass and polycrystalline ceramic materials are
transparent to 3.8 µm laser irradiation. This generally increases
the failure times of these materials quite substantially over those
at wavelengths to which these materials are opaque (e.g. 10.6 µm).
In these experiments, for irradiances greater than 3 kw/cm^2, all the
transparent materials tested (ZnS, MgF_2, calcium aluminate, and
calcium aluminosilicate glass) failed by either fracture or burnthrough in times less than two seconds (Figs. 8-11). The time required to fracture disks and spherical sections of MgF_2 irradiated
(3 - 10 kw) at United Technologies Research Center (UTRC) is shown
in Fig. 8. The solid line is the prediction of Eq. (7) for
0.25 cm thickness and an absorption coefficient of ~ 0.2/cm. The
thicker disks failed in slightly less time than the thinner, as is
predicted by Eq. (7) for transmitting materials, in contrast to the
predicted and observed (thickness squared) behavior at 10.6 µm (Eq.
(3)) where the material is opaque. The major difference of Eqs.
(3) and (7) is that the thickness dependence is drastically different
in the two cases because of the difference in the amount of beam
transmission. Limited data of ZnS show similar agreement with prediction (Eq. (7)) in Fig. 9. Figure 10 shows the current results
for calcium aluminate glass. The limited, small spot size data at
United Technologies Research Center (indicated by +) had indicated a
behavior that was quite different than expected by either Eq. (3) or
Eq. (7). However, the rest of the data are in accordance with the
prediction of Eq. (7), as shown. The one burnthrough datum shown at
United Technologies Research Center is misleading because of the
small beam diameter to thickness ratio. One would expect longer
times to fracture because of this, and, thus, burnthrough occurs
before fracture. At high power densities (≥ 40 kw/cm^2) burnthrough
is both predicted (Eq. (10)) and observed. However, there must be a
transition between Eq. (9) (fracture) and Eq. (10) (burnthrough) as
shown by the dashed line in Fig. 10. Whether this phenomenon is due
to surface absorption occurring at high power density, a bulk diffusion process, or radiation, is not known at this time. One sus-
pects that thermal diffusion is probably important, just as in the
opaque ceramic irradiation at high irradiance levels.

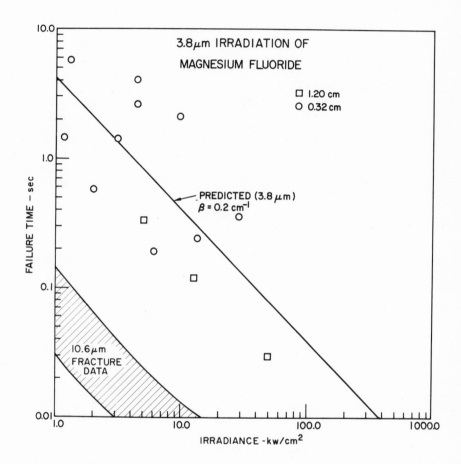

Figure 8.   Fracture time of MgF_2 as a function of irradiance. The
solid line is Eq. (7) for 0.32 cm thickness. Because of
the high degree of transmission, the fracture time of
MgF_2 is much longer than at 10.6 μm where the material
is opaque.

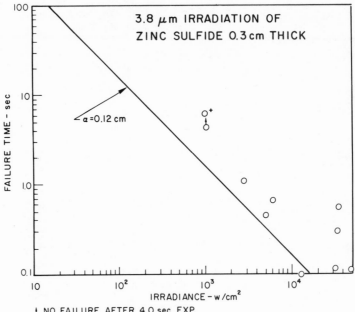

ᴉ NO FAILURE AFTER 4.0 sec EXP.

╋ POST FRACTURE AFTER 6.0 sec EXP.

Figure 9.   Fracture time of ZnS disks as a function of irradiance.
The solid line represents Eq, (7).

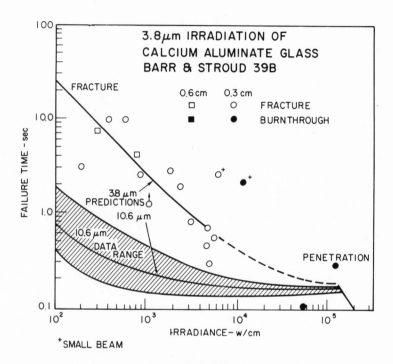

Figure 10   Failure time of Calcium Aluminate Glass (BS39B) as a
             function of irradiance. The solid lines for 3.8 µm and
             10.6 µm are the predictions for fracture of Eqs. (7) and
             (3) respectively. The solid line above 10^5 w/cm^2
             represents the prediction of Eq. (10) for penetration at
             both 10.6 and 3 8 µm. The dashed line represents the
             transition between fracture and burnthrough.

Predictions of failure for calcium aluminosilicate glass (Fig. 11) agree well with available data within the scatter typically encountered for glass and ceramics. Notice the intermixing of data for different thickness samples at 3.8 μm as opposed to 10.6 μm. The solid line shown in the figure is calculated for 0.44 cm thickness but essentially is applicable for 0.25 and 0.64 cm also since there is little thickness dependence. As with the calcium aluminate glass, a transition region (dashed line) is required to connect the fracture and burnthrough predictions.

Comparison of CW Laser Irradiation with Quench Test

An analysis of thermal stress theory shows that the approaches to analyzing CW laser irradiation and thermal quench data are basically the same.[14] High intensity ($\gtrsim 1$ kw/cm^2) CW laser irradiation effectively involves a large heat transfer coefficient, h, which is approximately greater than 10 cal/sec/°C/cm^2, whereas the water quench test depends on the exact condition of heat transfer and varies from h = 0.1 to 1.0 cal/sec/°C/cm^2. However, for most materials, this difference is not significant to final fractute behavior.

Ranking of materials for resistance to quench down, thermal stress fracture is generally made on the basis of the maximum temperature difference, ΔT_C, that can be tolerated such that the original room temperature strength, σ_I, is maintained. An important consideration to these tests is the residual strength (σ_R) that is left after the critical ΔT_C is reached. Depending on the particular application, ΔT_C and/or $\sigma_I - \sigma_R$ may be important. Rapid heating (quench up) is analytically similar to a quench down, except that the location of the maximum stress is on the inside of the sample instead of exterior as in the quench down, and the magnitude of the stress is not as severe for the same ΔT_C.[12]

Ranking of materials for resistance to thermal stress failure from laser irradiation can be made on the basis of time to failure (fracture or burnthrough), i.e. Eqs. (3) or (10). The time to reach 0.95 of the maximum stress in a thermally quenched sample is equivalent to the time to fracture under high intensity irradiation. It can be shown analytically that this time to fracture under laser irradiation and the time to reach 95% of the maximum stress is proportional to the thickness squared and inversely proportional to the thermal diffusivity,[11,12] as found for Eq. (3). The difference between the two cases is that the maximum stress reached in the quench test will not always result in a degradation of strength or fracture because the maximum ΔT_C has not been reached. Whereas in the CW laser irradiation case for the materials that fracture, we irradiate them until the fracture stress is reached, provided burnthrough does not occur first. A comparison between resistance to water quench and CW laser irradiation is given in Table I. The

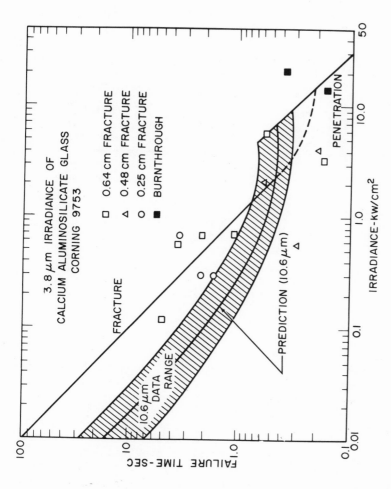

Figure 11. Failure time of Calcium Aluminosilicate Glass Plates as a function of irradiance. The solid line (fracture) is Eq. (7), and the solid line (prediction) within the scatter-band representing data is Eq. (3). The solid line above about 8 kw/cm^2 is Eq. (10) for burnthrough. The dashed line represents a transition between fracture and burnthrough.

conditions for the quench down and laser irradiation were selected
such that Biot's modulus was approximately equivalent for the two
cases. Notice that the rankings are equivalent if one compares the
time to fracture to the most commonly used criterion, ΔT_C, but are
not necessarily equivalent if σ_R/σ_I is used as the criterion for the
quench. In general, laser irradiation may be used to rank materials,
and these rankings can be compared to other heating environments provided that the geometry, i.e. primarily thickness, wavelength, i.e.
opaque or transparent, and heating rates are equivalent in the two
cases, e.g. by comparison of Biot's modulus.

TABLE I

THERMAL SHOCK OF CERAMICS

Material	Laser Induced t_f (sec)	$\dfrac{\sigma_R}{\sigma_I}\left(\dfrac{\text{ksi}}{\text{ksi}}\right)$	(Water) Quench ΔT_C ($^\circ$C)
50BN-50 Al$_2$O$_3$	20*	12/12	1200
70BN-30 Al$_2$O$_3$	15*	6/25	950
SCFS	8*	4/8	850
Pyroceram 9606	3	5/35	600
H P Si$_3$N$_4$	1.5	60/140	600
R.S. Si$_3$N$_4$	1	17/30	350
Alumina	0.3	15/45	225
MgF$_2$	0.1	12/26	150

* Burnthrough time

SUMMARY

Simple models have been presented which can predict the behavior
of nearly opaque and transparent ceramics exposed to high power laser
irradiation. These models show that the mechanism of damage can be
fracture or penetration, depending mainly on power density, beam
size, thickness, and thermal diffusivity. At high power densities,
the time to fracture of opaque materials is directly proportional to
the square of the thickness and inversely proportional to the thermal
diffusivity and independent of power density. Fracture time of
materials that are transparent to the incident irradiation is nearly
independent of thickness and inversely dependent on the power density
The penetration time of transparent and opaque materials is linear
with thickness and inversely proportional to power density. If the
beam diameter is of the order of $\sqrt{6Dt}$ or less or is larger than the
sample, longer times to fracture than predicted are observed. Laser
irradiation of opaque materials may be used to rank thermal stress
resistance and these rankings can be compared to other heating environments provided that geometry, wavelength, and heating rates are
equivalent in the two cases.

ACKNOWLEDGEMENTS

The authors thank G. DeLauder and his colleagues at Naval Research Laboratory for excellent photographic recording of the experiments. J. J. Mecholsky wishes to especially thank Professor C. T. Moynihan of Catholic University for helpful discussions of the thermal fracture models.

REFERENCES

1. E. L. Baardsen, D. J. Schmatz, and R. E. Bisaro, "High Speed Welding of Sheet Steel with a CO_2 Laser," Weld. J., 52(4): 227-229 (1973).
2. E. V. Locke and R. A. Hella, "Metal Processing with a High Power CO_2 Laser," IEEE/OSA Conf. on Laser Engineering, May 30-June 1, 1973, Washington, DC.
3. Un-Chul Paek and F. P. Gagliano, "Thermal Analysis of Laser Drilling Process," IEEE J. of Quantum Elec., 8:112-19 (1972).
4. G. K. Chu, "Laser Cutting of Hot Glass," J. Am. Cer. Soc., 54: 514-18 (1975).
5. R. M. Lumley, "Controlled Separation of Brittle Materials Using a Laser," Am. Cer. Soc. Bull. 48:850-54 (1969).
6. F. J. Grove, D. C. Wright, and F. M. Hamer, "Cutting of Glass with a Laser Beam," U.S. Pat. 3,543,979, Dec. 1970.
7. J. J. Mecholsky, S. W. Freiman, and R. W. Rice, "Fracture Surface Analysis of Ceramics," J. Mat. Sci., 11(4) (1976).
8. T. R. Goodman, "The Heat-Balance Integral and Its Application to Problems Involving a Change of Phase," Trans. ASME, (80) 335-342 (1958).
9. T. R. Goodman and J. J. Shea, "The Melting of Finite Slabs," Trans. ASME, J. Appl. Mech., (82) 16-24 (1960.
10. B. A. Boley and J. H. Weiner, "Theory of Thermal Stresses," J. Wiley and Sons (1960).
11. S. Timoshenko and J. N. Goodier, "Theory of Elasticity," McGraw Hill, Inc., 2nd Ed., 403 (1951).
12. D. P. H. Hasselman, "Thermal Shock by Radiation Heating," J. Am. Cer. Soc. (46) 229-34 (1963).
13. American Institute of Physics Handbook, D. E. Gray, Ed., 3rd Ed. A.I.P., 6-2 (1972).
14. J. J. Mecholsky, "A Summary of the Comparison of CW Laser Irradiation of Ceramics with Quench Tests," in Summary of Proceedings of the Workshop of Thermal Shock of Ceramics, Eds. Becher, Freiman, and Diness, Office of Naval Research, Arlington, VA, 22217, Dec. 1977.

INVESTIGATION OF THERMAL SHOCK RESISTANCE OF CERAMIC

MATERIALS UNDER PROGRAMMED HEATING

G. A. Gogotsi and Ya. L. Groushevsky

Institute for Problems of Strength
The Academy of Sciences of the Ukrainian SSR
Kiev, USSR

ABSTRACT

The report describes the equipment and test method for the thermal stress resistance of ceramic hollow cylindrical specimens under programmed heating conditions. The experimental data are recorded with the use of a computer. The analysis of the state of thermal stress of the specimens for elastic and nonelastic materials has been conducted, and the dependence of thermal shock resistance of ceramic materials upon their brittleness has been demonstrated. The relationships between the optimum thermal loading condition and specimen size have been established. The expedience of using the described technique in statistical investigations of ceramic materials is demonstrated.

INTRODUCTION

The limited deformability, insignificant stress relaxation, if any, and high brittleness of ceramics cause difficulties in ensuring serviceability of structures made from these materials. This provides the stimulation to carry out research aimed not only at improving these materials, but also at developing the experimental technique and analytical evaluations of their strength, particularly under conditions of thermal shock.

EXPERIMENTAL TECHNIQUE

The experiments were conducted on equipment for measuring thermal shock resistance[1] using a test method[2,3] involving a radial thermal loading of hollow cylindrical specimens. Such a method had been selected since it provides a great flexibility in temperature nonuniformities within the specimens resulting in occurrence of thermal stresses of sufficient magnitude to fracture almost any ceramic material.

In the tests, as shown in Fig. 1, the specimens (1) heated by a cylindrical heater (2), were stacked in a pile with a height equal to 3 to 4 outside diameters of the specimens. This arrangement ensured a uniform axial temperature field in two or three inner specimens in the stack, where all the measurements were taken. The equipment incorporated devices (3,4) which assured concentric alignment of the specimens and the heater, in order to assure radially symmetric heat flow within the specimens.

The heating of the specimens was controlled by an automatic temperature controller (units 6-8, Fig. 1) which allowed a thermal loading of the specimens according to any preset program. The program permits a steady state heating of the inside surface of the specimens at a rate of 0.01 to 50°C/min. The maximum test temperature in air is 1,400°C; in vacuum and in an inert atmosphere as high as 2,500°C. A cylindrical resistance furnace was used to raise the initial temperature of the specimens; special water coolers (4), were provided to increase the nonuniformity of the radial temperature field in specimens and to lower the temperature on their outside surface.

The temperature at different points in the specimens was measured with thermocouples radially arranged inside the specimens, and the moment of fracture of the specimens was determined from the measurements of strains on their outside surface and from acoustic signals that accompany the development of cracks.

The experimental data were recorded by potentiometers (9,10) and a device connected to a computer (15). The block diagram of the program for analysis of the data on the temperature and stressed state of the specimen is shown in Fig. 2. The results of the temperature measurements make up an experimental data file (EDF). Along with the parameters characterizing the operating mode of the device, calibration tables for thermocouples, and some data on physical properties of ceramics (modulus of elasticity, coefficient of linear thermal expansion, etc.) the EDF is entered into the computer.

The program allows recording of the emf's of all thermocouples and the corresponding temperatures, to find out a fracturing moment t_f, to plot temperature field as a function of time and radial coordinate, to calculate thermophysical properties, to compute a stressed state, etc. The results are presented in a

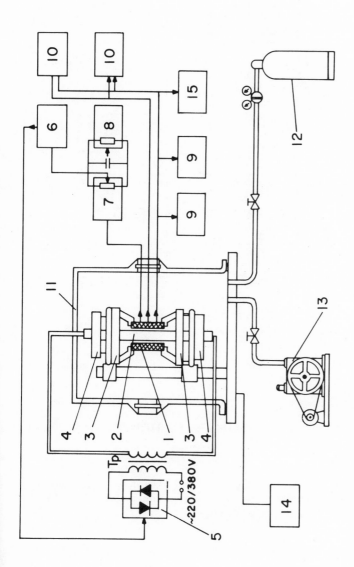

Fig. 1. Diagram of equipment for measuring thermal shock resistance: 1 – Stack of specimens; 2 – heater; 3,4 – alignment devices; 5 – transformer and thyristors; 6 – regulator; 7 – temperature measurement device; 8 – temperature controller; 9 – potentiometers; 10 – x–y recorders; 11 – vacuum chamber; 12 – pressure controller; 13 – vacuum pump; 14 – inert gas; 15 – signal converter for computer.

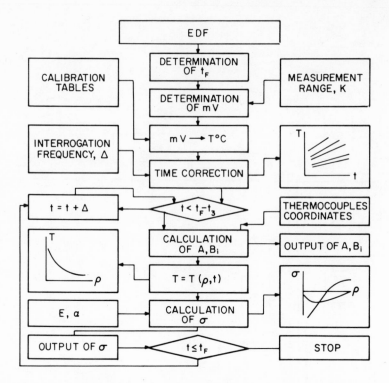

Fig. 2.   Block diagram for computer program for analysis of data
          on temperature and stress state in specimen.

form of tables and graphs and can also be displayed on the
computer's screen.

In order to establish the validity of the test method the
following ceramic materials were selected: alumina, yttria,
silicon-nitride and aluminum-boron nitride.   The main properties
of the materials are presented in Table 1.

ANALYSIS OF TEMPERATURE AND STRESS STATE OF SPECIMENS

It has been shown experimentally, that when the temperature
on the inside surface of specimens is raised at a constant rate,
a moment comes when the temperature in the specimens begins to
vary linearly.   This mode has been termed the Linear Heating
Mode.[4]   The temperature field of a specimen at the stage of the

Table 1. PROPERTIES OF CERAMIC MATERIALS UNDER STUDY

Material	Density, (g/cm^3)	Open porosity, (%)	Coefficient of linear thermal expansion, $(°C^{-1})$ $0 - 1,000°C$	Bending strength, kgf/cm^2	Limiting strains $x\ 10^2$ (%)	Elasticity modulus, $x\ 10^{-6}\ kgf/cm^2$
Alumina	3.97	0	10	1,904	5.9	3.39
Yttria	4.75	0.5	9.6	596	3.8	1.55
Silicon nitride	2.44	25	3.2	1,636	9.4	1.67
Aluminum–boron nitride	1.94	20	3.5	188	10.0	0.24

linear heating mode can be described as an asymptotic component of
the equation determining a linear unsteady-state heat conductivity,
or, in other words, a component which only depends on the boundary
conditions, whereas the other part of the solution, defined by the
initial temperature, tends to zero with increasing time.[5]   Thus,
the temperature field of a specimen can be described to a very
good approximation:

$$T = T_o + Vt \cdot \frac{1 - Bi\ \ln(x)}{1 - Bi\ \ln(x_o)} + \frac{V\ r_2^2}{A}\ F(x) \tag{1}$$

$$\text{where } F(x) = \frac{1}{1 - Bi\ \ln(x_o)} \cdot \left[\frac{x^2}{4}\ (1 + Bi) \right.$$

$$\left. - \frac{Bi\ x^2}{4}\ \ln x + L\ \ln(x) + M \right]$$

Here:

T = $T(\rho,t)$ temperature of specimen for distance ρ from its
center at time t, x = ρ/r_2 relative radius, x_o = r_1/r_2 relative
dimension, r_1 and r_2 inside and outside radii, V = rate of temperature rise at inner surface, A = thermal diffusivity, Bi = hr_2/λ =
Biot number, where h is the heat-transfer coefficient, and λ = the
thermal conductivity, T_o = initial temperature of specimen, L and
M = coefficient depending on parameters x_o and Bi.

The first two terms in Eq. (1) correspond to a steady-state
temperature distribution that would have been observed in a
specimen, in case the reached inside surface temperature had been
fixed at moment t.   The last term in Eq. (1) denotes the lag in
the temperature distribution behind the steady-state one, with the
lag being the greater, the higher is the rate of thermal loading,
and the lower is the thermal conductivity of the test specimen.

It follows from Eq. (1) that at the stage of the linear
heating mode, the function:

$$\eta(x) = \frac{T(x,t_2) - T(x,t_1)}{V(t_2 - t_1)} \tag{2}$$

calculated from the measurement of temperature, can be represented
by a linear plot in semi-logarithmic coordinates.   The function
describes the variation of the reduced temperature rise at a given
point of the specimen.

Fig. 3 shows as an example, the function $\eta(x)$ for various
conditions of thermal loading of alumina specimens whose outside
surface was cooled by free convection.   In this graph, the point
of intersection of the straight line with the ordinate axis

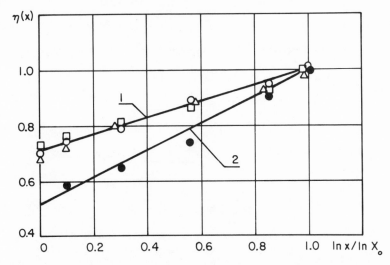

Fig. 3. Distribution of relative rates of heating in specimen: 1 - heat transfer by free convection; 2 - intensified heat transfer. ($\square$, V = 340°C/min; $\triangle$, V = 2°C/min; $\circ\bullet$, V - 170°C/min).

represents the ratio S between the rates of temperature variation on the outside and inside surfaces of the specimen, the ratio being related to the value of the Biot number by:

$$Bi = \frac{S - 1}{S \ln(x_0)} \tag{3}$$

derived from Eq. (1).

In the data represented in Fig. 3 (straight line 1) it is assumed that S = 0.73, which corresponds to the Biot number, Bi = 0.5. When the heat exchange on the outside surface of the specimen is intensified by a cooler with a copper heat-transfer wall (straight line 2 in Fig. 3), the value of Biot number equals 1.25. The use of a cooler with a rubber heat-transfer wall (6), ensuring a better contact with the specimen, raises the Biot number nearly by an order of magnitude.

The average thermal diffusivity for the test range is calculated from the data of the temperature distribution in a specimen by means of Eq. (1). Thus, for the yttria within a temperature range of 800-900°C, A = 0.011 cm^2/s. This is close to a thermal diffusivity calculated from the heat capacity, thermal conductivity and density shown in Fig. 4, which were determined from the same experimental data with the aid of an algorithm for

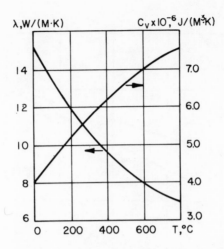

Fig. 4. Specific heat and thermal conductivity
of yttrium-oxide ceramic as a function
of temperature.

solving an inverse heat conductivity problem.

After the values of A and Bi have been found, Eq. (1) is
used to calculate the temperature fields as a function of time and
the radius as shown in Figs. 5a and 5b, as well as the rate of
rise of the inside surface temperature and the conditions of the
heat exchange between the outside surface and the surroundings.
Figs. 5b and 5c show, as an example, the temperature fields corresponding to heating rates of 170 and 2°C/min, the fields having
been calculated from the data of Fig. 5a.

The nonuniformity of the temperature field in a thermally
loaded specimen results in the generation of thermal stresses;
the stresses in linearly elastic material can be explained by the
well-known thermal elasticity relationships. It follows from
Eq. (1) that at the stage of the linear heating mode, the thermal
stresses σ_θ and σ_r, acting in circumferential and radial directions, are linear functions of time. Also, the stresses are
proportional to the rate of thermal loading.

The analysis of the stressed state of specimens made of
materials, which do not follow Hooke's law, was based on the
model of a nonlinearly elastic body, described by Kauderer.[7]

Assuming that to a first approximation a nonelastic deformation of a material, which results in fracture of its structure, occurs under the action of tensile stresses, the functions of volume compression and shear will be expressed in the form:

$$\frac{1}{K(\sigma_o)} = \frac{1}{K_o} \left[1 + K_1 \cdot H(\sigma_o) \cdot \sigma_o \right]; \quad G(\sigma_i) = G_o = \text{const} \qquad (6)$$

where

K_o = bulk modulus
G_o = shear modulus
K_1 = curvature of stress–strain curve
$H(\sigma_o)$ = Heaviside function defined by the expression:

$$H(\sigma_o) = \left\{ \begin{array}{l} 1 \\ \\ 0 \end{array} \right\} \quad \begin{array}{l} \text{at } \sigma_o \geq 0 \\ \\ \text{at } \sigma_o < 0 \end{array}$$

Then the calculation of the thermal stresses is reduced to solving the equations:

$$\rho \frac{d^2\sigma_r}{d\rho^2} + 3 \frac{d \sigma_r}{d\rho} + \alpha_T E \frac{dT}{d\rho} = \frac{E \cdot K_1}{K_o} \frac{d}{d\rho} [H(\sigma_o) \cdot \sigma_o^2]$$

$$\sigma_\theta = \frac{d\sigma_r \rho^2}{d\rho} \qquad (7)$$

which result from the equations of equilibrium and of compatibility of strains.

To determine the curvature K_1, we write down the stress-strain correlations for the case of a single-axis stressed state, which from Eqs. (4) to (6) are:

$$\varepsilon_1 = \left\{ \begin{array}{ll} \dfrac{\sigma_1}{E} + \dfrac{K_1}{9K_o} \cdot \sigma_1^2 & \text{at } \sigma_1 \geq 0 \\ \\ \dfrac{\sigma_1}{E} & \text{at } \sigma_1 < 0 \end{array} \right. \qquad (8)$$

From the definition of the material brittleness measure value:[8]

$$\chi = \frac{\sigma_{1\lim}^2}{2E \displaystyle\int_0^{\varepsilon_{1\lim}} \sigma \, d\varepsilon} \qquad (9)$$

and from Eq. (9), it follows that:

$$\frac{K_1}{K_o} = \frac{27}{4} \quad \frac{1 - \chi}{E \, \sigma_{lim} \, \chi}$$
(10)

where

σ_{lim} = strength,
ε_{lim} = limiting strain of material, corresponding to σ_{lim},
E = Young's modulus.

It is worth noting that the assumptions expressed by the correlations (6) and (9) are valid for ceramics for which, as is well known, the strength and proportional limits in compression exceed those in tension 5 to 10 times.

Comparing the stresses built up in the wall of specimens made of materials having the same elastic moduli, but possessing different degrees of nonelasticity expressed by a brittleness level value, indicates that with a decrease in the brittleness level from 1 down to 0.6, maximum circumferential stresses are reduced by 25-30%. The thermal stresses in specimens of both nonelastic and elastic materials are linear functions of time at the stage of the linear heating mode. Similarly, a proportional dependence of the thermal stresses on the rate of heating of the specimens is noted.

It has been assumed that fracture of a specimen occurs at the moment when the maximum tensile stress equals the value of tensile strength. At this moment the thermal state is characterized by a critical temperature difference for fracture and by a mean temperature of the specimen. The dependence of the temperature difference in the specimen wall on test condition and material properties for the case of a linearly elastic ceramics ($\chi = 1$) can be represented by the function:

$$\Delta T = \frac{\Delta J}{\overline{\Delta J}} R + \frac{V r_2^2}{A}\left(\Delta F - \frac{\overline{\Delta F \cdot \Delta J}}{\overline{\Delta J}}\right)$$
(11)

where

$\Delta T = T(r_1) - T(r_2)$ and R = limiting value of ratio $\dfrac{\sigma_\theta (r_2)}{\alpha_T E}$
for the material. Also, the following relation is valid for all the functions $J(x)$, $F(x)$, included in Eq. (11):

$$J(x) = \frac{1 - Bi \cdot \ln(x)}{1 - Bi \cdot \ln(x_o)} \quad \text{and} \quad F(x) = \text{function from Eq. (1).}$$

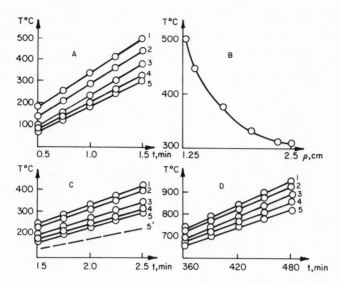

Fig. 5. Calculated and experimental temperature distributions as a function of time (A,C,D) and position (B). (Solid lines represent experimental data; dashed lines – calculated data).

In accordance with the model, the correlation between the components of the stress-strain tensor is represented by the expressions:

$$\varepsilon_\theta = \frac{\sigma_o}{K(\sigma_o)} + \frac{1}{2G(\sigma_i)} \, (\sigma_\theta - \sigma_o) + \alpha_T T \qquad (4)$$

$$\varepsilon_r = \frac{\sigma_o}{K(\sigma_o)} + \frac{1}{2G(\sigma_i)} \, (\sigma_r - \sigma_o) + \alpha_T T \qquad (5)$$

where

$\varepsilon_\theta, \varepsilon_r$ = strain of specimens in circumferential and radial directions, respectively
$\sigma_o = 1/3(\sigma_r + \sigma_\theta)$ = mean stress
$K(\sigma_o)$ = function of volume compression
$G(\sigma_i)$ = function of shear
σ_i = intensity of tangential stresses
α_T = coefficient of linear thermal expansion.

The relation thus obtained explicitly expresses the dependence of the critical temperature difference for fracture in terms of the rate of heating; it also shows its dependence upon the specimen size and conditions of heat transfer, and the properties of the material. A similar relation is valid also for the mean temperature of a specimen.

Figure 6 shows the results for specimens made of materials possessing various degrees of nonelasticity, (providing that the rest of their mechanical and thermophysical properties being identical). As can be seen, the critical temperature difference for fracture increases with an increase in degree of non-linearity. It should be noted that, as an example, the calculated data for aluminum-boron nitride ($\chi = 0.59$), is close to the one obtained from the experimental data.

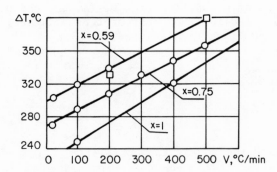

Fig. 6.   Dependence of critical temperature difference for fracture on rate of heating and degree of non-elasticity of ceramics: o - calculated; □ - experimental.

In general, all the above discussions allow one to draw the conclusion that it is possible to apply the data obtained from the results of tests at one rate of heating to evaluation of these obtained at another rate of thermal loading or at different conditions of heat transfer on the outside surface of the specimen. This makes it possible to reduce the scope of experimental work.

INVESTIGATION OF EFFECTS OF INDIVIDUAL FACTORS

The problem of selecting specimen size and conditions of thermal loading to be used in the tests can also be solved in the course of the above-mentioned analysis when the thermal stress state of a hollow cylindrical specimen has been determined. Consequently, it is necessary to determine inside and outside radii as well as a rate of thermal loading in a specimen in a way, so that in its most stressed zone the nature of temperature and stress variations as a function of time would approach that of the piece in natural conditions. Using Eq. (1), the temperature on the outside surface and the tensile stresses acting on the surface can be presented in a form:

$$T(r_2) = T_0 + T_0 \cdot \nu \cdot f_T\left(r_1, r_2, \frac{hr_2}{\lambda}, t_k\right) \tag{12}$$

$$\sigma_\theta(r_2) = \alpha_T \cdot E \cdot \nu \cdot f_\sigma\left(r_1, r_2, \frac{hr_2}{\lambda}, t_k\right) \tag{13}$$

where

t_k = time during which temperature T and stress σ in the piece are reached (the explicit expressions for the functions f_T and f_σ are not presented, as being too cumbersome).

The problem set forth as a problem of minimizing the function will be formulated like this

$$W(\nu, r_1, r_2) = q(\sigma^* - \nu\, f_\sigma)^2 + (1 - q)(T^* - \nu\, f_T)^2 \tag{14}$$

which represents a departure of calculated values from the given ones.

Here, $\sigma^* = \dfrac{\sigma}{\alpha_T E\, T_0}$; $T^* = \dfrac{T - T_0}{T_0}$; and q is a weight parameter varying from 0 to 1; the best approximation to the given temperature is when $q = 0$, and the best approximation to the given stress, is when $q = 1$.

Inasmuch as the relations (12) and (13) are linear with respect to the rate of heating, the value of ν, which minimizes the function W at the given values r_1 and r_2, is defined by the expression

$$\nu(r_1, r_2) = \frac{q \cdot \sigma^* \cdot f_\sigma + (1 - q) \cdot T^* \cdot f_T}{q \cdot f_\sigma^2 + (1 + q) \cdot f_T^2}$$

Here, the values of r_1 and r_2 vary within a region whose area is confined by the dimensions and wall thickness of the specimen; these in turn depend upon the representative dimensions of a ceramic piece and the capabilities of the equipment. The required values of r_1 and r_2 are determined by using any one of the known methods for finding the extremum of functions. As an example of the above-discussed approach, presented below is the calculation of the parameters ν, r_1 and r_2 for simulating the working conditions of a silicon-nitride piece which alternatively undergoes 1.5 minutes heating up to 600°C followed by 30 minutes cooling. Preliminary tests of hollow cylindrical specimens have shown that with cooling of the outside surface due to a free convection the ratio of the heat transfer coefficient to the value of thermal conductivity of the material is about 0.12 cm^{-1}. Minimizing Eq. (14) has yielded the required rate of heating, $V = 915°C/min$, the inside radius $r_1 = 1.25$ cm, and the outside radius $r_2 = 3.75$ cm; the value of outside surface temperature and maximum thermal stresses, at the time $t_k = 1.5$ minutes, were found to be 599°C and 1,090 kgf/cm^2 respectively. Thus the problem of simulating a thermal stressed state can be considered to be solved.

The method of a radial heating of hollow cylindrical specimens proves to be useful, not only for simulating the actual conditions of loading in investigating the structural strength of ceramics, but also for studying the effect of process factors on serviceability of these ceramic materials. In particular, the dependence of thermal shock resistance of ceramics upon mechanical working conditions is found to be of interest. As an example, data obtained for alumina ceramics are listed in Table 2. The data show that under conditions of thermal shock, just as is the case of mechanical loading, the scattering of the results is fairly described by the Weibull distribution.[9] The change in the shape of the distribution curve, due to grinding of specimens (if compared to the testing data of specimens before grinding) is similar to that in four-point bending. It is also appropriate to point out that when hollow cylindrical specimens are tested to thermal loading, it is possible to observe particular features of the fracture process running under various conditions of heating and to determine the state of stress at the specimen surface.

CONCLUSIONS

The data presented in this report indicate that the technique and analysis of measuring the thermal shock resistance of hollow cylindrical specimens as described in this report considerably extends the potentialities of this conventional method for studying the serviceability of ceramics under conditions of thermal shock.

Table 2. EFFECT OF MACHINING ON STRENGTH OF CERAMIC MATERIALS

Type of specimen	State of specimens	Average fracturing difference, $\Delta t°C$	Standard deviation, °C	Variation factor, %	Weibull modulus
Alumina thermal shock specimen	before grinding (1)	193	36	19	6.3
	ground (2)	204	25	12	8.9

Type of specimen	State of specimens	Average bending strength, (mor) kgf/cm^2	Standard deviation, kg/cm^2	Variation factor %	Weibull modulus
Silicon nitride bend specimen	as-supplied (1)	1,464	394	27	4.0
	ground (2)	1,636	205	13	8.4

REFERENCES

1. G. A. Gogotsi, "Ogneupory" ["Refractories" USSR], 9(1976), 50.
2. Proc. of Symp. on Thermal Fracture, Journ. Amer. Cer. Soc.,
    38(1955), 1-44.
3. G. A. Gogotsi, "Problemy prochnosti" ["Problems of strength"],
    5(1974), 64.
4. G. A. Gogotsi, "Problemy prochnosti" ["Problems of strength"],
    1(1970), 96.
5. G. S. Pisarenko, G. A. Gogotsi, Ja. L. Groushevsky, "Problemy
    prochnosti" ["Problems of strength"], 4(1978), 36.
6. G. A. Gogotsi, A. G. Gashchenko, N. N. Radin, Inventor's
    Certificate No. 3,94,707 with priority of October 22, 1971.
7. G. Kauderer, Nelinejnaja mekhanika [Nonlinear Mechanics].
    Moscow, "IL", 1961.
8. G. A. Gogotsi, "Problemy prochnosti" ["Problems of strength"],
    10(1973).
9. G. J. Hahn and S. S. Shapiro, Statistical Models in
    Engineering. John Wiley and Sons, N.Y. (1967).

THERMAL DEFORMATIONS AND STRESSES

IN COMPOSITE MATERIALS

I.M. Daniel

IIT Research Institute

10 W. 35th Street, Chicago, Illinois   60616

ABSTRACT

Composite materials are highly anisotropic thermally with the
coefficient of thermal expansion in the fiber direction much lower
than that in the transverse to the fiber direction. Coefficients
of thermal expansion in unidirectional and multidirectional laminates
can be calculated by using the properties of the constituents and
lamination theory. Residual stresses are introduced in multidirectional laminates during curing as a result of thermal anisotropy.
These stresses have been investigated analytically and experimentally.
It was found that the significant strains recorded during the cooling
stage of curing correspond to thermal expansion of the laminate.
Residual or restraint strains are computed from measured restrained
and unrestrained thermal expansions. Residual stresses are computed
using appropriate orthotropic constitutive relations. Results have
been obtained from a variety of materials including boron, graphite,
Kevlar, S-glass and hybrids with epoxy or polyimide matrices, for
a variety of lamination angles. It was found that residual stresses
do not relax appreciably with time. Results show that, for graphite
and Kevlar laminates, residual stresses at room temperature are high
enough to have caused damage in the transverse to the fiber direction.

INTRODUCTION

Composite laminates are made by bonding and curing together
a number of unidirectionally reinforced plies. The basic lamina
or ply consists of an array of parallel fibers in a matrix. Two
types of residual thermal stresses are induced during curing or
subsequent thermal variations, micro-residual stresses produced
by matrix shrinkage around the reinforcing fibers, and macro-residual

or lamination stresses caused by anisotropic thermal deformations of the various plies.

Micro-residual stresses have been analyzed by means of two- and three-dimensional photoelastic models.[1-7] In some cases these stresses can be beneficial as they tend to increase the bond strength between fibers and matrix. Shrinkage stress distributions are similar to thermal stress distributions arising from the difference in thermal expansion in the fiber and matrix materials when the composite is exposed to a uniform change in temperature.

Lamination stresses are induced during curing or by changes in temperature. They are a function of many parameters, such as the ply orientation and stacking sequence, fiber content, curing temperature and other variables. They can reach values comparable to the transverse strength of the ply and thus induce cracking. Residual stresses in each ply are equilibrated with interlaminar shear stresses transmitted from adjacent plies and thus may result in ply separation. In the design and evaluation of composite structures one must take these residual stresses into account and superimpose them onto those produced by subsequent external loading and environmental fluctuations. Lamination residual stresses have been analyzed theoretically using lamination theory and properties of the constituent materials.[8-12] Experimentally, they have been measured and determined using embedded strain gage techniques.[13-18] Residual stresses were computed from measured restrained and unrestrained thermal strains using appropriate orthotropic constitutive relations. Results have been obtained for a variety of material including boron, graphite, Kevlar, S-glass and hybrids with epoxy or polyimide matrices, for a variety of lamination angles.

THERMAL DEFORMATIONS

Theoretical Results

The coefficient of thermal expansion α for a single ply or a unidirectional laminate, can be calculated by knowing the coefficient of the fiber and matrix, their geometric arrangement and mechanical properties.[19-20] Schapery's[1] energy theory yields the following expressions for a unidirectional fibrous composite:

$$\alpha_{11} = \frac{\alpha_f E_f V_f + \alpha_m E_m V_m}{E_f V_f + E_m V_m} \tag{1}$$

$$\alpha_{22} = \alpha_f V_f (1+\nu_f) + \alpha_m V_m (1+\nu_m) - \alpha_{11}(\nu_f V_f + \nu_m V_m) \tag{2}$$

where α_{11}, α_{22} = coefficients of thermal expansion for composite lamina in the fiber and transverse to the fiber directions, respectively.

α_f, α_m = coefficients of thermal expansion for fiber and matrix, respectively.

E_f, E_m = fiber and matrix moduli, respectively.

V_f, V_m = fiber and matrix volume fractions, respectively.

ν_f, ν_m = fiber and matrix Poisson's ratios, respectively.

For realistic graphite/epoxy composites with very low α_f and high E_f the expression for the transverse coefficient of expansion can be approximated by

$$\alpha_{22} \cong \alpha_f V_f + \alpha_m V_m (1+\nu_m) \tag{3}$$

The expansion coefficients along an arbitrary coordinate system x-y rotated by an angle θ with respect to the fiber direction can be obtained by a linear coordinate transformation as

$$\alpha_x = \alpha_{11}\cos^2\theta + \alpha_{22}\sin^2\theta$$

$$\alpha_y = \alpha_{11}\sin^2\theta + \alpha_{22}\cos^2\theta \tag{4}$$

$$\alpha_{xy} = (\alpha_{22} - \alpha_{11}) \sin\theta\cos\theta$$

The effective coefficients for multidirectional laminates can be evaluated from the properties of the single lamina by means of lamination theory.[21-24] For a balanced symmetric laminate the effective coefficients are:

$$\alpha_x = \frac{A_{22}N_x^T - A_{12}N_y^T}{(A_{11}A_{22} - A_{12}^2)\Delta T}$$

$$\alpha_y = \frac{A_{11}N_y^T - A_{12}N_x^T}{(A_{11}A_{22} - A_{12}^2)\Delta T} \tag{5}$$

where

$$A_{ij} = h \sum_{k=1}^{n} \bar{Q}_{ij}^k \tag{6}$$

h = lamina thickness

$\bar{Q}_{ij}^k$ = stiffness matrix for k^{th} lamina referred to x-y axes of laminate

N_x^T, N_y^T = thermal force resultants defined as follows:

$$N_x^T = h\Delta T \sum_{k=1}^{n} (\bar{Q}_{11}^k \alpha_x^k + \bar{Q}_{12}^k \alpha_y^k + 2\bar{Q}_{16}^k \alpha_{xy}^k)$$

$$N_y^T = h\Delta T \sum_{k=1}^{n} (\bar{Q}_{12}^k \alpha_x^k + \bar{Q}_{22}^k \alpha_y^k + 2\bar{Q}_{26}^k \alpha_{xy}^k)$$

(7)

$\alpha_x^k, \ \alpha_y^k, \ \alpha_{xy}^k$ = coefficients of thermal expansion of kth lamina, referred to x-y axes of laminate.

ΔT = temperature change.

The expressions above are linear elastic and do not take into consideration the viscoelastic response of the matrix at elevated temperatures. Coefficients of thermal expansion for a general [0/±45/90]$_c$ boron/epoxy laminate at room temperature and at the curing temperature of 450 degK (350°F) are shown in Fig. 1 as a function of ply orientation and ply composition of the laminate.[22] Thus, for a ply configuration of [0₂/±45]$_c$ the coefficients of thermal expansion at room temperature would be

$$\alpha_{11} = 4.0 \times 10^{-6} K^{-1} (2.2 \mu\varepsilon/°F)$$

$$\alpha_{22} = 10.8 \times 10^{-6} K^{-1} (6.0 \mu\varepsilon/°F)$$

These coefficients can be compared with the unrestrained single-ply coefficients for the 0-deg ply

$$\alpha_{11} = 4.5 \times 10^{-6} K^{-1} (2.5 \mu\varepsilon/°F)$$

$$\alpha_{22} = 23.4 \times 10^{-6} K^{-1} (13 \mu\varepsilon/°F)$$

The differences between the coefficients of the angle-ply laminate and those of the single ply give a measure of the residual strains present in that ply within the laminate.

Experimental Methods

Thermal expansion in unidirectional and angle-ply laminates has been measured experimentally for various materials and over wide temperature ranges. Freeman and Campbell[25] used a Leitz dilatometer and strain gages to measure thermal expansion in graphite fiber composites over a temperature range of 78 to 561°K (-320 to 550°F). Freund[26] used a vacuum interferometric dialatometer to measure thermal expansion in high-modulus graphite/epoxy laminates for applications to mirror mounts requiring high dimensional stability. An accuracy of ±1 x 10⁻⁸/°K was mentioned. Thermal expansion has also been measured by Wang et al.[27] and the author[13-18] for a variety of composite materials using strain gages. Commercial strain gages have been found suitable for measuring small thermal strains in composites. To properly interpret the strain gage

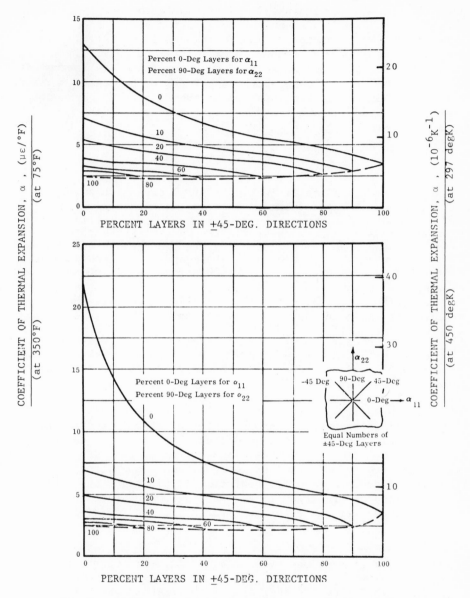

Fig. 1 Coefficients of Thermal Expansion for $[0/90/\pm45]_c$ Boron/
Epoxy Laminates at Room Temperature and 450 $\overline{\text{deg}}$ $\overset{\circ}{\text{K}}$ (350°F)
(Ref. 22)

output ε_a (apparent strain), it is necessary to separate this output into the component ε_t due to the deformation of the specimen (thermal strain) and the component ε_g due to changes in resistivity and gage factor of the gage with temperature (thermal output). To determine ε_g a reference specimen of known thermal expansion is instrumented with the same type of gage as the specimens to be tested. The true thermal strain ε_t is then obtained by subtracting algebraically from the recorded apparent strains ε_a the output of the gage on the reference material ε_r and adding the known expansion of the reference material ε_{tr}

$$\varepsilon_t = \varepsilon_a - \varepsilon_r + \varepsilon_{tr} \tag{8}$$

Fused quartz with a coefficient of thermal expansion of 0.7 x $10^{-6}K^{-1}$ (0.4$\mu\varepsilon$/°F) and titanium silicate with a coefficient of 0.03 $10^{-6}K^{-1}$ (0.017$\mu\varepsilon$/°F) have been used as reference materials. Many types of gages have been used. In the case of graphite/epoxy composites and especially when gages are embedded between plies it is necessary to use fully encapsulated gages with insulated leads. Micro-Measurements gages of the WK-00 series were found most suitable as they have a very low purely thermal output. The apparent strain recorded with a typical gage of this series on a titanium silicate specimen is shown in Fig. 2 as a function of temperature.

In one recent application of strain gages to measurement of thermal deformations,[15] encapsulated gages (WK-00-125TM-350, Option B-157) were embedded between the plies during laminate assembly. The attached ribbon leads were sandwiched between thin (0.013 mm; 0.005 in.) polyimide strips. A thermocouple was also embedded in each laminate. The instrumented specimens, including a reference titanium silicate specimen, were subjected to the curing and postcuring cycles in the autoclave. Strain gage and thermocouple readings were taken throughout. Subsequently, the same specimens were subjected to a thermal cycle from room temperature to 444 degK (340°F) and down to room temperature. Strain gages and thermocouple were recorded at 5.5 degK (10°F) intervals. The true thermal strain were obtained from the recorded apparent strains as discussed above.

Thermal Strains

Thermal strains measured on eight-ply unidirectional graphite/ epoxy, Kevlar 49/epoxy and S-glass/epoxy specimens are shown in Fig. 3. Both Kevlar 49/epoxy and graphite/epoxy exhibit negative thermal strains in the longitudinal (fiber) direction. The Kevlar 49/epoxy exhibits the largest positive transverse and negative longitudinal strains. The S-glass/epoxy undergoes the lowest thermal deformation in the transverse direction and the highest (positive) in the longitudinal direction. Coefficients of thermal expansion computed from such data are tabulated in Table 1 for eight composite materials.[13,15] Coefficients are listed at room temperature, 297 degK (75°F), and at the elevated temperature of

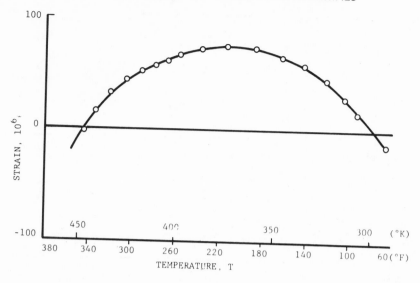

Fig. 2   Apparent Strain as a Function of Temperature of WK-00-
125TM-350 Gage Bonded on Titanium Silicate

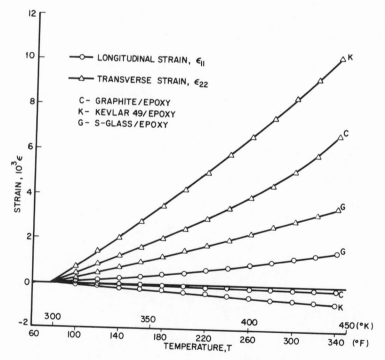

Fig. 3   Thermal Strains in Unidirectional Composites as a Function
of Temperature

Table 1

Thermal Expansion Coefficients of Unidirectional
Composite Materials

Material	Longitudinal Coefficient of Thermal Expansion, α_{11}, 10^{-6}K^{-1} ($\mu\varepsilon$/°F)		Transverse Coefficient of Thermal Expansion, α_{22}, 10^{-6}K^{-1} ($\mu\varepsilon$/°F)	
	297degK(75°F)	450degK(350°F)	297degK(75°F)	450degK(350°F)
Boron/Epoxy (Boron/AVCO 5505)	6.1 (3.4)	6.1 (3.4)	30.3 (16.9)	37.8 (21.0)
Boron/Polyimide (Boron/WRD 9371)	4.9 (2.7)	4.9 (2.7)	28.4 (15.8)	28.4 (15.8)
Graphite/Epoxy (Modmor I/ERLA 4289)	-1.1 (-0.6)	3.2 (1.3)	31.5 (17.5)	27.0 (15.0)
Graphite/Epoxy (Modmor I/ERLA 4617)	-1.3 (-0.7)	-1.3 (-0.7)	33.9 (18.8)	83.7 (46.5)
Graphite/Polyimide (Modmor I/WRD 9371)	-0.4 (-0.2)	-0.4 (-0.2)	25.3 (14.1)	25.3 (14.1)
S-Glass/Epoxy (Scotchply 1009-26-5901)	3.8 (2.1)	3.8 (2.1)	16.7 (9.3)	54.9 (30.5)
S-Glass/Epoxy (S-Glass/ERLA 4617)	6.6 (3.7)	14.1 (7.9)	19.7 (10.9)	26.5 (14.7)
Kevlar/Epoxy (Kevlar 49/ERLA 4617)	-4.0 (-2.2)	-5.7 (-3.2)	57.6 (32.0)	82.8 (46.0)

450 degK (350°F). All graphite fiber composites exhibit negative
thermal expansion in the fiber direction. The polyimide matrix
composites do not show any variation of thermal coefficients with
temperature. This is true at least up to the postcuring temperature
of 589 degK (600°F).

Freeman and Campbell[25] measured the longitudinal and transverse
dilatation of five graphite composites over a temperature range of 78
to 561 degK (-320 to 550°F). The moisture content was found to have
a significant effect on thermal deformation with a permanent shrinkage
resulting after thermal cycling. Expansion coefficients were also
measured in ±θ angle-ply graphite laminates. The coefficient in the
0-deg. direction decreases with increasing θ up to θ = 30-deg,
thereafter the influence of the resin becomes predominant as shown
by a rapid increase in the coefficient α with θ. Thermal deformation
varies fairly linearly with temperature for angle-ply laminates with
lamination angles less than 45-deg. Above this angle the increasing
influence of the resin makes thermal deformations nonlinear. Experimental results were in good agreement with theoretical ones obtained
by using lamination theory.[21] Orthotropic laminates such as $[0/90]_c$,
$[\pm45]_c$ and $[0/90/\pm45]_c$ have equal coefficients along two principal
material axes. Theoretically, all three types of laminates above
are supposed to have the same coefficients along the 0-deg and 90-deg
directions. However, experimental measurements show that the $[0/90]_c$
laminate has slightly lower thermal expansion than the other two
because of the higher percentage of fibers in the directions of
measurement. High modulus fibers produce laminates with low thermal
coefficients. These coefficients are reduced further by increasing
the fiber volume ratio. Theoretically, a quasi-isotropic laminate
of $[0/\pm60]_c$ layup with high modulus fibers and a fiber volume ratio
of 0.62 would result in zero thermal expansion. With the exception
of unidirectional composites, the thermal coefficient was found
to increase nonlinearly with temperature.

A review of thermal and other properties of composites in the
cryogenic temperature range has been given by Kasen[28] Results
by Toth et al.[29] for glass/epoxy in the 5 to 295 degK (-450 to 71°F)
range show that the coefficient of thermal expansion decreases with
temperature approaching zero near absolute zero. Longitudinal
thermal deformations in the cryogenic temperature range were
discussed for a variety of advanced composites.[28] Borsic/Aluminum
and boron/epoxy show contraction with decreasing temperature.
Graphite/epoxy shows very little thermal expansion (small negative
coefficient) in the fiber direction down to 77 degK (-417°F).
Thereafter there seems to be a slight reversal in contraction.
Kevlar 49/epoxy shows appreciable thermal expansion with decreasing
temperature with the (negative) coefficient of thermal expansion
approaching zero near absolute zero.

THERMAL STRESSES

Residual Strains

An important result of the anisotropic thermal expansions of
composite plies is the introduction of lamination residual or thermal
stresses in angle-ply laminates during curing. These stresses have
been investigated recently both analytically and experimentally.[8-18]
Residual stresses during curing have been measured in a variety
of angle-ply laminates using embedded strain gage techniques.
Unidirectional and angle-ply specimens were instrumented with surface
and embedded gages and thermocouples and the output recorded during
curing and postcuring. The unidirectional specimen was used as
a reference to determine the unrestrained stress-free thermal
expansions of an individual ply.

It was found that apparent strains recorded during the heating
stage of the curing cycle are not significant as they correspond to
the fluid state of the matrix resin. Residual stress buildup
occurs only upon solidification of the matrix resin at the peak
curing temperature and during subsequent cooldown. Strains measured
during the cooldown stage of curing as well as those measured during
postcuring correspond to the thermal expansion of the laminate.

Thermal strains measured in a unidirectional graphite/epoxy
laminate are shown in Fig. 4a with room temperature as the reference
level. Thermal strains measured in a $[0_2/\pm45]_s$ graphite/epoxy
angle-ply laminate during the cooling stage of posturing are shown
in Fig. 4b. The residual stresses induced in each ply correspond to
the so-called restraint strains, i.e., the difference between the
unrestrained thermal expansion of that ply (obtained from the
unidrectional specimen) and the restrained expansion of the ply
within the laminate (obtained from the angle-ply specimen). Restrain
or residual strains in the 0- and 45-deg plies of the $[0_2/\pm45]_s$
graphite/epoxy laminate are plotted in Figs. 5a and 5b as a function
of temperature with room temperature as the reference level. The
stress-free level can be shifted to 444 degK (340°F), the temperature
at which the matrix solidifies. Other investigators have claimed
that the stress-free temperature level might be lower as indicated
by comparing experimental and theoretical results.[12] In the case
in question the maximum residual strain at room temperature is
6.43×10^{-3} in the ±45-deg. plies in the transverse to the fiber
direction. The corresponding maximum residual strain in the 0-deg.
plies is 5.95×10^{-3}.

Determination of Residual (Thermal) Stresses

Residual or thermal stresses in any given ply within a laminate
can be computed from the thermal strains using the appropriate
thermoelastic orthotropic constitutive relations such as

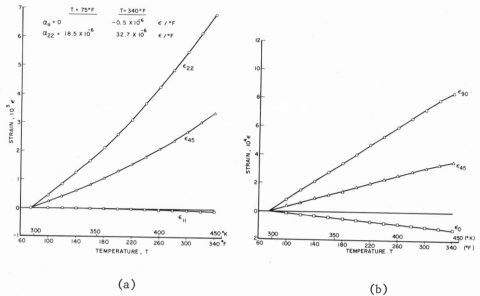

(a)   (b)

Fig. 4 Thermal Strains in Graphite/Epoxy Specimens, (a) $[0_8]$
Unidirectional Specimen, (b) $[0_2/\pm45]_s$ Specimen

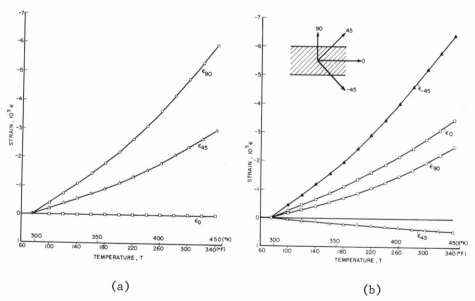

(a)   (b)

Fig. 5 Residual Strains in $[0_2/\pm45]_s$ Graphite/Epoxy Specimen,
(a) 0-Degree Plies, (b) 45-Degree Plies

$$\left\{ \begin{array}{c} \sigma_x \\ \sigma_y \\ \sigma_{xy} \end{array} \right\} = \left[\begin{array}{ccc} \bar{Q}_{11} & \bar{Q}_{12} & \bar{Q}_{16} \\ \bar{Q}_{12} & \bar{Q}_{22} & \bar{Q}_{26} \\ \bar{Q}_{16} & \bar{Q}_{26} & \bar{Q}_{66} \end{array} \right] \left\{ \begin{array}{c} \varepsilon_x - \alpha_x \Delta T \\ \varepsilon_y - \alpha_y \Delta T \\ 2\varepsilon_{xy} - 2\alpha_{xy} \Delta T \end{array} \right\} \qquad (9)$$

where,

$[\bar{Q}]$ is the stiffness matrix of the lamina referred to the x-y axes of the laminate.

The strain components in the equation above represent the difference between the actual strains in the laminate and the unrestrained thermal expansion in the ply, assuming constant coefficients of thermal expansion.

In a somewhat different approach, residual stresses in a given ply are obtained directly from the measured restraint strains ε_{ij}^r discussed before. Assuming linear elastic behavior, the residual stresses in any given ply referred to the ply principal directions are given by

$$\sigma_{11} = \frac{E_{11}}{1-\nu_{12}\nu_{21}} [\varepsilon_{11}^r + \nu_{21} \varepsilon_{22}^r]$$

$$\sigma_{22} = \frac{E_{22}}{1-\nu_{12}\nu_{21}} [\varepsilon_{22}^r + \nu_{12} \varepsilon_{11}^r] \qquad (10)$$

$$\sigma_{12} = 2 G_{12} \varepsilon_{12}^r$$

where subscripts 1 and 2 refer to the fiber and the transverse to the fiber directions.

Residual stresses at room temperature computed for the 0-deg. and ±45-deg. plies of the $[0_2/\pm45]_s$ graphite/epoxy laminate at room temperature are tabulated in Table 2. The transverse to the fibers stress in the ±45-deg. plies seems to exceed somewhat the measured transverse tensile strength of the unidrectional material which is 42 MPa (6.1 ksi). This means that these plies are probably damaged in their transverse direction upon completion of curing.

Table 2

Residual Stresses at Room Temperature in
$[0_2/\pm45]_s$ Graphite/Epoxy Laminate

Ply (deg)	Stress, MPa (ksi)		
	σ_{11}	σ_{22}	σ_{12}
0	23 (3.3)	42 (6.1)	0
±45	-52 (-7.5)	45 (6.5)	6 (0.9)

Comparable results for glass/epoxy and boron/epoxy show that residual tensile stresses exhaust a significant portion of the transverse tensile strength of the ply. In the case of Kevlar 49/epoxy computed residual stresses, assuming linear elastic behavior, far exceed the transverse strength of the ply. It has been shown also that the amount of relaxation of residual stresses is fairly small.[30]

Effect of Laminate Construction

The effects of ply orientation and ply stacking sequence on residual stresses were investigated for graphite/polyimide specimens.[17] The effects of ply orientation were evaluated with specimens of $[0_2/\pm15]_s$, $[0_2/\pm45]_s$, and $[0_2/90_2]_s$ layups. The effects of ply stacking sequence were studied with specimens of $[0_2/\pm45]_s$, $[\pm45/0_2]_s$, $[0/+45/0/-45]_s$, and $[+45/0_2/-45]_s$ layups.

Residual strains obtained for the first three sets of laminates are shown in Figs. 6-8. The highest residual strains occur in the $[0_2/\pm90_2]_s$ laminate and the lowest, approximately one-fourth in magnitude, occur in the $[0_2/\pm15]_s$ laminate. The maximum residual strains in the $[0_2/\pm45]_s$ laminate are slightly lower than those in the $[0_2/90_2]_s$ laminate. The ply stacking sequence was found to have no effect on the magnitude of residual strains.

Residual stresses at room temperature were computed for all plies assuming linear elastic behavior. Results are tabulated in Table 3. Computed residual stresses in the transverse to the fiber direction in the $[0_2/\pm45]_s$ and $[0_2/90_2]_s$ laminates exceed the transverse strength of the unidirectional material. This implies that the plies in these laminates may have already been damaged in the transverse direction upon completion of curing. Stresses in the fiber direction are tensile, except in the 15-degree plies, and are equal to a fraction of the longitudinal tensile strength of the material. The highest residual shear stress, occurring in the 15-degree plies, is approximately equal to one-third the in-plane shear strength of the unidirectional material.

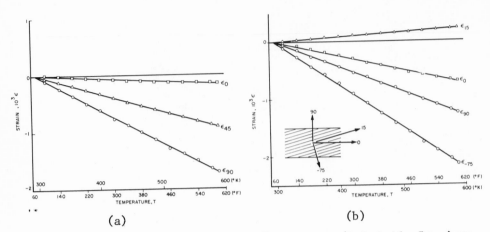

Fig. 6   Residual Strains in $[0_2/\pm15]_s$ Graphite/Polyimide Specimen.
(a) 0-Degree Plies, (b) 15-Degree Plies

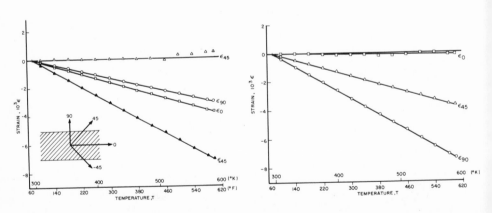

Fig. 7   Residual Strains in 45-
Degree Plies of $[0_2/\pm45]_s$
Graphite/Polyimide Specimen

Fig. 8   Residual Strains In 0-
Degree Plies of $[0_2/90_2]_s$
Graphite/Polyimide Specimen

Table 3

Residual Stresses at Room Temperature in Angle-
Ply Graphite/Polyimide Laminates

Laminate	Ply (deg)	Stress, MPa (ksi)		
		σ_{11}	σ_{22}	σ_{12}
$[0_2/\pm45]_s$				
$[\pm45/0_2]_s$	0	37.1(5.38)	17.9(2.59)	0
$[0/45/0/-45]_s$	45	16.3(2.36)	18.7(2.71)	1.2(0.17)
$[+45/0_2/-45]_s$				
$[0_2/\pm15]_s$	0	23.5(3.40)	4.6(0.67)	0
	15	−21.2(−3.07)	5.5(0.79)	7.2 (1.04)
$[0_2/90_2]_s$	0	29.8(4.32)	19.4(2.81)	0
	90	29.8(4.32)	19.4(2.81)	0

Effect of Hybridization

The influence of hybridization with stacking sequence variation on residual stresses was investigated for angle-ply hybrid graphite/Kevlar/epoxy and graphite/S-glass/epoxy laminates of the following layups:[18]

$$[0^K/\pm45^C/0^C]_s, \quad [\pm45^C/0^K/0^C]_s, \quad [0^K/\pm45^C/0^K]_s, \quad [\pm45^C/0^K_2]_s$$

$$[0^G/\pm45^C/0^C]_s, \quad [\pm45^C/0^G/0^C]_s, \quad [0^G/\pm45^C/0^G]_s, \quad [\pm45^C/0^G_2]_s$$

where superscripts C, K, and G denote graphite, Kevlar and S-glass, respectively.

Residual stresses at room temperature were computed for the various plies of the laminates studied assuming linear elastic behavior. Results are tabulated in Table 4. For laminates of the same composition, stacking sequence variations have no influence

on residual stresses. Hybridizing the basic $[0_2^C/\pm45^C]_s$ graphite/
epoxy laminate by substituting Kevlar or S-glass plies for 0-degree
graphite plies has a relatively small influence, a small reduction,
on residual stresses in the remaining graphite plies. This is due
in part to the fact that the thermal deformations in the angle-ply
laminates are an order of magnitude lower than the unrestrained
strains in the unidirectional material and in part to the relatively
lower stiffness of Kevlar and S-glass. Computed transverse to the
fiber residual stresses in the S-glass plies are approximately
seventy-five percent of the static transverse strength of the
unidirectional material. Computed residual stresses in the graphite
plies exceed the static strength of these plies by approximately
ten percent, indicating that these plies may have already failed
transversely. In the case of Kevlar the computed stresses indicate
that these plies must have failed in the early stages of cool down
before reaching room temperature.

Table 4

Residual Stresses at Room Temperature in
Angle-Ply Graphite and Hybrid Laminates

Laminate	Ply	Stress, MPa (ksi)		
		σ_{11}	σ_{22}	σ_{12}
$[0_2^C/\pm45^C]_s$	0^C	13.1 (1.9)	42.5 (6.2)	0
	45^C	-73.5(-10.6)	45.0 (6.5)	5.4 (0.8)
$[0^K/\pm45^C/0^C]_s$ or $[\pm45^C/0^K/0^C]_s$	0^K	-11.2(-1.6)	40.8 (5.9)	0
	0^C	12.3 (1.8)	39.6 (5.7)	0
	45^C	-97.0(-14.0)	43.2 (6.3)	6.9 (1.0)
$[0^K/\pm45^C/0^K]_s$ or $[\pm45^C/0_2^K]_s$	0^K	2.3 (0.4)	39.4 (5.7)	0
	45^C	-85.5(-12.4)	44.0 (6.4)	9.3 (1.4)
$[0^G/\pm45^C/0^C]_s$ or $[\pm45^C/0^G/0^C]_s$	0^G	103.5(15.0)	58.0 (8.4)	0
	0^C	6.9 (1.0)	39.2 (5.7)	0
	45^C	-117.8(-17.1)	42.8 (6.2)	9.0 (1.3)
$[0^G/\pm45^C/0^G]$ or $[\pm45^C/0_2^G]_s$	0^G	88.3(12.8)	57.2 (8.3)	0
	45^C	-151.3(-22.1)	41.0 (5.9)	6.9 (1.0)

Note: Superscripts K, C and G denote Kevlar, graphite and S-glass,
respectively.

The influence of residual stresses on failure patterns of hybrid laminates is illustrated in Fig. 9. The graphite/glass/epoxy $[\pm 45^C/0^G/0^C]_s$ specimen (where superscripts C and G denote graphite and glass, respectively) failed in a "brooming" fashion after the 0-deg. graphite ply failed first and the isolated layer $(\pm 45^C/0^G)$ curled as shown because of the residual tensile stresses in the 0-deg. glass ply. The same residual stresses caused the outer $(0^G/\pm 45^C)$ layer in the $[0^G/\pm 45^C/0^G]_s$ specimens to curl outwards after the central 0-deg. glass plies had delaminated.

SUMMARY AND CONCLUSIONS

The basic lamina of a filamentary composite is highly aniso-tropic thermally with the coefficient of thermal expansion in the fiber direction much lower than that transversely to the fiber direction. In the case of graphite and Kevlar composites the thermal coefficient in the fiber direction is negative.

This thermal anisotropy and the corresponding mechanical anisotropy allow the fabrication of angle-ply laminates with near-zero thermal expansion, a fact of great importance in structures requiring exceptional dimensional stability.

Coefficients of thermal expansion in unidirectional and angle-ply laminates can be calculated by using the properties of the constituents and lamination theory. Most of these theories, however, are linear and do not account for nonlinear and viscoelastic effects in matrix dominated properties at high temperatures.

Thermal expansion in composite laminates has been measured for various materials over a wide temperature range using dilato-metric and strain gage techniques. Of the most commonly used composites, Kevlar/epoxy has the highest coefficient of thermal expansion in the transverse to the fiber direction and the lowest (negative) coefficient in the fiber direction. S-glass/epoxy has the lowest transverse and highest longitudinal coefficients, i.e., it is the least anisotropic thermally. In general, especially in epoxy matrix dominated response, the coefficient of thermal expansion increases with temperature. This trend exists also in the cryogenic range where the coefficient approaches zero near absolute zero. In the case of polyimide matrix composites thermal deformation varies linearly with temperature. The moisture content has a significant effect on thermal deformation with some shrinkage resulting after thermal cycling at elevated temperature. Thermal expansion is stabilized after sufficient drying.

Residual stresses are produced during curing in angle-ply laminates as a result of anisotropic thermal deformations of the variously oriented plies. Residual strains have been measured experimentally using embedded strain gage techniques and residual stresses were computed using orthotropic stress-strain relations. Results show that, for graphite and Kevlar laminates residual stresses at room temperature are high enough to have caused damage in the plies in the transverse to the fiber direction. It has

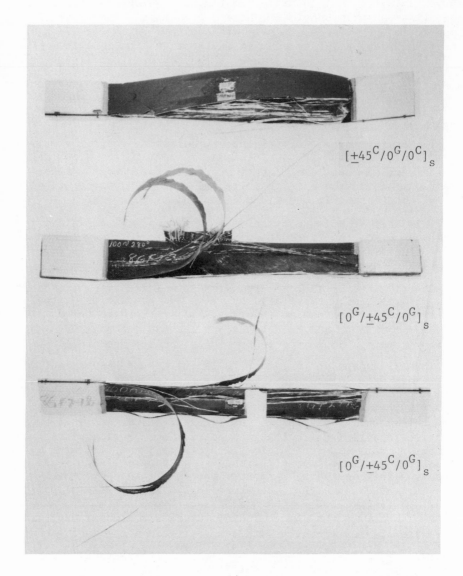

Fig. 9   Characteristic Failure Patterns of Three Graphite/S-Glass/
         High Modulus Epoxy Specimens Under Uniaxial Tensile Loading

also been shown that residual stresses do not relax appreciably. The ply stacking sequence has no effect on the magnitude of average residual stresses. Residual stresses and susceptibility to cracking during curing are strongly dependent on ply layup. In laminates of $[0_2/\pm\theta_2]_s$ layup, for example, the transverse tensile residual stress at room temperature in the $\pm\theta$-deg plies increases from zero to a maximum as θ varies between 0-deg. and 90-deg. The laminate is susceptible to microcracking when this stress exceeds the transverse strength of the unidirectional lamina. Interply hybridization of laminates has only a small influence on residual stresses in the various plies.

ACKNOWLEDGEMENTS

Most of the work discussed here was sponsored by NASA-Lewis Research Center, Cleveland, Ohio. The author is grateful to Dr. C. C. Chamis of NASA-Lewis for his encouragement and cooperation; to Dr. T. Liber and Mr. T. Niiro of IIT Research Institute for their assistance; and to Ms. I. Pakos for typing this manuscript.

REFERENCES

1. I.M. Daniel and A.J. Durelli, "Photoelastic Investigation of Residual Stresses in Glass-Plastic Composites," Proc. of 16th Conf. of Reinf. Plastics Div., Soc. of Plastics Industry, 19-A:1 (1961).

2. I.M. Daniel and A.J. Durelli, "Shrinkage Stresses Around Rigid Inclusions," Exp. Mechanics, 2:240 (1962).

3. T. Koufopoulos and P.S. Theocaris, "Shrinkage Stresses in Two-Phase Materials," J. Composite Materials, 3:308 (1969).

4. I.M. Daniel, "Photoelastic Studies of Mechanics of Composites" in IIT Research Institute Report No. M6132 to General Dynamics: Application of Advanced Fibrous Reinforced Composite Materials, Contract AF33(615)-3323, (1966).

5. I.M. Daniel, "Micromechanics" in "Structural Airframe Application of Advanced Composite Materials," Tech. Report AFML-TR-69-101, Vol. II (1969).

6. A.J. Durelli, V.J. Parks, H.C. Feng and F. Chiang, Strains and Stresses in Matrices with Inserts in "Mechanics of Composite Materials," J.W. Wendt, H. Liebovitz, and N. Perrone, eds., Pergamon Press, New York, (1970).

7. R.H. Marloff and I.M. Daniel, "Three-Dimensional Photoelastic Analysis of a Fiber-Reinforced Composite Model," Exp. Mechanics, 9:156 (1969).

8. C.C. Chamis, "Design and Analysis of Fiber Composite Structural Components," NASA Report SP227, (1970).

9. C.C. Chamis, "Lamination Residual Stresses in Cross-Plied Fiber Composites," Proc. of 26th Annual Conference of SPI, Reinforced Plastics/Composites Division, 17-D, (1971).

10. C.C. Chamis, and T.L. Sullivan, "A Computational Procedure to Analyze Metal Matrix Laminates with Nonlinear Lamination Residual Strains," Composite Reliability, ASTM STP 580, American Society for Testing and Materials, 327 (1975).

11. H.T. Hahn and N.J. Pagano, "Curing Stresses in Composite Laminates," J. Composite Materials, 9:91 (1975).

12. H.T. Hahn, "Residual Stresses in Polymer Matrix Composite Laminates," J. Composite Materials, 10:266 (1976).

13. I.M. Daniel and T. Liber, "Lamination Residual Stresses in Fiber Composites," IITRI Report D6073-I, for NASA-Lewis Research Center, NASA CR-134826, (1975).

14. I.M. Daniel, T. Liber and C.C. Chamis, "Measurement of Residual Strains in Boron/Epoxy and Glass/Epoxy Laminates," Composite Reliability, ASTM STP 580, American Society for Testing and Materials, 340, (1975).

15. I.M. Daniel and T. Liber, "Lamination Residual Stresses in Hybrid Composites," IITRI Report D6073-II, NASA CR-135085, (1976).

16. I.M. Daniel and T. Liber, "Measurement of Lamination Residual Strains in Graphite Fiber Laminates," Proc. of Second International Conference on Mechanical Behavior of Materials, ICM-II, Boston (1976).

17. I.M. Daniel and T. Liber, "Effect of Laminate Construction on Residual Stresses in Graphite/Polyimide Composites," Exper. Mechanics, 17:21, (1977).

18. I.M. Daniel and T. Liber, "Lamination Residual Strains and Stresses in Hybrid Laminates," Composite Materials: Testing and Design (Fourth Conference), ASTM STP 617, American Society for Testing and Materials, 331, (1977).

19. R.A. Schapery, "Thermal Expansion Coefficients of Composite Materials Based on Energy Principles," J. Composite Materials, 2:380 (1968).

20. B.W. Rosen and Z. Hashin, International Journal of Engineering Science, 8:157 (1970).

21. J.E. Ashton, J.C. Halpin and P.H. Petit, "Primer on Composite Materials:Analysis, Technomic, Stamford, (1966).

22. Grumman Aerospace Corp., "Advanced Composite Wing Structures-Boron/Epoxy Design Data," Vol. II, Analytical Data, Tech. Report AL-SM-ST-8085, (1969).

23. J.C. Halpin and N.J. Pagano, Consequences of Environmentally Induced Dilatation in Solids, in "Recent Advances in Engineering Science," A.C. Eringen, ed., Gordon and Breach, London (1970).

24. J.M. Whitney, I.M. Daniel and R.B. Pipes, "Experimental Mechanics of Fiber Reinforced Composite Materials," Society for Experimental Stress Analysis Monograph, to be published by Iowa State University Press.

25. W.T. Freeman and M.D. Campbell, "Thermal Expansion Characteristics of Graphite Reinforced Composite Materials," Composite Materials: Testing and Design (Second Conference), ASTM STP 497, American Society for Testing and Materials, (1972).

26. N.P. Freund, "Measurement of Thermal and Mechanical Properties of Graphite/Epoxy Composites for Precision Applications," Composite Reliability, ASTM STP 580, American Society for Testing and Materials, (1975).

27. A.S.D. Wang, R.B. Pipes and A. Ahmadi, "Thermoelastic Expansion of Graphite-Epoxy Unidirectional and Angle-Ply Composites," Composite Reliability, ASTM STP 580, American Society for Testing and Materials, (1975).

28. M.B. Kasen, "Properties of Filamentary-Reinforced Composites at Cryogenic Temperatures," Composite Reliability, ASTM STP 580, American Society for Testing and Materials, (1975).

29. L.W. Toth, B.R. Lloyd and R.L. Tennant, "Determination of the Performance of Plastic Laminates at Cryogenic Temperatures," ASD-TDR-62-794, Part II (N64-24212), Wright-Patterson Air Force Base, (1964).

30. I.M. Daniel and T. Liber, Relaxation of Residual Stresses in Angle-Ply Composite Laminates, in "Composite Materials: The Influence of Mechanics of Failure on Design," Army Symposium on Solid Mechanics, (1976).

RESIDUAL STRESSES AND MICROCRACKING INDUCED BY

THERMAL CONTRACTION INHOMOGENEITY

A.G. Evans[*] and D. R. Clarke[†]

University of California, Berkeley[*] and
Rockwell International Science Center
Thousand Oaks, CA 91360[†]

ABSTRACT

Brittle materials are subject to microcrack formation at grain boundaries and at second phase particles. These cracks are induced by residual stress that results from incompatibilities in thermal contraction. The development of residual stress and its partial relaxation by diffusion (at elevated temperatures) are described. The evolution of microcracks within the residual stress fields are then examined. Particular attention is devoted to considerations of the critical microstructural dimension at the onset of microcracking.

INTRODUCTION

Many properties of ceramic materials are dependent upon the incidence of microcracking. The most notable physical character-istics that exhibit a strong dependence on microcrack formation are certain mechanical (fracture toughness[1] and fracture strength[2]) and thermal (thermal diffusivity[3]) properties. The formation of stable microcracks is primarily related to localized residual stresses that develop because of thermal contraction mismatch or anisotropy (the former in multiphase materials[4] and the latter in single phase materials[5]). Significant progress has recently been achieved in the analysis of microcracking events, using a combination of stress analysis (based on the Eshelby concept) and fracture mechanics.[5,6,7] The intent of this paper is to examine the microcracking phenomenon in order to emphasize both the progress that has been achieved and the limitations of the available analyses.

One of the dominant characteristics of microcracking is its dependence on the scale of the microstructure. Typically, there is a 'critical' microstructural dimension ℓ_c below which microcracking is not generally observed and above which a significant density of microcracks becomes evident.[4,8,9] The development of a capability for predicting ℓ_c is a primary objective of microcracking analyses. A critical comparison with measured values of ℓ_c is also a demanding test of the validity of such analyses.

The amplitude of residual stress fields produced by thermal contraction mismatch is independent of the scale of the microstructure. A criterion for microfracture based on the peak tension would not, therefore, yield a size dependence. This dilemma was first addressed by suggesting[8] that the onset of microfracture be dictated by an equality of the loss of strain energy and the increase in surface energy associated with the microfracture event. The former is a volume dependent term and the latter is a surface area term and hence, a critical size emerges in a natural way. A reasonable correspondence with experimental observation was achieved by specifying the ratio of the final crack size to the dominant dimension of the microstructure. A conceptual difficulty with the approach arises because only the thermodynamics of the initial and final stages of the fracture event are considered; whereas, fracture is dictated by the rate of energy change at the critical condition for unstable crack extension.[5,9]

Subsequently, since size effects in brittle fracture often derive from statistical considerations,[10] a potential role of flaw statistics was suggested.[11] Notably, since fracture initiates from small inhomogeneities (pores, inclusions, etc.), the spatial and size distribution of these fracture initiating sites can influence the incidence of fracture. However, if the size distribution of these inhomogeneities is independent of the scale of the microstructure the fracture probability, for a constant volume fraction of the responsible microstructural phase, would be either independent of size (volume distributed inhomogeneities)[12] or decrease with increase in size (interface distributed inhomogeneities).[13] A statistically based argument must, therefore, invoke inhomogeneities that increase in size as the microstructure enlarges. This effect is a plausible possibility, because inhomogeneities (such as pores) tend to increase in size during sintering, in direct proportion to the size of the grains[14] (or other microstructural entities). However, the absence of well-substantiated distribution functions to describe the inhomogeneity size limits the quantitative utility of the statistical approach.

More recently, it has been recognized that a size effect can stem directly from considerations related to either the gradient of the residual stress field[5,6] or to stress relaxation.[7] For example, if the fracture initiating inhomogeneity is of sufficient size that it experiences an appreciable gradient of stress, then dimensional considerations demand that the fracture be size dependent. Specifically, the stress intensity factor, K, is given by,[5]

$$K \sim \sqrt{a} \int_0^1 \sigma(x/\ell) \; F(x/a) \; d(x/a) \qquad (1)$$

where a is the inhomogeneity size, σ is the stress and F(x/a) is the appropriate Green's function. Since the stress can always be expressed in the form,

$$\sigma(x/\ell) = \hat{\sigma} \; \Omega(x/a, \; a/\ell) \qquad (2)$$

where $\hat{\sigma}$ is the peak residual tension, the stress intensity factor can be written

$$\frac{K}{\hat{\sigma} \sqrt{\ell}} = \sqrt{\frac{a}{\ell}} \int_0^1 \Omega(x/a, \; a/\ell) \; F(x/a) \; d(x/a) \equiv \kappa(a/\ell) \qquad (3)$$

where $\kappa(a/\ell)$ is the function determined by integration. Now, if the stress intensity factor is equated to its critical value for crack extension, K_c, Eq. (3) yields a 'critical size' given by,

$$\ell_c = \left(\frac{K_c}{\hat{\sigma} \; \kappa(a/\ell)} \right)^2 \qquad (4)$$

An additional size influence derives from the diffusive stress relaxation that can occur at elevated temperatures.[15] The rate of relaxation will be more rapid in fine-grained materials, because of the enhanced diffusive fluxes. Smaller residual stresses will thus obtain and the tendency for microcrack formation will be reduced.

The considerations of microcracking developed in this paper relate primarily to the size effect that derives from residual stress gradients and relaxation phenomena. Beyond the scope of this review are the residual stresses[6,17] produced by phase transformations during cooling (as in the ZrO_2 based alloys) and the effect of externally applied stress fields upon the residual stresses and the onset of microcracking.

RESIDUAL STRESSES

The residual stresses typically encountered in ceramic materials derive from differences in thermal contraction (anisotropy of the thermal expansion coefficient, α, for a single phase material, and contraction mismatch for multiphase systems). Thermal contraction differences are important because ceramics are fabricated at elevated temperatures (by hot pressing or sintering) and, during cooling, stress relaxation (by diffusion or viscous flow) becomes sufficiently inoperative below a temperature T_g that appreciable local stresses must develop from the contraction mismatch. The elastic stresses that evolve below T_g can be computed using adaptations of the Eshelby approach.[16] Several such calculations will be presented below. A more difficult problem to address is the definition of T_g; an issue also examined in the following section.

Elastic Stresses

The stresses that develop below T_g can generally be calculated using the Eshelby approach, illustrated for the anisotropic contraction of a hexagonal grain in Fig. 1. This method of calculation firstly extracts the microstructural entity (or entities) subject to shape deformation and its shape is allowed to change (as characterized by the unconstrained transformation strain ε_{ij}^T). Subsequently, its shape is restored to the shape of the matrix cavity (by exerting a uniform stress) and then reinserted into the cavity. Finally, interface forces are applied (of equal magnitude, but opposite sign, to the restoring stress) to achieve continuity of stress. For isolated particles of ellipsoidal geometry this process yields a uniform stress within the particle and hence, stress analysis is relatively straightforward. More complex behavior is expected for other geometries, such as individual grains within a polycrystalline aggregate.

Multiphase Materials

The stresses within a spherical particle subject to transformation strains ε^T (hydrostatic) and $'\varepsilon_{ij}^T$ (deviatoric) are[16,17]

$$\sigma^I = - \frac{\varepsilon^T}{(1 + \nu_m)/2E_m + (1 - 2\nu_p)/E_p} \tag{5a}$$

$$'\sigma_{ij}^I = - \frac{'\varepsilon_{ij}^T}{(1 + \nu_p)/E_p + 2(1 + \nu_m)(4 - 5\nu_m)/E_m(7 - 5\nu_m)} \tag{5b}$$

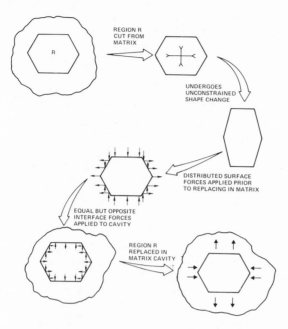

Fig. 1.   A schematic indicating the Eshelby method for
calculating the residual stresses and strains
generated by anisotropic thermal expansion of a
hexagonal grain.

whose σ^I is the hydrostatic stress and σ^I_{ij} is the deviatoric
stress. The same hydrostatic stress level pertains for ellipsoidal particles, irrespective of their shape; but, the deviatoric
stress is sensitive to the particle shape.[16] Two extremes are of
interest. Firstly, if the particles and matrix have isotropic
thermal contraction coefficients, the resultant stress is exclusively hydrostatic, and hence,

$$\sigma^I = - \frac{(\alpha_m - \alpha_p)(T_g - T)}{(1 + \nu_m)/2E_m + (1 - 2\nu_p)/E_p} \qquad (6)$$

where $\alpha_m - \alpha_p$ is the thermal contraction mismatch between matrix
and particle and T is the temperature. For a particle and matrix
with similar elastic constants, Eq. (6) becomes,

$$\sigma^I = - \frac{(\alpha_m - \alpha_p)(T_g - T)E}{3(1 - \nu)} \qquad (7)$$

Secondly, if the particle exhibits anisotropy of thermal expansion (e.g., α_1 and α_2), such that the average expansion matches that of the matrix, then the stress within the particle is purely deviatoric, and given by

$$\sigma_{ij}^I = - \frac{(\alpha_m - \alpha_2)(T_g - T)}{(1 + \nu_p)/E_p + 2(1 + \nu_m)(4 - 5\nu_m)/E_m(7 - 5\nu_m)} \tag{8}$$

which for uniform elastic properties reduces to;

$$\sigma_{ij}^I = - \frac{(7 - 5\nu)(\alpha_m - \alpha_2)(T_g - T) E}{15(1 - \nu^2)} \tag{9}$$

In general, therefore, the resultant stress, σ_{ij}, for uniform elastic constants is;

$$\frac{\sigma_{ij}}{E(T_g - T)} = - \frac{(7 - 5\nu)(\alpha_m - \alpha_2)}{15(1 - \nu^2)} - \frac{(\alpha_m - \alpha_p)}{9(1 - \nu)} \delta_{ij} \tag{10}$$

The equivalent result for ellipsoidal geometry is,

$$\frac{\sigma_{ij}}{E(T_g - T)} = - \frac{(\alpha_m - \alpha_p)}{9(1 - \nu)} \delta_{ij} - \frac{\gamma}{(1 + \nu)}(\alpha_m - \alpha_2) \tag{11}$$

where γ is a function of the particle shape.[16] For a needle, $\gamma = 1/2$: for a sphere $\gamma = (7 - 5\nu)/15(1 - \nu)$: and for a disc, $\gamma = \pi(c/b)(2 - \nu)/4(1 - \nu)$, where c is the disc thickness and b is the disc radius.

   The stresses within the matrix are more difficult to analyze and hence, with the exception of spherical[18] and cylindrical particles[19-21] have not been computed exactly. Eshelby shows that if the harmonic and biharmonic potentials of the particle (however arbitrary its shape) are known then the displacements in the matrix are related to the transformation strain ε_{ij}' in the particle by

$$u_i = \frac{\sigma_{jk}^T \Psi_{,ijk}}{16\pi\mu(1 - \nu)} = \frac{\sigma_{ik}^T \phi_{,jk}}{4\pi\mu} \tag{12}$$

where σ_{ik}^T is the stress derived by Hooke's law from the strain ε_{ij}^T. The matrix stresses can then be obtained from the displacement derivatives. In the case of a spherical particle subject to hydrostatic strain the matrix stresses are particularly simple,

$$\sigma_{rr} = \sigma(r_0/r)^3, \qquad \sigma_{\theta\theta} = -(\sigma/2)(r_0/r)^3 \tag{13}$$

where r_0 is the particle radius and r is the distance from the particle center.

The harmonic and biharmonic potentials for a very long cylindrical particle (fiber) have recently been calculated[20] and may be used to compute the matrix stresses from Eq. (12),

$$\phi = 2\pi r_0^2 \ln (r/r_0) \tag{14a}$$

$$\psi = \pi r_0^2 (r^2 - r_0^2) - \pi r_0^2 r^2 + \left(r_0^2/2\right) \ln (r/r_0) \tag{14b}$$

For the case of thermal contraction mismatch between the fiber and matrix, described by a hydrostatic strain ε^T the matrix stresses are

$$\sigma_{11}^c = E \varepsilon^T \frac{(3 - 2\nu)}{2(1 - \nu)} \frac{r_0^2 (x^2 - y^2)}{r^4}$$

$$\sigma_{22}^c = -\sigma_{11}^c \tag{15}$$

$$\sigma_{12}^c = E \varepsilon^T \frac{(3 - 2\nu)}{2(1 - \nu)} \frac{r_0^2 \, xy}{r^4}$$

A case of particular interest is that of the cylindrical particle exhibiting an anisotropy of thermal contraction in the plane perpendicular to the long axis of the cylinder such that the average contraction matches that of the matrix. The stresses generated within the matrix are of the form,[21]

$$\sigma_{11}^c = \frac{E \varepsilon_{11}^T}{(1 + \nu)(1 - 2\nu)} \left[\frac{r_0^2 (x^2 - y^2)}{r^4} + \frac{\nu}{1 - \nu} \frac{(x^4 - y^4)}{r^2} \right]$$

$$+ \frac{E \varepsilon_{11}^T}{4(1 - \nu^2)} \left[\frac{4r_0^2 y^2 (3x^2 - y^2)}{r^6} + \frac{3r_0^2 (x^4 + y^4 - 6x^2 y^2)}{r^2} \right] \tag{16}$$

where

$$\varepsilon_{11}^T = (\alpha_1 - \alpha_m)(T_g - T)$$

However, since the stresses immediately adjacent to the
interface are of greatest interest for microfracture problems, some
pertinent information can be obtained by deriving the stresses in
the matrix just outside the inclusion. For an ellipsoidal inclusion subject to dilatation, the stresses in the matrix can be
written quite generally as:[16]

$$\sigma_{ij}^{I} = \sigma^{I}(n_i n_j - 1/3\delta_{ij}) \tag{17}$$

where n_{ij} are the normals to the ellipsoid surface. Particular
values for the stresses at the particle matrix interface have been
calculated for disc and needle shaped particles.[22] The stresses
are a maximum near the termination of the major axis of the ellipsoid. However, as described earlier, the gradient in stress (in
addition to the stress level at the particle interface) is of
importance in dictating the size effect. As far as the authors
are aware the stress gradients around ellipsoidal particles subject to a transformation strain have not been calculated and
remains a subject for further work.

Single Phase Materials

Grains in single phase materials exhibit relatively complex
geometric configurations, and stress analysis is more complex than
for the isolated ellipsoidal particle. However, some useful approximations can be obtained quite straightforwardly. The general
level of residual stress within the grains can be obtained by
simply requiring a grain to be contained within an isotropic matrix, with the average properties of the polycrystalline
aggregate, and inserting the anisotropic contraction coefficients
into Eq. (10). However, this simplification neglects the stress
enhancing influence of grain junctions, an effect which has
important consequences for microfracture.

Estimates of the stresses that develop in the vicinity of
grain junctions can be obtained by adopting two-dimensional ana-
logues, such as an array of hexagonal grains. The stresses that
develop in such an array can be determined by firstly establishing
the resultant body forces at each grain boundary facet (see
Fig. 1). These body forces, p, generate non-uniform stresses that
superimpose upon the uniform stresses developed during shape res-
toration. Of principal interest are the stresses at the grain
boundaries, because these are the dominant sites for microfracture. The stresses at a site (x,z) inclined at an angle β to the
boundary, (Fig. 2) are of the form[5]

$$\frac{4\pi\sigma_{xx}}{p\cos\beta} = \int_0^1 \frac{(z + \alpha\sin\beta)}{[\alpha^2 - 2\alpha(z\sin\beta - x\cos\beta) + (x^2 + z^2)]}$$

$$\left\{(1 - \nu) - \frac{2(1 + \nu)(\alpha\cos\beta - x)^2}{[\alpha^2 + 2\alpha(z\sin\beta - x\cos\beta) + (x^2 + z^2)]}\right\} d\alpha$$

(18)

Equation (18) should be used to obtain the stresses on boundary AB (Fig. 3) from the body forces of the four adjacent boundaries (AA', AA", BB', BB"). For more remote boundaries sufficient accuracy can be achieved by placing a single force at the grain facet center,[23] which represents the total body force on that boundary (Fig. 3). These stresses superimpose on a uniform stress (equal in magnitude to half the initial stress) which derives from the body forces on AB coupled with the initial stress.

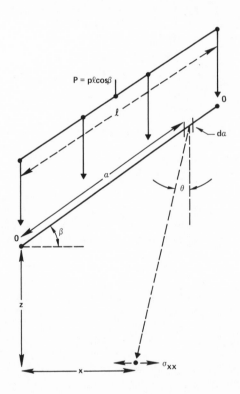

Fig. 2. The linear boundary segment used to compute the relaxation stresses, showing the coordinate system (x, z).

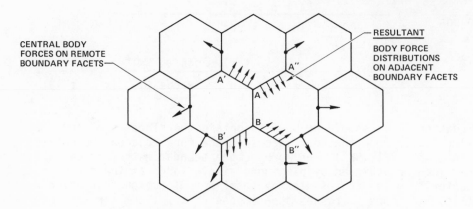

Fig. 3.   An hexagonal grain array showing the body forces
used to calculate the stress at the central facet AB.

The component of the stress from the four adjacent boundaries
dominates the behavior in the vicinity of the grain junction.
This stress component is of the form;[5]

$$\frac{4\pi\sigma}{p\cos\beta} = A_1(\beta,\nu) \; \ell n \left[\frac{\ell^2 + x^2 + z^2 + 2\ell(z\sin\beta - x\cos\beta)}{x^2 + z^2} \right]$$

$$+ A_2(\beta,\nu,x,z) \left[\tan^{-1} \frac{2\ell + z\sin\beta - x\cos\beta}{2(z\cos\beta + x\sin\beta)} \right]$$

$$- \tan^{-1} \left[\frac{z\tan\beta - x}{2(z + x\tan\beta)} \right] + A_3(\ell, \beta, \nu, x, z) \qquad (19)$$

where A_1, A_2, and A_3 are relatively complex functions in the range
$\pm 2\pi$. The logarithmic term is singular at the grain junction and
is thus the most influencial with regard to microcrack formation.[5]

The specific stress magnitudes that develop depend on the relative orientations of the grains circumventing the boundary of interest. Preliminary calculations have been conducted for the orientation that has been assumed to yield the maximum stress, depicted
in Fig. 4a. The results are shown in Fig. 4b. Calculations for
more general grain orientation relations are now in progress.[23]
These results will provide a full perspective of residual stress
distributions in polycrystalline aggregates in which there is a
random distribution of contraction anisotropy orientations.

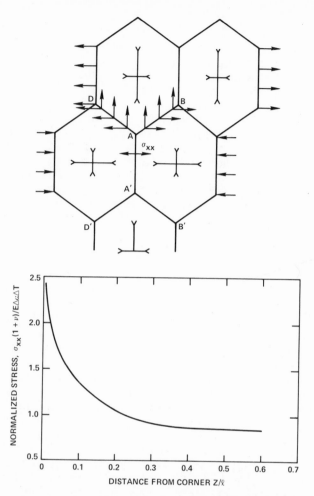

Fig. 4.   The grain configuration that yields large values
          of the residual stress of facet AA' and the stresses
          calculated to exist along that facet.

STRESS RELAXATION EFFECTS

Stress relaxation in ceramics occurs primarily by diffusion
(or by viscous flow in the presence of an amorphous phase). These
relaxation processes are usually motivated by local gradients in
hydrostatic stress and thus, occur in response to localized
thermal contraction stresses, while the material is at elevated
temperatures.

Multiphase Materials

For isotropic multiphase materials, there is no gradient of hydrostatic stress within the isolated phase (Eq. 5). However, large shear stresses exist within the surrounding matrix (note that the hydrostatic stress is zero). The shear stresses within the matrix cause grain boundary sliding, and diffusive deformation will occur in response to local normal stresses induced by sliding.[24] The deformation field will be similar to that of a cavity subject to internal pressure. Initially, radial flow in the matrix will redistribute the residual stress. Reduction of the stresses will initiate when the redistribution extends across the sample, following the onset of interaction between the stress fields around adjacent particles. The authors are not aware of solutions for this problem, although the analysis is relatively straightforward.

Polycrystalline Single Phase Aggregates

It has already been demonstrated that, anisotropic thermal contraction in polycrystalline aggregates develops tensile or compressive stresses on grain boundaries. A gradient of chemical potential suitable for diffusive relaxation (Fig. 5) is thus established. The 'initial' stress involves singularities near grain junctions (Fig. 4). But the singularities are weak (logarithmic) and should be rapidly dispersed by localized diffusive fluxes. The rate controlling relaxation process involves diffusion between adjacent grain facets (Fig. 5), such that atoms are removed from the boundaries subject to compression and deposited on boundaries under tension. If it is assumed that the relaxation times are sufficiently rapid that atom deposition and removal occurs uniformly along the respective grain boundaries, a parabolic 'steady-state' stress distribution must develop along the boundaries during the relaxation process. The extent of strain relaxation can then be deduced using well-established mathematical procedures for diffusive flow. This mode of analysis is only permissable when the diffusivities are large, notably at the highest temperatures. The stress evolution at intermediate temperatures requires 'transient' solutions involving more complex formulations. Such analyses have not yet been performed. Presently, therefore, it is only possible to obtain approximate solutions by permitting 'steady-state' relaxation above a 'freezing' temperature T_s and invoking fully elastic stress development below T_g.

The stress relaxation problem can be posed by firstly establishing the normal elastic displacement δ_1 of the boundaries during cooling (the driving force for the diffusive flow, Fig. 5) and the displacement relaxation δ_2 due to diffusion. Then, the resultant displacement δ (= $\delta_1 - \delta_2$), that determines the level of the

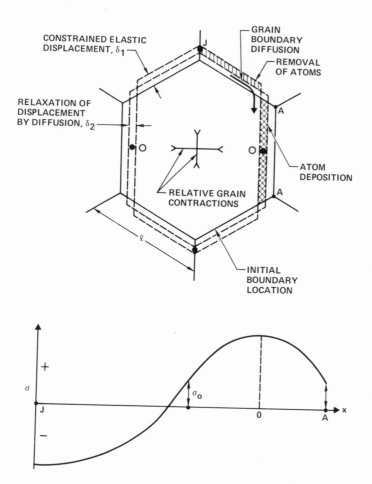

Fig. 5. The elastic and diffusion displacement that occurs
during cooling, indicating the direction of the diffusive flux. Also shown are the stresses that develop
during steady-state diffusive flow.

relaxed stress, can be derived. The solution for a constant
cooling rate T will be presented. [25]

The elastic stress level on a grain boundary normal to the
direction of maximum contraction is;

$$\sigma_{xx} = \frac{\beta \, E(T_0 - \dot{T}t) \Delta\alpha}{(1 + \nu)} \tag{20}$$

where T_0 is the initial temperature, $\Delta\alpha$ is the deviation of the contraction coefficient from the average, and β is a coefficient that depends on the orientations of the adjacent grains. The corresponding elastic displacement is,

$$\delta_1 = \frac{\sqrt{3} \, \ell \, \beta \, \dot{T} \, t \, \Delta\alpha}{2 \, (1 + \nu)} \tag{21}$$

The relaxations of these displacements by diffusion are governed by the relation,[26]

$$\frac{d^2 \sigma(x,t)}{dx^2} = - \frac{kT \dot{\delta}_2(t)}{\Omega \, D_b \, \delta_b} \tag{22}$$

where $D_b \delta_b$ is the diffusion parameter, Ω the atomic volume and $\dot{\delta}$ is assumed to be uniform (as noted above). Integration of Eq. (22) gives the stress distribution,

$$\sigma(x,t) = \frac{-\xi x^2}{2} + C_1 x + C_2 \tag{23}$$

where $\xi = kT\dot{\delta}_2(t)/\Omega \, D_b \, \delta_b$ and C_1 and C_2 are the integration constants. The positions of zero flux ($d\sigma/dx = 0$) in the system are at the grain facet center 0 and at the grain junction J (Fig. 5). Hence, since the flux must be continuous at the grain corner A, the constant C_1 must be equal to $\xi\ell/2$. If σ_0 is the stress at A, the stress distribution becomes;

$$\sigma(x,t) = - \frac{\xi x^2}{2} + \frac{\xi x \ell}{2} + \sigma_0 \tag{24}$$

The equivalent average stress is:

$$<\sigma>(t) = \frac{\xi\ell^2}{12} + \sigma_0 \tag{25}$$

Volume conservation requires that the volume of material deposited on the tensile boundary must equal the volume removed from the compressed boundary. The stress at the grain corner thus becomes;

$$\sigma_0 = \xi \ell^2 / 12 \qquad (26)$$

and the average stress reduces to;

$$<\sigma>(t) = \frac{kT\ell^2 \dot{\delta}_2}{6\Omega D_b \delta_b} \qquad (27)$$

The average stress on each boundary must also be related to the resultant displacement of the grain;

$$\frac{\sqrt{3}\ell}{2E} \frac{d<\sigma>}{dt} = \dot{\delta}_1 - \dot{\delta}_2 \qquad (28)$$

Substituting $\dot{\delta}_1$ from Eq. (21) and $\dot{\delta}_2$ from Eq. (27), and noting that D_b is temperature dependent

$$D_b = D_o e^{-Q/kT} \qquad (29)$$

where Q is the activation energy for boundary diffusion, the following differential equation obtains,

$$\frac{d<\sigma>}{dT} - \frac{12\Omega\delta_b D_o E}{\sqrt{3}k\ell^3 \dot{T}} \frac{e^{-Q/kT}}{T} <\sigma> = - \frac{\beta E \Delta\alpha}{(1 + \nu)} \qquad (30)$$

The solution to this equation must be conducted numerically. However, an approximate series expansion may be derived[25] for comparison with the elastic result expressed in terms of a "freezing" temperature T_g;

$$\frac{<\sigma>(1 + \nu)}{\beta \cdot E \cdot \Delta\alpha} = T_g - T$$

From this T_g is given approximately by

$$T_g \sim \frac{Q/k}{\ell n[12\Omega D_o \delta_b E/\sqrt{3}nk\ell^3 \dot{T}]} \qquad (31)$$

The trends in T_g with the influential variables are immediately apparent. Importantly, T_g increases as the grain size and cooling rate increase and as the diffusivity and modulus decrease. This behavior is exemplified for Al_2O_3 ($D_o\delta_b = 10^{-9}$ m^3 s^{-1},

Q = 100 kcal/mole, $\Omega \sim 10^{-29}$ m³, n = 30, E = 420 GN m²), for which:

$$T_g = \frac{50,000}{29.5 - \ell n \, \ell^3 \, \dot{T}}$$

where ℓ is in microns, T in °K s^{-1} and T_g is in °K. Specifically, for ℓ = 1 μm and $\dot{T}$ = 1°K s^{-1}, T_g = 1695°K; whereas for ℓ = 10 μm and $\dot{T}$ = 1°K s^{-1}, T_g = 2210°K.

MICROCRACK FORMATION

The formation of microcracks at grain boundaries or at second phase particles has been considered to depend upon the existence of a distribution of small inhomogeneities that pre-exist at the boundaries (especially at three grain junctions) or interfaces.[5,6,7] These inhomogeneities have been proposed because the stress intensification levels associated with the residual stress fields do not appear to be of sufficient magnitude to induce fracture in defect free material (although further study is needed to establish whether this possibility can be discounted). The role of the proposed inhomogeneities is to further enhance the stress intensification to a level suitable for microcrack evolution. It is certainly reasonable to presume that inhomogeneities do exist at boundaries or interfaces in ceramics, e.g., small pores remaining at grain triple points. However, little effort has yet been devoted to the elucidation of the inhomogeneities that induce microfracture in specific miocrostructural situations. It is thus generally assumed that the inhomogeneities exhibit the stress concentrating properties of small microcracks: a presumption that is evidently an over-simplification. Thereafter, stress intensity factors can be calculated (from the residual stress levels) and compared with the critical values for grain boundary separation. Approximate stress intensity factors are conveniently calculated using a superposition method, based on the prior stress field.[5,6] A typical example is illustrated in Fig. 6; wherein a microcrack develops along two symmetrically stressed grain facets, initiating at the common triple junction. The stress intensity factor for such a crack is given by:

$$\frac{K(1 + \nu)}{E \Delta\alpha\Delta T \sqrt{\ell}} = \kappa(a/\ell) \tag{32}$$

where $\kappa(a/\ell)$ is the function plotted in Fig. 6. It is noted that the stress intensity factor exhibits a maximum, K. This is typical of crack extension in residual or spatially varying stress fields. The principal maximum in the present analysis essentially

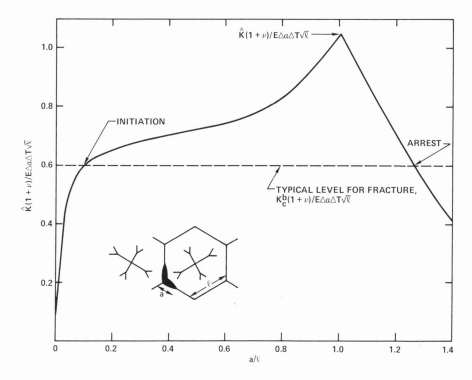

Fig. 6. The variation of the normalized stress intensity
factor with crack length for the grain, crack configuration indicated on the inset.

coincides with a crack front located at the first triple junction,
where the residual stress changes sign (i.e., the residual stress
becomes compressive along the impinging boundaries). Equating K
to the boundary separation resistance, K_c^b, yields an <u>absolute
minimum</u> condition for the formation of microcracks. This corresponds to an <u>upper bound</u> for the critical grain facet size;

$$\hat{\ell}_c = \frac{K_c^b (1 + \nu)}{\hat{\kappa} \, E \Delta\alpha\Delta T}^2 \tag{33}$$

where $\hat{\kappa}$ is the magnitude of the normalized stress intensity factor
at the maximum. Estimates of specific values of the grain facet
sizes that induce microcracking involve statistical considerations
based on flaw distributions.

More exact calculations of the stress intensity factor can be
obtained using numerical (finite element or finite difference)

methods. A recent example[7] is the use of a finite difference scheme for calculating the stress intensity factor for a crack at the interface of a spherical second phase particle. A convenient use of the finite difference method involves the calculation of the strain energy, U, as a function of crack length, a. The stress intensity factor is then deduced from the crack length derivative of the strain energy. A maximum value, K, is obtained. The corresponding absolute minimum requirement for microcrack initiation is;

$$\hat{R}_c = \frac{1.2(K_c^I/\langle\sigma\rangle)^2}{(1 - \nu)} \tag{34}$$

where K_c^I is the resistance of the interface to fracture and $\langle\sigma\rangle$ is the stress in the uncracked particle.

Further progress in the elucidation of microfracture is achieved by incorporating defects that reduce the critical size below the upper bound values. Little progress has yet been achieved in selecting appropriate defects and defect size distributions; although results could clearly be obtained by selecting arbitrary distributions (such as one of the extreme value distributions). Careful obsevations coupled with pertinent stress intensity factor calculations are needed to establish a more fundamental appreciation of microcracking. Comparison of the available calculations of K with experimental observations of microcracking suggest[5,6,7] that triple point defects in the size range, 0.1 < 2a/ℓ < 0.3, are typically involved in the microcrack initiation process. However, more direct observations and further calculations are needed to substantiate and refine this result.

CONCLUSION

This paper has described methods for calculating the residual stresses that develop at the microstructural level due to thermal contraction inhomogeneity. The stresses are typified by locally large amplitudes (with some singularities) and rapid gradients. These characteristics are central to the onset of microcracking: both microcrack initiation and arrest.

The stress level is also shown to depend upon the rate of stress relaxation at elevated temperatures, by diffusion or viscous flow. The relaxation rate is a strong function of microstructure: rapid relaxation rates obtain in fine grained materials or in materials containing an amorphous boundary phase. These relaxation effects have been demonstrated to be manifest in an effective freezing temperature: a temperature at which elastic residual stresses begin to develop.

The onset of microcracking within the residual stress field has been considered to depend on the presence of small microstructural inhomogeneities (such as voids) at the susceptible interfaces. These features certainly exist, but have not yet been uniquely correlated with the onset of microcracking. By treating these pre-existent inhomogeneities as crack-like entities, stress intensity factors have been calculated. The level and variation in stress intensity factors indicate the potential for microcrack initiation and arrest at grain boundaries and interfaces.

In particular, a lower bound for the critical microstructural size needed to initiate microcracks has been identified (no microcracks can be observed at size levels below this bound). The actual formation of microcracks above the lower bound depends upon the statistical characteristics (size and spatial) of the pre-existent inhomogeneities. This issue has not been addressed.

ACKNOWLEDGEMENT

The authors wish to thank the Office of Naval Research under Contract N-0014-79-C-0159. (AGE) and the Rockwell International Independent Research and Development Program (DRC) for their financial support.

REFERENCES

1. R.G. Hoagland, G.T. Hahn and A.R. Rosenfield, Rock Mechanics 5, 77 (1973).
2. F.A. McClintock and H.J. Mayson, ASME Applied Mechanics Conf. (June 1976).
3. H.J. Siebenbeck, D.P.H. Hasselman, J.J. Cleveland and R.C.Bradt, Jnl. Amer. Ceram. Soc. 59, 241 (1976).
4. F.F. Lange, Fracture Mechanics of Ceramics (Ed. R.C. Bradt, D.P.H. Hasselman and F.F. Lange) Plenum, N.Y. Vol. 4 (1977).
5. A.G. Evans, Acta Met. 26, 1845 (1978).
6. D.R. Clarke, Acta Met., in Press.
7. Y.M. Ito, M. Rosenblatt, L.Y. Cheng, F.F. Lange and A.G. Evans, Int. J. Fract. (In Press).
8. R.W. Davidge and G. Tappin, Jnl. Mater. Sci 3, 297 (1968).
9. J.A. Kuszyk and R.C. Bradt, Jnl. Amer. Ceram. Soc. 56, 420 (1973).
10. W. Weibull, Ingenioers Vetenskaps Akad., 151, (1939).
11. A.G. Evans, Jnl. Mater. Sci., 9, 210 (1974).
12. O. Vardar, I. Finnie, D.S. Biswas and R.M. Fulrath, Intl. Jnl. Frac. 13, 215 (1977).
13. A.G. Evans, D.S. Biswas and R.M. Fulrath, Jnl. Amer. Ceram. Soc. 62, 95 (1979)
14. W.D. Kingery and B. Francois, in "Sintering and Related Phenomena," edited G.C. Kuczynski, Gordon and Breach, 1967.

15. J.E. Blendell, R.L. Coble and R.J. Charles, in "Ceramic Microstructure '76," edited R.M. Fulrath and J.A. Pask, Westview Press, Boulder, 1977.

16. J.D. Eshelby, Proc. Roy. Soc., A241, 376 (1957).

17. A.G. Evans and A.H. Heuer, Jnl. Amer. Ceram. Soc. 63, (May/June 1980), to be published.

18. J. Selsing, Jnl. Amer. Ceram. Soc., 44, 419 (1961).

19. L.M. Brown and D.R. Clarke, Acta. Met. 25, 563 (1977).

20. D.R. Clarke, PhD Thesis, University of Cambridge (1974)

21. D.R. Clarke to be published.

22. G.C. Weatherly, Phil. Mag. 17, 791 (1968).

23. Y. Fu and A.G. Evans, to be published.

24. R. Raj and M.F. Ashby, Acta Met. 23, 653 (1975).

25. D.R. Clarke and A.G. Evans, to be published.

26. A.G. Evans and A.S. Rana, Acta Met., 28, 129 (1980).

27. R.M. Cannon and R.L. Coble, Deformation of Ceramics (Ed., R.E. Tressler and R.C. Bradt), Plenum, N.Y. (1975) p. 61.

THERMAL MICROCRACKING IN CELION 6000/PMR-15 GRAPHITE/POLYIMIDE

C.T. Herakovich - Engineering Science & Mechanics
Virginia Tech., Blacksburg, Va. 24061

J.G. Gavis, Jr. - National Aeronautics & Space Admin.
Langley Research Ctr., Hampton, Va. 23665

J.S. Mills[1] - McDonnell Douglas Astronautics Company
Huntington Beach, California 92647

ABSTRACT

Six laminate configurations were subjected to five different thermal exposures in the temperature range 78K to 603K (-320°F to 625°F), and then studied using microscopy and x-ray to determine the characteristics of microcracks formed during the thermal loadings. The laminates studied were: $[0_3/90_3]_s$, $[0_2/90_2]_s$, $[(0/90)_3]_s$, $[45/-45/0/90]_s$, $[0/45/90/-45]_s$, and $[0/60/0/-60]_s$. The material system investigated was found to be free of cracks after curing, but microcracks did develop in most laminates when cooled from 603K (625°F) by quenching in ice water or liquid nitrogen. Crack density was dependent on laminate configuration and rate of cooling. Microcracks present at free edges extended across the entire width of the specimens. The $[45/-45/0/90]_s$ laminate proved to be very resistant to microcracking for all thermal loadings. The thermal load required to initiate microcracking, determined using laminate analysis with stress and temperature dependent material properties, compared reasonably well with experimental results.

INTRODUCTION

Utilization of graphite/polyimide (Gr/Pi) composite materials in primary structures is viewed as one of the most promising approaches for reducing the structural mass and improving the per-

---

[1] Formerly, graduate student at Virginia Tech.

formance of future space vehicles. Such structures may experience
temperatures as low as 117K(-250°F) while in space and as high as
589K (600°F) during re-entry. Preliminary efforts to fabricate Gr/
Pi laminates showed that significant numbers of microcracks could
be developed when laminates were cooled rapidly from the final cure
temperature of 603K (625°F) to room temperature. Figure 1 shows
transverse microcracks (TVM) in an "as fabricated" cross-plied
HTS1/PMR-15 graphite/polyimide laminate. The cracks are uniformly
distributed throughout the laminate with most crack spacing approximately double the layer thickness. The presence of such microcracks
may have deleterious effects on the performance of structures.

Thermal microcracking has been studied previously by several
investigators [1-4] with the most comprehensive treatment being that
of Spain [1]. He investigated a number of factors affecting the
formation of microcracks in carbon fiber/resin composites including
resin properties, cure temperature, thermal loading history, fiber
surface roughness, shape and bundle form, and laminate configuration.

This investigation was undertaken to experimentally determine
the effect of thermal exposures between 78K and 603K (-320°F and
625°F) on microcracking of Gr/Pi. Crack densities were determined
for six different laminates after they were subjected to five
different thermal exposures. Ply residual stresses and the thermal
load required to initiate microcracking were determined using
laminate analysis with stress and temperature dependent material
properties. The stress-free temperature of the material was determined experimentally.

EXPERIMENTAL PROGRAM

The graphite/polyimide selected for use in this investigation
consisted of Celion 6000 graphite fibers and PMR-15 polyimide matrix.
Fiber volume fraction, glass transition temperature and the thermal
stress-free temperature range for the Gr/Pi were 57 percent, 603K
(625°F) and 572K to 630K (570°F to 625°F), respectively. (The range
for the stress-free temperature indicates batch to batch and specimen
to specimen variation.) Test specimens 2.54 cm (1.0 inch) square
were machined from laminated plates. The six laminate configurations
investigated were: $[0_3/90_3]_s$, $[0_2/90_2]_s$, $[(0/90)_3]_s$, $[45/-45/0/90]_s$,
$[0/45/90/-45]_s$ and $[0/60/0/-60]_s$. The five thermal exposures used
were:

        Type 1 - cool from 603K (625°F) to room temperature
                 at 3K (5°F) per minute, the cure cycle cool-down
                 (as fabricated condition),
        Type 2 - cool from room temperature to 78K (-320°F) by
                 immersing in liquid nitrogen,

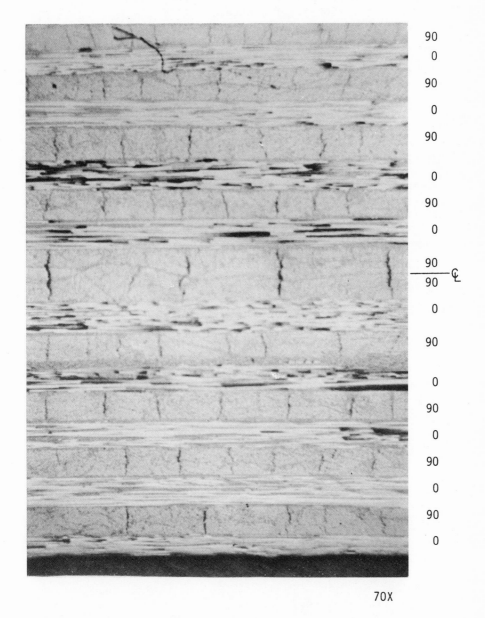

90
0
90
0
90
0
90
0
90
90 — ℄
0
90
0
90
0
90
0
90
0

70X

Figure 1.   Microphotograph of $[(0/90)_5]_s$ Laminate After Cure.
HTS1/PMR-15 Composite.

Type 3 - cool from 603K (625°F) to 273K (32°F) by
        immersing in ice water,
Type 4 - cool from 603K (625°F) to 78K (-320°F) at
        11K (20°F) per minute,
Type 5 - cool from 603K (625°F) to 78K (-320°F) by
        immersing in liquid nitrogen.

After exposure, the specimen edges were polished and observed
through a microscope to detect microcracks. A magnification of 50X
or greater was generally necessary to discern crack details. Crack
depth in from free edges was measured using a liquid dye penetrant
(tetra-bromo-ethane) x-ray technique.

Theoretical Considerations

Laminate analysis with stress and temperature dependent
material properties was used to predict the residual stresses
in laminates after cooling from the cure temperature. All calculations were based upon a stress-free temperature of 603K (625°F).
The temperature-dependent analysis follows that of Hahn and
Pagano [5]. The total strains in the kth layer $\{\varepsilon\}^k$ are assumed to
be the sum of the mechanical strains $\{\varepsilon^\sigma\}^k$ and the free thermal
strains $\{\varepsilon^T\}$

$$\{\varepsilon\}^k = \{\varepsilon^\sigma\}^k + \{\varepsilon^T\}^k \tag{1}$$

The free thermal strains for a temperature change T_o to T are

$$\{\varepsilon^T(T)\}^k = \int_{T_o}^{T} \{\alpha(\xi)\}^k d\xi \tag{2}$$

where $\{\alpha(\xi)\}^k$ are the temperature dependent coefficients of thermal
expansion for the kth layer. If T_o is the stress-free temperature,
the stresses induced in a symmetric laminate for the temperature
change $T_o \rightarrow T$ are

$$\{\sigma(T)\}^k = [\bar{Q}(T)]^k(\{\varepsilon^\circ\} - \{\varepsilon^T(T)\}^k) \tag{3}$$

where $[\bar{Q}(T)]$ is the stiffness matrix at temperature T, and $\{\varepsilon^\circ\}$
are the total midplane strains which are determined from the thermo-
mechanical laminate analysis equations

$$\{\varepsilon^\circ\} = [A(T)]^{-1}\{N^T\} \tag{4}$$

The exact integration of (2) is accomplished using the best fit polynomial of the available temperature dependent experimental data. The stress and temperature dependent analysis was based upon an incremental form of (3) with $[\bar{Q}(\sigma,T)]$ constant for each temperature increment. The material properties used in the stress and temperature dependent analysis were expressed as polynomial equations of the two independent variables stress, σ, and temperature, T.

RESULTS AND DISCUSSION

Viewing Direction

For this study, the viewing direction has been defined as the acute angle between the 0° fiber (x laminate) direction and the normal to the surface under observation (Fig. 2). The cross-sections of the fibers in a 0° ply are evident when the laminate is observed from the 0° viewing direction and a side view of the fiber is observed from the 90° viewing direction. The effect of viewing direction on TVM detection is clearly shown in Figure 3. Cracks are visible in the 0° plies when viewed from the 0° direction, in the 90° plies when viewed from the 90° direction, and in both 0° and 90° plies when viewed from the 45° direction.

The primary viewing direction used in this investigation was the 0° direction. All laminates were viewed from the 0° direction for all five types of thermal loadings. In addition, the specimens subjected to the Type 5 thermal loading were also viewed from the 90° viewing direction.

Crack Characteristics

As shown in Figure 4, in some cases, resin-rich regions act as crack arresters. This figure suggests that the cracks form due to the stress concentrations, present at the micro-mechanics level, in the fiber/matrix region and propagate to the resin-rich region. The crack may be arrested at the boundary of the resin-rich region (Fig. 4) or may propagate through it (Fig. 3). The orientations of the fibers in the layers adjacent to the resin-rich region have a significant influence on the crack propagation. Whereas the cracks are usually arrested by the resin-rich region between layers of different fiber angle (Fig. 4), the cracks always propagated through the resin-rich region between adjacent plies of the same fiber orientation (Fig. 3) for the five thermal loading considered in this investigation. Complimentary work on TVM due to tensile loading has indicated that cracks do not propagate through the resin-rich region if the load is only marginally large enough to initiate cracking.

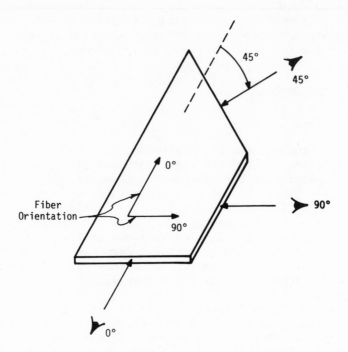

Figure 2.   Illustration of Viewing Directions.

Hence, it would appear that the thermal loadings in this investigation were more than marginally sufficient to initiate microcracking.
This ability of the resin-rich region to arrest cracks is a result
of the transverse strength of the neat PMR-15 being as much as 33
percent higher than the transverse strength of the unidirectional
composite [6].

     It is also apparent from Figures 3 and 4 that the the 90°
plies adjacent to a cracked 0° layer have a constraining influence
on crack development. The crack width is largest at mid-layer and
tapers in width near the resin-rich region.

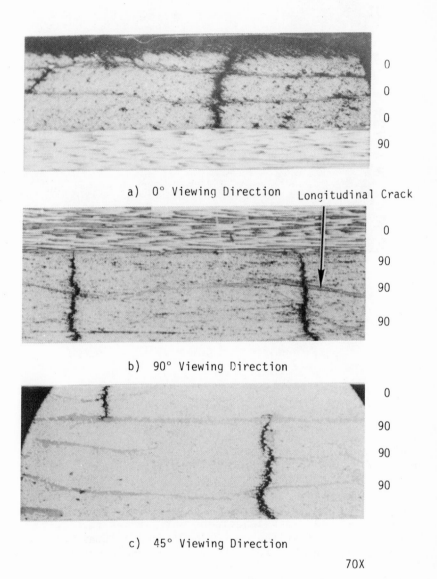

a) 0° Viewing Direction   Longitudinal Crack

b) 90° Viewing Direction

c) 45° Viewing Direction

70X

Figure 3.   Effect of Viewing Direction on TVM Observation
in Type 5 Thermally Loaded $[0_3/90_3]_s$ Laminate.

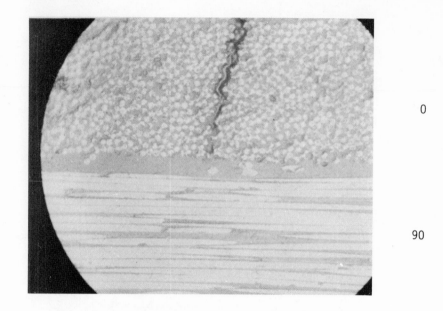

0

90

200X

Figure 4.   TVM Near an Interface of a $[0_2/90_2]_s$ Laminate.

The extent of transverse microcracks in from the free edge of specimens subjected to Type 5 thermal loading is shown in Figure 5. In all but a very few cases, the cracks extend across the entire width of the laminate. In some cases it appears that cracks emanate from opposite edges and "meet" to form the complete crack. In at least one case, a crack is evident only at the free edge. This is further evidence that the cracks do indeed initiate at the free edge and then propagate across the width of the specimen. The dark circular shapes in Figure 5c are delaminations which, interestingly, do not extend to the free edges.

X-rays also showed that TVM were found only in the 0° and 90° plies of the $[0/45/90/-45]_s$ laminate, not at all in the $[\pm45/0/90]_s$ laminate, and primarily only in the outer 0° plies of the $[0/60/0/-60]_s$ laminate.

a) $[0_2/90_2]_s$

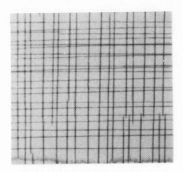

b) $[0_3/90_3]_s$

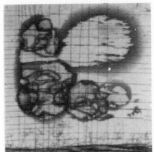

2X

c) $[(0/90)_3]_s$

Figure 5.   X-rays of Specimens Subjected to Type 5
Thermal Loading.

TVM Densities

As mentioned previously, the laminates under investigation did
not exhibit microcracks after curing (Type 1 thermal loading). TVM
densities for all thermal loadings are summarized in Table 1.   The
absence of TVM in Celion 6000/PMR-15 after curing is in sharp
contrast to the densely microcracked HTS1/PMR-15 (approximately
100 TVM/inch - Fig. 1). This difference is attributed to the
improved cure cycle and possibly the higher transverse strength
of the Celion 6000/PMR-15 material.

Table 1. Effect of Thermal Exposure on Microcracking in Celion/PMR-15 Graphite/Polyimide

Thermal Exposure	Laminate Orientation		
	$[0/45/90/-45]_s$	$[45/-45/0/90]_s$	$[0/60/0/-60]_s$
Type 1 625°F→RT @ 5°F/min.	No TVM	No TVM	No TVM
Type 2 RT→ -320°F, by LN$_2$ Quench store RT air	No TVM	No TVM	No TVM
Type 3 RT→625°F @ 15°F/min., 625°F→32°F by ice water quench, store RT air	2 TVM/inch in 0° plies	No TVM	No TVM
Type 4 RT→625°F @ 5°F/min., 625°F→ -320°F @ 20°F/min., store RT air	2 TVM/inch in 0° plies	No TVM	No TVM
Type 5 RT→625°F @15°F/min., 625°F→ -320°F by LN$_2$ quench store RT air	19 TVM/inch in 0° plies and 4 cracks/inch in +45° plies	No TVM	21 TVM/inch in outer 0° plies and 9 cracks/inch in -60° plies

Thermal Exposure	Laminate Orientation		
	$[(0/90)_3]_s$	$[0_3/90_3]_s$	$[0_2/90_2]_s$
Type 1 625°F→RT @ 5°F/min.	No TVM	2.5 TVM/inch in 0° plies on one side	No TVM
Type 2 RT→ -320°F by LN$_2$ quench, store RT air	No TVM	2 TVM/inch in 0° plies	1.5 TVM/inch in 0° plies
Type 3 RT→625°F @15°F/min. 625°F→32°F by ice water quench, store RT air	2 TVM/inch in outer 0° plies	8 TVM/inch in 0° plies	4 TVM/inch in 0° plies
Type 4 RT→625°F @ 5°F/min., 625°F→ -320°F @ 20°F/min., store RT air	5 TVM/inch in outer 0° plies	7 TVM/inch in 0° plies	5 TVM/inch in 0° plies
Type 5 RT→625°F @ 15°F/min., 625°F→ -320°F by LN$_2$ quench, store RT air	28 TVM/inch in outer 0° plies and 16 cracks/inch in midplane 90° plies	11 TVM/inch in 0° plies and 8 cracks/inch in all 90° plies	15 TVM/inch in 0° plies and 26 cracks/inch in all 90° plies

The results in Table 1 show that TVM density is laminate dependent and generally increases with increased rate of cooling and range of temperature drop. Only the $[0_3/90_3]_s$ and $[0_2/90_2]_s$ laminates exhibited microcracking after the room temperature to liquid nitrogen quench (Type 2 thermal loading). The TVM density is quite low for both laminates. Since the densities represent at most two cracks per specimen for the one inch (2.54 cm) square specimens, the results may be influence by the statistical distribution of stress risers and strength.

Post heating to the maximum cure temperature and then quenching in ice water (Type 3 thermal loading) resulted in microcracks in four of the six laminates. The crack densities were generally higher than those from Type 2 loading, but still quite low with the maximum density of 8 TVM/inch (3.2 TVM/cm) in the $[0_3/90_3]_s$ laminate.

The influence of rate of loading on crack density is clearly shown when the results for Type 4 loading (slow cool) are compared with the results for Type 5 loading (rapid cool). Five of the six laminates developed significant microcracking after exposure to the Type 5 loading with the TVM densities being much higher than the densities from Type 4 loading. The ratio of Type 4 to Type 5 TVM densities are presented in Table 2.

It is important to note that the TVM densities for Type 5 thermal loading (the most severe case investigated) are well below the TVM density of the HTS1/PMR-15 after a much less severe loading. This fact clearly demonstrates the superior thermal resistance of the materials and curing process used in this investigation.

As indicated by the results presented in Table 1, the $[\pm45/0/90]_s$ laminate proved to be very resistant to microcracking. This laminate was completely free of cracks for all five types of thermal loadings. The substantially different behavior of the two quasi-isotropic laminates clearly shows the influence of stacking sequence and edge effects on crack formation.

Residual Stress Predictions

Room temperature residual stresses in a $[0/90]_s$ laminate were predicted using constant material properties, temperature dependent properties, and stress and temperature dependent properties. The residual stresses and the percent of transverse failure stress are presented in Table 3. Results based upon a stress free temperature of 603K (625°F) indicate that the residual stresses range from 95-115 percent of the ultimate stress depending upon the theory employed. The temperature dependent results predict 20 percent higher stresses than does the constant property

Table 2.   Ratio of TVM Densities for Type 4 and
           Type 5 Thermal Loadings

Laminate	TVM Density Type 5:Type 4
$[0_3/90_3]_s$	11:7
$[0_2/90_2]_s$	15:5
$[(0/90)_3]_s$	28:5
$[0/45/90/-45]_s$	19:2
$[0/60/0/-60]_s$	21:0

Table 3.   Predicted Transverse Residual Curing
           Stresses for $[0/90]_s$ Gr/Pi Laminates

Form of Mechanical Properties	Residual Stresses ksi (MPa)	Percent of Transverse Failure Stress
Constant	7.15 (49.3)	95
Temperature Dependent	8.65 (59.6)	115
Stress- and Temperature- Dependent	7.98 (55.0)	106

NOTE: Stress-Free Temperature = 625°F (603K)
        Y_T = 7.523 ksi (51.9 MPa)

assumption. The stress and temperature dependent assumption
predicts intermediate stresses of 106 percent of ultimate.

The stress-free temperature dependence of the predicted residual
stresses is shown in Figure 6 for temperature dependent, and stress
and temperature dependent properties. For the measured range of
stress-free temperature (570-625°F), the temperature dependent
theory always predicts stresses above the failure strength, but
the stress and temperature dependent theory predicts that failure
does not occur for a stress-free temperature below 598°F.

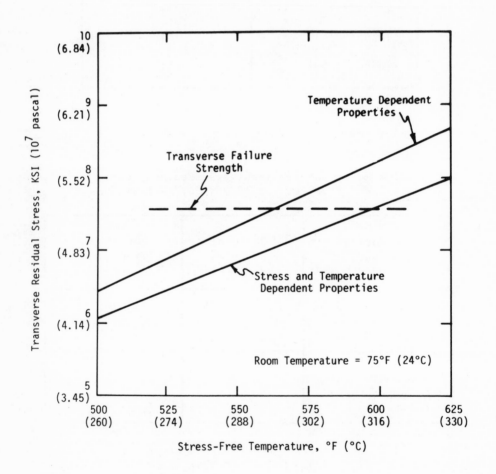

Figure 6.   Residual Stresses for [0/90]$_s$ and [0/90/±45]$_s$
            Laminates as a Function of Stress-Free Temperature

These predictions indicate that the stresses are marginally
high enough to initiate microcracking. The temperature change
required to initiate microcracking at room temperature may be
predicted more accurately if viscoelastic and/or nonlinear effects
are included in the analysis. Weitsman [7] predicts a 20 percent
decrease in residual stresses when viscoelastic behavior is included
for an epoxy resin composite. The higher TVM densities found in
this investigation for the Type 5 rapid thermal cooling gives
further evidence to the importance of rate effects on the development of residual stresses in polyimide composites.

CONCLUSIONS

It has been shown that Celion 6000/PMR-15 graphite-polyimide
is free of cracks after curing. However, exposure to 78K (-320°F)
produced microcracks in five of the six laminates investigated.
The $[\pm45/0/90]_s$ laminate proved to be very resistant to microcracking. This laminate was free of microcracks for all five
thermal loadings considered. The density of microcracks was found
to be dependent upon fiber orientations, stacking sequence, edge
effects and rate of cooling. Microcracks produced by thermal
exposures extended across the entire width of one inch (2.54 cm)
square specimens.

REFERENCES

1. Spain, R. G., "Thermal Microcracking of Carbon Fibre/Resin
    Composites," Composites, pp. 33-37, March, 1971.

2. Novak, R. C. and DeCrescente, M. A., "Fabrication Stresses in
    Graphite-Resin Composites," J. of Engineering for Power, Vol.
    92, pp. 377-380, October, 1970.

3. Molcho, A. and Ishai, O., "Thermal Cracking of C.F.R.P.
    Laminates," Society for the Advancement of Material and Process
    Engineering, Vol. 10, pp. 255-262, October, 1978.

4. Lee, B. L. and McGarry, F. J., "Study of Processing and
    Properties of Graphite Fiber/High Temperature Resin Composites,"
    AMMRC CTR 76-10, April, 1976.

5. Hahn, H. T. and Pagano, N. J., "Curing Stresses in Composite
    Lamiantes," Journal of Composite Materials, Vol. 9, p. 91, 1975.

6. Cavano, P. J. and Winters, W. E., "PMR Polyimide/Graphite Fiber
    Composite Fan Blades," NASA-CR-135113, December 15, 1976.

7. Weitsman, Y., "Residual Thermal Stresses Due to Cool-Down of
    Epoxy-Resin Composites," J. Applied Mechanics, Vol. 46,
    September, 1979.

ACKNOWLEDGEMENTS

This work was supported by the NASA-Virginia Tech Composites
Program, NASA CA NCCI-15. Mr. Mills was in residence at Langley
Research Center during the investigation and the assistance of
numerous technical staff members there is greatfully acknowledged
as is the fine typing of Ms. Fran Carter at Virginia Tech.

EXTERNAL CRACK DUE TO THERMAL EFFECTS IN AN INFINITE ELASTIC SOLID

WITH A CYLINDRICAL INCLUSION

Ranjit S. Dhaliwal

Department of Mathematics and Statistics
The University of Calgary
Calgary, Alberta, Canada   T2N 1N4

ABSTRACT -- This paper deals with the state of stress in an infinite
elastic solid with an external crack which is subjected to a
prescribed temperature distribution. The infinite elastic medium
consists of two materials which are separated by a cylindrical
surface. It is assumed that there is perfect bonding at the common
cylindrical surface. By assuming a suitable representation for the
temperature function, the heat conduction problem is reduced to the
solution of a Fredholm integral equation of the second kind. Then,
using suitable biharmonic functions as thermoelastic potentials,
the thermoelastic problem is also reduced to the solution of a
Fredholm integral equation of the second kind. Both the integral
equations are solved numerically. The numerical values of the
stress intensity factor are displayed graphically.

1.   INTRODUCTION

In recent years, considerable effort has been devoted to the
problems of calculating thermal stresses in infinite and semi-
infinite solids, thick plates and infinitely long cylinders containing penny-shaped or external cracks. Adequate references to
this type of work may be found in Kassir and Sih [1]. The
axisymmetric external crack problems for a media with cylindrical
cavities have been discussed by Srivastav and Lee [2].

In this paper, we consider the problem of determining the
distribution of thermal stresses in an infinite medium containing an
external crack and bonded to an infinitely long circular cylinder
(inclusion). We assume that the plane of the crack is z = 0, the

axis of the cylinder being the z-axis and (r,θ,z) is a cylindrical polar coordinate system. The external crack occupying the region $a < r < \infty$, $z = 0\mp$ is subjected to a prescribed temperature distribution $T(r)$. Both the materials are assumed to be homogeneous, isotropic and elastic such that μ_i, ν_i and k_i are respectively the shear modulus, the Poisson ratio and the thermal conductivity of the region i (i = 1,2) where the region 1 is $b < r < \infty$, $-\infty < z < \infty$ and the region 2 is $0 \le r < b$, $-\infty < z < \infty$. It is assumed that there is a perfect bonding of the two materials at the common surface $r = b(b < a)$. Since the geometry of the regions is symmetric about the crack plane $z = 0$, the problem is reduced to a mixed boundary value problem in heat conduction and a mixed boundary value problem of thermoelasticity for the region $z \ge 0$, $r \ge 0$.

The boundary conditions for the heat conduction problem may be taken as:

$$T^1(r,0) = T(r), \qquad\qquad\qquad a < r < \infty, \qquad (1.1)$$

$$\frac{\partial}{\partial z} T^1(r,z) = 0, \qquad z = 0, \qquad b < r < a, \qquad (1.2)$$

$$\frac{\partial}{\partial z} T^2(r,z) = 0, \qquad z = 0, \qquad 0 \le r < b, \qquad (1.3)$$

and the continuity conditions are

$$T^1(r,z) = T^2(r,z), \ k_1 \frac{\partial}{\partial r} T^1(r,z) = k_2 \frac{\partial}{\partial r} T^2(r,z), \ r = b, \quad (1.4)$$

where the superscript i (i = 1,2) denote quantities for the region i. The temperature function $T^i(r,z)$ for the region i (i = 1,2) satisfies the Laplace's equation

$$\left(\frac{\partial^2}{\partial r^2} + \frac{1}{r}\frac{\partial}{\partial r} + \frac{\partial^2}{\partial z^2}\right) T^i(r,z) = 0. \qquad (1.5)$$

The boundary conditions for the thermoelastic problem may be taken as

$$\sigma^1_{zz}(r,0) = p(r), \qquad\qquad\qquad a < r < \infty, \qquad (1.6)$$

$$u^1_z(r,0) = 0, \qquad\qquad\qquad b < r < a, \qquad (1.7)$$

$$\sigma^1_{rz}(r,0) = 0, \qquad\qquad\qquad b < r < \infty, \qquad (1.8)$$

$$\sigma^2_{rz}(r,0) = 0, \ u^2_z(r,0) = 0, \qquad\qquad 0 \le r < b. \qquad (1.9)$$

The continuity conditions are:

$$u_z^1(b,z) = u_z^2(b,z), \quad u_r^1(b,z) = u_r^2(b,z),$$

$$\sigma_{rr}^1(b,z) = \sigma_{rr}^2(b,z), \quad \sigma_{rz}^2(b,z) = \sigma_{rz}^2(b,z) \tag{1.10}$$

where $(u_r^i, 0, u_z^i)$ denote the displacement vector and $(\sigma_{rr}^i, \sigma_{\theta\theta}^i, \sigma_{rz}^i, \sigma_{zz}^i)$ is the stress tensor for the region i.

2. THE HEAT CONDUCTION PROBLEM

We assume suitable temperature functions for the regions 1 and 2 in the following form:

$$T^1(r,z) = \int_0^\infty \frac{1}{s} \theta_1(s) \, e^{-sz} \, J_0(rs)ds + \int_0^\infty \frac{1}{s} \theta_2(s) \, K_0(sz) \, \cos(sz) \, ds, \tag{2.1}$$

$$T^2(r,z) = \int_0^\infty \frac{1}{s} \theta_3(s) \, I_0(rs) \, \cos(sz)ds, \tag{2.2}$$

where the functions $\theta_1(s)$, $\theta_2(s)$ and $\theta_3(s)$ are to be determined from the boundary and the continuity conditons (1.1) – (1.4), and $J_\nu(rs)$, $K_\nu(rs)$ and $I_\nu(rs)$ denote the Bessel function of the first kind, the modified Bessel function of the first kind and the Macdonald's function of order $\nu \geq 0$, respectively.

The boundary conditions (1.1), (1.2), along with the equations (2.1) and (2.2) lead to the following dual integral equations

$$\int_0^\infty \frac{1}{s} \theta_1(s) \, J_0(rs)ds + \int_0^\infty \frac{1}{s} \theta_2(s) \, K_0(rs)ds = T(r), \quad a < r < \infty, \tag{2.3}$$

$$\int_0^\infty \theta_1(s) \, J_0(rs)ds = 0, \qquad\qquad b < r < a. \tag{2.4}$$

For solving the above dual integral equations (2.3) and (2.4) we follow the method discussed by Lowengrub and Sneddon [3] and assume

$$\theta_1(s) = s \int_a^\infty g(t) \, \sin(st)dt = g(a) \, \cos(as) + \int_a^\infty g'(t) \, \cos(st)dt, \tag{2.5}$$

where the prime denotes differentiation with respect to t and
$g(\infty) = 0$. Substituting for $\theta_1(s)$ in equation (2.3) from (2.5) and
using the result

$$\int_0^\infty J_0(rs)\ \sin(st)ds = \begin{cases} (t^2 - r^2)^{-\frac{1}{2}}, & 0 < r < t \\ 0, & r > t \end{cases} \qquad (2.6)$$

we find that

$$\int_r^\infty \frac{g(t)dt}{(t^2 - r^2)^{\frac{1}{2}}} + \int_0^\infty \frac{1}{s}\ \theta_2(s)\ K_0(rs)ds = T(r),\ a < r < \infty, \qquad (2.7)$$

and the equation (2.4) is identically satisfied.

The above equation (2.7) is of Abel type and its solution is
given by

$$g(t) + \int_0^\infty \frac{\theta_2(s)}{s}\ e^{-ts}\ ds = -\frac{2}{\pi}\frac{d}{dt}\int_t^\infty \frac{rT_0(r)dr}{(r^2 - t^2)^{\frac{1}{2}}},\ a < t < \infty. \qquad (2.8)$$

It is easy to see from (2.2) that the boundary condition (1.3) is
identically satisfied. The continuity conditions (1.4) and the use
of the Fourier inversion theorem yield

$$\theta_2\ K_0(bs) - \theta_3\ I_0(bs) = -\frac{2}{\pi}\ s \int_0^\infty \frac{\theta_1(u)\ J_0(bu)du}{u^2 + s^2}, \qquad (2.9)$$

$$\theta_2\ K_1(bs) + R\theta_3\ I_1(bs) = -\frac{2}{\pi}\int_0^\infty \frac{u\ \theta_1(u)\ J_1(bu)du}{u^2 + s^2}, \qquad (2.10)$$

Solving equations (2.9) and (2.10) we find that:

$$\theta_2 = -\frac{2}{\pi\Delta}\left[s\ R\ I_1 \int_0^\infty \frac{\theta_1(u)\ J_0(bu)du}{u^2 + s^2} + \right.$$

$$\left. + I_0 \int_0^\infty \frac{u\ \theta_1(u)\ J_1(bu)du}{u^2 + s^2} \right], \qquad (2.11)$$

$$\theta_3 = \frac{2}{\pi\Delta}\left[s\ K_1 \int_0^\infty \frac{\theta_1(u)\ J_0(bu)du}{u^2 + s^2} - K_0 \int_0^\infty \frac{u\ \theta_1(u)\ J_1(bu)du}{u^2 + s^2} \right], \qquad (2.12$$

where

$$\Delta = K_1 I_0 + R K_0 I_1, \quad K_j = K_j(bs), \quad I_j = I_j(bs), \quad j = 0,1,2. \quad (2.13)$$

Making use of (2.5), we find that

$$\int_0^\infty \frac{u \, J_1(bu) \, \theta_1(u) du}{u^2 + s^2} = \int_a^\infty g(t) dt \int_0^\infty \frac{u^2 \, J_1(bu) \, \sin(ut) du}{u^2 + s^2} . \quad (2.14)$$

With the help of Erdelyi ([4], Vol. I, p. 101, Entry (16)), we find that

$$\int_0^\infty \frac{u^2 \, \sin(ut) \, J_1(bu) du}{u^2 + s^2} = -\frac{\pi s}{2} I_1 e^{-st} , \quad (2.15)$$

and hence

$$\int_0^\infty \frac{u \, J_1(bu) \, \theta_1(u) du}{u^2 + s^2} = -\frac{\pi s}{2} I_1 \int_a^\infty g(t) e^{-st} dt . \quad (2.16)$$

Again using the following integral from Erdelyi ([4], Vol. I, p. 100, Entry (12))

$$\int_0^\infty \frac{u \, J_0(bu) \, \sin(ut) du}{u^2 + s^2} = \frac{\pi}{2} e^{-st} I_0(bs) , \quad (2.17)$$

we find that

$$\int_0^\infty \frac{\theta_1(u) \, J_1(bu) du}{u^2 + s^2} = \frac{\pi}{2} I_0 \int_a^\infty e^{-st} g(t) dt . \quad (2.18)$$

Using the results (2.16) and (2.18), we find from (2.11) and (2.12) that

$$\{\theta_2, \theta_3\} = \frac{s}{\Delta} \int_a^\infty e^{-st} g(t) dt \{I_0 I_1 (1 - R), \Delta_1\} \quad (2.19)$$

where $\Delta_1 = K_1 I_0 + I_1 K_0$. Substituting the value of $\theta_2(s)$ from equation (2.19) into (2.8) we obtain the following Fredholm integral equation of the second kind for the determination of $g(t)$:

$$g(t) + \int_a^\infty g(u) \, K_1(u,t) du = -\frac{2}{\pi} \frac{d}{dt} \int_t^\infty \frac{r \, T(r) dr}{(r^2 - t^2)^{\frac{1}{2}}} \, , \quad a < t < \infty,$$

$$(2.20)$$

where

$$K_1(u,t) = (1 - R) \int_0^\infty \frac{I_0(bs) \, I_1(bs) \, e^{-(u+t)s}}{\Delta} \, ds \, . \tag{2.21}$$

If $T(r) = T_0/r^2$, where T_0 is constant, we find that

$$-\frac{2}{\pi} \frac{d}{dt} \int_t^\infty \frac{r \, T(r) dr}{(r^2 - t^2)^{\frac{1}{2}}} = -\frac{2T_0}{\pi} \frac{d}{dt} \int_t^\infty \frac{dr}{r(r^2 - t^2)^{\frac{1}{2}}} = T_0/t^2 \, . \tag{2.22}$$

3.   THE THERMOELASTIC PROBLEM

3a.   Basic Equations

It is well known that the displacement components of u_i are given in terms of the potential function $\psi(r,z)$ by the following equations (see Nowacki [5], p. 45):

$$u_r(r,z) = \frac{\partial \psi}{\partial r} \, , \quad u_z(r,z) = \frac{\partial \psi}{\partial z} \, , \quad \sigma_{ij} = 2\mu(\psi,_{ij} - \delta_{ij} \, \psi,_{kk}) \, , \tag{3.1}$$

where

$$\psi,_{kk} = mT_i \, , \quad m = \frac{1 - \nu}{1 + \nu} \alpha_t \, , \tag{3.2}$$

ν being the Poisson's ratio and α_t is the coefficient of linear thermal expansion.   Equations (1.5) and (3.2) give

$$\left(\frac{\partial^2}{\partial r^2} + \frac{1}{r} \frac{\partial}{\partial r} + \frac{\partial^2}{\partial z^2} \right)^2 \psi(r,z) = 0 \, . \tag{3.3}$$

A suitable biharmonic function for the region $1(b < r, \, z \geq 0)$ may be taken as

$$\psi_1(r,z) = \frac{m_1}{2} \int_0^\infty s^{-2} \, \theta_2(s) \, rK_1(rs) \, \cos(sz) ds \, -$$

$$- \frac{m_1}{2} \int_0^\infty s^{-3} \theta_1(s)(1 + sz) \, \bar{e}^{sz} \, J_0(rs)ds, \quad b < r, \ z > 0, \tag{3.4}$$

where $m_1 = (1 + \nu_1)(\alpha_t)_1/(1 - \nu_1)$ and $(\alpha_t)_1$ represents the coefficient of linear thermal expansion for the region 1. With the help of the relations (3.1) we find that the displacement and stress components for the region 1 due to thermal effects are given by:

$$u_r^*(r,z) = \frac{m_1}{2} \int_0^\infty s^{-2} \theta_1(s)(1 + sz) \, \bar{e}^{sz} \, J_1(rs)ds -$$
$$- \frac{m_1 r}{2} \int_0^\infty s^{-1} \theta_2(s) \, K_0(sr) \, \cos(sz)ds \ , \tag{3.5}$$

$$u_z^*(r,z) = \frac{m_1 z}{2} \int_0^\infty s^{-1} \theta_1(s) \, J_0(rs)ds -$$
$$- \frac{m_1 r}{2} \int_0^\infty s^{-1} \theta_2(s) \, K_1(sr) \, \sin(sz)ds \ , \tag{3.6}$$

$$\frac{\sigma_{rz}^*(r,z)}{2\mu_1} = -\frac{m_1 z}{2} \int_0^\infty \theta_1(s) \, \bar{e}^{sz} \, J_1(rs)ds +$$
$$+ \frac{rm_1}{2} \int_0^\infty \theta_2(s) \, K_0(rs) \, \sin(sz)ds \ , \tag{3.7}$$

$$\frac{\sigma_{zz}^*(r,z)}{2\mu_1} = -\frac{m_1}{2} \int_0^\infty s^{-1} \theta_1(s)(1 + sz) \, \bar{e}^{sz} \, J_0(rs)ds -$$
$$- \frac{m_1}{2} \int_0^\infty s^{-1} \theta_2(s)[2K_0(rs) + rs \, K_1(rs)] \, \cos(sz)ds \ , \tag{3.8}$$

$$\frac{\sigma_{rr}^*(r,z)}{2\mu_1} = \frac{m_1}{2} \left[- \int_0^\infty \left\{ \frac{J_0(rs)}{s} + \frac{J_1(rs)}{rs^2} \right\} \bar{e}^{sz} \, \theta_1(s)ds \right.$$
$$\left. - z \int_0^\infty \left\{ \frac{J_1(rs)}{rs} - J_0(rs) \right\} \bar{e}^{sz} \, \theta_1(s)ds \right. \tag{3.9}$$

$$+ \int_0^\infty \theta_2(s) \left\{ r\, K_1(rs) - \frac{3}{s}\, K_0(rs) \right\} \cos(sz)ds \Bigg] \ .$$

Let us now assume a suitable biharmonic function for the region $2(0 < r < b, \ z \geq 0)$ in the following form:

$$\psi_2 = \frac{m_2 r}{2} \int_0^\infty s^{-2}\, \theta_3(s)\, I_1(rs)\, \cos(sz)ds \ . \tag{3.10}$$

With the help of equations (3.1) and (3.10), we find that the displacement and stress components for the region 2 due to thermal effects are

$$u_r^{**}(r,z) = \frac{m_2 r}{2} \int_0^\infty s^{-1}\, \theta_3(s)\, I_0(rs)\, \cos(sz)ds \ , \tag{3.11}$$

$$u_z^{**}(r,z) = - \frac{m_2 r}{2} \int_0^\infty s^{-1}\, \theta_3(s)\, I_1(rs)\, \sin(sz)ds \ , \tag{3.12}$$

$$\frac{\sigma_{rz}^{**}}{2\mu_2} = - \frac{m_2 r}{2} \int_0^\infty \theta_3(s)\, I_0(rs)\, \sin(sz)ds \ , \tag{3.13}$$

$$\frac{\sigma_{zz}^{**}}{2\mu_2} = - \frac{m_2}{2} \int_0^\infty s^{-1}\, \theta_3(s)\, [2I_0(rs) + rs\, I_1(rs)]\, \cos(sz)ds \ , \tag{3.14}$$

$$\frac{\sigma_{rr}^{**}}{2\mu_2} = \frac{m_2}{2} \int_0^\infty s^{-1}\, \theta_3(s)\, [rs\, I_1(rs) - I_0(rs)]\, \cos(sz)ds \ , \tag{3.15}$$

where $m_2 = (1 + \nu_2)(\alpha_t)_2/(1 - \nu_2)$ and $(\alpha_t)_2$ represents the coefficient of linear thermal expansion for the region 2. Equations (3.5) – (3.9) and (3.11) – (3.15) give the particular solution for both the regions.

We shall now find the displacement and the stress components for the isothermal elastic problem. From Keer [6], we find that the axially symmetric displacement field in an infinite solid with an external crack subjected to normal loading is given by:

$$u_r(r,z) = (1 - 2\nu) \frac{\partial \phi}{\partial r} + z \frac{\partial^2 \phi}{\partial r \partial z} \ , \quad u_\theta(r,z) = 0 \ ,$$

$$u_z(r,z) = -2(1 - \nu) \frac{\partial \phi}{\partial z} + z \frac{\partial^2 \phi}{\partial z^2} . \tag{3.16}$$

The corresponding stress components are then given by:

$$\sigma_{rr}(r,z) = 2\mu \left[(1 - 2\nu) \frac{\partial^2 \phi}{\partial r^2} - 2\nu \frac{\partial^2 \phi}{\partial z^2} + z \frac{\partial^3 \phi}{\partial r^2 \partial z} \right] ,$$

$$\sigma_{zz}(r,z) = 2\mu \left[-\frac{\partial^2 \phi}{\partial z^2} + z \frac{\partial^3 \phi}{\partial z^3} \right] , \quad \sigma_{rz}(r,z) = 2\mu \left[\frac{\partial^3 \phi}{\partial r \partial z^2} \right] , \tag{3.17}$$

where

$$\phi(r,z) = \int_0^\infty \frac{1}{s^2} F(s) J_0(rs) e^{-sz} ds , \tag{3.18}$$

and μ and ν represent the shear modulus and the Poisson ratio, respectively.

For a circular cylindrical inclusion containing no external crack, the displacement components can be taken in the following form (Collins [7]):

$$u_r(r,z) = \frac{\partial \phi_1}{\partial r} + (3 - 4\nu) \phi_2 - r \frac{\partial \phi_2}{\partial r} ,$$

$$\tag{3.19}$$

$$u_\theta(r,z) = 0 , \quad u_z(r,z) = \frac{\partial \phi_1}{\partial z} - r \frac{\partial \phi_2}{\partial z} .$$

With the help of the stress-displacement relations, it is found that

$$\frac{\sigma_{rr}}{2\mu} = \frac{\partial^2 \phi_1}{\partial r^2} + \frac{2\phi_2}{r} + 2(1 - \nu) \frac{\partial \phi_2}{\partial r} - r \frac{\partial^2 \phi_2}{\partial r^2} ,$$

$$\frac{\sigma_{zz}}{2\mu} = \frac{\partial^2 \phi_1}{\partial z^2} + \frac{2\nu \phi_2}{r} + 2\nu \frac{\partial \phi_2}{\partial r} - r \frac{\partial^2 \phi_2}{\partial z^2} ,$$

$$\tag{3.20}$$

$$\frac{\sigma_{\theta\theta}}{2\mu} = \frac{1}{r} \frac{\partial \phi_1}{\partial r} + (3 - 2\nu) \frac{\phi_2}{r} - (1 - 2\nu) \frac{\partial \phi_2}{\partial r} ,$$

$$\frac{\sigma_{rz}}{2\mu} = \frac{\partial^2 \phi_1}{\partial r \partial z} + (1 - 2\nu) \frac{\partial \phi_2}{\partial z} - r \frac{\partial^2 \phi_2}{\partial r \partial z} .$$

The functions $\phi_1(r,z)$ and $\phi_2(r,z)\cos\theta$ are harmonic and can be found from the following equations:

$$\left(\frac{\partial^2}{\partial r^2} + \frac{1}{r}\frac{\partial}{\partial r} + \frac{\partial^2}{\partial z^2}\right)\phi_1(r,z) = 0 \ ,$$

$$\left(\frac{\partial^2}{\partial r^2} + \frac{1}{r}\frac{\partial}{\partial r} + \frac{\partial^2}{\partial z^2} - \frac{1}{r^2}\right)\phi_2(r,z) = 0 \ . \tag{3.21}$$

For the region $b < r$, $z \geq 0$, we take

$$\phi_1(r,z) = \int_0^\infty \frac{1}{s} A(s) \, K_0(rs) \, \cos(sz)ds$$

$$\phi_2(r,z) = \int_0^\infty B(s) \, K_1(rs) \, \cos(sz)ds \tag{3.22}$$

Due to isothermal effect the displacement field for the region 1 obtained by adding the equations (3.16) and (3.19) and using the results of equations (3.18) and (3.22) is given by:

$$\left[u_r(r,z)\right]_1 = \int_0^\infty \frac{1}{s}(2\nu_1 - 1 + sz)\, F(s)\, J_1(rs)\, \bar{e}^{sz}\, dz \tag{3.23}$$

$$+ \int_0^\infty \left\{-K_1(rs)\, A(s) + [4(1 - \nu_1)\, K_1(rs) + rs\, K_0(rs)]\, B(s)\right\}$$

$$\cos(sz)ds \ ,$$

$$\left[u_z(r,z)\right]_1 = \int_0^\infty \frac{1}{s}(2 - 2\nu_1 + sz)\, F(s)\, J_0(rs)\, \bar{e}^{sz}\, ds \tag{3.24}$$

$$+ \int_0^\infty [-K_0(rs)\, A(s) + rs\, B(s)\, K_1(rs)]\, \sin(sz)ds \ .$$

In the same way, by making use of (3.17), (3.18), (3.20) and (3.22) we find that the stress components for the region 1 are given by

$$\frac{\left[\sigma_{rz}\right]_1}{2\mu_1} = -z \int_0^\infty s\, F(s)\, J_1(rs)\, \bar{e}^{sz}\, ds \tag{3.25}$$

$$+ \int_0^\infty s \left\{ K_1(rs) \, A(s) - [2(1 - \nu_1) \, K_1(rs) + rs \, K_0(rs)] \, B(s) \right\} \sin(sz) ds,$$

$$\frac{[\sigma_{zz}]_1}{2\mu_1} = - \int_0^\infty (1 + sz) \, F(s) \, J_0(rs) \, \bar{e}^{sz} \, ds$$

$$- \int_0^\infty s \left\{ K_0(rs) \, A(s) + [2\nu_1 \, K_0(rs) - rs \, K_1(rs)] \, B(s) \right\} \cos(sz) ds \; ,$$

(3.26)

$$\frac{[\sigma_{rr}]_1}{2\mu_1} = \int_0^\infty \left[(-1 + sz) \, J_0(rs) + (1 - 2\nu_1 - sz) \, \frac{J_1(rs)}{rs} \right] F(s) \, \bar{e}^{sz} \, ds$$

$$+ \frac{1}{r} \int_0^\infty \left[\left\{ K_1(rs) + rs \, K_0(rs) \right\} A(s) - \left\{ (4 - 4\nu_1 + r^2 s^2) \, K_1(rs) \right. \right.$$

$$\left. \left. + (3 - 2\nu_1) \, rs \, K_0(rs) \right\} B(s) \right] \cos(sz) ds \; .$$

(3.27)

For the region 2 $(0 < r < b)$, we take

$$\phi_1 = \int_0^\infty \frac{1}{s} \, C(s) \left[I_0(rs) \, \cos(sz) - 1 \right] ds \; ,$$

(3.28)

$$\phi_2 = \int_0^\infty D(s) \, I_1(rs) \, \cos(sz) ds \; ,$$

(3.29)

which are solutions of equations (3.21). Then, by making use of (3.19), (3.20), (3.28) and (3.29) we find that

$$\left[u_r(r,z) \right]_2 = \int_0^\infty \left[I_1(rs) \, C(s) + \left\{ 4(1 - \nu_2) \, I_1(rs) - \right. \right.$$

$$\left. \left. - rs \, I_0(rs) \right\} D(s) \right] \cos(sz) ds,$$

(3.30)

$$\left[u_z(r,z) \right]_2 = \int_0^\infty \left[rs \, I_1(rs) \, D(s) - I_0(rs) \, C(s) \right] \sin(sz) ds, \quad (3.31)$$

$$\frac{\left[\sigma_{rz}(r,z)\right]_2}{2\mu_2} = -\int_0^\infty s\left\{I_1(rs)\ C(s) + \left[2(1-\nu_1)\ I_1(rs) -\right.\right.$$

$$\left.\left. - rs\ I_0(rs)\right]\ D(s)\right\}\ \sin(sz)ds\ , \tag{3.32}$$

$$\frac{\left[\sigma_{zz}(r,z)\right]_2}{2\mu_2} = -\int_0^\infty s\left\{I_0(rs)\ C(s) - \left[2\nu_2\ I_0(rs) +\right.\right.$$

$$\left.\left. + rs\ I_1(rs)\right]\ D(s)\right\}\ \cos(sz)ds\ , \tag{3.33}$$

$$\frac{\left[\sigma_{rr}(r,z)\right]_2}{2\mu_2} = -\frac{1}{r}\int_0^\infty \left\{I_1(rs) - rs\ I_0(rs)\right\}\ \cos(sz)\ C(s)ds$$

$$- \frac{1}{r}\int_0^\infty \left\{(4 - 4\nu_2 + r^2s^2)\ I_1(rs) - (3 - 2\nu_2)\ rs\ I_0(rs)\right\}D(s)$$

$$\tag{3.34}$$
$$\cos(sz)ds\ .$$

The general solution of the crack problem can be written by combining the particular solution with the auxiliary solution. Therefore, we find that

$$u_z^1(r,z) = \left[u_z(r,z)\right]_1 + u_z^*(r,z)\ ,\quad u_r^1(r,z) = \left[u_r(r,z)\right]_1 + u_r^*(r,z)\ ,$$

$$\sigma_{rr}^1(r,z) = \left[\sigma_{rr}(r,z)\right]_1 + \sigma_{rr}^*(r,z)\ ,\quad \sigma_{zz}^1(r,z) = \left[\sigma_{zz}(r,z)\right]_1 + \sigma_{zz}^*(r,z)$$

$$\sigma_{rz}^1(r,z) = \left[\sigma_{rr}(r,z)\right]_1 + \sigma_{rz}^*(r,z),\quad u_z^2(r,z) = \left[u_z(r,z)\right]_2 + u_z^{**}(r,z)$$

$$u_r^2(r,z) = \left[u_r(r,z)\right]_2 + u_r^{**}(r,z)\ ,\quad \sigma_{rr}^2(r,z) = \left[\sigma_{rr}(r,z)\right]_2 + \sigma_{rr}^{**}(r,z)$$

$$\sigma_{zz}^2(r,z) = \left[\sigma_{zz}(r,z)\right]_2 + \sigma_{zz}^{**}(r,z),\quad \sigma_{rz}^2(r,z) = \left[\sigma_{rz}(r,z)\right]_2 + \sigma_{rz}^{**}(r,z)$$

$$\tag{3.35}$$

3b. Reduction of the Thermoelastic Problem to a
     Fredholm Integral Equation of the Second Kind

From the solution obtained in equations (3.35), we find that
the boundary conditions (1.8) and (1.9) are identically satisfied
and the boundary conditions (1.6) and (1.7) yield the following
dual integral equations:

$$\int_0^\infty F(s) \ J_0(rs)ds + \int_0^\infty s\left\{K_0(rs) \ A(s) + \left[2\nu_1 \ K_0(rs) - rs \ K_1(rs)\right]B(s)\right\}ds$$

$$+ \frac{m_1}{2} \int_0^\infty s^{-1}\theta_1(s)J_0(rs)ds + \frac{m_1}{2} \int_0^\infty s^{-1} \ \theta_2(s)\left[2K_0(sr) + sr \ K_1(rs)\right]ds = 0,$$

$$a < r < \infty , \tag{3.36}$$

$$\int_0^\infty \frac{F(s)}{s} \ J_0(rs)ds = 0 , \quad b < r < a . \tag{3.37}$$

If we make the integral representation

$$F(s) = s \int_a^\infty h(t) \ \sin(st)dt = -\sin(sa) \ h(a) - \int_a^\infty \sin(st) \ h(t) \ dt , \tag{3.38}$$

such that $h(\infty) = 0$, the equation (3.37) is satisfied identically
whatever be the form of $h(t)$. Making use of (2.3), the equation
(3.36) can be written in the following form:

$$\int_0^\infty F(s) \ J_0(rs)ds + \int_0^\infty s\left\{K_0(rs) \ A(s) + \left[2\nu_1 \ K_0(rs) - rs \ K_1(rs)\right]B(s)\right\}ds$$

$$+ \frac{m_1T_0}{2} + \frac{m_1}{2} \int_0^\infty s^{-1} \ \theta_2(s)\left[K_0(rs) + rs \ K_1(rs)\right]ds = 0 , \quad a < r < \infty . \tag{3.39}$$

Substituting for $F(s)$ from equation (3.38) into equation (3.39) we
obtain

$$h(t) + \frac{2}{\pi} \int_0^\infty \left[s\left\{A(s) + 2\nu_1 \ B(s)\right\} \int_t^\infty \frac{r \ K_0(rs)dr}{(r^2 - t^2)^{\frac{1}{2}}}\right.$$

$$- s^2 B(s) \int_t^\infty \frac{r^2 K_0(rs)dr}{(r^2 - t^2)^{\frac{1}{2}}} \bigg] ds + \frac{m_1}{\pi} \int_t^\infty \frac{r T_0(r)dr}{(r^2 - t^2)^{\frac{1}{2}}}$$

$$+ \frac{m_1}{\pi} \int_0^\infty s^{-1} \theta_2(s) \left[\int_t^\infty \frac{r K_0(rs)dr}{(r^2 - t^2)^{\frac{1}{2}}} + s \int_t^\infty \frac{r^2 K_1(rs)dr}{(r^2 - t^2)^{\frac{1}{2}}} \right] ds = 0 .$$

(3.40)

Equation (3.40) may be written in the form:

$$h(t) + \int_0^\infty \left[\left\{ A(s) + 2\nu_1 B(s) \right\} - B(s)(1 + st) \right] e^{-st} ds$$

(3.41)

$$+ \frac{m_1}{2} \int_0^\infty \frac{\theta_2(s)}{s^2} (2 + st) e^{-st} ds + \frac{m_1}{\pi} \int_t^\infty \frac{r T(r) dr}{(r^2 - t^2)^{\frac{1}{2}}} = 0 , \quad a < t < \infty .$$

If $T(r) = T_0/r^2$, we find

$$\int_t^\infty \frac{r T(r)dr}{(r^2 - t^2)^{\frac{1}{2}}} = \frac{T_0 \pi}{2t} .$$

(3.42)

Making use of equation (3.35) and the Fourier inversion theorem, we find that the continuity conditions (1.10) yield

$$K_0 A - bsB K_1 + [bs I_1 D - I_0 C]$$

$$= - \frac{m_1 b}{2} \left[s^{-1} \theta_2(s) K_1(bs) \right] + \frac{m_2 b}{2} \left[s^{-1} \theta_3(s) I_1(bs) \right]$$

$$+ \frac{2}{\pi} \int_0^\infty \frac{1}{u} \left\{ 2(1 - \nu_1) f_1 + u f_2 \right\} F(u) J_0(bu)du$$

$$+ \frac{m_1}{\pi} \int_0^\infty u^{-1} f_2 \theta_1(u) J_0(bu)du = L_1 ,$$

(3.43)

$$I_1 C + \left\{ 4(1 - \nu_2) I_1 - bs I_0 \right\} D + K_1 A - \left\{ 4(1 - \nu_1) K_1 + bs K_0 \right\} B$$

$$= \frac{2}{\pi} \int_0^\infty \frac{1}{u} \left\{ (2\nu_1 - 1) f_3 + u f_4 \right\} F(u) J_1(bu)du$$

$$+ \frac{m_1}{\pi} \int_0^\infty \frac{1}{u^2} \theta_1(u) (f_3 + u f_4) J_1(bu)du$$

$$- \frac{m_2 b}{2} \left[\overset{-1}{s} \; \theta_3(s) \; I_0 \right] - \frac{m_1 b}{2} \left[\overset{-1}{s} \; K_0 \; \theta_2(s) \right] = L_2 \; , \tag{3.44}$$

$$G \left[(K_1 + bs \; K_0) \; A - \left\{ (4 - 4\nu_1 + b^2 s^2) \; K_1 + (3 - 2\nu_1) \; bs \; K_0 \right\} B \right]$$

$$+ \left[(I_1 - bs \; I_0) C + \left\{ (4 - 4\nu_2 + b^2 s^2) \; I_1 - (3 - 2\nu_2) bs \; I_0 \right\} D \right]$$

$$= \frac{m_2 b}{2} \left(b \; I_1 - \frac{I_0}{s} \right) \theta_3(s) - \frac{b \; G \; m_1 \; \theta_2(s)}{2} \left(b \; K_1 - \frac{3}{s} K_0 \right)$$

$$- \frac{2 \; G \; b}{\pi} \int_0^\infty \left[(-f_3 + u \; f_4) \; J_0(bu) + \left\{ (1 - 2\nu_1) \; f_3 - u \; f_4 \right\} \frac{J_1(bu)}{bu} \right]$$

$$F(u) du$$

$$+ \frac{m_1 \; G \; b}{\pi} \int_0^\infty \left[f_3 \left\{ \frac{J_0(bu)}{u} + \frac{J_1(bu)}{bu^2} \right\} + \left\{ \frac{J_1(bu)}{bu} - J_0(bu) \right\} f_4 \right] \theta_1(u) du$$

$$= L_3 \; , \tag{3.45}$$

$$G \left[s \left\{ K_1 A - [2(1 - \nu_1) \; K_1 + bs \; K_0] B \right\} \right] + s \left[I_1 C + \left\{ 2(1 - \nu_1) \; I_1 - bs \; I_0 \right\} D \right]$$

$$= - \frac{b \; m_1 \; G}{2} \; \theta_2(s) \; K_0 - \frac{m_2 b}{2} \; I_0 \; \theta_3(s)$$

$$+ \frac{m_1 \; G}{\pi} \int_0^\infty \theta_1(u) \; J_1(bu) \; f_2 \; du + \frac{2}{\pi} \; G \int_0^\infty u \; F(u) \; J_1(bu) \; f_2 \; du = L_4 \; , \tag{3.46}$$

where $G = \mu_1 / \mu_2$ and

$$f_1 = \int_0^\infty \sin(sz) \; \overset{-uz}{e} \; dz = \frac{s}{s^2 + u^2}$$

$$f_2 = \int_0^\infty z \; \sin(sz) \; \overset{-uz}{e} \; dz = \frac{2su}{(s^2 + u^2)^2}$$

$$f_3 = \int_0^\infty \cos(sz)\, e^{-uz}\, dz = \frac{u}{s^2 + u^2}$$

$$f_4 = \int_0^\infty z \cos(sz)\, e^{-uz}\, dz = \frac{u^2 - s^2}{(u^2 + s^2)^2}\,. \tag{3.47}$$

Solving equations (3.43) to (3.46) we obtain

$$A = \frac{-1}{(a_2 a_6 - a_3 a_5)(a_{12} I_0 + a_7 I_1)} \Bigg[L_1 \Big\{ (a_{13} I_0 + a_8 I_1)$$

$$(a_1 a_6 - a_3 a_4) + a_3 I_1 (a_{12} I_0 + a_7 I_1) \Big\}$$

$$+ L_2 \Big\{ a_3 I_0 (a_{12} I_0 + a_7 I_1) + (a_9 I_1 + a_{14} I_0)$$

$$(a_1 a_6 - a_3 a_4) \Big\} + L_3 \Big\{ (a_1 a_6 - a_3 a_4)(a_{10} I_1 + a_{15} I_0)$$

$$- I_1 a_6 (a_{12} I_0 + a_7 I_1) \Big\} + L_4 \Big\{ (a_{11} I_1 + a_{16} I_0)$$

$$(a_1 a_6 - a_3 a_4) + \frac{a_6}{s} (a_{12} I_0 + a_7 I_1)(I_1 - bs\, I_0) \Big\} \Bigg]\,, \tag{3.48}$$

$$B = \frac{1}{(a_2 a_6 - a_3 a_5)(a_{12} I_0 + a_7 I_1)} \Bigg[L_1 \Big\{ a_2 I_1 (a_{12} I_0 + a_7 I_1)$$

$$- (a_2 a_4 - a_5 a_1)(a_{13} I_0 + a_8 I_1) \Big\}$$

$$+ L_2 \Big\{ a_2 I_0 (a_{12} I_0 + a_7 I_1) - (a_2 a_4 - a_5 a_1)(I_1 a_9 + I_0 a_{14}) \Big\}$$

$$- L_3 \Big\{ a_5 I_1 (a_{12} I_0 + a_7 I_1) + (a_{10} I_1 + a_{15} I_0)(a_2 a_4 - a_5 a_1) \Big\}$$

$$- L_4 \Big\{ (a_2 a_4 - a_1 a_5)(a_{11} I_1 + a_{16} I_0)$$

$$- \frac{a_5}{5} (I_1 - bs\, I_0)(a_{12} I_0 + a_7 I_1) \Big\} \Bigg]\,, \tag{3.49}$$

$$C = \frac{1}{(I_1 a_7 + I_0 a_{12})} \Bigg[L_1 (a_7 a_{13} - a_8 a_{12}) + L_2 (a_7 a_{14} - a_{12} a_9)$$

$$+ L_3 (a_7 a_{15} - a_{10} a_{12}) + L_4 (a_7 a_{16} - a_{11} a_{12}) \Bigg]\,, \tag{3.50}$$

$$D = \frac{1}{(a_{12} \, I_0 + a_7 \, I_1)} \left[L_1(a_{13} \, I_0 + a_8 \, I_1) + L_2(I_1 \, a_9 + I_0 \, a_{14}) \right.$$

$$\left. + L_3(a_{10} \, I_1 + a_{15} \, I_0) + L_4(a_{11} \, I_1 + a_{16} \, I_0) \right] , \qquad (3.51)$$

where

$$a_1 = b^2 s^2 (I_1^2 - I_0^2) + 2(1 - \nu_2) \, I_1^2 \, , \quad a_2 = bs \, G(K_0 \, I_1 + K_1 \, I_0) \, ,$$

$$a_3 = -G[K_1 \, I_1(2 - 2\nu_1 + b^2 s^2) + 2(1 - \nu_1) \, bs \, I_0 \, K_1$$

$$+ 2bs(1 - \nu_1) \, K_0 I_1 + K_0 \, I_0 \, b^2 s^2] \, ,$$

$$a_4 = bs(I_1^2 - I_0^2) + 4(1 - \nu_2) \, I_1 \, I_0 \, ,$$

$$a_5 = a_2/bs \, G = K_0 \, I_1 + K_1 \, I_0 \, ,$$

$$a_6 = -[bs(I_1 \, K_1 + K_0 \, I_0) + 4(1 - \nu_1) \, K_1 \, I_0] \, ,$$

$$a_7 = bs \, I_1 + \frac{sb \, K_1(a_2 a_4 - a_1 a_5) + K_0(a_3 a_4 - a_1 a_6)}{a_2 a_6 - a_3 a_5} \, ,$$

$$a_8 = 1 + \frac{I_1(a_2 \, sb \, K_1 + a_3 \, K_0)}{a_2 a_6 - a_3 a_5} \, ,$$

$$a_9 = \frac{I_0(sb \, K_1 \, a_2 + K_0 \, a_3)}{a_2 a_6 - a_3 a_5} \, , \quad a_{10} = \frac{- \, I_1(a_6 \, K_0 + bs \, a_5 \, K_1)}{a_2 a_6 - a_3 a_5} \, ,$$

$$a_{11} = \frac{1}{s} \frac{I_1 - bs \, I_0}{a_2 a_6 - a_3 a_5} \, (a_6 \, K_0 + a_5 \, bs \, K_1) \, ,$$

$$a_{12} = 2(1 - \nu_2) \, I_1 - bs \, I_0 + \frac{G(a_3 a_4 - a_1 a_6) \, K_1}{a_2 a_6 - a_3 a_5}$$

$$+ G \frac{a_2 a_4 - a_1 a_5}{a_2 a_6 - a_3 a_5} \, [2(1 - \nu_1) \, K_1 + bs \, K_0]$$

$$a_{13} = \frac{G}{a_2 a_6 - a_3 a_5} \left[a_3 \, K_1 \, I_1 + a_2 \, I_1 \left\{ 2(1 - \nu_1) \, K_1 + bs \, K_0 \right\} \right]$$

$$a_{14} = \frac{G}{a_2 a_6 - a_3 a_5} \left[a_3 \, K_1 \, I_0 + a_2 \left\{ 2(1 - \nu_1) \, K_1 + bs \, K_0 \right\} I_0 \right]$$

$$a_{15} = \frac{-G \, I_1}{a_2 a_6 - a_3 a_5} \left[a_6 \, K_1 + a_5 \left\{ 2(1 - \nu_1) \, K_1 + bs \, K_0 \right\} \right]$$

$$a_{16} = \frac{1}{s} \left[1 + \frac{G(I_1 - bs \, I_0)}{a_2 a_6 - a_3 a_5} \left\{ a_6 \, K_1 + [2(1 - \nu_1) \, K_1 + bs \, K_0] a_5 \right\} \right] .$$

$$(3.52)$$

Substituting the values of A and B from the equations (3.48) and (3.49) into the equation (3.41) and making use of (3.42) we find that (3.41) simplifies to

$$h(t) + \int_0^\infty [L_1 \, (s) \, B_1(s,t) + L_2(s) \, B_2(s,t) + L_3(s) \, B_3(s,t)$$

$$+ L_4 \, (s) \, B_4(s,t)] \, e^{-st} \, ds + \frac{m_1}{2} \int_0^\infty \frac{\theta_2(s)}{s^2} \, (2 + st) \, e^{-st} \, ds$$

$$= - \frac{m_1 T_0}{2t} , \qquad 0 < t < a , \qquad (3.53)$$

where

$$B_1(s,t) = D_1[a_{13} \, I_0 + a_8 \, I_1)(a_1 a_6 - a_3 a_4) + a_3 \, I_1(a_{12} \, I_0 + a_7 \, I_1)$$

$$+ W\{a_2 \, I_1(a_{12} \, I_0 + a_7 \, I_1) - (a_2 a_4 - a_5 a_1)(a_{13} \, I_0 + a_8 \, I_1)\}]$$

$$(3.54)$$

$$B_2(s,t) = D_1[a_3 \, I_0(a_{12} \, I_0 + a_7 \, I_1) + (a_9 \, I_1 + a_{14} \, I_0)(a_1 a_6 - a_3 a_4)$$

$$+ W\{a_2 \, I_0(a_{12} \, I_0 + a_7 \, I_1) - (a_2 a_4 - a_1 a_5)(I_1 \, a_9 + I_0 \, a_{14})\}]$$

$$(3.55)$$

$$B_3(s,t) = D_1[(a_1 a_6 - a_3 a_4)(a_{10} \, I_1 + a_{15} \, I_0) - I_1 \, a_6(a_{12} \, I_0 + a_7 \, I_1)$$

$$- W\{a_5 \, I_1(a_{12} \, I_0 + a_7 \, I_1) + (a_{10} \, I_1 + a_{15} \, I_0)(a_2 a_4 - a_5 a_1)\}]$$

$$(3.56)$$

$$B_4(s,t) = D_1[(a_{11} \, I_1 + a_{16} \, I_0)(a_1 a_6 - a_3 a_4) + \frac{a_6}{s} \, (I_1 - bs \, I_0)$$

$$(a_{12} I_0 + a_7 I_1) - W\{(a_2a_4 - a_1a_5)(a_{11} I_1 + a_{16} I_0)$$

$$- \frac{a_5}{s} (I_1 - bs\ I_0)(a_{12} I_0 + a_7 I_1)\}] \tag{3.57}$$

with

$$D_1 = - \frac{1}{(a_2a_6 - a_3a_5)(a_{12} I_0 + a_7 I_1)} , \quad W = 1 - 2\nu_1 + st . \tag{3.58}$$

Let us take

$$L_j = X_j + P_j \qquad (j = 1,2,3,4) \tag{3.59}$$

where X_j is the isothermal part and P_j is the contribution due to the temperature field.

After a lot of manipulations, we find that

$$X_1 = [2(1 - \nu_1) I_0(bs) + I_0(bs) + bs\ I_1(bs)] \int_a^\infty h(u)\ e^{-su}\ du$$

$$- s\ I_0(bs) \int_a^\infty u\ h(u)\ e^{-su}\ du . \tag{3.60}$$

$$P_1 = \left[\frac{m_1b\ I_1}{2} + \frac{m_2b\ \Delta_1 I_1 - m_1b\ I_0 I_1 K_0(1 - R)}{2\Delta} \right]$$

$$\int_a^\infty g(u)\ e^{-us}\ du + \frac{I_0\ m_1}{2} \int_a^\infty u\ g(u)\ e^{-us}\ du . \tag{3.61}$$

$$X_2 = -\left[\frac{bs}{2} (I_0 + I_2) + 2\nu_1\ I_1\right] \int_a^\infty h(u)\ e^{-su}\ du + s\ I_1 \int_a^\infty u\ h(u)\ e^{-su}\ du. \tag{3.62}$$

$$P_2 = \left[\frac{m_1}{2} \left\{\frac{b}{2} (I_0 + I_2) + \frac{I_1}{s}\right\} - \frac{m_2\ b\ \Delta_1 I_0}{2\Delta}\right.$$

$$\left. - \frac{m_1\ I_0 I_1 K_0(1 - R)}{2\Delta}\right] \int_a^\infty g(u)\ e^{-us}\ du - \frac{I_1\ m_1}{2} \int_a^\infty u\ g(u)\ e^{-us}\ du . \tag{3.63}$$

$$X_3 = G \int_0^\infty h(t) \, \bar{e}^{st} \left[sb\left(\frac{I_0}{2} + bs \, I_1 - st \, I_0\right) + I_1(st - 2\nu_1) - \frac{sb}{2} I_2 \right] dt.$$

$$\tag{3.64}$$

$$P_3 = \int_a^\infty \bar{e}^{su} \, g(u) \left[\frac{m_2 b}{2} (bs \, I_1 - I_0) \frac{\Delta_1}{\Delta} - \frac{G \, b \, m_1}{2\Delta} (bs \, K_1 - 3K_0) \right.$$

$$I_0 \, I_1(1 - R) + \frac{m_1 \, G}{2s} \left\{ I_1(1 - us) + \frac{bs}{2} (I_0 + I_2) - bs^2(b \, I_1 - \right.$$

$$\left. \left. - u \, I_0) \right\} \right] du \, .$$

$$\tag{3.65}$$

$$X_4 = sG \left[(st - 2) \, I_1 - \frac{sb}{2} (I_0 + I_2) \right] \int_a^\infty h(u) \, \bar{e}^{su} \, du \, .$$

$$\tag{3.66}$$

$$P_4 = \int_a^\infty \left[\frac{m_1 \, G}{2} \left\{ I_1(1 - us) + \frac{bs}{2} (I_0 + I_2) \right\} \right.$$

$$\left. - \frac{sb}{2\Delta} \left\{ m_1 \, G \, K_0 \, I_0 \, I_1(1 - R) + m_2 \, I_0 \, \Delta_1 \right\} \right] \bar{e}^{su} \, g(u) \, du \, .$$

$$\tag{3.67}$$

Substituting the values of L_1, L_2, L_3, L_4 and θ_2 from equations (3.59) - (3.67) and (2.19) into equation (3.53), we find that

$$H(t) + \int_a^\infty H(u) \, K_2(u,t) du + \int_a^\infty G(u) \, K_3(u,t) du = \frac{1}{t} \, , \quad a < t < \infty, \tag{3.68}$$

where

$$K_2(u,t) = \int_0^\infty \bar{e}^{s(u+t)} \left[\left\{ (3 - 2\nu_1) \, I_0 + bs \, I_1 - su \, I_0 \right\} B_1 \, (st) \right.$$

$$+ \left\{ - \frac{bs}{2} (I_0 + I_2) - 2\nu_1 \, I_1 + su \, I_1 \right\} B_2(s,t)$$

$$+ G \left\{ sb\left[\frac{1}{2} I_0 + bs \, I_1 - us \, I_0 \right] \right.$$

$$\left. + I_1(su - 2\nu_1) - \frac{sb}{2} I_2 \right\} B_3(s,t)$$

$$+ G\ B_4(s,t)\ \left\{s(su - 2)\ I_1 - \frac{bs^2}{2}\ (I_0 + I_2)\right\}\right]ds, \qquad (3.69)$$

$$K_3(u,t) = \int_0^\infty e^{-s(u+t)}\left[\left\{\frac{M\ b\ I_1\ \Delta_1}{2\Delta} - \frac{b\ I_0\ I_1\ K_1(1 - R)}{2\Delta} - \frac{b}{2}\ I_1 + \frac{I_0\ u}{2}\right\}\right.$$

$$B_1(s,t) + B_2(s,t)\ \left\{\frac{b}{4}\ (I_0 + I_2) + \frac{I_1}{2s} - \frac{M\ b\ I_0\ \Delta_1}{2\Delta} - \right.$$

$$- \frac{I_0\ I_1\ K_0(1 - R)}{2\Delta} - \frac{I_1\ u}{2}\right\}$$

$$+ B_3(s,t)\ \left\{\frac{M\ b\ \Delta_1}{2\Delta}\ (bs\ I_1 - I_0) - \frac{bG}{2\Delta}\ (bs\ K_1 - 3K_0)\right.$$

$$I_0\ I_1(1 - R)$$

$$+ \frac{G}{2s}\left[I_1(1 - us) + \frac{bs}{2}\ (I_0 + I_2)\ - s^2\ b(b\ I_1 - u\ I_0)\right]\right\}$$

$$+ B_4(s,t)\ \left\{\frac{G}{2}\left[I_1(1 - us) + \frac{bs}{2}\ (I_0 + I_2)\right] - \frac{sb}{2\Delta}\right.$$

$$[G\ K_0\ I_0\ I_1(1 - R) + M\ I_0\ \Delta_1]\right\}\right]ds$$

$$+ \frac{(1 - R)}{2}\int_0^\infty \frac{I_0\ I_1(2 + st)}{s\Delta}\ e^{-s(t+u)}\ ds\ , \qquad (3.70)$$

$$M = \frac{m_2}{m_1}\ , \quad h(t) = -\frac{T_0\ m_1}{2}\ H(t)\ , \quad g(t) = -\frac{T_0}{2}\ G(t)\ . \qquad (3.71)$$

With the help of (3.71), the Fredholm integral equation (2.20) can be written in the form

$$G(t) + \int_a^\infty G(u)\ K_1\ (u,t)du = -\frac{2}{t^2}\ , \quad a < t < \infty\ . \qquad (3.72)$$

Now G(t) can be determined from equation (3.72) and then H(t) can be obtained from equation (3.68). In the next section, we will obtain the expression for the stress intensity factor in terms of H(t).

4. EXPRESSION FOR THE STRESS INTENSITY FACTOR

The expression for normal stress for the region $b < r$ can be written in the form

$$\frac{\sigma^1_{zz}(r,0)}{2\mu_1} = -\int_0^\infty F(s) \, J_0(rs)ds - \int_0^\infty s\{K_0(rs) \, A(s)$$

$$+ [2\nu_1 \, K_0(rs) - rs \, K_1(rs)] \, B(s)\} \, \cos(sz)ds$$

$$-\frac{m_1}{2}\int_0^\infty s^{-1} \, \theta_1(s) \, J_0(rs)ds - \frac{m_1}{2}\int_0^\infty s^{-1} \, \theta_2(s)$$

$$[2K_0(rs) + rs \, K_1(rs)] \, \cos(sz)ds \,, \quad b < r \,. \tag{4.1}$$

The expression for the stress intensity factor K is

$$K = \underset{r\to a^-}{\text{Lim}} \, \{[2(a - r)]^{\frac{1}{2}} \, \sigma^1_{zz}(r,0)\} \,. \tag{4.2}$$

Substituting the value of $F(s)$, $A(s)$, $B(s)$, $\theta_1(s)$ and $\theta_2(s)$ from (3.38), (3.48), (3.49), (2.19) and (2.20) into equations (4.1) and (4.2), we find that

$$K = 2\mu_1 \, h(a)/\sqrt{a} \,. \tag{4.3}$$

From Kassir and Sih [1], we find that the expression for the stress intensity factor for an external crack, in an infinite medium, when the thermal boundary conditions are given by (1.1) - (1.3) is given by

$$K_\infty = \frac{-\alpha E_1}{\pi(1 - \nu_1)a^{\frac{1}{2}}} \int_a^\infty \frac{rT(r)dr}{(r^2 - a^2)^{\frac{1}{2}}} \,, \tag{4.4}$$

where E_1 is Young modulus of the material. For $T(r) = T_0/r^2$, we find that

$$K_\infty = - \mu_1(1 + \nu_1)T_0 / \left[1 - \nu_1\right) a^{3/2}\right] ,$$
(4.5)

and hence

$$K|K_\infty = - \frac{2a\ h(a)}{T_0\ m_1} = a\ H(a) .$$
(4.6)

5. NUMERICAL SOLUTION OF THE INTEGRAL EQUATIONS

First of all we have solved numerically the integral equation
(3.72) for $G(t)$. These numerical values of $G(t)$ have been used to
obtain the numerical values of $H(t)$ from (3.68). Using these
numerical values of $H(a)$ we have obtained numerical values of
$K|K_\infty$ from (4.6). For the numerical solution we have taken
$\nu_1 = \nu_2 = \frac{1}{4}$, b = 1 and a = 1.2,1.3,1.4,...,2.0, and various com-
binations of values for G, R and M. The numerical values of $K|K_\infty$
have been graphed in Figures 1-6.

REFERENCES

1. M. K. Kassir and G. C. Sih, "Three Dimensional Crack Problems,"
   Noordhoff International Publishing, Lyden (1975).
2. R. P. Srivastav and D. Lee, Axisymmetric external crack
   problems for media with cylindrical cavities, Int. J. Eng.
   Sci. 10:217 (1972).
3. M. Lowengrub and I. N. Sneddon, The solution of a pair of
   dual integral equations, Proceedings of Glasgow Mathematical
   Association 6:14 (1963).
4. A. Erdelyi, ed., "Tables of Integral Transforms, Vol. I,"
   McGraw-Hill Inc., New York (1954).
5. W. Nowacki, "Thermoelasticity," Pergamon Press, Oxford, England
   (1962).
6. L. M. Keer, A class of non-symmetrical punch and crack
   problems, Q. J. Mech. Appl. Math. 27:423 (1963).
7. W. D. Collins, Some axially symmetric stress distributions in
   elastic solids containing penny-shaped cracks I, Proceedings of Royal Society, Series A 266:359 (1962).

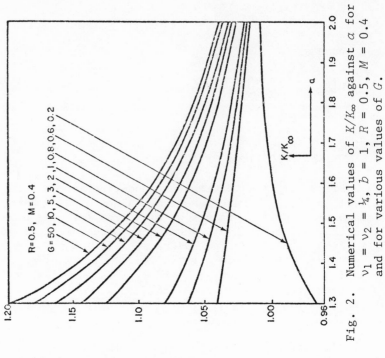

Fig. 2. Numerical values of K/K_∞ against α for $\nu_1 = \nu_2 = \tfrac{1}{4}$, $b = 1$, $R = 0.5$, $M = 0.4$ and for various values of G.

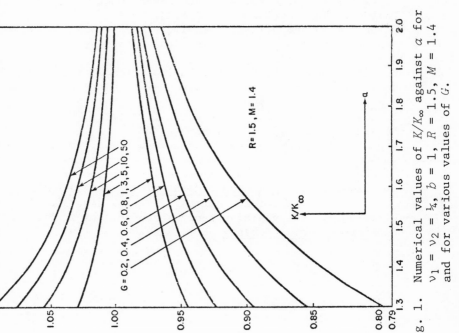

Fig. 1. Numerical values of K/K_∞ against α for $\nu_1 = \nu_2 = \tfrac{1}{4}$, $b = 1$, $R = 1.5$, $M = 1.4$ and for various values of G.

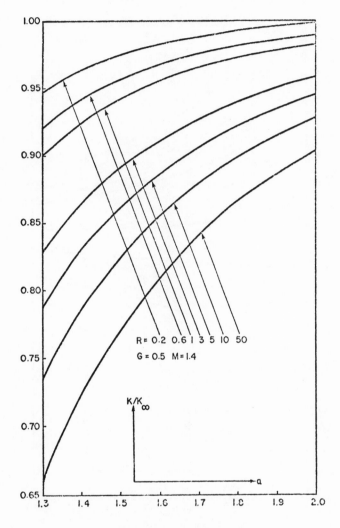

Fig. 3. Numerical values of K/K_∞ against a for $\nu_1 = \nu_2 = \frac{1}{4}$, $b = 1$, $G = 0.5$, $M = 1.4$ and for various values of R.

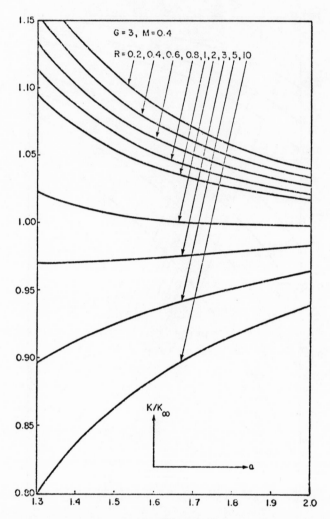

Fig. 4.   Numerical values of K/K_∞ against a for $\nu_1 = \nu_2 = \frac{1}{4}$,
$b = 1$, $G = 3.0$, $M = 0.4$ and for various values of R.

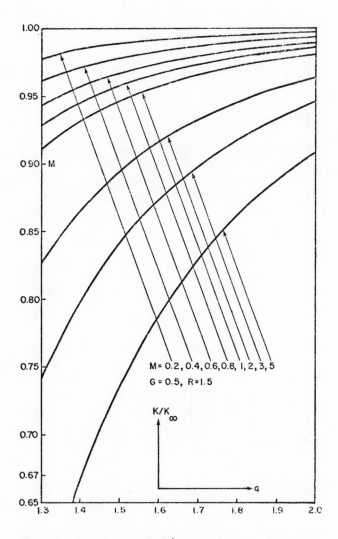

Fig. 5.   Numerical values of K/K_∞ against a for $\nu_1 = \nu_2 = \frac{1}{4}$, $b = 1$, $G = 0.5$, $R = 1.5$ and for various values of M.

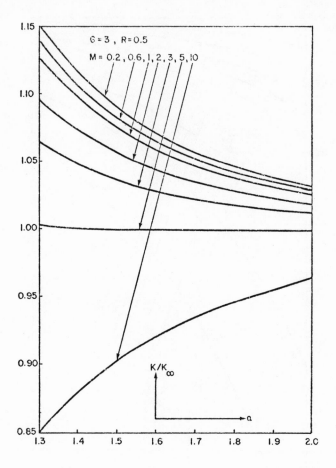

Fig. 6.   Numerical values of K/K_∞ against a for $\nu_1 = \nu_2 = \frac{1}{4}$,
$b = 1$, $G = 3.0$, $R = 0.5$ and for various values of M.

THE VISCOELASTIC BEHAVIOR OF A

COMPOSITE IN A THERMAL ENVIRONMENT

D. H. Morris,[*] H. F. Brinson,[*] W. I. Griffith[*] and
Y. T. Yeow[**]

INTRODUCTION

The merits of composite materials for potential use in
structural design are well established. Their high strength to
weight ratio make them attractive in aerospace and automotive applications where improved fuel economy by weight reduction is
desirable. Unfortunately, a number of factors have inhibited the
ready acceptance of such materials. First, costs are high compared to conventional materials. With increased usage, however,
costs are likely to become increasingly competitive in the future.
Another, perhaps more serious limitation, is the current lack of
understanding of the mechanical behavior of polymer based laminates
under long term environmental exposure.

It is well known that the epoxy resins which are now often
used as the polymer matrix component exhibit viscoelastic or time
effects which are significantly affected by exposure to both
temperature and humidity. Epoxies soften as temperatures are increased with resulting loss of both moduli and strength.[1-4] In
addition, they absorb moisture and swell giving rise to residual
stresses.[5-8]

Polymer based composite laminates will be similarly time
dependent and affected by moisture and temperature under certain
circumstances. Fiber dominated composites are not likely to

*Department of Engineering Science and Mechanics, Virginia Poly-
technic Institute and State University, Blacksburg, VA 24061

**Allied Chemical Corp., Morristown, NJ 07960

suffer large reductions of either moduli or strength in the fiber
direction. In other directions, time, temperature, and moisture
dependent losses of both strength and modulus are likely.

Because of the effects of environment, there is concern that
time dependent properties such as creep, relaxation, creep ruptures,
etc., may be important long-term design considerations for the
temperature and moisture levels anticipated in current structural
applications. It would be desirable to be able to measure the
environmental effects with short-term laboratory tests rather than
perform long-term prototype studies. In addition, it would be
desirable to be able to predict these effects with analytical techniques for either short- or long-term situations. As a result, it
is clear that there is a need for accelerated characterization
techniques for laminates similar to those used for other structural
materials.

For metals and polymers a variety of techniques are available
such as linear elastic stress analysis, empirical extrapolative
equations such as the Larson-Miller parameter method, and the
time-temperature superposition principle. Several procedures have
been proposed for the purpose of making such lifetime or viscoelastic predictions of composite materials. Some of these are the
"wear out model" proposed by Halpin, Jernia, and Johnson,[9] a nonlinear viscoelastic technique proposed by Lou and Schapery,[10] and
a combined viscoelastic-lamination theory model proposed by
De Runtz and Crossman.[11] The former[9] is a statistically based
method for the prediction of fatigue lifetimes and the latter[10-11]
are methods to predict environmental degradations of moduli or
compliances.

The purpose of the work reported herein was to develop an
accelerated characterization method by which design information
or predictions of long time moduli and strength could be made from
short-term tests on graphite/epoxy laminates. In general, the
procedure is based upon the time-temperature superposition
principle and the widely used lamination theory for composite
materials.

ACCELERATED CHARACTERIZATION AND FAILURE PREDICTION

The procedures for accelerated characterization and life time
predictions are outlined in Fig. 1. Using these proposed procedures a designer can systematically incorporate predictions of
long term viscoelastic failures in the initial design process.
That is, by use of the method shown in Fig. 1, delayed failures
during the life time of a structural component can be avoided.
Thus, the ideas of Fig. 1 are for an arbitrary polymer based
composite laminate. In the discussion to follow, the particular

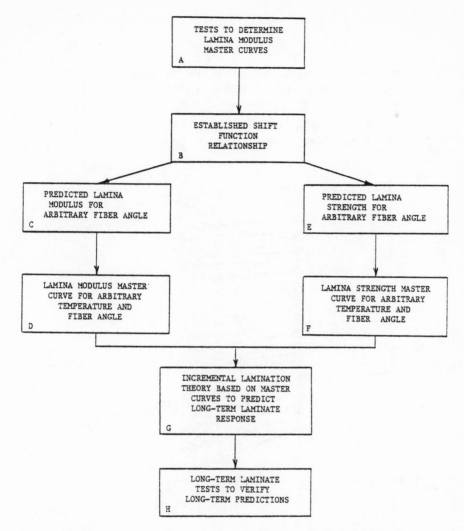

Fig. 1.   Flow Chart of the Proposed Procedures for Laminate
          Accelerated Characterization and Failure Prediction

laminate being studied is a 350°F cured graphite/epoxy system,
T300/934.

    The various experimental and analytical procedures under-
taken to verify the proposed methodology given in Fig. 1 are outlined below.   The letters of the items below agree with those
identifying each task in Fig. 1.

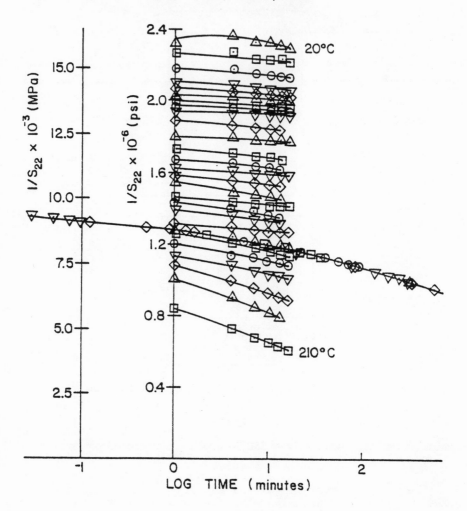

△ 20°C, 60°C, 100°C, 145°C, 180°C, 205°C
⊡ 30°C, 65°C, 110°C, 155°C, 185°C, 210°C
⊙ 40°C, 70°C, 120°C, 160°C, 190°C
▽ 50°C, 76°C, 127°C, 165°C, 195°C
◇ 55°C, 85°C, 135°C, 175°C, 200°C

Fig. 2. Reduced Reciprocal of Compliance, $1/S_{22}$, and Portion of
180°C Master Curve for $[90°]_{8s}$ T300/934 Graphite/Epoxy
Laminate

ITEM A: Tests to Determine Lamina Modulus Master Curves

As previously mentioned, the proposed accelerated characterization method is based upon the time-temperature superposition
principle (TTSP) for polymeric materials. The validity of the TTSP
for the T300/934 graphite/epoxy system has been established. [12]

Figure 2 shows the results of short term (15 min.) creep tests
on a $[90°]_{8S}$ specimen at various temperatures. The ordinate of
this figure is reciprocal of reduced compliance, which is calculated
using[12] $S_{22} = \dfrac{T}{T_O}\dfrac{\epsilon(t)}{\sigma}$ where T = temperature (°R), T_O = reference
temperature (taken to be the glass transition temperature of 453°R),
$\epsilon(t)$ is the time dependent axial strain, and σ is the applied axial
stress. Figure 2 also shows a portion of the master curve, while
Fig. 3. shows the complete master curve. The master curve is obtained by horizontally shifting the short-time data until a smooth
curve is obtained. The amount of horizontal shift at each
temperature is the shift factor, a_T.

Testing was performed at other fiber angles than 90°, and
master curves similar to Fig. 3 were constructed.[13] The need for
such master curves will be subsequently discussed. It is assumed
that reciprocal of compliance, such as shown in Figs. 2 and 3, is
equal to modulus. Such an assumption appears to be justified.[14]

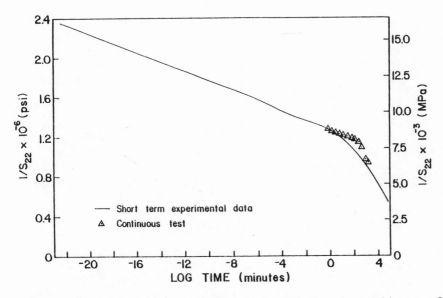

Fig. 3.   Master Curve of the Reciprocal of Reduced Compliance of
          $[90°]_{8S}$ Laminate at 180°C

ITEM B: Established Shift Function Relationship

Graphical shifting of short-time (15 min.) data to determine
shift factors as a function of temperature for various fiber angles
gave the results shown in Fig. 4. Examination of this figure
reveals that the' shift function is relatively insensitive to fiber
orientation. This result is important when calculating lamina
modulus for an arbitrary fiber angle using anisotropic transformation equations.

ITEM C: Predicted Lamina Modulus for Arbitrary Fiber Angle

The viscoelastic compliance of an arbitrary ply in a general
laminate (which is assumed to be the same as a unidirectional
laminate of the same fiber orientation) for any time t may be found
using the orthotropic transformation equation

$$S_{xx}(t) = m^4 S_{11} + 2m^2 n^2 S_{12} + n^4 S_{22}(t) + m^2 n^2 S_{66}(t) \tag{1}$$

where $S_{xx}(t)$ is the time dependent compliance in the load direction
for a specimen with the fibers oriented θ degrees from the loading
axis, $m = \cos \theta$, $n = \sin \theta$, and S_{11}, S_{12}, S_{22}, and S_{66} are
components of the principal compliance matrix.   A previous study[12]
found S_{11} and S_{12} to be independent of time for our graphite/epoxy
material. The time dependent compliances $S_{22}(t)$ and $S_{66}(t)$ are
found from creep tests of specimens with fiber orientations of 90°
and 10°, respectively.[12]

Figure 5 shows a comparison between the predicted master
curve using equation (1) and those obtained using the time-
temperature superposition principle with short-term (15 min.) tests
for a specimen whose fibers are at an angle of 30° with the load
axis. Similar comparisons were made for other fiber orientations
and in all cases agreement was moderate to good.[13]

To further validate the predictive ability of equation (1) and
the applicability of the time-temperature superposition principle,
medium-term, 25-hour creep tests were also run for a number of
fiber angles. Figures 3 and 5 show the results for 90° and 30°,
respectively. Favorable comparisons exist between the results
using the time-temperature superposition principle (15-min. tests),
the transformation equation (1), and the medium-term, 25-hour data.
Similar results were obtained for other orientations.[13]

ITEM D: Lamina Modulus Master Curve for Arbitrary Temperature
          and Fiber Angle

Given the information in A, B and C, a master curve for the
modulus of a lamina or ply of arbitrary fiber orientation can be
found for an arbitrary reference temperature. This is needed input

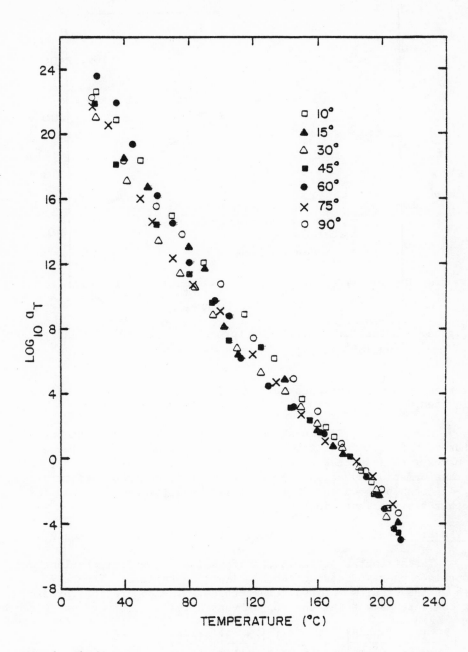

Fig. 4. Shift Factors Versus Temperature for Creep of Off-Axis
        Tensile Coupons (T300/934 Graphite/Epoxy)

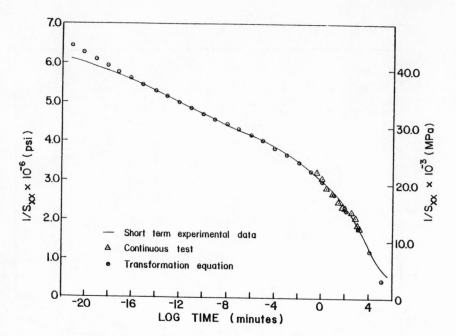

Fig. 5. Master Curve of the Reciprocal of Reduced Compliance
$1/S_{xx}$, of $[30°]_{8s}$ Laminate at 180°C

for any computational scheme to predict laminate failure, as shown
in Item G.

For example, Fig. 6 shows master curves, at three temperatures, for a laminate with a fiber orientation of 30°. These
curves were generated using the 180°C master curve (Fig. 5), the
transformation equation (1), and the shift function-temperature
relationship shown in Fig. 4. Similar results may be found for
other fiber orientations.

ITEM E: Predicted Lamina Strength for Arbitrary Fiber Angle

The strengths of laminates of various fiber orientations were
obtained by ramp loading the specimens to failure. Figure 7
shows the results for two temperatures. The theoretical predictions were made using the Puppo-Evensen failure criterion[16] given by

$$\frac{1}{\sigma_x^2} = \left[\frac{\cos^2\theta}{X_{11}}\right]^2 - \gamma \left[\frac{X_{11}}{X_{22}}\right]\left[\frac{\cos^2\theta}{X_{11}}\right]\left[\frac{\sin^2\theta}{X_{22}}\right]$$

$$+ \gamma \left[\frac{\sin^2}{X_{22}}\right]^2 + \left[\frac{\cos\theta \ \sin\theta}{X_{66}}\right]^2 \qquad (2)$$

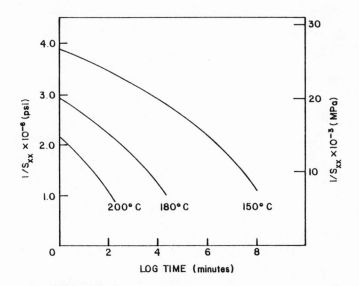

Fig. 6. Predicted Master Curves of the Reciprocal of Reduced
        Compliance, $1/S_{xx}$, of $[30°]$ 8s Laminate at Different
        Temperatures

$$\frac{1}{\sigma_x^2} = \gamma \left[\frac{\cos^2\theta}{X_{11}}\right]^2 - \gamma \left[\frac{X_{22}}{X_{11}}\right]\left[\frac{\cos^2}{X_{11}}\right]\left[\frac{\sin^2\theta}{X_{22}}\right]$$

$$+ \left[\frac{\sin^2\theta}{X_{22}}\right]^2 + \left[\frac{\cos\theta \ \sin\theta}{X_{66}}\right]^2 \qquad (3)$$

where $\gamma = \left[\frac{3X_{66}^2}{X_{11} \ X_{22}}\right]^n$

and σ_x = applied uniaxial stress, X_{11} is the strength of a 0°
specimen, X_{22} is the strength of a 90° specimen, X_{66} is the
strength of a 10° specimen, and n is a material parameter. Using
a value of n = 1, it was found that equation (2) gave good correlations for $\theta < 45°$, and equation (3) gave good correlations for
$\theta > 45°$.

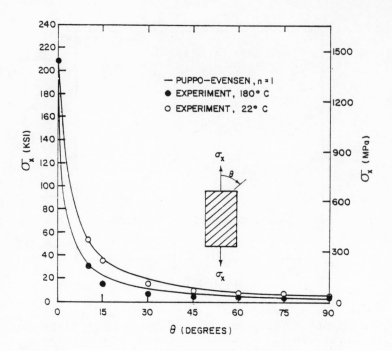

Fig. 7. Comparison of Experimental and Predicted Off-Axis
Strengths (Ramp Loaded to Failure)

<u>ITEM F</u>: Lamina Strength Master Curve for Arbitrary Temperature
and Fiber Angle

Due to the length of time needed to perform enough testing to
experimentally determine strength master curves for an arbitrary
temperature and for arbitrary fiber angles, it was assumed that
strength master curves would have the same shape with the same
shift function as the corresponding compliance master curve. As a
result strength master curves were determined by using the corresponding ramp loaded strengths[17] and the known shape of the
compliance master curves and shift function from Items A, B and D.
Portions of strength master curves so generated are shown in Figs.
8a and b for fiber orientations of 30° and 60°, respectively. Also
shown in Figs. 8a and b are the results of creep rupture tests.

Deviations between predicted and measured creep rupture
stresses are less than 25% over the range of data. However, predictions on creep to rupture times differ with measurements from

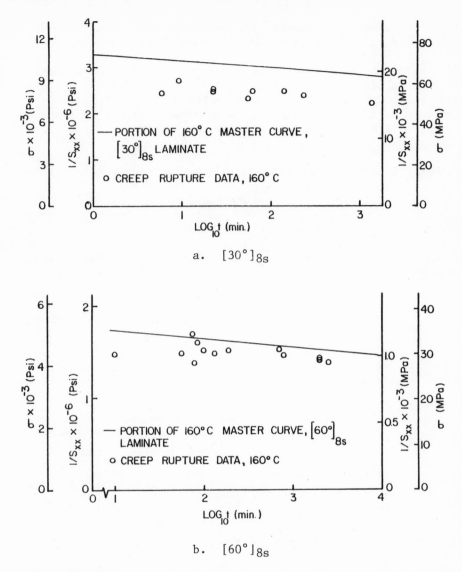

a. $[30°]_{8s}$

b. $[60°]_{8s}$

Fig. 8. Comparison of Creep-Rupture Predictions and Experimental Results

one to three orders of magnitude. Such large errors appear to be inherent in a creep rupture process.[18]

ITEM G: Incremental Lamination Theory Based on Master Curves to
          Predict Long-Term Laminate Response

     Using the well-known time independent lamination theory, an
incremental lamination theory could be developed which would predict the moduli and strength of arbitrary laminates based upon the
lamina modulus and strength master curves developed in Items D and
F. Efforts are underway to develop the incremental lamination
theory. The results will be presented at a later date.

ITEM H: Long-Term Laminate Tests to Verify Long-Term Predictions

     The results of creep rupture tests for a $[90°/\pm60°/90°]_{2s}$
laminate are shown in Fig. 9. There is a large amount of scatter
in the data. As previously stated, such scatter appears to be
inherent in a creep rupture process. Comparison between predicted
results (Item G) and experimental results are not available at
this time.

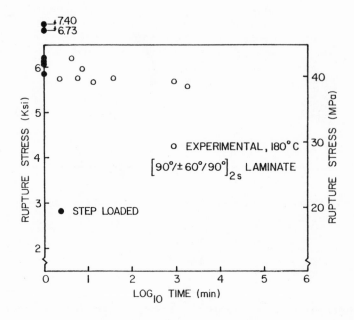

Fig. 9.   Creep-Rupture of $[90°/\pm60°/90°]_{2s}$ Laminate at 180°C
          (Solid Circles Represent Step, or Ramp Loaded Tests)

SUMMARY AND CONCLUSIONS

The purpose of the work reported herein was to develop an accelerated characterization method by which long-term predictions of time dependent moduli and strength could be made on the basis of short-time laboratory tests. The method was based on the time-temperature superposition principle and lamination theory.

Several key assumptions were made regarding the accelerated characterization method shown in Fig. 1. For example, the ortho-tropic transformation equation for composites was assumed to be valid for time dependent moduli (Item C). Figure 5 shows that this assumption is reasonably correct.

The TTSP was assumed to be valid for both moduli and strengths, and strength master curves were assumed to have the same shape as modulus master curves. These assumptions led to predictions of lamina strength which were in error less than 25% from measured strengths, as shown in Figs. 8a and b.

It was also assumed that classical lamination theory, in incremental form, was valid. The validity of this assumption awaits development of an incremental lamination theory, and comparison with the experimental results shown in Fig. 9.

It should be noted that all data gathered herein was for small stress and strain levels such that linear viscoelastic concepts be applicable. When failures such as creep ruptures occur, stresses and strains at the point of failure are high, and non-linear processes are likely to be involved. Thus, it is reasonable to assume that master curves and shift functions are also likely to be stress dependent. Without further experimental evidence, it is likely that non-linear processes may result in large variations between predictions and the experimental results shown in Fig. 9.

Finally, a time independent failure law of Puppo and Evensen was assumed to be valid by simply including time dependent data in determining the necessary constants. Further work is needed to validate this assumption.

Efforts are underway to investigate the non-linear effects, to find a time dependent failure law, and to incorporate these into the lamination theory process for moduli and strength predictions.

REFERENCES

1.   J. D. Ferry, "Viscoelastic Properties of Polymers", John Wiley and Sons, Inc., New York (1970).

2. H. F. Brinson, Mechanical and optical characterization of
    Hysol 4290, Experimental Mechanics.8(12):561-566 (Dec.,
    1968).
3. H. F. Brinson, M. P. Renieri and C. T. Herakovich, Rate and
    time dependent failure of structural adhesives, ASTM STP
    593, American Society for Testing and Materials, pp. 177-
    199 (1975).
4. M. P. Renieri, C. T. Herakovich and H. F. Brinson, Rate and
    time dependent behavior of structural adhesives, VPI&SU
    Report, VPI-E-76-7 (April, 1976).
5. J. R. Vinson and R. B. Pipes, The effects of relative humidity
    and elevated temperature on composite structures, AFOSR
    Workshop, Univ. of Delaware, Newark, Del. (March, 1976).
6. C. E. Browning, G. E. Husman and J. M. Whitney, Moisture
    effects in epoxy resin matrix composites, ASTM STP 617,
    American Society for Testing and Materials, pp. 481-496
    (1977).
7. N. R. Adsit, The effect of environment of the compressive
    behavior of graphite/epoxy, presented at 4th ASTM Conference
    on Composite Materials: Testing and Design, Valley Forge,
    PA (May, 1976).
8. D. Roylance, An experimental method for determining environment-
    fatigue interactions in composite materials, SESA paper no.
    WR-23-1975, presented at SESA meeting, Silver Spring, MD
    (May, 1976).
9. J. C. Halpin, K. L. Jernia and T. A. Johnson, Characterization
    of composites for the purpose of reliability evolution,
    ASTM STP 521, American Society for Testing and Materials,
    pp. 5-64 (1975).
10. Y. C. Lou and R. A. Schapery, Viscoelastic characterization
    of a nonlinear fiber-reinforced plastic, J. of Composite
    Materials, 5:208-234 (April, 1971).
11. J. A. De Runtz and F. W. Crossman, Time and temperature ef-
    fects in laminated composites, Proc. of Computer Simulation
    of Materials Application, Nuclear Materials, Vol. 20 (1976).
12. Y. T. Yeow, D. H. Morris and H. F. Brinson, The time-
    temperature behavior of a unidirectional graphite/epoxy
    composite, ASTM STP 674, American Society for Testing and
    Materials, pp. 117-125 (1979).
13. Y. T. Yeow, The time-temperature behavior of graphite epoxy
    composites, Ph.D. Dissertation, Virginia Polytechnic
    Institute and State University, Blacksburg, VA (May, 1978).
14. R. A. Schapery, Viscoelastic behavior and analysis of
    composite materials, in: "Mechanics of Composite
    Materials," Vol. 2, G. P. Sendeckyj, ed., Academic Press,
    New York (1974).
15. D. H. Morris, Y. T. Yeow and H. F. Brinson. The viscoelastic
    behavior of the principal compliance matrix of a uni-
    directional graphite/epoxy composite, VPI&SU Report, VPI-
    E-79-9 (Feb., 1979). Also, to appear in Polymer Engineering
    and Science.

16. A. H. Puppo and H. A. Evensen, Strength of anisotropic materials under combined stresses, AIAA Journal 10(4):468-474 (1972).
17. Y. T. Yeow and H. F. Brinson, A comparison of simple shear characterization methods for composite laminates, Composites, pp. 49-55 (Jan., 1978).
18. S. Goldfein, General formula for creep and rupture stresses in plastics, Modern Plastics 37:127 (April, 1960).
19. R. M. Jones, "Mechanics of Composite Materials," McGraw-Hill Book Co., New York (1975).

STRESS RELATED THERMAL EMISSION

Kenneth L. Reifsnider and Edmund G. Henneke, II

Engineering Science and Mechanics Department
Virginia Polytechnic Institute and State University
Blacksburg, Virginia 24061

ABSTRACT

Thermal nondestructive testing and evaluation of materials has attracted increased attention in recent years with the appearance of commercially available, real-time infrared scanning devices. Previous work in our laboratory and others has indicated that material damage can be studied in situ, as well as in initiation and growth stages during the application of load. To develop the thermal techniques quantitatively, appropriate physical and mathematical models which will delineate the interdependence of damaged regions and the observed thermal patterns need to be found. This paper investigates some of the current physical models which have been discussed in the literature for such phenomena as internal friction, viscoelasticity, thermoelasticity, etc., in relation to their possible utility in explaining "Stress Related Thermal Emission." The parametric dependence of stress related thermal emission on variables such as wavelength, size of the damage zone, thermal conductivity, material inhomogeneity, and others are discussed along with factors affecting the possible observed heat patterns.

INTRODUCTION

Thermal nondestructive testing methods have existed, in principle, for over one hundred fifty years. It has only been with the relatively recent advent of commercially available real-time, infrared scanning devices, however, that a truly major interest has appeared in developing these techniques for qualitative and quantitative material inspection. The older temperature detection techniques were limited in that they were either contact methods or they were slow and gave only point information rather than areal. The

newer real-time scanning equipment presents visual pictures of the
temperature isotherms existing across the surface of the inspected
object. With proper calibration, the actual values of the temperatures can be measured.

Recent work has shown that real-time thermography can be very
useful in the investigation of damage in materials [1-5]. Provocative techniques [3] use an external heating or cooling source to
cause heat flow in the object under investigation. Variations in
heat flow caused by local inhomogeneities (flaws, holes, etc.) cause
thermal gradients to be established that can be viewed by the thermographic camera. On the other hand, active techniques [1,2,4,5]
apply a source of mechanical energy to the specimen. The mechanical
energy, supplied by fatigue loading or small vibrations, is transformed partially into heat. This heat transformation has been found
in the cited references to be preferential in regions where internal
material damage is located or is developing. In order to utilize
this technique for quantitative evaluation of materials, appropriate
physical and mathematical models must be found for interpretation of
the mechanical to heat transformation process.

We interpret "Stress Related Thermal Emission" to mean energy
converted from mechanical excitation to heat by various material
(and occasionally structural) phenomena. While classical areas of
activity such as "internal friction" and "damping" are thought to be
included in our Stress Related Thermal Emission definition, a new,
more general and more clearly defined (at least from the standpoint
of mechanics) field of study is desirable, for reasons to be expressed shortly.

The subject of energy dissipation is a truly interdisciplinary
field. It has attracted the attention of physicists, mechanical
engineers, electrical engineers, mechanicians, applied mathematicians,
and artists — especially those who have been making cymbals and
bells — for centuries. The related literature is extensive. Review
articles, however, are sparse; McClintock [6] and Nowick [7] have
made fairly recent worthwhile review attempts. If one were to
characterize the general field at this point in a word, it would be
something like "disjunct." As a body of work, existing literature is
more multidisciplinary than interdisciplinary. This is especially
true when one views the field from the perspective of engineering
applications. The rather active field of structural damping has
developed quite apart from the large body of literature from the
metallurgy and physics community which deals with the mechanisms of
dissipation. Within the latter group there is a considerable amount
of fragmentism, attempts to split off parts of the subject as being
unique, distinct and unrelated to the remainder. Of course, careful
classification and isolation of various physical behavior is
essential, but the lack of a unified approach to the field has, in
many respects, been a serious impediment to progress. Major

collections of work can be found in the classical book by Zener [8]
and the more recent and complete work by Nowick and Berry [9]. Our
concern in the present paper, will be to deal with a part of the
subject which is of particular engineering interest, namely Stress
Related Thermal Emission (hereafter called SRTE), by addressing the
mechanics of the engineering situation as a unified approach to the
dissipation problem and related phenomena.

DEFINITION OF THE PROBLEM

To set our problem and to establish a nomenclature, we will
outline the major aspects of the general field of SRTE, indicate their
relevance to the engineering aspects of such behavior, and focus (by
elimination) our attention on the part of the problem we will address
in detail. The criterion for "relevance" in this case will be the
degree to which each aspect can be expected under stress excitations
common to engineering applications, and the extent to which each
aspect can be used to gain information about the integrity or the
nature of the mechanical response of the material.

We can begin by breaking the general subject into two categories,
the first we will call reversible SRTE and the second we will call
irreversible SRTE. The first category is the most extensively treated
and includes the classical literature from metallurgy, chemistry and
physics disciplines. Reversible SRTE usually involves rate dependent
materials response and includes the classical "anelastic" and "viscoelastic" classes of behavior. Irreversible SRTE is less well understood mechanistically, but is more intuitively clear in the sense that
such things as plastic deformation and fracture are more obviously
dissipative processes to the engineer who has a clear concept of
mechanical work. Irreversible SRTE is generally rate-independent in
the strict sense, i.e., the response of the material itself is not a
function of time derivatives of stress or strain. We shall examine
these two categories in the order of introduction above.

Our reversible category refers to thermodynamically reversible
mechanical behavior in the sense that there is no inherent history
dependence involved. Stress-strain laws for this case are usually
assumed to be linear [6]. At low stress levels anelastic effects
including the various relaxation phenomena (atomic diffusion, order-
disorder processes, thermal diffusion, etc.) are common. Over a
larger range of stress, viscoelastic effects are observed. While the
mechanics of these two types of behavior are closely related, anelasticity has developed as a rather special part of the subject
having to do with physical and chemical phenomena in crystalline
solids. For the most part, the object of that development was, and
is, to study the material itself as a bulk substance, and to establish
the character of certain internal features by their relaxation spectra.
Viscoelasticity is an older and more general field having closer ties
to engineering situations wherein the inherent response of a material

to an applied state of stress or strain is rate dependent. Viscoelastic creep under quasi-static or static loading is a common consequence of this type of behavior although cyclic response is also an
active part of this sub-field. We will discuss these two types of
behavior (anelastic and viscoelastic) together since the basic
descriptive concepts involved are not separable.

From our present point of view, some of the dissipative
mechanisms associated with anelasticity are not of practical importance. As a group, this is especially true of the atomic diffusion
events that are activated by stress, including such things as grain
boundary motion, single or paired solution atom motion, and twin
boundary activity. While these events may be prominent in the small
strain range, the heat produced by them is not significant because of
the low ratios of dissipated to input energy, the low levels of input
energy (for small strains) and the low frequencies at which these
events are commonly excited [8].

The last point requires a momentary diversion to explain the
implication that there is a maximum in the dissipation as a function
of the frequency of cyclic excitation. This implication follows from
first principals, as noted by Zener [8]. At a given temperature, for
our reversible SRTE, if an amount of heat, δQ, is added to a (unit)
volume of material, the change in entropy is given by

$$\delta S = \frac{\delta Q}{T} .$$

(1)

For an equilibrium process, the energy produced by the dissipation
(and the resultant heat) must be exactly balanced by the heat flow
out of the specimen (or away from a process zone) since the total
entropy change must be zero. Hence, for the equilibrium situation,

$$\Delta Q = \frac{\overline{\delta Q}}{T_o} \, \delta T$$

(2)

where ΔQ is the net heat flow, $\dfrac{\overline{\delta Q}}{T_o}$ is the entropy change per cycle and

T_o is the mean specimen temperature. The product $\overline{\delta Q} \delta T$ dictates that
the heat flow will diminish when the process is driven so rapidly
that heat exchange is precluded (adiabatic) or so slowly that the
temperature change is precluded (isothermal). The heat produced will
be maximized at some intermediate frequency defined by the physical
process involved. We will return to this concept when we have focused
on a more specific situation.

By comparison to the reversible SRTE literature, much less attention has been given to irreversible SRTE. Experimental data have
been generated which relates plastic hysteresis energy loss to specimen temperature during cyclic loading (c.f. references 10 and 11).
Heat generation by internal flaws through mechanisms such as the

rubbing of flaw surfaces or energy release during micro-crack formation has received virtually no attention.

For the present case we focus on SRTE which can be used to provide information about the integrity and mechanical response of a material specimen or component. Hence we address two practical questions. How is heat produced under practical situations, and how do we discern variations in integrity or response from the temperature differences caused by point to point variations in the heat produced? For purposes of illustrative discussion, we imagine that these variations of interest occur over local regions of material. These regions may be flaw neighborhoods, material phase boundaries, or mechanical "process zones" of various types involving such things as local concentrations of plastic slip, local concentrations of microcracking, etc. The characteristic size of these regions is taken to be of the order of 1-10 mm. Our problem, then, is how to characterize SRTE from such regions and how to maximize the temperature differences produced by such regions so that the most information about the regions can be obtained. Finally, we also need to develop a philosophy for interpreting such information.

BASIC CONCEPTS AND MODELING

We can extract a useful bit of practical information from an argument borrowed from the physics community [12]. If our regions are sufficiently small, say with characteristic dimension D, then we can deduce the essence of the behavior associated with thermal diffusion from the heat conduction equation:

$$\frac{\partial T}{\partial t} = \frac{k}{\rho c} \nabla^2 T \tag{3}$$

where k is the thermal conductivity, c is the specific heat, ρ is the density and T is the absolute temperature. The spatial derivatives can be written in difference form as:

$$\nabla^2 T \simeq \frac{T - T_o}{D^2} \tag{4}$$

where D is the characteristic dimension of the region over which the gradient of temperature exists. Then equation (3) becomes:

$$\frac{\partial T}{\partial t} = \frac{k}{\rho c} \frac{T - T_o}{D^2} \quad , \tag{5}$$

from which;

$$(T - T_o) = A e^{-kt/\rho c D^2} \tag{6}$$

This is a relaxation phenomenon with relaxation time given by

$$\tau = \frac{\rho c D^2}{k} \tag{7}$$

Fig. 1 shows an interpretation of this equation for some materials of
interest for a range of region sizes of practical importance. Two
types of practical information are illustrated by Fig. 1. If energy
dissipation is due to thermal currents within the region of interest,
then the frequencies of oscillation for a range of materials which
will maximize the dissipative loss is given directly by the figure.
It is important to notice the strong dependence on material properties
in Fig. 1. (The ordinate is a logarithmic scale.) The material
property that contributes the most to this dependence is the thermal
conductivity. The density and specific heat rarely change by an order
of magnitude but the thermal conductivity may change by four (or more)
orders of magnitude. This will be a recurring situation as we continue our characterization: the thermal conductivity has a dominant

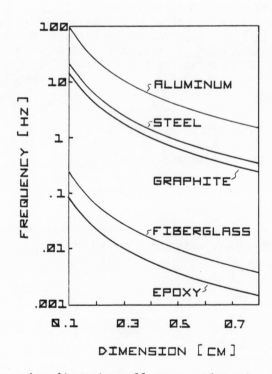

Fig. 1.   Region size dimension effect on relaxation frequency.

influence over SRTE material interrogation schemes in many cases. Dependence on geometry (the region size in this example) is not as great as the material property influence.

Fig. 1 is also of interest for a second practical situation. If thermal currents within the regions of interest are not to be used as dissipative mechanisms, then we will have to choose frequencies of excitation that are significantly different than the relaxation frequencies shown in the figure. For modeling purposes, if we choose frequencies significantly higher than those shown we can assume that the regions of interest have constant temperature distributions since conduction is not a major influence, i.e., this is essentially the adiabatic condition.

In this regard a few words need to be said concerning the evolution of heat by stress waves, either traveling or standing. The adiabatic condition is not quite satisfied for longitudinal waves since heat energy flows from the hot, compressed regions to the cooler, rarefied regions. This interchange of heat energy will by necessity remove energy from the mechanical wave, causing general heating of the body and eventual extinction of the stress wave. One might expect there to be preferential heating due to this effect in regions where the stresses are intensified due to material discontinuities. Upon closer examination, however, it is found that this effect is minimal at frequencies of practical interest. This heat loss will occur most readily at frequencies for which the wave length becomes small enough that the compressed and rarefied regions are so close that heat exchange can most easily take place. In fact, for very high frequencies, one approaches an isothermal state where heat exchange takes place nearly instantaneously. In this region, the heat loss from the stress wave is a maximum. The frequency at which this transition from the adiabatic to the isothermal condition occurs can be estimated from [13]

$$ f \gg \frac{1}{2\pi} \rho \frac{c_v v^2}{K} \tag{8} $$

where K is the bulk modulus, ρ is the mass density, c_v is the specific heat at constant volume and v is the longitudinal phase speed. For quartz, for example, the transition frequency is of the order of 10^{11} Hz. Hence in all practical cases, this transition will never be observed. The wave attenuation coefficient for this phenomenon in normal frequency ranges has been estimated to be [13].

$$ A = \frac{2\pi^2 f^2}{\rho v^3} \left[\frac{K}{c_v} \frac{C_{11}^{\sigma} - C_{11}^{\theta}}{C_{11}^{\sigma}} \right] \tag{9} $$

where f is the frequency, and $C_{11}{}^{\sigma}$ and $C_{11}{}^{\theta}$ are the isentropic and isothermal elastic moduli, respectively. For quartz, this value is of the order of 12×10^{-4} nepers/cm for f = 100 MHz. Hence, only at the very high frequencies normally used in ultrasonic testing does this heat generation effect become non-negligible. Much of our interest is in an intermediate frequency range, especially in view of the results of earlier work which shows that heat images of defects in several materials can be observed at frequencies of excitation between about 30 Hz and the mega-hertz range [1,4,5]. For these frequencies, especially in the kilohertz range, the wavelength of traveling stress waves in most engineering materials is greater than the order of a centimeter. We can then assume that the stress or strain due to the external excitation is uniform over the regions of interest (1-10 mm). Situations for which the wavelength is of the order of the inhomogeneities or discontinuities produce what is often called a loss to propagating stress waves. One can refer, for example, to scattering losses incurred in polycrystalline materials when the wavelength is comparable to the grain size [13]. This, however, is only an apparent loss due to the scattering of wave energy in all directions, and hence away from the primary propagation direction. This loss does not directly result in a preferential transformation of mechanical to thermal energy at the scattering site, at least as far as presently accepted physical models predict.

For the "long wavelength case" for which we are still well above frequencies which correspond to thermal current losses within the regions, we can examine another important physical situation of practical importance. The situation of interest is the case for which hysteresis losses occur because of material inhomogeneity. We take inhomogeneity to mean that there are regions of material separated by phase boundaries or discontinuous property boundaries which dominate the local physical response and control the flow of thermal energy from one region to another. To get a physical feel for the parameters which have the greatest influence over this situation we examine the following simple model.

For illustrative purposes we consider a bi-material having regularly spaced-regions of alternating material, as shown in Fig. 2. We assume that heat transfer is dominated by the region boundaries. Each region is represented by a single dependent variable (temperature). Then for each region:

$$\frac{dQ}{dt} = k\nabla^2 T - \delta(T_1 - T_2) \tag{10}$$

where k is a conductivity and δ is a surface transfer coefficient. If the gradients in temperature exist only over the semi-discontinuous boundary regions, and if a heat source of strength U per unit volume exists in each region (due to dissipation), we can write,

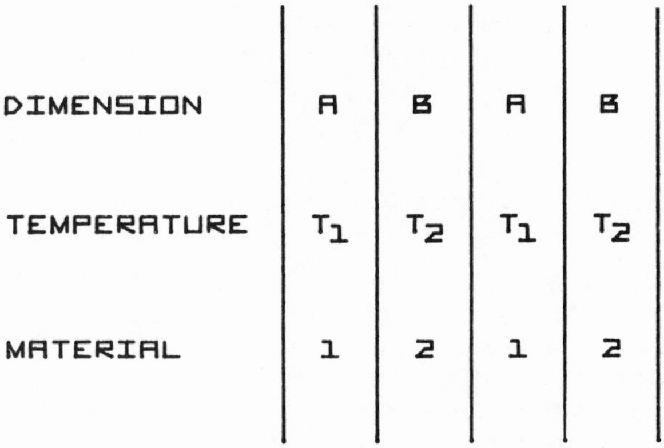

DIMENSION	A	B	A	B
TEMPERATURE	T_1	T_2	T_1	T_2
MATERIAL	1	2	1	2

Fig. 2.   Schematic diagram of regions in an inhomogeneous material.

$$\rho_1 c_1 V_1 \frac{dT_1}{dt} = V_1 U_1 - 2\gamma A(T_1 - T_2) \tag{11}$$

$$\rho_2 c_2 V_2 \frac{dT_2}{dt} = V_2 U_2 - 2\gamma A(T_2 - T_1) \tag{12}$$

for regions 1 and 2, respectively, where V_i are the volumes of the regions (per unit thickness), A is the surface area (per unit thickness) between the regions, ρ_i are the densities and c_i are the specific heats of the respective regions. If we make the common assumption (c.f. reference 6) that the dissipative energy sources can be represented as,

$$U_i = \omega\pi\delta_i\sigma_a\varepsilon_a \qquad i=1,2 \tag{13}$$

for each region where δ_i is the classical phase lag between stress and strain for each material, ω is the frequency of excitation and σ_a and ε_a are the stress and strain amplitudes of the excitation, then our balance equations become:

$$\rho_1 c_1 V_1 \frac{dT_1}{dt} = V_1(\omega\pi\delta_1\sigma_a\varepsilon_a) - 2\gamma A(T_1 - T_2) \tag{14}$$

$$\rho_2 c_2 V_2 \frac{dT_2}{dt} = V_2(\omega\pi\delta_2\sigma_a\varepsilon_a) - 2\gamma A(T_2 - T_1) \tag{15}$$

where we have assumed that our excitation wavelength is sufficiently large so that the stress and strain amplitudes are common to both region types. The steady state temperature difference is then given by:

$$(T_1 - T_2) = \frac{\sigma_a \varepsilon_a}{4\gamma} \; [\; \frac{V_1}{A} \delta_1 - \frac{V_2}{A} \delta_2 \;] \tag{16}$$

The influence of excitation amplitude and surface transfer coefficient is independent of our specific region configuration, but the surface-to-volume ratios and the phase lag (which is proportional to the anelastic dissipative loss) enter in a special way. The temperature difference is proportional to the difference in the product of the respective phase lag and the inverse of the surface-to-volume ratio for each region. From a practical standpoint, we would expect the observed temperature difference, which we want to use to provide a heat-image of the regions, to be greater for material regions which have the greatest difference in dissipative loss. Materials which have the greatest loss generate the most heat, but region definition depends on the difference in loss factors, not a suprising result.

Equation (16) also indicates that surface to volume ratios are important. For the above example for which the surface is an interface between the regions (common to both), the volume difference is controlling. However, we also note that the general level of temperature produced is greater for smaller surface-to-volume ratios, i.e., more nearly sphere-like regions, and less for higher ratios, i.e., more nearly plate-like regions.

A related effect that can occur in inhomogeneous materials is caused by the difference in elastic properties of adjacent materials. As far as we are aware, this effect has not been studied for inhomogeneous materials but has been discussed for polycrystalline homogeneous materials. In the latter case the variation in elastic properties arises due to the anisotropy and varying orientations of neighboring grains. This effect, called the Zener effect [8] arises due to thermal currents established between adjacent grains that are strained differently because of their anisotropy. This loss is again a relaxation type of loss with the peak frequency depending upon grain size and the thermal properties of the material. For most homogeneous materials, the peak frequency is less than 100 KHz and it is found that the greater the anisotropy, the greater the heat generation. It is easy to visualize the extension of this effect to heterogeneous materials which have neighboring phases of greatly differing elastic properties. However, as yet no estimation has been made of the pertinent frequency range or magnitude of the effect.

OBSERVATIONAL FACTORS

From a practical standpoint, we are limited in our use of

thermal emission by the sensitivity of our instrumentation and the feasibility of the excitation methods required. We examine a few pertinent factors which affect our practical ability to observe and produce heat images below.

First, we return to a point that we made earlier regarding the importance of thermal conductivity to relaxation spectra, and demonstrate a second situation for which it is true. This second situation is much more obvious but no less important. We ask, for a local source of heat of fixed strength (fixed flow per unit time out of a region of fixed size), how do the material properties affect the nature of the heat image of that source that we observe? Again, for demonstration purposes we idealize our situation so that our region of interest is a circular area with radius r_1 in an infinite plate of unit thickness. The classical solution of Fourier's equation yields,

$$(T(r) - T_1) = \frac{Q}{2\pi k} \ln \frac{r}{r_1} \qquad (17)$$

where k is the conductivity and Q is the source strength. The temperature difference per unit source strength (scaled in an unusual way for convenience of data comparison) is plotted in Fig. 3 for several materials having a wide range of properties. Data for that figure are shown in Table 1. The thermal conductivity controls the behavior, of course, but the nonlinear dependence on the distance, r, introduces some interesting behavior. First, as expected, the magnitude of the temperature difference for insulating materials is three orders of magnitude greater than for good conductors. Second, the sharpness of the temperature gradients is also different, i.e., for good conductors the distance over which the temperature change is spread is less. Both of these factors affect the acuity of the heat image. The temperature difference must be of the order of 1°C for most of the optical image forming thermographic instrumentation available for engineering purposes, especially the video devices of current interest. For a source of, say, 10 BTU per hour (175.8 watts) which is a value which could be typical for local plastic flow at a defect in a metal, from Fig. 3, materials with conductivities greater than or equal to steel or graphite would be at or below such instrument sensitivity. The temperature gradient widths must also be greater than the spatial resolution of available equipment. If an imaging camera can resolve 150 lines (75 line pairs in common optical terminology) across its imaging plane, the lenses and optical paths then limit the resolution for a given field of view. For a common video camera, for example, a field of view of about 5 cm might result in a resolution of about 0.33 mm or about 1.5 line pairs per millimeter.

Figure 4 presents another interpretation of equation (17) which can be used to estimate image acuity and spatial resolution. The abscissa of Fig. 4 is the percent difference of the temperature between two points on any curve in Fig. 3 separated by a change of

Table 1.

$\Delta T/q$ (°F/BTU)

r/r_1	Epoxy	Glass-Epoxy	Graphite	Steel	Aluminum
2	1.0	3.8×10^{-1}	7.9×10^{-3}	4.2×10^{-3}	9.3×10^{-4}
3	1.6	6.0×10^{-1}	1.2×10^{-2}	6.7×10^{-3}	1.4×10^{-3}
4	2.0	7.6×10^{-1}	1.6×10^{-2}	8.5×10^{-3}	1.8×10^{-3}
5	2.3	8.8×10^{-1}	1.8×10^{-2}	9.8×10^{-3}	2.2×10^{-3}
7	2.8	1.1×10^{0}	2.2×10^{-2}	1.2×10^{-2}	2.8×10^{-3}
10	3.3	1.3×10^{0}	2.6×10^{-2}	1.4×10^{-2}	3.1×10^{-3}
15	3.9	1.5×10^{0}	3.1×10^{-2}	1.6×10^{-2}	3.6×10^{-3}
20	4.3	1.6×10^{0}	3.4×10^{-2}	1.8×10^{-2}	4.1×10^{-3}
25	4.6	1.8×10^{0}	3.7×10^{-2}	1.9×10^{-2}	4.3×10^{-3}
30	4.9	1.9×10^{0}	3.9×10^{-2}	2.0×10^{-2}	4.6×10^{-3}
35	5.1	1.9×10^{0}	4.1×10^{-2}	2.2×10^{-2}	4.8×10^{-3}
40	5.3	2.0×10^{0}	4.2×10^{-2}	2.3×10^{-2}	4.9×10^{-3}

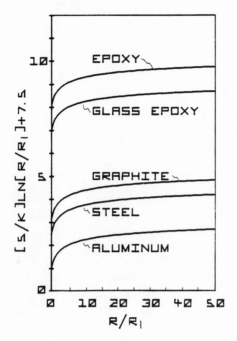

Fig. 3. Temperature difference vs. distance from a dissipative
source of fixed strength for several materials.

0.1 in the normalized distance from the source, r/r_1, where r_1 is
the source (or region) size. By plotting the information this way
we can reduce all of the curves down to a single one, separate out
of the logarithmic function the influence of the conductivity (which
does not affect this plot), and extract a "rule of thumb." To
interpret Fig. 4 we notice that if we wish to resolve a defect region
with a 0.1 in (2.5 mm) characteristic radius and we determine from
Fig. 3 that our instrument will resolve 1 percent of the temperature
range to be expected, Fig. 4 tells us that the temperature gradient
that we can resolve for this case will appear to be spread over a
distance of about 5.8 times 0.1 or 0.58 in (14.5 mm) which can easily
be resolved. Fig. 3 also indicates that only 10 percent changes will
be occurring at about 0.2 in (5 mm), i.e., most of the apparent heat
image will be quite close to the source. It bears repeating that as

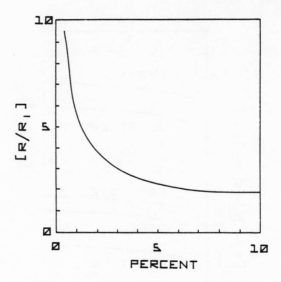

Fig. 4. Percent difference between successive points on the
temperature distribution at a source for a change of 0.1 in
r/r_1.

long as we talk about percentage changes, our conclusions hold for all
materials. In fact, from Fig. 4 we could state a "rule of thumb"; the
apparent size of the heat image produced by a source, normalized by
the source size, will be in the neighborhood of 5 or less for many
practical situations. This rule can be very helpful in the interpretation of heat patterns.

"PRACTICAL" DISPERSIVE MECHANISMS

SRTE occurs by means of many mechanisms, as we have seen.
However, if we are to use SRTE to locate material defects and characterize the local response in the neighborhood of a defect or flaw only
certain mechanisms are of "practical" interest. A few of the most
important ones will be briefly noted.

Anelastic effects are nearly always present. One major reason
for this is that the magnitude of temperature changes due to adiabatic
straining at practical excitation frequencies is often significant,
especially for the more conductive materials. The temperature change
for adiabatic elastic deformation is given by [14],

$$\Delta T = - \frac{3T\alpha Ke}{\rho c} \tag{18}$$

where α is the coefficient of thermal expansion, K is the bulk modulus, e is the volume dilalation, ρ is the mass density and c is the specific heat of the material being strained. For aluminum loaded over a range between the tensile and compressive yield strengths in a uniaxial test at 1 Hz the entire specimen may oscillate over a temperature range of 0.5°C or more, depending upon the type of aluminum, the conductive, convective and radiation losses and other experimental factors. Such large temperature oscillations which cycle in phase with the excitation are easily observed in many materials and have been observed for several metals and plastics in our laboratory. However, for homogeneous materials the dissipative loss associated with thermal currents induced by these temperature changes is generally rather small. In general, the losses are proportional to the difference between the adiabatic and isothermal elastic compliances of a given material, i.e.

$$S_{ij}\bigg|_{adiabatic} - S_{ij}\bigg|_{isothermal} = - \frac{\alpha_i \alpha_j}{c} T \tag{19}$$

where S_{ij} are the elastic compliance tensor components, α_i are the vector components of thermal expansion, T is the temperature and c is the specific heat of the material. The physical reason for the strength of the dissipative loss being controlled by expression (19) is simply the fact that the difference between the adiabatic and isothermal behavior controls the maximum area in the ideal hysterasis loop associated with what is commonly called the "elastic after-effect," shown in Fig. 5. As an example of this situation, for "transverse thermal currents in a reed" (beam) of rectangular cross section Nowick gives the internal friction as [7],

$$\tan \delta = \frac{E\alpha^2 T}{c} \left[\frac{\omega\tau}{1 + \omega^2\tau^2} \right] \tag{20}$$

where E is the elastic modulus, ω is the frequency of excitation, α, T and c are defined in Eqn. (17) and the relaxation time is

$$\tau = \frac{a^2}{\pi^2 D} \tag{21}$$

where a is the reed thickness and D is the thermal diffusivity. However, it should be emphasized that these expressions are for homogeneous materials. Virtually no basic characterization of thermal current losses in inhomogeneous materials such as composites has been carried out. Values of relative damping given by Eqn. (20) usually fall in the range of 0.001 to 0.01.

"Viscoelastic" effects and "anelastic" effects include, as noted

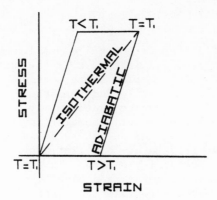

Fig. 5.   Idealized hysteresis loop due to elastic after-effect.

earlier, several other mechanisms such as grain boundary or phase
boundary effects, eddy current effects, and various molecular motions
in long-chain-molecule materials. Only the last of these is of
practical importance in the present context. These types of dissipative mechanisms can produce relative damping values of 0.1 to 1.5 [6].
Furthermore, since long-chain (polymeric) materials are generally poor
conductors, temperature changes at local sources are generally higher
and easier to observe as noted earlier. Also, using Eqn. (20), losses
due to thermal currents are essentially negligible.

    If history-dependent behavior is present, especially so-called
plastic deformation and localized slip, very great temperature differences may be produced in both good and poor conductors. The
plastic zone at the tip of a propagating (fatigue) crack is easily
resolved, even in aluminum [15]. However, long-term steady-steate
heat emission due to this type of deformation is not a practical
material interrogation method, especially if a nondestructive scheme
is required. Thermography is an excellent way of observing such
detail, however.

    In our earlier discussions, very little has been said about what
might be called structural dissipation. We refer, here, to dissipation
that comes about because of the interference or interaction of two
regions in a structural way rather than dissipation due to a material
micro-response. The most obvious, and, it would appear, most important
of such mechanisms for present purposes is the interaction of neighboring surfaces. Commonly, such surfaces rub, flap and pound against
each other under cyclic excitation and may cause considerable heat

release. This is almost purely a mechanical process in the sense that vibrations of the surfaces and the local deformation fields can be analyzed with a fairly standard mechanics approach. However, the manner in which such response produces heat is not well understood and, to our knowledge, has not been modeled. Much work is yet to be done in this area.

ACKNOWLEDGEMENT

This work was sponsored by the Army Research Office, Durham, Grant No. DAAG29-79-G-0037.

REFERENCES

1. K. L. Reifsnider and W. W. Stinchcomb, Investigation of Dynamic Heat Emission Patterns in Mechanical and Chemical Systems, Proceedings of 2nd Biennial Infrared Information Exchange, AGA Corp, St. Louis, MO, pp. 45-58 (1974).

2. J. A. Charles, F. J. Appl and J. E. Francis, Using The Scanning Infrared Camera in Experimental Fatigue Studies, Experimental Mechs. 15, pp. 133-138 (April 1975).

3. G. J. Trezek and S. Balk, Provocative Techniques in Thermal NDT Imaging, Materials Evaluation 34, pp. 172-176 (August 1976).

4. E. G. Henneke, II and T. S. Jones, Detection of Damage in Composite Materials by Vibrothermography, "Nondestructive Evaluation and Flaw Criticality for Composite Materials," STP 696, Amer. Soc. for Testing and Materials, Phila. (1979).

5. E. G. Henneke, II, K. L. Reifsnider and W. W. Stinchcomb, Thermography - An NDI Method for Damage Detection, Journal of Metals 31, pp. 1115 (1979).

6. F. A. McClintock and A. S. Argon, "Mechanical Behavior of Materials," Addison Wesley, pp. 471-486 (1966).

7. A. S. Nowick, Internal Friction in Metals, Prog. in Metal Physics 4, pp. 1-70 (1953).

8. C. Zener, "Elasticity and Anelasticity of Metals," Univ. of Chicago Press (1948).

9. A. S. Nowick and B. S. Berry, "Anelastic Relaxation in Crystalline Solids," Academic Press (1972).

10. H. T. Hahn and R. Y. Kim, "Fatigue Behavior of Composite Laminate," J. Comp. Materials 10, pp. 156-180 (April 1976).

11. E. H. Jordan and B. I. Sandor, "Stress Analysis from Temperature Data," J. of Testing & Evaluation, Vol 6, No. 6, pp. 325-331 (1978).

12. I. Malecko, "Physical Foundation of Technical Acoustics," Pergamon Press, New York, (1969).

13. W. P. Mason, "Piezoelectric Crystals and Their Application to Ultrasonics," D. Van Nostrand Co., Inc., New York (1950).
14. M. A. Biot, "Thermoelasticity and Irreversible Thermo-dynamics," J. Applied Physics, Vol 27, No. 3, pp. 240-253 (1956).
15. James C. Hsieh, "Temperature Distribution Around a Propa-gating Crack," Doctoral Dissertation, College of Engineering, Virginia Polytechnic Institute and State University, (April 1977).

CONTRIBUTORS

Co-Chairmen

D. P. H. Hasselman, Whittemore Professor of Materials
     Engineering, Virginia Polytechnic Institute and State
     University, Blacksburg, Virginia 24061, U.S.A.

R. A. Heller, Professor, Department of Engineering Science and
     Mechanics, Virginia Polytechnic Institute and State
     University, Blacksburg, Virginia 24061, U.S.A.

Conference Session Chairmen

H. Abe, Asahi Glass, Company, Yokohama, Japan

K. Bills, Aerojet General Corp., P. O. Box 13400,
     Sacramento, California 94813

O. W. Dillon, Professor, Dept. of Engineering Mechanics,
     University of Kentucky, Lexington, Kentucky 40506

R. J. Gottschall, Division of Materials Sciences, Department
     of Energy, Washington, DC 20545

R. B. Hetnarski, Professor, Department of Mechanical
     Engineering, Rochester Institute of Technology,
     Rochester, New York 14623

D. Post, Professor, Department of Engineering Science and
     Mechanics, Virginia Polytechnic Institute and State
     University, Blacksburg, Virginia 24061, U.S.A.

V. J. Tennery, Department of Energy, Oak Ridge National
     Laboratory, Oak Ridge, Tennesses 37830

AUTHORS

H. Awaji, Fukushima Technical College, Iwaki, Fukushima, 970
      Japan

Z. P. Bazant, Civil Engineering Department, Northwestern Univer-
      sity, Evanston, Illinois 60201

P. F. Becher, Naval Research Laboratory, Washington, DC 20374

L. Bentsen, Montana Energy & Research Institute, Butte, MT 59701

H. Blauel, Fraunhofer-Institute fur Werkstoffmechanik, D-7800
      Freiburg, West Germany

B. A. Boley, The Technological Institute, Northwestern University,
      Evanston, Illinois 60201

R. C. Bradt, Materials Science and Engineering Department,
      Pennsylvania State University, University Park, Pennsylvania
      16802

H. F. Brinson, Engineering Science and Mechanics Department,
      Virginia Polytechnic Institute and State University,
      Blacksburg, Virginia 24061

E. P. Cernocky, Mechanical Engineering Department, University
      of Colorado, Boulder, Colorado 80301

F. S. Chau, Unilever Research, Colworth House, Sharnbrook,
      Bedford MK44 1LQ, Great Britain

F. H. Chou, Department of Engineering Mechanics, San Diego
      State University, San Diego, CA 92182

N. Claussen, Max-Planck Institute fur Metallforschung,
      7000 Stuttgart - 80, West Germany

D. R. Clarke, Rockwell Science Center, Thousand Oaks, Cali-
      fornia 91360

T. L. Cost, Aerospace Engineering Department, University of
      Alabama, University, Alabama 35486

W. J. Craft, School of Engineering, North Carolina A. and T.
      State University, Greensboro, North Carolina 27411

I. M. Daniel, Illinois Institute of Technology Research
      Institute, Chicago, Illniois 60616

A. Das Gupta, U. S. Army Ballistic Research Laboratory, Aberdeen
      Proving Grounds, Maryland   21005

J. G. Davis, NASA Langeley Research Center, Hampton, Virginia
      23665

R. S. Dhaliwal, Mathematics and Statistics Department, University
      of Calgary, Calgary, Canada T2N 1N4

A. F. Emery, Mechanical Engineering Department, FU-10, University
      of Washington, Seattle, Washington 98195

A. G. Evans, Materials and Engineering Department, University of
      California, Berkeley, California 94720

E. C. Francis, Chemical Systems Division, United Technologies,
      Sunnyvale, California   94088

S. W. Freiman, National Bureau of Standards, Washington, DC   20234

R. Gardon, Ford Motor Company, Dearborn, Michigan   48121

N. Ghadiali, Batelle Columbus Laboratories,
      Columbus, Ohio   43201

G. A. Gogotsi, Institute for Problems of Strength, Academy of
      Sciences of the UkrSSR, Kiev, USSR

W. I. Griffith, Michelin Applied Research Center, Greenville,
      South Carolina   29602

Y. L. Groushevsky, Institute for Problems of Strength, Academy of
      Sciences of the UkrSSR, Kiev, USSR

T. K. Gupta, Westinghouse Research & Development Center,
      Pittsburgh, PA   15235

D. Hartmann, Institute for Structural Mechanics, University of
      Dortmund, Dortmund 50, West Germany

D. P. H. Hasselman, Department of Materials Engineering, Virginia
      Polytechnic Institute and State University, Blacksburg,
      Virginia   24061

R. A. Heller, Department of Engineering Science and Mechanics,
      Virginia Polytechnic Institute and State University,
      Blacksburg, Virginia   24061

E. G. Henneke, Engineering Science and Mechanics Department,
    Virginia Polytechnic Institute and State University,
    Blacksburg, Virginia 24061

C. T. Herakovich, Engineering Science and Mechanics Department,
    Virginia Polytechnic Institute and State University,
    Blacksburg, Virginia 24061

J. Homeny, Department of Materials Science and Engineering, The
    Pennsylvania State University, University Park, PA 16802

A Hopper, Batelle Columbus Laboratories, Columbus, Ohio 43201

W. L. Hufferd, University of Utah, Salt Lake City, Utah 84102

G. Hulbert, Batelle Columbus Laboratories, Columbus, Ohio 43201

J. Ignaczak, Polish Academy of Sciences, Warsaw, Poland

Y. Imamura, University of Ibaraki, Hitachi, 316, Japan

C. Jaske, Batelle Columbus Laboratories,
    Columbus, Ohio 43201

M. P. Kamat, Department of Engineering Science and Mechanics,
    Virginia Polytechnic Institute and State University,
    Blacksburg, Virginia 24061

K. Kawamata, University of Ibaraki, Hitachi, 316, Japan

D. E. Klett, School of Engineering, North Carolina A. and T.
    State University, Greensboro, North Carolina 27411

K. Kokini, Department of Mechanical & Aerospace Engineering
    Syracuse University, Syracuse, New York 13210

E. Krempl, Mechanical Engineering Department, Rensselaer
    Polytechnic Institute, Troy, New York 12181

G. Krishnamoorthy, Department of Engineering Mechanics, San
    Diego State University, San Diego, California 92182

E. Kuznetsov, Applied Solid Mechanics Section, Batelle Columbus
    Laboratories, Columbus, Ohio 43201

B. Leis, Batelle Columbus Laboratories, Columbus, Ohio 43201

D. Lewis III, Naval Research Laboratory, Washington, DC 20375

C. Libove, Department of Mechanical and Aerospace Engineering,
        Syracuse University, Syracuse, New York 13210

J. Margetson, Research Division, Propellants, Explosive and Rocket
        Motor Establishment, Westcott, Aylesbury, Buckingham, United
        Kingdom

D. L. Martin, Jr., Propulsion Directorate, U.S. Army Missile
        Command, Redstone Arsenal, Alabama   35809

W. J. McDonough, Naval Research Laboratory, Washington, DC   20374

J. J. Mecholsky, Sandia Laboratory, Albuquerque, NM   87185

J. S. Mills, McDonnel Douglas Astronautics Corporation, W.
        Huntington Beach, CA   92647

D. H. Morris, Engineering Science and Mechanics Department,
        Virginia Polytechnic Institute and State University
        Blacksburg, Virginia   24061

S. Musikant, Re-Entry and Space Division, General Electric
        Company, Philadelphia, Pennsylvania   19101

T. Oku, Japan Atomic Energy Research Institute, Ibaraki, 319-11,
        Japan

R. W. Perkins, Mechanical and Aerospace Engineering Department,
        Syracuse University, Syracuse, New York 13210

W. H. Peters, III, Department of Engineering Science and Mechanics,
        Virginia Polytechnic Institute and State University,
        Blacksburg, Virginia 24061

J. O. Pilcher, U. S. Army Ballistic Research Laboratory, Aberdeen
        Proving Grounds, Maryland   21005

K. L. Reifsnider, Engineering Science and Mechanics Department,
        Virginia Polytechnic Institute and State University,
        Blacksburg, Virginia 24061

R. W. Rice, Naval Research Laboratory, Washington, DC   20375

S. Sato, University of Ibaraki, Hitachi, 316, Japan

K. Satyamurthy, Department of Materials Engineering, Virginia
        Polytechnic Institute and State University, Blacksburg,
        Virginia   24061

S. G. Schwille, Re-Entry and Space Division, General Electric
    Company, Philadelphia, Pennsylvania 19101

L. D. Simmons, Multnomah School of Engineering, University of
    Portland, Portland, Oregon 97203

J. P. Singh, Department of Materials Engineering, Virginia
    Polytechnic Institute and State University, Blacksburg,
    Virginia 24061

M. P. Singh, Department of Engineering Science and Mechanics,
    Virginia Polytechnic Institute and State University,
    Blacksburg, Virginia 24061

J. R. Spann, Naval Research Laboratory, Washington, DC 20375

D. Stahn, Fraunhofer-Institut fur Werkstoffmechanik, D-7800
    Freiburg, West Germany

P. Stanley, Simon Engineering Laboratories, University of
    Manchester, Oxford Road, Manchester M13 9PL, England

M. Stern, Aerospace Engineering and Engineering Mechanics
    Department, University of Texas at Austin, Austin,
    Texas 78712

T. R. Tauchert, Engineering Mechanics Department, University of
    Kentucky, Lexington, Kentucky 40506

R. A. Tanzilli, Re-Entry and Space Division, General Electric
    Company, Philadelphia, Pennsylvania 19101

J. R. Thomas, Mechanical Engineering Department, Virginia Poly-
    technic Institute and State University, Blacksburg,
    Virginia 24061

T. R. Trafton, U.S. Army Ballistic Research Laboratory,
    Aberdeen Proving Grounds, MN 21005

G. G. Trantina, Corporate Research and Development, General
    Electric Company, Schenectady, NY 12301

J. H. Wiener, Division of Engineering, Brown University,
    Providence, Rhode Island 02912

R. W. Wierman, Westinghouse Hanford Company, Richland, Washington
    99352

Y. T. Yeow, Allied Chemical Corporation, Morristown, NJ 07960

G. E. Youngblood, Montana Energy & MHD Research Institute,
     Butte, Montana   59701

INDEX